SECOND EDITION

ENVIRONMENT AND SOCIETY

Human Perspectives on Environmental Issues

Charles L. Harper
Creighton University

Upper Saddle River, New Jersey 07458

Library of Congress Cataloging-in-Publication Data

Harper, Charles L.
Environment and society : human perspectives on environmental issues / Charles L. Harper.—2nd ed.
p. cm.
Includes bibliographical references and index.
ISBN 0-13-016555-7
1. Human ecology. 2. Environmental policy. 3. Environmentalism. I. Title.

GF49 .H373 2000
304.2'8—dc21

00-033644

For *Anne*
. . . my mate and best friend, who continues to make my life beautiful by loving me, blemishes and all.

VP, Editorial Director: Laura Pearson
Publisher: Nancy Roberts
Managing Editor: Sharon Chambliss
Project Manager: Merrill Peterson
Prepress and Manufacturing Buyer: Mary Ann Gloriande
Cover Director: Jayne Conte
Cover Design: Joseph Sengotta
Cover Illustration: Jose Ortega/Stock Illustration Source, Inc.
Director of Marketing: Beth Gillett Mejia
Part opening photos: page 1: Tim McCabe/U.S. Department of Agriculture; page 79: Dave Van de Mark/USDA/NRCS/NCGC/National Cartography and Geospatial Center; page 169: U.S. Public Health Service; page 305: NASA Headquarters.

This book was set in 10/12 Palatino by DM Cradle Associates and was printed and bound by Courier Companies, Inc.
The cover was printed by Phoenix Color Corp.

A Division of Pearson Education
Upper Saddle River, New Jersey 07458

Printed in the United States of America

10 9 8 7 6 5 4 3 2 1

ISBN 0-13-016555-7

Prentice-Hall International (UK) Limited, *London*
Prentice-Hall of Australia Pty. Limited, *Sydney*
Prentice-Hall Canada Inc., *Toronto*
Prentice-Hall Hispanoamericana, S.A., *Mexico*
Prentice-Hall of India Private Limited, *New Delhi*
Prentice-Hall of Japan, Inc., *Tokyo*
Pearson Education Asia Pte. Ltd., *Singapore*
Editora Prentice-Hall do Brasil, Ltda., *Rio de Janeiro*

Contents

PART III THE HUMAN CAUSES OF ENVIRONMENTAL PROBLEMS

Chapter Five

Population, Environment, and Food *171*

Chapter Six

Energy and Society *220*

Preface

Environment and Society: Human Perspectives on Environmental Issues is intended to provide college students and other interested readers with an introduction to environmental problems and issues. More specifically, it is about the human connections and impacts on the environment—and vice versa. There are many specialized research reports and monographs about particular environmental topics and issues, but I intend this book to work as an integrative vehicle for many different human and environmental issues. It is intended to be usable in a variety of settings that are seriously concerned with the connections between human societies, ecosystems, and the geophysical environment. It is appropriate for upper division undergraduates and, with appropriate supplements, for beginning graduate students.

Stimulated by the enormous growth of interest in environmental issues and problems in higher education, the book is addressed to the diverse backgrounds of students in classes and programs that attend to environmental and ecological topics. My own classes have a yeasty mix of students from biology, environmental sciences, the social sciences, and sometimes others from education, philosophy, or marketing. I tried to write a book that is at least understandable to them all. Its social science perspective is mostly sociological, but readers expecting a narrow disciplinary treatise will be disappointed. I hope it will be intellectually challenging for students, but perceptive readers will note that in some places the book alternates between more advanced and more elementary topics. This is deliberate, because social science students know some things that natural science students do not, and vice versa.

The book treats blocks of material that recognizably constitute contemporary environmental concerns, controversies, and discourses that you can see from the table of contents. The second edition has new data in many places, new material about human ecology and world political economy that connects human environmental issues to the evolution of ecosystems; that material frames later, more particular issues. This edition also has new material in many places—about, for instance, the economic costs of declining biodiversity, energy transitions at the end of the fossil age in the coming century, community resource management, environmental movements, and global issues.

As with most such books, some chapters can be omitted or rearranged, but I have tried to write a book that is truly developmental and ties the topics of different chapters together. One pervasive theme is that disciplinary scholars bring very different intellectual views (*paradigms*) to the understanding of human-environmental issues. I argue that these different views are not ultimately irreconcilable. But if you do not like attention given to different points of view, this is probably not the book for you.

This is a book about "big issues," but it is, I hope, written in a way that engages individual readers. I had intended to include an epilogue to examine the connections between big issues and the personal life, but reviewers suggested that I do so in smaller installments at the end of each chapter instead of at the end of the book. So each chapter is followed by some questions and issues ("Personal Connections") that attempt to make macro-micro links between large-scale issues and the lives of persons. These are *not* "review questions" that summarize chapter content, but rather they provide opportunities for dialogue between the book and its readers. They may provide points of departure for discussion and argumentation. I hope they are useful, but they are clearly not everybody's cup of tea, nor will they be useful for every setting in which the book is used.

Every intellectual work is in some sense autobiographical. My early college education (of many years ago!) was in biology and the physical sciences. But I subsequently pursued graduate studies in sociology, and for years I have been engaged in a professional life that dealt only peripherally with environmental and ecological issues. This book attempts to put together the chronological pieces of my education into a coherent whole, and to do so in a way that addresses important intellectual and social concerns of our times.

This book is also dedicated to George Perkins Marsh, Aldo Leopold, Rachael Carson, Lois Gibbs, Karen Silkwood, Jaime Lerner, Chico Mendez, and Wangari Maathai—in different ways, all pioneers in consciousness and concern about the connections between humans and the natural world. All appear briefly in these pages. Some were stigmatized by powerful people and agencies. Some paid with their lives.

Intellectual works are not just autobiographical. They involve the insights, encouragement, and constructive criticism of many others. I am indebted to many persons for helping to bring the idea for this book to completion, and I need to thank them. I thank my colleagues and students at Creighton University, who contributed substantially to this work and who also tolerated me while I was working on it. Thanks especially to Tom Mans, who fed me a constant stream of relevant articles and material for several years, and to James T. Ault, who had the patience to read and critically comment on many parts of the book. Thanks to Dean Barbara Braden of the Creighton University Graduate College for her important material support. Finally, thanks to Karen Prescott, our reliable departmental secretary, who suffered through the formidable task of helping to get this manuscript ready to send to the publisher.

I also want to thank a truly amazing network of environmental social scientists at other institutions who supported the first or second edition. They include Riley Dunlap (Washington State University), William Freudenburg (University of Wisconsin), Eugene Rosa (Washington State University), Thomas Dietz (George Mason University), Robert Brulle (George Washington University), J. Allen Williams (University of Nebraska–Lincoln), Andrew Szasz (University of California at Santa Cruz), Paul Stern (National Research Council), and Bruce Podobnik (Lewis and Clark College).

These colleagues sent me, sometimes unsolicited, an incredible collection of their research papers and reports that inform various parts of the book. I do not, of course, hold them responsible for errors or omissions. They are mine alone. I thank the reviewers of the manuscript at different stages of completion, who were critical but universally encouraging, especially Victor Agadjanian, Arizona State University. I owe an enormous debt of gratitude to Publisher Nancy Roberts and Managing Editor Sharon Chambliss, who have suffered with me through several projects and who have been patient, supportive, and encouraging. Through the years they have been the "human faces" of Prentice Hall.

If you would like to contact me, I would be happy to hear your comments and reactions to the book and its uses. I look forward to improving it.

Charles L. Harper
Department of Sociology and Anthropology
Creighton University
Omaha, Nebraska, 68178
charper@creighton.edu

PART I

INTRODUCTION

CHAPTER ONE

Environmental Problems and Ecosystems

Assuming that your parents paid attention to the news, how many of these words or phrases do you think they would have found familiar when they were your age?

> Acid rain, air pollution, smog, thermal inversion, deforestation, global warming/greenhouse effect, indoor air pollution, landfill overcrowding, low level nuclear wastes, meltdown, eutrophication, urban sprawl, landfill overcrowding, ozone depletion, global warming, Kyoto treaty, radiation from power lines, species extinction, sustainable development, biodiversity, toxic waste dump, desertification, green politics, green consumerism, NIMBY syndrome

My guess is that they would have been familiar with two or three of them at the most (probably air pollution/smog, and toxic waste dump). My guess is also that you have at least heard of many of them. That, I think, is one measure of how rapidly and pervasively environmental issues and problems have entered the popular consciousness and political discourse of our times. This book is about those problems, their human causes and implications.

In the last several decades, just behind the headlines about politics, jobs, crime, and other perennial media topics, there has been a steady barrage of news items about environmental problems and controversies. I'm sure the environmental news is not really news to you. But, just for openers—to frame but not analyze issues—here's a quick and selective laundry list of some particularly newsworthy environmental issues of our times and some controversies surrounding them.

ENVIRONMENTAL NEWS

The Vanishing Wilderness

The world's forests are gradually being destroyed. In 1999, half of the forests that once blanketed the earth—3 million hectares—were gone. And the loss has accelerated since the 1960s, so that each year another 16 million hectares are converted to "other uses." Deforestation is not the only problem: secondary (regrown) forests are very different from the original ones and the fragmentation is a serious problem (Abramovitz, 1998: 124–125). There are now few virgin or "old-growth" forests left anywhere in the world. The rainforests of the tropics and subtropics are being rapidly depleted as is the planet's great conifer forest on the Pacific Northwest coast in the United States and Canada, stretching from the Olympia Peninsula in Washington to the Tongas forest in Alaska. In parts of the world, trees are cut for fuel wood by villagers, but mostly they have been destroyed commercially for lumber or to clear land for agriculture. Trees and forests are particularly important because they hold soil, maintain water tables, and recycle the gases that maintain the chemical balance of the atmosphere. Virgin rainforests, or "old-growth" forests, are particularly important in that they provide habitats for species of living things that live nowhere else (Raven, 1990; Rohr, 1992).

Along with the coastal wetlands, marshes, and mangroves swamps—also diminishing rapidly because of pollution and the encroachment of human settlements—the rainforests are the great genetic "biological storehouses" of the planet. But at the rate of destruction of the forests and wetland habitats, at least fifty to one hundred species disappear each day—for good (Rohr, 1992). Why worry? Because time after time species thought useless have proven essential to the ecological systems that support humans. Living things in little known corners of the world have also provided invaluable sources of human medicines, including those that treat cancer, malaria, leukemia and Hodgkin's disease, high blood pressure, and multiple sclerosis (Miller, 1998: 335–344).

Agricultural Resources under Stress

The resources that produce human food are also threatened. Crop and range land is threatened by soil erosion and degradation resulting from overuse. More land has been brought into agricultural production—some of it fragile and marginal land—and irrigation, fertilizers, and new seed hybrids have steadily increased *total* food production. But *per capita* food production has declined in the last decade, so that in the context of growing human population an era dominated by food surpluses may be coming to an end (Brown, 1998:13). Furthermore, a 1992 United Nations report alarmed experts about the degree to which soils are eroding and being degraded around the planet. The problem is that each year the world's farmers must feed millions more people with billions of tons less topsoil.

The consequences of overharvesting and poor resource management is not limited to the land: The world's fish catch peaked in the 1970s and has steadily declined. In spite of the catches by Mexican and Indonesian fishers that resulted in an exceptional world catch in 1996, the United Nations Food and Agriculture Organization (FAO) estimates that eleven of the world's fifteen major ocean fishing grounds are seriously depleted (Strauss, 1998: 34).

But the most critical resource for agriculture as well as industrial and household use—clean, fresh water—is coming to be in short supply on a global basis. As agriculture, populations, and human economies grew, global water use increased dramatically since the 1950s, and the consequences of such expanding demand are becoming apparent in falling water tables, shrinking lakes and wetlands, and dwindling streams and rivers. Around the world water shortages result in economic and legal conflict, and they amplify international tensions in water-short areas (Livernash and Rodenberg, 1998: 20).

Pollution and Other Garbage

Environmental problems are not limited to the overuse and poor management of physical and biological resources. They also involve the *pollution and wastes* (or "effluents") that result from human social and economic activity. In the United States, 25 percent to 75 percent of the groundwater is polluted by some combination of seepage from underground storage tanks, hazardous wastes, sewage and landfills drainage, or from accumulated nitrates, pesticides, and herbicides from farming. Hazardous wastes are stored in *thousands* of contaminated sites that may threaten human health or the environmental system (Miller, 1998: 534, 566).

Americans generate 200 million tons of solid municipal waste per year, commonly called *garbage* (Miller, 1998: 566). Many municipal landfills are overflowing or reaching capacity, even though local communities resist building more landfills and garbage dumps "in their back yards." The problem is so visible and severe that garbage has been dumped at sea or "exported" away from cities, sometimes to less developed countries, and in wealthier nations many communities are experimenting with recycling programs to deal with its vast flow (Stark, 1999).

Various forms of air pollution also present problems. *Acid rain* produced from automobile and industrial emissions has killed or damaged forests in Appalachia and the New England states as well as in Canada, the Black Forest in Germany, Poland, and Central Europe. In Eastern Europe and Russia, these conditions have produced shortened life expectancy, soaring cancer rates, and a host of other maladies (French, 1990: 11–20).[1] Dangers from urban *smog*, produced by auto and industrial exhausts trapped under atmoshperic thermal inversion layers, have been improved in the industrial companies by modern air pollution controls. But the hazardous health experiences of London, Los Angeles, and Pittsburgh are now being repeated in Mexico City, New Delhi, São Paulo, and many other places in the industrial world. Each year, coal burning kills prematurely an estimated 178,000 persons in China alone (Flavin and Dunn, 1999: 25).

The high-drama media events of chemical pollution are, of course, the *ocean oil spills*, which have visible and environmentally devastating effects. Oil spills from offshore production rigs and tankers have not been rare events. North Americans remember them from Santa Barbara in 1969 to the Gulf of Mexico near the Mexican and Texas coastlines in 1997, with the megaspill being the 1989 wreck of the tanker Exxon *Valdez* in the pristine northern environment of Prince William Sound in Alaska. Hitting submerged rocks, the tanker created an oil slick that coated more than 1,600 kilometers (1,000 miles) of shoreline. The full number of wildlife lost will never be fully known because most died and decomposed without being counted. The good news is that many wildlife populations are recovering, but long-terms effects on such natural ecosystems are unknown. The company was required to spend billions on a cleanup that in some ways did more harm than good. *But* you need to know that more oil is released during the normal operations of offshore wells—washing tankers and releasing oily water into the seas—and from pipeline and storage tank leaks. A 1993 study found that U.S. oil companies spilled or leaked as much as 1,000 huge Exxon *Valdez* tankers—*more oil than Australia uses* (Miller, 1998: 527–529)!

A frightening potential form of toxic waste is the *radioactive material* associated with nuclear power plants. Doses of strong nuclear radiation can kill quickly, but low doses have the potential to produce, over time, much higher rates of genetic mutations, birth disorders, and fatal cancers. Several nuclear plant "mishaps" have almost occurred around the world, (notably Three Mile Island, Pennsylvania, in 1979), and the Chernobyl reactor in Kiev, USSR (now Ukraine) actually did explode in 1986. Because of their long-lasting toxicity, however, even in the absence of such a catastrophic event the radioactive waste materials from such power plants are problems in themselves: Where and how to store these materials undisturbed by leaks, geological disturbances, human sabotage or error for hundreds of years? But such accumulating nuclear wastes *have* to be stored somewhere, as long as we produce electricity in such a way. In your back yard, maybe?

The Spectre of Climate Change

I've saved the broadest and most controversial environmental threat of the millenium till last. Climatologists have noted that for several decades carbon dioxide and other gases have been accumulating in the atmosphere. As they accumulate, those gases increase the amount of heat retained in the atmosphere, just like the glass panes in a greenhouse or your car window on a hot summer day.

The earth's temperature *is* measurably warming, and such fluctuations have happened "naturally" before in history and prehistory, but over *very long* time spans. By 1997, the Intergovernmental Panel on Climate Change (IPCC) had taken the position that the rapid increase in greenhouse gases is the gradual result of the combustion of carbon fuels and other emissions produced by modern societies, which now emit literally billions of tons of carbon dioxide and other gases into the atmosphere per year. This would likely result in a "wide array of dislocation to human and natural systems" (Dunn, 1998: 66). Most relevant scientists agree, but some do not. As you may know, there not only scientific uncertainties here, but also intense public and political controversies about dealing with their implications. One hundred and seventy-one nations, including the United States, agreed to do so at Kyoto, Japan in 1997. But, strongly influenced by American petrochemical industries, the U.S. Senate pledged not to ratify the treaty if it were sent to them by the president. A second international meeting about the problem in Buenos Aires, Argentina, ended inconclusively. Since there is an entire chapter about this topic later on, I will say no more now.

Taken together, this is quite a gory environmental bill of particulars, and by now you have probably grown a bit jaded just by reading it. I have presented it to you as unsystematic bits of information without much depth or background—the written equivalent of news sound bites, if you will. I could go on and on. But you get the picture. In modern times, human social and economic behavior has resulted in broad, multidimensional assaults on the integrity of the biophysical earth.

ECOCATASTROPHE OR ECOHYPE?

Is all that just alarmist stuff? How *real* are these problems? Sure, everyone knows that there are environmental problems—with pollution and the rainforests, nuclear energy, and the possibility of global warming. But is ecocatastrophe really around the corner, or are the problems greatly exaggerated?

Like me, you probably don't spend much time or energy thinking about these problems. The world seems okay: I get up and go to work and enjoy my family life, farmers continue to grow food that is plentiful and normally tasty, and drinking tap water has not made me ill (not yet, anyway). In a similar reflection about a Thanksgiving holiday, environmental social scientist Lester Milbrath mused that he had much to be thankful for. His house was warm and stocked with food for a feast; and dinner guests were soon to arrive. His personal world and society were functioning reasonably well. Yet Milbrath had a nagging sense of unease that all was not really well (1989: 1). To many of us in the richer nations, the biophysical world still seems okay. Perhaps, if you are like me, it is hard to experience directly the environmental devastation depicted here. We are aware, of course, that there *is* human suffering, poverty, and disease in the world, but to most of us the economic, political, and individual causes of human problems and misery seem more direct and obvious than the environmental ones.

Pollyannas and Cassandras

We do, of course, have the capacity for abstract and long-range thinking that transcends our concrete experience. Yet we live mostly in the concrete present, so we worry more about losing our jobs, or about crime or family problems, than about the hazardous pollutants we come in contact with or the possibility of a core meltdown at a nearby nuclear power facility. Indeed, particularly for those of us in the world's more developed nations, the whole notion of ecocatastrophe may seem hazy, abstract, and hypothetical. Are environmental activists romantic tree-hugging *"Cassandras,"* habitually painting unreal gloom-and-doom scenarios, but who, in fact advocate policies that impede human progress?

But the problem is not simply that our life experience and awareness are often concrete and short term, while the environmental threats described are mostly abstract and long term. There *is* a body of scholarly opinion that argues—on the basis of broader arguments and evidence like that just cited—that fears of an impending global ecocatastrophe are greatly exaggerated. And while such thinkers do not deny the reality of *all* environmental problems, they generally argue that actively "doing something about them" as urged by environmental activists is not only misguided but would involve costs and blunders worse than "letting nature take its course." Fixing the environment, they argue, is fixing something that "ain't really broke." (For a sampler of such views, see Samuelson, 1992; Simon, 1990, 1996, 1998; Simon and Kahn, 1984; Singer, 1992, 1997; McKenzie,1992; Lee, 1992.)

These ecological *"Pollyannas"*—to continue the caricature—have been in continuous debate and contention with the Cassandras who paint ominous scenarios of impending ecocatastrophe.[2] Given the environmental bill of particulars we have elaborated, how *can* it be argued that "Things are okay. Don't worry."? There are at least three ways.

First, the Pollyannas argue that this bill of particulars is selective and distorted. There is evidence of environmental degradation, but also of improvements that get left out of the threatening scenarios. There are, for example, more trees now in the Hudson Valley in New York State than at the turn of the twentieth century (Mann, 1993: 59). Large regions of the Midwest, in Minnesota, Wisconsin and Missouri, were clear cut for lumber by the 1920s, where today national forests have been regrown and are protected. We have exotic new forms of pollution but fail to appreciate the health and aesthetic insults of open sewers and piles of horse manure that plagued urban dwellers at the turn of the century.

Salmon have been reintroduced into the Thames and Lake Erie is no longer a biologically dead lake, as it was in the 1960s. White-tailed deer, almost extinct in 1900, now plague New England gardens and Iowa cornfields. American bison and alligators are certainly no longer endangered species. Lead pollution has gone out of American gasoline and the atmosphere. Smog—still a problem—has remained level while the economy has grown, and in some cities (e.g., Tokyo) it is markedly less a problem. In general, the argument of the Pollyannas is that while some things get worse, other things improve, and that there is no global trend of environmental degradation (Mann, 1993: 59).

Second, it has been argued, most relentlessly by economist Julian Simon, that resources—whether soil, water, energy, or biological—are never really scarce, because human ingenuity keeps finding more of them and human inventiveness has always come up with ways to get around or resolve existing scarcities. Simon and others note the historic relationship between scarcity and inventiveness. For example, the English used coal to replace England's decimated wood fuel supplies in the 1700s, and in America in the 1800s kerosene was substituted as fuel for lamps for refined whale oil, when the New England whale hunters hunted sperm whales almost to extinction. Simon argues that the most important resource is *human inventiveness* and that impending shortages typically produce market conditions that encourage resource conservation, technological innovation, and progress. Simon points out that as human populations have grown, many resources—such as food—have become cheaper and more plentiful and that more people are living better, healthier, and longer lives today than at any other time in human history. These facts, he argues, should persuade us that ecocatastrophe is not lurking just around the corner (1990; 1994; 1996; 1998).

Third, environmental Pollyannas argue that the forecasters of global collapse, even so-called scientific ones, have a miserable track record about this because they oversimplify and misapprehended the way the world works. Such predictions—including the ideas of Malthus about the disastrous consequences of overpopulation in 1798, the predictions by demographers of impending world famine in the 1975 (Paddock and Paddock, 1975), and the prediction by the prestigious "Club of Rome" researchers of an impending collapse of the world ecological system that would be visible by the turn of the century (Meadows et al., 1972)—have not come true. Furthermore, some scientists contest the global warming theory, based on mathematical models with many unknowns, or predict that natural processes will counter its effects (Singer, 1992, 1997). Kenneth E. F. Watt, an environmental scientist from the University of California–Davis, has gone so far as to call the global warming theory the "laugh of the century" (Sanction, 1989: 28). The point: If the Cassandras have always been wrong, why should we believe them now?

Veteran Cassandras (Brown and Flavin, 1999; Ehrlich and Ehrlich, 1992; Hardin, 1993; Lovins, 1977; Meadows et al., 1992; Ophuls and Boyan, 1992) concede many specific points made by their critics without giving away the main thrust of their arguments. It is true, for instance, that the earth is like a huge recycling system that dilutes, soaks up, transforms, and restores the damage done to it by humans or other species. So there are success stories, and environmental problems can often be cleaned up, or left to restore themselves.

However, these contemporary success stories are all local or regional cases. Data about environmental problems aggregated over time for most nations—and the world—are unremittingly depressing. The Hudson River Valley forests rebounded because farmers abandoned them in order to wipe out the native grasslands of the Great Plains. Lake Erie has come back to life but barely so; meanwhile, many other lakes, such as Lake Superior, are in a state of profound chemical and biological degradation. And the lower Mississippi River is so polluted by chemical industries that it has become famous as America's premiere "toxic waste corridor." It is true that the lumber industry has moved away from Wisconsin and Missouri; they have moved on to the old-growth conifer forests of the Pacific Northwest, which, as I previously noted, they are cutting with a vengeance. Furthermore, the ecological "success" stories are mostly in the richer, more developed nations, which can develop more effective responses that at least buy time in a way that less developed nations cannot. But even in the more developed nations difficulties remain: soil erodes, wetlands shrink, more species disappear than are successfully protected, and toxic wastes keep accumulating (Meadows et al., 1992: 59).

Projections about human and environmental futures are highly speculative, and they have failed *insofar as they have forecast a general environmental collapse*. For centuries, humans caused no irreparable environmental damage when "population was sparse, factories small, and products few in number compared to today. The environment's dilutive capacity was rarely exceeded and was perceived as infinite in its ability to absorb waste" (Buchholz, 1993: 8). But environmental Cassandras point to abundant evidence that contemporary ecological circumstances are without historical precedent. They have a different and more sinister character because (1) there are *many* more people making demands on the environmental resource system, (2) they have dramatically more powerful technologies to disrupt, and alter, the environment than preindustrial people did, (3) people around the world want an "improved standard of living," which means higher levels of material consumption and environmental damage, and (4) the volume of garbage derived from human activity is much greater and much more chemically difficult for nature to absorb and recycle than were preindustrial effluents.

Increasingly, there is less wilderness to migrate to from overpopulated and environmentally degraded areas, and fewer places to throw it away and forget it. Unprecedented human numbers and technical capabilities mean that we are increasingly forced to live in a "closed system" (more about this concept later), and compelled to live with the ecological consequences of our behavior in ways that humans never have been before. Pollyannas, according to the Cassandras, fail to appreciate the ecological implications of the unprecedented increase in the scale of human activities.

I'm sure you realize by now that we have been talking about an extremely complicated set of issues and controversies. And I hope you also recognize that they are terribly important for the human future—if not for you, then certainly for your children and grandchildren. Not "merely" scientific and academic debates, they have become political issues and policy dilemmas. The American political process is infused at all levels by arguments and political conflict between those who want to protect nature from impacts and keep it "clean" for reasons of heath and well-being and those who argue that further environmental protection itself represents threats to individual liberty, jobs, a healthy economy, and the material security and aspirations of citizens. This political dialogue reverberates not just in the United States but around the world. With the more traditional political controversies, environmental issues have unquestionably entered the world's political arenas.

I will return to these controversies throughout the book because they are the defining problems of contemporary intellectual and political discourse about humans and the environment. But I won't use the terms *Cassandra* and *Pollyanna* very often because both the issues and the contemporary debates are more complex than those labels suggest.[3]

But now I step back from the smoke and fire of these scholarly and popular debates to examine more basic ideas and concepts about environments and ecosystems—and how they work. The following chapter does the same for human social systems and the relation of humans to the environment. My purposes here and in Chapter Two are (1) to introduce concepts and an agenda of topics in the rest of the book and (2) to enable you to put topics and controversies about human-environment interaction in more intellectually rigorous contexts.

ECOSYSTEMS

Many physical sciences, including physics, geology, chemistry, and atmospheric sciences, have contributed to our current understanding of the biophysical environment, but the most sustained and focused interest in environmental questions has come from *ecology*, which has been a specialty field within biology for a long time. The term *ecology* was coined in 1868 by German botanist Ernest Haeckel, who thereby provided a conceptual framework for the study of the relationships between organisms and their environments. Ecologists are particularly interested in (1) the process by which species adapt to environmental changes, and (2) the exchange of matter and energy among functionally interrelated species of plants and animals in a given habitat (Humphrey and Buttel, 1982: 39). Historically, the ecological perspective was a manifestation of the broader scientific revolution that took place in the eighteenth and nineteenth centuries. It was stimulated by the reports of world traveler-naturalists such as Alexander Von Humbolt, but more importantly by the seminal scientific works of James Hutton (1726–1797) in geology; Thomas Malthus (1766–1834) in population studies; and Charles Darwin (1804–1882) in biological evolution theory. In America ecology was stimulated by the late nineteenth century conservation movement, which led to the establishment of national parks and forests, and by concern about overgrazing and desertification of the Western states rangelands. An early and particularly sophisticated ecological study about these problems was written by George Perkins Marsh, titled *The Earth as Modified by Human Action* (1874). In the 1920s, Austrian-born American biophysicist Alfred J. Lotka could have been writing for a present-day government report when he declared:

> Whatever may be the ultimate course of events, the present is an eminently atypical epoch. Economically, we are living on our capital; biologically we are changing radically the complexion of our share in the carbon cycle by throwing into the atmosphere from coal fires and metallurgical furnaces, ten time as much carbon dioxide as in the natural biological process of breathing. (1924/1956: 222)

Lotka's ideas helped to found the modern science of ecology. I will return to them later in this chapter.

But it was really studies of the 1960s, such as those of Rachel Carson about the impact of insecticides on wildlife (*Silent Spring*, 1962) and zoologist Paul Ehrlich on human population problems (*The Population Bomb*, 1968), that ensured the growth and popular support for ecology as a scientific discipline as well as an intellectual cause. I won't say more here about the development of ecology; there are many excellent books and texts that can provide you with more information if you wish (e.g., Miller, 1998; Odum, 1971, 1983; Southwick, 1995).

Ecosystem Concepts

The most fundamental concept for ecological understanding is the notion of a *system* as a *network of interconnected and interdependent parts*. Not everything is a system, of course; there are also *aggregates*, which are assemblages of things or elements in adjacent time and/or space, but these elements are not connected by "systemic links." The concept of system is important to many sciences, including most of the social sciences, but it is particularly important for ecology, which views the "sum" as greater—or at least different from—the parts taken separately. In other words, the science of ecology is based on *holistic analysis*, which seeks to understand the interconnections between organisms in an environmental field of forces.

An *ecosystem* is the most basic unit of ecological analysis, which includes all the varieties and populations of living things that are interdependent in a given environment. The parts of an ecosystem are operationally inseparable from the whole. An entity, such as a pond, lake, or tract of forest, may be considered an ecosystem as long as the major components of the system are present and operate together to achieve some sort of stability in the way that they function, if only for a short period of time (Odum, 1971: 9). All systems have a degree of stability (or *equilibrium*) in the configuration of parts, but systems are also dynamic and changing, so that the parts are also continuously *reconfiguring* themselves, as it were. There is a *balance of nature*, but it is a dynamic and changing rather than a static one. Finally, it is important to note that there are systems within systems (or *subsystems*) at many different levels. While each system and subsystem must have identifiable boundaries, most such boundaries are in fact zones or *gradients* across which matter and energy flow or across which species sometimes migrate. So far this is all pretty abstract, so now let me be more concrete.

Ecosystem Units

Ecological systems are composed of structural units that form a progressively more inclusive hierarchy:

Organism	Any individual form of life, including plants and animals (Felix and Fido; you and me)
Species	Individual organisms of the same kind (e.g., dolphins, oak trees, corn, domestic cats, humans)
Population	The group of individual organisms of the same species living within a particular area
Community	Populations of different plants and animals living and interacting in an area at a particular time (e.g., the interacting life forms in the Monterey Bay estuary in California)
Ecosystem	The community of organisms and populations interacting with one another *and* with the chemical and physical factors making up the inorganic environment (e.g., a lake; the Amazon basin rainforest; the High Plains grasslands in the United States)
Biome	Large life and vegetation zones made up of many different smaller ecosystems (e.g., tropical grasslands or savannas, northern conifer forests)

In addition to these, there are two other important terms you should know that ecologists use. The *biosphere* is the entire realm where life is found. It consists of the lower part of the atmosphere, the hydrosphere (all the bodies of water) and the lithosphere, (the upper region of rocks and soil). Combined, the biosphere is a relatively thin, 20-kilometer (12-mile) zone of life extending from the deepest ocean floor to the tops of the highest mountains (Miller, 1998: 92). Some ecologists use the term *ecosphere* to mean the earth's total collection of living things found in the biosphere. The goal of global-level ecology is to learn how this thin layer of life interacts with the earth and maintains itself. Exchanges (or *cycles)* form the interconnections that bind the components of ecosystems and subsystems with the inorganic environment.

Cycles

Matter, energy, and nutrients circulate in predictable paths from the biosphere to living organisms and back. Among the more important *such cycles* are the flows of carbon, nitrogen, oxygen, phosphorus, and water (the *hydrological cycle*). The carbon and water cycles are illustrated in Figures 1.1 and 1.2.

The carbon cycle illustrates the interdependent *system connections* between living things, since green plants (*primary producers*) absorb carbon dioxide and water from geochemical reservoirs and—with energy inputs in the form of sunlight—manufacture complex carbohydrates and emit oxygen as a waste product. This complex reaction, *photosynthesis*, can be summarized as follows:

$$6CO_2 + 6H_2O + \text{sunlight} \text{ ----> } C_6H_{12}O_6 \text{ (glucose)} + 6O_2$$

Plants then use the energy stored in glucose and other organic nutrient compounds to drive their life processes, as do the animals (*consumers*), which eat the green plants and, in turn, each other. This process, *aerobic respiration*, can be summarized as follows:

$$C_6H_{12}O_6 + 6O_2 \text{ -----> } 6CO_2 + 6H_2O + \text{energy}$$

With vast oversimplification, you can see that these two processes, *photosynthesis* and *aerobic respiration*, are the virtual biogeochemical basis of life, in which oxygen is used to release energy stored in the chemical bonds of carbohydrates and other organic nutrient compounds. There are, of course, purely geochemical cycles, such as the hydrological cycle, in which water is cycled from the earth to the atmosphere and back without involving living things. Note that in these cycles no matter is destroyed but only rearranged in its chemical forms. In fact—aside from nuclear reactions—no matter *can* be destroyed. The *law of conservation of matter* is that matter cannot be created

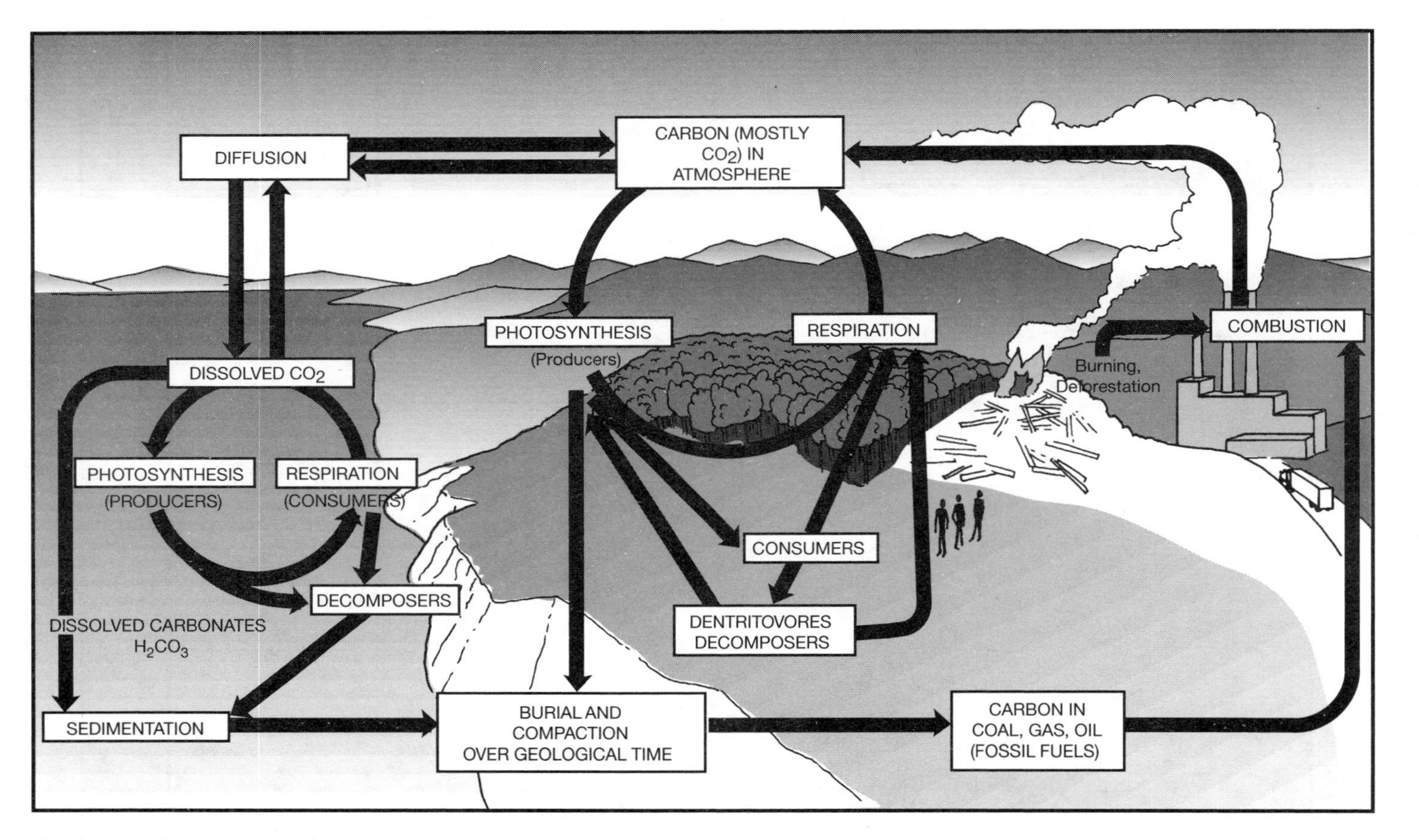

Figure 1.1 The Carbon Cycle
Source: Adapted from T. G. Miller, Jr., 1998: 114–115.

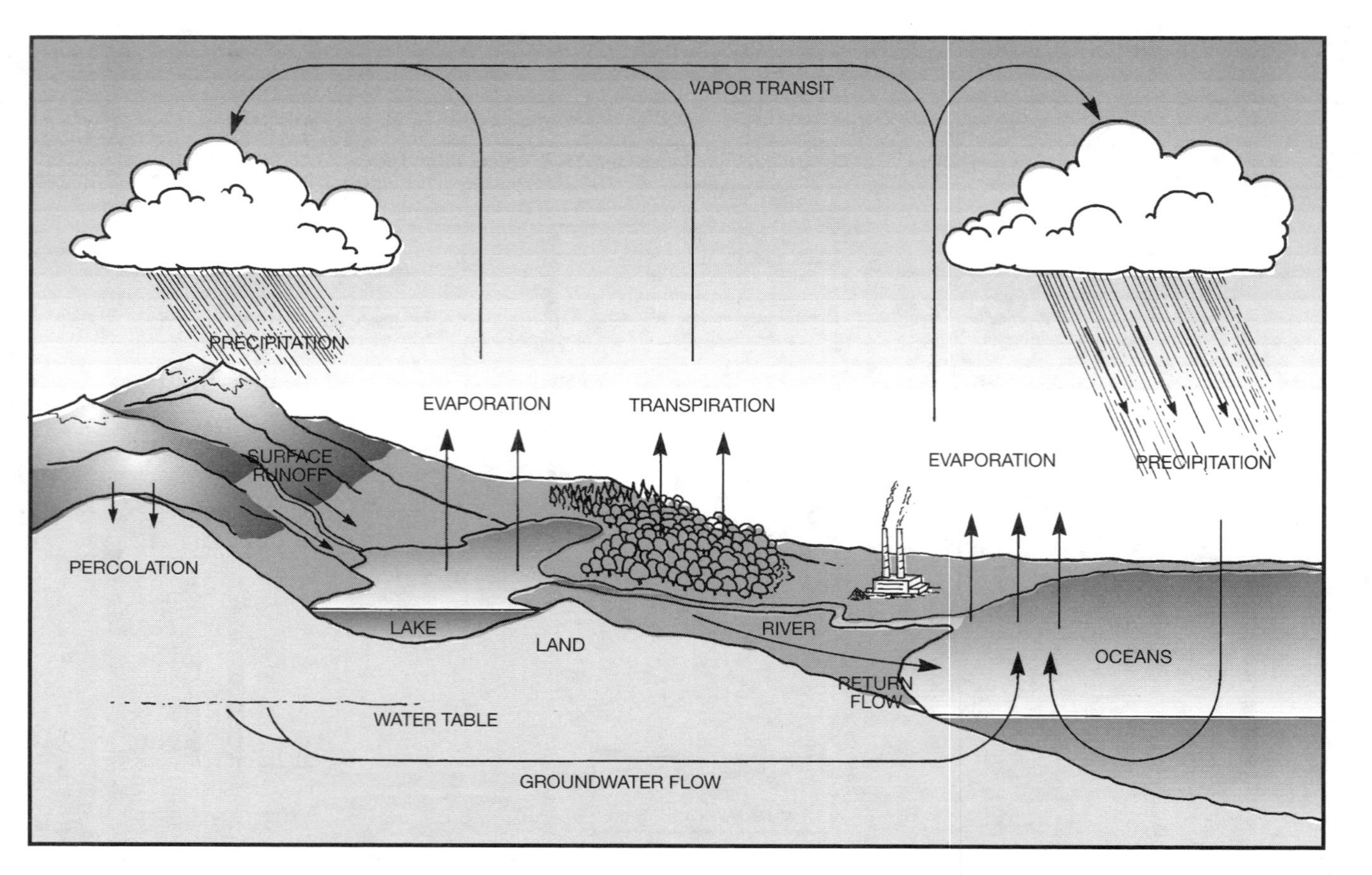

Figure 1.2 The Water Cycle
Source: Adapted from J. W. Maurits La Riviere, 1990: 39.

or destroyed, only changed in form (and the practical environmental principle is, taking the earth as a whole, there is really no "away" to throw anything into!).

Also note that these cycles are symmetrical in terms of energy. The ultimate source of the earth's energy is solar radiation, which is built up into complex forms and used by living things, then eventually emitted into the environment, mostly as low-quality heat energy near the earth's surface. Similar to the conservation of matter, the *first law of thermodynamics* says that energy cannot be created or destroyed, only changed into different forms. So, like matter, you can't get something for nothing—energy input always equals energy output. But wait! When you use energy, you can't even break even. Unlike matter, energy can't be recycled over and over. Respiration or burning gasoline in your car permanently degrades useful complex forms of energy (such as that stored in complex carbohydrates or petrochemicals) to low-grade forms, such as heat that can't be reused. The *second law of thermodynamics* states that we can't recycle or reuse high-quality energy to perform useful work again. This tendency for energy to run downhill is called *entropy*. The heat produced from combustion and from the respiration processes of living things is eventually diffused over the earth and radiates back into space. All of our usable energy comes from solar radiation that has been stored in various forms (I will have more to say about energy in Chapter Six).

Food Chains

The transfer of food energy from its source in plants (primary producers) through a series of consumer organisms where eating and being eaten is repeated a number of times is called a *food chain*. For example, *humans* eat *big fish*, which eat *small fish*, which eat *zooplankton* (tiny to microscopic aquatic invertebrate animals such as brine shrimp), which feed on *phytoplankton* (small and microscopic drifting plants, mostly algae and bacteria) which produce carbohydrate compounds by photosynthesis. A special ecological class of consumers, *detritovores*, feeds on and decomposes dead organisms and the wastes of living organisms and return important nutrient chemicals to the soil (examples include earthworms, termites, fungi, and bacteria). Food chains do have distinct feeding levels, but really they are food *webs* since they are far more complex than the simple chain in our illustration.

At each transfer point (the feeding or *trophic level*), some energy is lost or transformed into heat, and so the high-quality energy available for respiration diminishes at each level. This effect illustrates the process of entropy just described. The greater the number of trophic levels, the greater the cumulative loss of usable energy, which explains why larger populations at lower trophic levels are required to support smaller populations at higher levels, and particularly at the top of food chains. This means that food chains or webs are also *food pyramids*. For example, a million phytoplankton in a small pond may support 10,000 zooplankton, which in turn may support 100 perch, which might feed one person for a month or so. This energy-flow pyramid explains why larger populations of people can be maintained if people eat mostly at lower levels on the food chain (by directly eating grains such as wheat or rice) than at higher levels (by eating cattle, which have been fed on grains) (Bender and Smith, 1997: 14–16).

Understanding food chains is important for another reason. Pollutants and toxins are also transferred in this process, and rather than losing their effect, they generally become more concentrated as they move from lower to higher trophic levels. By the time they reach the top of the food chains in the flesh of animals that humans consume, toxins such as pesticides and pollutants can become so concentrated that they are harmful to human health. The term for this process is *bioaccumulation* (Miller, 1998: 260). The toxic chemical Mirex has not been dumped into Lake Ontario for more than three decades, but it is still found in the flesh of fish at rates as high as seven parts per million. To put this number in perspective, in the early 1990s you would have had to drink *half the water* in Lake Ontario to get as much Mirex as you get from eating just one fish (Boyer, 1991).

Limiting Factors and Habitats

The number of organisms an ecosystem can support depends on a very complex set of *limiting factors* or conditions for their success. Limiting factors for terrestrial species are certain amounts of water, heat, light, and soil nutrients. For example, if corn is planted in a field where there is too little phosphorus, corn will not grow well, even if other nutrient conditions—for water, nitrogen, potassium, and the like—are okay. Limiting factors are really a range in between minimum (too little) and maximum (too much) of necessary things within which living things can thrive. Thus too much as well as too little water, sunlight, and nutrients can kill (Every home gardener knows you can kill plants with too much fertilizer). Important limiting factors in aquatic ecosystems include salinity (how much salt is dissolved in water) and at different depths, the amount of sunlight, temperature, water pressure, and the dissolved oxygen content of the water.

Ecological communities adapt for different combinations of limiting factors. Some organisms can tolerate a quite narrow range of limiting factors; for example, the saguaro and organ pipe cacti occupy distinct and narrow temperature and altitude zones in the Arizona desert. Others can tolerate a broad range of limiting factors, such as ground squirrels, cockroaches, English sparrows, starlings, silver maple trees, and ragweed plants. These combinations of limiting factors produce different *habitats*, meaning simply the places where organisms live. The major categories of habitats include *marine, freshwater, estuarine* (a semiclosed coastal body of water that has a free connection with the open sea), and *terrestrial*. Each of these habitats can be divided into subtypes, according to the amount of sunlight, temperature, rainfall, and the like. In cold regions, there are, depending on the amount of moisture, cool deserts (in Mongolia and Chile), grasslands (the Arctic tundra regions in Canada and Siberia), and forests (the northern conifer forests in Canada and Alaska). At more temperate climates there are also—again depending on the amounts of rainfall—temperate deserts (in northern Arizona and Utah), grasslands (in Kansas, Nebraska, and Ukraine), and forests (in Kentucky and in Central Europe). There are also hot (tropical) deserts, grasslands, and forests, in Saudi Arabia, Kenya, and Brazil, respectively.

It is important to understand that many species are adapted to very specific combinations of limiting factors and habitats. Many are changed by human encroachments on natural ecosystems (by the addition of pollutants, draining wetlands or clearing land for agriculture). This is also why some scientists are worried about the ability of some species to adapt to rapidly changing global climate conditions. Some may be able to, but others, such as hardwood trees, which take decades to grow and reproduce, may suffer dieback because they are unable to migrate rapidly enough.

Niches

An *ecological niche* is an organism's functional role in the community, or its status. Its habitat is its address, whereas its niche is its role in a community of organisms that comprise an ecosystem. Sometimes niches overlap and two species compete for the same resources. But often different kinds of *resource partitioning* make it possible for different species to share the same habitat without much competition. For instance, species inhabit and feed from different layers of the Brazilian rainforest: some birds (the antpitta) are ground feeders, tapirs feed on short shrubs, opossums live in the shady understory, toucans live in the high canopy, and harpy eagles live and feed in the tallest outcroppings. The droppings of all of these species feed the detritovores that recycle nutrients to the otherwise fragile tropical soil.

There are other ways that species "share the wealth" in a given ecosystem. Hawks and owls feed on similar prey, but hawks hunt during the day and owls hunt at night. Where lions and leopards occur together, lions take mostly larger animals as prey and leopards take mostly smaller ones. But it is not only that species manage to share environmental resources in various ways; the notion of niche also refers to the role that species play in the ecosystem maintenance. The detritovores just mentioned play a critical role in decomposing wastes and the bodies or dead organisms to soil nutrients, enabling primary producers to grow without exhausting the soil. At the other end of the food chain, top carnivores, such as wolves and hawks, act as checks on population size of reindeer and rabbits, preventing "overshoot" and keeping the system from exceeding its carrying capacity. One important implication was already mentioned: that the loss of seemingly unimportant species (unimportant to humans, anyway) can destabilize entire ecosystems by removing the occupants of critical niches. Conversely, the addition of "nonresident" species by migration or by introduction by humans often destabilizes an ecosystem. These interlopers may outbreed or outcompete resident species because they have no established controls of predators.

Carrying Capacity

Thus, every ecosystem has limits in terms of the size of various populations that it can support (whether we are talking about plants, animals, or humans). Every organism has nutrient needs that the ecosystem and its physical environment must provide for it to thrive. If any population gets too large, the ecosystem is overloaded and cannot provide the basic needs of every organism. If this overload occurs, populations become stressed and may begin dieback. The concept of ecosystem *carrying capacity* and the possibility that population growth can produce an *overshoot* of available resources is illustrated by William Clark's analogy of bacteria in a petri dish. When bacteria are introduced into a nutrient-rich petri dish, exuberant growth follows. But in the limited world of the petri dish, such growth is not sustainable forever. "Sooner or later, as the bacterial populations deplete available resources and submerge in their own wastes, their initial blossoming is replaced by stagnation and collapse" (1990: 1). But we don't have to rely on analogies such as this; there are many real cases in which species have outgrown ecosystem carrying capacity, and after such overshoot, population size has collapsed. For example, David Klein's study of reindeer tells of the introduction of twenty-nine animals, minus wolves—their natural predators—to remote Matthew Island off the coast of Alaska. In the next nineteen years, they had multiplied to 6,000 animals and then, through starvation, had crashed to forty-two in the following three years. When discovered, the 42 reindeer were in miserable condition, all probably sterile (1968: 350–367).

Like other species, humans need space, clean air, water, food, and other essential nutrients to survive and maintain a quality existence. If human population gets too large relative to its environment, however the carrying capacity of that ecosystem may be overtaxed and human welfare may be threatened (Buchholz, 1993: 34). And like animal species, there are numerous real cases of human local and regional overshoot disasters and population crashes in various countries throughout history (I will return to some of these cases later). The human consequences have included widespread malnutrition, disease, starvation, all kinds of social stress, outmigration, and sometimes war as people compete for scarce resources.

But does the idea of a finite carrying capacity apply to humans as to other species? In spite of local and regional disasters, the overall human population has continued to grow. We have made technological, cultural, and social changes that have extended the earth's carrying capacity for us (while often *reducing* it for many other species). Are we finally approaching the earth's limits as population and material consumption have increased exponentially during the twentieth century? Some scholars believe so and describe how we are approaching or may have already passed the earth's sustainable limits. Some are in sympathy, but conclude that carrying capacity is not very useful as concept. Some, including many industry leaders, reject this projection out of hand in order to rationalize continued growth and consumption. Whether or not it really applies to humans is the big question about which tomes have been written over the last few centuries (Cohen, 1995). This is a highly contentious and highly politicized issue to which I will return in greater depth in Chapter Seven.

Explaining Ecosystem Dynamics: Change and Evolution

Today most people believe that biological species evolve and that natural selection and rare genetic mutations are important mechanisms for the evolution of species. But ecosystems also change and have done so since long before humans arrived on the scene. *How do ecosystems change and evolve?* Alfred J. Lotka, one of the founders of ecological science, provided important leads to this question beginning in the 1920s. Viewed ecologically, the competition among species is fundamentally about sources of energy. That competition triggers changing relationships among different species, often causing them to evolve into more inclusive systems. *When* energy is available in the environment, the species with the most efficient energy-capturing mechanisms has a survival advantage. The principle is that organisms with superior energy-capturing devices will be favored by *natural selection*, increasing their mass and also their total energy flux throughout the ecosystem (Lotka, 1922; 1945: 172–185).

These processes result in *ecological succession*, which is a process in which species replace one another in gradual changes. Populations tend to modify the physical environment making conditions favorable for each other until an equilibrium between the physical environment and biological organisms is reached (Odum, 1971: 257–358). Ecological systems evolve in stages from young, simpler ecosystems to older, more complex, and more mature ones. You could watch ecological succession in action by doing nothing to the grass in your yard or vacant lot for about 15 years and observing the successive invasion of weeds, low shrubs, and eventually of trees (along with the forms of rodents, bugs, and other creatures that may come with them). If you live in an urban area, however, you are likely to get a call from the city or county weed control department long before that succession process reaches its climax!

Succession can produce three kinds of biological communities. First, young or "immature" stages have a few types of relatively small, rapidly breeding organisms, low species diversity, and mostly producers, with few consumers and detritovores. Such a simple, generalized structure of ecological niches results in *growth communities* of organisms that change rapidly. Second, older, more "mature" *climax communities* have much larger and more slowly breeding organisms, much greater diversity, and a much greater mixture of producers, consumers, and detritovores, with much more complex food webs or chains. Such mature, or climax, communities are much more efficient users and recyclers of available matter and energy. They are the final or stable stage in a developmental series. Climax communities are self-perpetuating and in relative stable equilibrium with the physical habitat. In general, young stages are simple and rapidly changing, while older stages appear to be in equilibrium but in fact change more slowly. Third are communities that exhibit *pulse stability*, where pervasive physical changes from without, such as forest fires, floods, and prairie fires destroy a more mature ecosystem that gradually regrows. There is stability from such changes or oscillations, but only when there is a complete community of organisms. Most such physical stresses introduced by humans are too sudden, too violent, or too arrhythmic for such pulse stability to occur (Odum, 1971: 269).

The Evolution of Ecosystems

The long-term evolution of ecosystems is shaped jointly by two factors: (1) outside factors such as geological and climate change, and (2) processes resulting from the activities of living components of the ecosystem, such as reproductive success and the competition for energy among various organisms. Over the earth's long 3-billion-year geological history, ecosystems have evolved by (1) *natural selection*, as described earlier, and (2) *coevolution*, or the reciprocal natural selection that forms relationships between different species, called symbiosis, and (3) *group or community selection*, which produces the maintenance of traits favorable to groups even when disadvantageous to individual genetic carriers within the group. Few ecologists doubt that such group selection occurs, although the process by which it does so is unclear. Some researchers believe that "genetic drift" within populations and the extinction of some species allowing others to survive are involved. *Symbiosis*, the special relationships between organisms in an ecosystem, can be mutually beneficial (mutualism), as when birds provide services for large wild animals by cleaning them of insects. In contrast, host-parasite relations (parasitism) are only beneficial to one species but are not mutually beneficial, as when fungi or micro-organisms infect humans and other species. Interestingly, there are micro-organisms that live in human digestive tracts that appear to be in a mutualistic relation with humans by aiding in the digestive process (Odum, 1971: 271–275).

The Relevance of Ecological Theory for Human-Environment Interactions

In discussing ecosystem change and evolution, I relied extensively on the work of renowned ecologist and ecological theorist E. P. Odum, and I end this discussion with a brief summary of some of his ideas about the relevance of evolution in the natural biophysical world and human-environment interactions. Odum understood the relationships between different kinds of environments and ecosystems as a *compartment model* in which four broad types of natural settings are partitioned according to the their biotic function and life cycle criteria. There are (1) environments with young, relatively immature and rapidly growing ecosystems; (2) ones with more mature, diverse, or climax ecosystems that tend toward protective equilibrium; (3) compromise or multiple-use environments and ecosystems that combine both types and functions; and (4) urban-industrial environments that are relatively *abiotic* in relation to the other types. You can see these four types represented schematically in Figure 1.3.

The important point is that the growth of human settlements and communities obviously decrease the proportion other types of environments and ecosystems at the expense of the more mature, protective ones. Human activity creates urban-industrial environments, with their vast sprawling growth and their great expansion of simplified growth ecosystems. This happens through the cutting of forests, the expansion of land for agriculture

Figure 1.3 Compartment Model of Environments and Ecosystems According to Function and Life Cycle Criteria

Source: Adapted from Odum, 1971: 269.

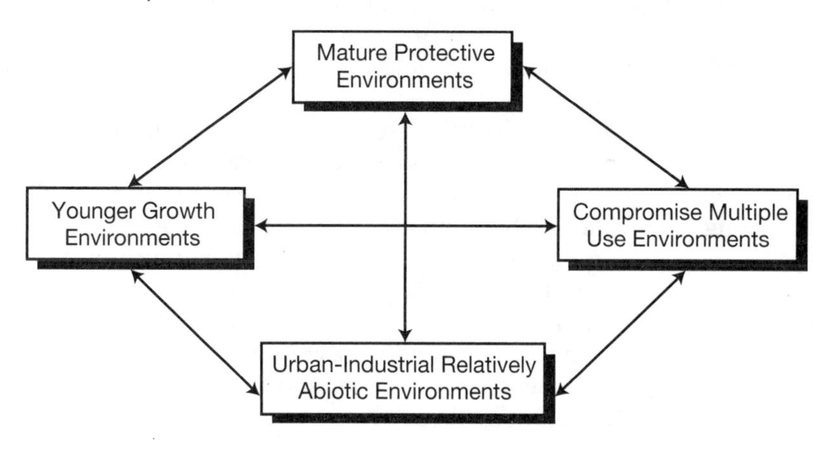

and other uses, and the increase of multiple-use ecosystems that combine some wilderness with fields, towns, highways, among other factors.

The impact of human activity usually creates simplified-growth ecosystems by producing virtual *monocultures* (areas where primarily one type of organism grows). Whether cutting trees, plowing prairies for crops, or cultivating grass in a lawn, humans reduce the biological diversity of living things that exist in "wild" ecosystems. A field of corn or soy beans is such a monoculture. If your lawn has mainly one kind of grass (blue grass, rye, Zoysia or such), it too is a monoculture. If you have had to maintain such a monoculture, you know that it takes a great deal of effort in weed pulling and requires herbicides and pesticides to keep other life forms from invading it. The *loss of biodiversity* in monocultures has its price, not only by the addition of chemicals that are very difficult for nature to recycle, but also by the fact that they are much less robust and hardy than more diverse systems. They are notoriously more susceptible to damage by drought and diseases, such as sod webworm that kills blue grass, or the whole range of insect, fungi, and microbe infections that can decimate grain crops and livestock monocultures. The Irish Potato Famine of the 1840s is an example of the devastation that can be caused by the collapse of an agricultural monoculture. A fungus ("blight") infection killed the Irish potato crop for several years, resulting in widespread starvation and civil disorders, and—importantly—triggering massive waves of Irish emigration to countries such as the United States, Canada, and Australia.

As a crude contemporary measure of the expanding human impact on protective climax ecosystems, consider that ecologists from Cornell University computed that human activity accounted for 40 percent of the *net primary production* on the world's land (the amount of energy captured from sunlight by green plants and fixed into living tissues, and the base of all food chains). Humans directly consumed only abort 3 percent (through food, animal food, and firewood), but indirectly another 36 percent went into crop wastes, forest burning, desert creation, and the conversion of natural areas into human settlements (Vitousek et al., 1986). What will this proportion be as the earth's expansive human population grows beyond 6 billion and world agricultural economic activity increases proportionately?

Odum's observations of the 1970s are still relevant: "Until we can determine more precisely how far we may safely go in expanding intensive agriculture and urban sprawl at the expense of the protective landscape, it will be good insurance to hold inviolable as much of the latter as possible" (1971: 270).

Is there a "saturation limit" for what how and how much of the biophysical environment can be appropriated for human use and still provide broadly positive conditions for social life for most of humanity? To what extent can we do this and still value and respect for its own sake the earth's rich and diverse genetic inheritance of species and ecosystems that resulted from 3 billion years of evolution? *Tough questions* but important ones.

WHAT YOU CAN EXPECT FROM THE REST OF THIS BOOK AND HOW IT IS ORGANIZED

Many writers put information like this in the preface, but since some readers—including me—are likely to skip lightly over a preface, I decided to put this material here because it is important that you get an idea of what kind of a book you are going to be reading and a preview of its contents.

I will use the best and most recent evidence about environmental problems themselves, but this is a *social science perspective.* So I will be even more concerned with how these problems are caused by human behavior, culture, and social institutions like political and economic systems. I will also examine current efforts to change the human-environment relationship and to promote a more "sustainable" society and world order. Finally, it is important for you to know that this book will provide a broad overview that focuses more on the interconnections among a variety of issues rather than on any particular issue in great depth. *Many* other books and research papers provide in-depth coverage of specific topics.

The next chapter in Part I introduces basic concepts about social systems, their historical development, and various ways that people in different kinds of societies have understood and interacted with their biophysical environments. It examines how human-environment relations have come to be understood by the social sciences. It ends with a summary of the driving forces of human activity that impact the biophysical environment. The chapters in Part II are a reading of the "vital signs" of the planet, examining more systematically various resource issues and the prospects for climate change. The chapter about climate change also discusses the notion of *risk* and the sources of uncertainty about human-environment issues as they affect science and policy. Part III focuses on the human impacts on the environment and ecosystems in terms of two particular forces: *population growth,* with special reference to food problems, and the *energy systems* that underlie all human economic activity. It concludes by analyzing the notion of sustainability and the prospects for the emergence of more sustainable societies in greater depth. Part IV examines the elements of such a possible transition: the possibility of transforming markets, politics, and environmentalism (ideology, collective action, and movements) to produce a more sustainable world of societies.

A central theme that I try to develop with progressive clarity is the importance of *paradigms,* implicit notions about the "way the world works" that people have, both as individuals in a culture and as scholars. I will be concerned with the social paradigms that people share by which they envision the environment and organize their behavior toward it as well as the intellectual paradigms embedded in scholarly disciplines that often make communication difficult.

I am a sociologist by training, and my outlook on environmental issues is informed by environmental sociology, a subdiscipline that has developed rapidly over the last 25 years. Even so, no single scholarly discipline has a corner on truth about such a multifaceted and important topic. I have therefore attempted to give attention to the works and perspectives of environmental economists, political scientists, anthropologists, geographers, and policy analysts, making this really as much a social science work than a treatise narrowly about environmental sociology. But of these fields, the book will draw most heavily on environmental sociology and economics, and I will say more about those two fields in the next chapter.

Science, Values, and Language

I have tried to write an objective book about how social scientists analyze the human causes of and reactions to environmental problems and issues. However, as you can probably tell from the introductory pages, I will not ignore scholarly controversy and disagreement about environmental issues. The book will address some outrageously difficult and multidimensional social and environmental issues as reasonably as possible but—obviously—will not do so to everyone's liking. You need to understand that all really good social science, indeed, all good science of any kind, sooner or later connects objective "facts" with things that people find important (values) and with criteria for making normative choices. As environmental sociologist Thomas Dietz put it, speaking about the prospects for a new "human ecology":

> We must become a normative as well as a positive science. I don't mean that human ecologists, as scientists, need continually to be engaged in advocacy. I do mean that we must use our analytical skills to develop arguments for the proper criteria for making decisions. We must help individuals and collectivities make better decisions by offering methods for handling value problems. (1994: 50)

There is, in truth, no completely value-free social or any other kind of science. So we will talk about facts and data, but this is also a book that exhibits my own thoughts, values, hopes, and fears about the human predicament. It is impossible (and I think undesirable) to eliminate one's own opinions and values from scholarly work. But they should be labeled as such, so I have tried to be careful to put "I think . . ." statements in front of those places where I am particularly aware that not all readers would agree with what I have written. The book will also be concerned with ideas about making choices in conjunction with considerable uncertainty. About this, I would be less than honest if I did not share with you up front some of what I believe to be true.

So here's what I think: Ecocatastrophe is *not* just around the corner. But I agree more with those who argue we are in a historically new and difficult situation. The *scope* and *scale* of human activity have grown so enormously, a fact not appreciated by Pollyannas, that they threaten to overwhelm the dilutive and absorptive capacity of the biophysical environment. Throughout most of human history on the earth, we were so few and our capabilities so meager in relation to the resources of the earth that the environmental messes we made really didn't matter much. Those conditions are rapidly changing in the new century in which we live. To deal with these problems, we will have to "invent" our way into a new and *sustainable relationship* with our biophysical environment, and that will be challenging, contentious, expensive, and fraught with many false turns and blunders. But I *do* believe evidence is mounting that the costs of doing nothing—and letting nature take its course while we do business as usual—are far higher and more ominous. Finally, I do not believe that we are somehow trapped by human nature, or business as usual, into doing nothing and destroying our own support systems. People have a capacity for rationality and adaptiveness, even on a large scale. That's my hunch, but about this, one hunch is a good as another. There is certainly a lot of room for honest disagreement with these ideas, and the views of those who do so are presented in a variety of places in the book that you will be reading.

It's fair to warn you that you will be reading a book that details a lot of bad news about human-environment interactions. Reading a sustained fare about problems can be very depressing and can generate fatalism. But it is also important to note that I find some compelling reasons for hope (if not optimism) about the possibilities for a positive future. Those reasons occur mainly in the later chapters that conclude the book, so if what you read initially depresses you, *read on*. The book moves, after the first two chapters, from the more physical to the more social dimensions of environmental problems, and from the more depressing litany of facts and problems to examining some possibilities for positive change. I discovered in writing the book, somewhat to my surprise, that in the final analysis I'm not a very true-blue Cassandra.

I should mention one other thing that should be obvious to you by now. As much as possible, this book is written in an informal and, I hope, unpretentious style. I have often tried to write as if I were carrying on an imaginary conversation with you as an individual rather than communicating with anonymous groups of people. It's the way I like to communicate, and I hope it makes the book more engaging to read.

PERSONAL CONNECTIONS

This book deals with large and important questions about the life of humans in relation to the environments and ecosystems in which they live. At the end of each chapter, I will try to engage you in thinking about personal implications and questions related to the material of each chapter. It may include some pointed comments about personal implications or questions to help you think about your own life (*"Consequences and Questions"*). It may also include an appreciation of things, products, attitudes, or activities that are environmentally good and personally good but that we take for granted (*"Real Goods"*). And it may include suggestions about things that you can consider or do to modify your own lifesyle while satisfying your needs or to work for change (*"What You Can Do"*).

Consequences and Questions

The discussion of ecosystems in this chapter suggests that you can never "do your own thing" without impacting other people or species, and the quality of the physical environment. That is because you are in a system relation with them. Humans are certainly as embedded in nature (biological ecosystems and physical resource systems) as any other species. Yet you live in a made environment, constructed by humans that organizes, processes, and mediates the natural world for your profit, safety, and comfort. By *made environment* I mean things like houses, central heating, and air conditioning as well as small items like the stereo headphones and the large constructions of urban centers that literally reshape the land to human uses. We like these things; they make our lives more comfortable, but ironically, they also blunt our awareness of human embeddedness and interdependence on the natural and biotic world. Think about this:

1. What are some of the layers of culture and civilization that tend to insulate you from the natural world? Look at the products in your home, or on your back, or on your dinner plate in a new light: How do they illustrate your embeddedness in nature? How do the processes by which they get to you tend to blunt your awareness of this? Here's an example: Buying food in a supermarket, which you normally understand as being a consumer, actually makes you a participant in vast food chains, energy and resource transfers in far-flung human constructed market systems that nature never knew. Being a consumer has ecosystem and environmental impacts,. but when you buy the luscious vegetables or convenience foods at the grocery store, you don't see the rest of the system. Nor do you see it when you throw away waste and packaging. What are some other examples of how your awareness of natural embeddedness is blunted by living in the made environment?
2. When do you think about the natural world? When you see it on TV or in books (you know, the breathtaking picture of distant mountains, fascinating wildlife, oceans, and so forth)? Or when you really participate in it? Does your daily routine include being in the natural world? Do you normally view nature with aesthetic appreciation, as a resource to be used, or as an intrusion to be minimized in an otherwise comfortable life?

What You Can Do

"Think globally, act locally" has become a ritual slogan (mantra?) of the environmental movement. If you are concerned about environmental problems, you do need to think globally about them. You also need to act locally, in your own corner of the world. But you also need to act in ways that have larger-scale relevance. Including a list of things you can do to "walk lighter on the earth" is almost a ritual in most books about environmental issues. I mention some of these ideas for lifestyle change in later chapters. Changing individual lifestyles is important, but it is not sufficient to address the environmental problems that beset us. Powerful institutions and organizations operate on structural levels beyond individual behaviors. But it does not follow that the actions or attitudes of individuals irrelevant for larger scale change. (I will address some questions about social change later.) For now, I want to leave you with the notion that *individuals matter*. That is another important theme of this book. It was well put in the novel *Middlemarch*, by the famous British writer, George Eliot: "The growing good of the world is dependent on individual acts."

Renowned anthropologist Margaret Mead had similar thoughts: "Never doubt that a small group of thoughtful, committed citizens can change the world; indeed, it is the only thing that ever has."

Real Goods

Let me tell you about something I have lately come to value, that I didn't for many years: *Anne's garden*. My wife, Anne, likes to grow things. We live in an ordinary older urban neighborhood with brick and wood frame houses and big established trees. Our backyard measures about 60 by 75 feet, which is pretty normal for Omaha but would be considered large in more densely populated cities. The trees that shade the backyard are not fancy ones; in fact, a landscaper would understand most as "weed trees." There's an alanthus (sometimes called a tree of heaven), a mulberry tree, several chinese elms, and a big cottonwood tree in the neighbor's yard. I cut the grass—whatever grows, some blue grass and rye grass, but also a variety of weeds and clover that have taken root. By contrast, some of my neighbors spend lots of money having their lawns regularly doused with fertilizer, herbicides and pesticides, and have beautifully manicured bluegrass and zoysia monocultures.

Since we first lived there, Anne has kept planting new flowers every summer while nurturing the old ones and tending a vegetable garden. Gradually, she pared the "lawn" back to a smaller space in the middle of the yard and paths to walk to other corners. This is not a manicured "English garden," mind you, because we both work and don't have the time for that. A variety of weeds comes with the flowers and vegetables. It is very different from our neighbor's manicured and open backyards. For a long time I just thought it was weird. My role was to help a bit but mainly to sit under a shade tree and watch all this encroaching vegetation with a bit of perplexity. But one summer I counted the variety of plants in our back yard. There are irises, day lilies, roses, delphiniums, crocuses, tulips, daffodils, black-eyed susans, salvia, impatiens, German ivy, elephant ears (colladium), horehound plants, asters, hausta lilies, larkspur, tiger lilies, dahlias, phlox, Queen Ann's lace, sunflowers, zinnias. . . . etc. (and more). There's a peach tree now as high as my head that sprouted from a peach seed someone threw on the compost pile. It doesn't have much of a fruit-bearing future in Nebraska. And we have some hazelnut bushes and apricot trees my wife planted.

In various years, the vegetable garden grows green beans, snow peas, strawberries, carrots, cabbage, radishes, tomatoes, broccoli, bok choy, peppers, and a variety of herbs (dill, mint, basil, sage). The yard has attracted a variety of creatures: a tribe of entertaining and contentious squirrels, a multitude of bees and pollinators, summer cicadas and other bugs, garter snakes that nest under an upturned corner of an old driveway slab, a variety of birds that nest and feast, bats that hunt bugs on summer evenings, and, until last winter when he died, the backyard was the home of an old beaglehound, Max, who had pretensions to being the top carnivore (he wasn't as good at climbing trees as the squirrels!). The squirrels have enjoyed all the apricots. Not a one has matured for a human.

What's the point? It dawned on me that our whole backyard has become Anne's garden. A mini-ecosubsystem of its own. A green, leafy, vibrant, buggy urban polyculture (compared to the backyards of our neighbors) where something is always blooming and dying in great variety. I have come to appreciate why the English word *paradise* derives from a word in an ancient Mideastern language meaning "a small green garden." It is a small corner of the world that I have come to cherish as very beautiful in its own right. Every winter I wait for its return.

ENDNOTES

1. In a trip to Poland in 1990, I observed firsthand that many people had chronic lung, eye, and skin diseases that could, in the broader sense, be understood as environmentally induced illness. And while driving from a rural district into an industrial region around Kracow and Nowa Huta, everyone in our car developed an asthmalike bronchial condition in the space of a few hours. We became so used to it that we didn't notice the difference until we left the region several days later.
2. You may be curious about the origin of the words. Pollyanna was a young girl in a novel by Eleanor H. Porter who saw the world through rose-colored glasses so she refused to see much that was ugly and wrong. Cassandra was a figure from

Greek mythology who saw impending catastrophes, which were real, but she announced them so often and so stridently that no one paid any attention to her.

3. These are not original with me and have been used by many others (e.g., Mann, 1993). In an odd way, the labels became real. In 1987, a number of environmentalists organized an official "Cassandra Conference" at Texas A&M University (see Elrlich and Holdren, 1988, for the proceedings of this conference).

CHAPTER TWO

Human Systems, Environment, and Social Science

Like Chapter One, this chapter has an important role in "setting the stage" for what follows. It will maintain some parallelism with Chapter One, about ecosystems. First, I spell out the basic elements of human sociocultural systems. Second, I describe in broad strokes historical change in human systems and their human-environment relations, from the small, scattered human groups to the large complex systems of today. Third, after describing this change, I turn to explanations of sociocultural evolution, using some ideas from human ecology and political economy that derive from many academic fields. Fourth, the chapter examines in more depth the emergence of environmental social science in two fields (economics and sociology) and illustrate applications of each. Fifth, it will conclude by examining, in a summary way, the proximate causes or driving forces by which humans alter their environments and the system connections between human and ecosystems.

SOCIOCULTURAL SYSTEMS

The last chapter noted the Irish Potato Famine of the 1840s to illustrate the biotic vulnerability of agricultural monocultures. The fact that a large number of persons of Irish descent are in the United States, Canada, and Australia partly because of this catastrophe demonstrates in a very graphic way the important connections between humans and the natural world. Humans and human societies are certainly embedded in the ecosphere but, as is often noted, humans are also unique creatures among all others.

Humans are social animals, a characteristic they share with other species, such as bees, gorillas, and dolphins. In other words, humans live in *groups*. But even though *group* is a common word, I won't often use it because it is imprecise: When we speak of groups, do we mean small groups? Families? Professional communities? Crowds at rock concerts? Organizations? Or what? For sociologists, a basic abstract organizing concept is the *social system*, which is like the concept of *ecosystem* for ecologists. I could begin, as I did in the previous chapter, by simply iterating the structural units of societies, from small to large and inclusive (e.g., individuals, small groups, communities, bureaucracies, societies, world order), but that wouldn't be very enlightening, particularly because it ignores a whole dimension of human systems that most differentiate *Homo sapiens* from other species: *culture*. Even though the social animals mentioned above live in social systems, they lack a cultural dimension. A *sociocultural system* is a network of interdependent actors (individuals, organizations, subsystems) that are in relatively stable patterns of interaction and intercommunication. They share cultural patterns (both material and symbolic), and which are distinguishable from those of other such systems. If you are suspicious that I am not exactly on new ground, you aren't wrong; a human system is another specific version of the general system concept introduced in the last chapter. A systems perspective is fundamental to both ecology and the social sciences. This is important, because it means that for humans as well as other species (1) everything is ultimately connected to everything else, and therefore, (2) you can't ever do *just* one thing without some consequences for other parts of the systems in which you live.

Here, in Table 2.1, are the components of human systems, distinguishing some clusters of related elements.

Table 2.1 Elements of Sociocultural Systems

Culture	worldviews paradigms ideologies knowledge, beliefs, values symbols, language
Social structure	world system society nation state complex organizations (bureaucracies) social stratification systems (based on economic class, ethnicity, kinship, or gender) small groups kinship systems status roles
Material infrastructure	wealth (tokens, wives, cattle, money) material culture, subsistence technologies (plows, computers) human population (size and characteristics) human-environment relations biophysical resources (land, forests, minerals, fish)

This is a useful and fairly conventional analytical scheme. As will become apparent, however, things are not divided so neatly; others do it a bit differently: See Lenski and Nolan (1999) and Sanderson (1995).

Since the relevance of these human system elements or subsystems may not be quite obvious to you, let me say a few things about them, particularly as they relate to understanding environmental issues. *First,* you may be wondering how some are different, particularly the difference between a nation state and a society. Today we usually think of them as the same, but they really are not. Real nation states did not even exist much before the 1500s, but *society,* the most inclusive structural unit of human systems, are as old as human civilizations.[1] There are people, such as the Berbers of North Africa, who comprise a coherent society but who live in several north African nation states (Algeria, Mauritania), as do the Mohawks (whose "territory" straddles the U.S.-Canadian border). *Second,* these elements are really not an evolutionary or developmental sequence. For the earliest known *Homo sapiens,* and among the few scattered indigenous peoples of the world today, there *is* no operating society beyond the level of families or kinship systems, no larger communities, and no stratification systems beyond elementary status roles based on age or perhaps gender. Furthermore, an authentic world order that has the potential to knit nations and societies into a truly global system of sorts has been emerging for only about the last five hundred years, and its features are not yet very clear. *Third,* there are some things left out. There are, obviously, *individual* human organisms, and there are *social networks,* that are somewhere in between populations and organized groups in the number and strength of the system bonds between actors.

Culture

Surely the most and important distinction between *Homo sapiens* and other species is the extent to which humans are cultural creatures. Nonhuman animal social behavior is more shaped by the behavioral instructions or codes carried in their genetic makeup—which interact with their environments in complex ways. Human behavior and environmental adaptation is more flexible, open ended, and shaped by learning; in other words, it is cultural. *Culture* is the total learned way of life that people in groups share. You can think of it as a sort of humanly constructed software (to use a computer analogy) for, for instance, what the world is like, how people should relate to each other, and how they ought to adapt and "make a living" in the biophysical environment. Since our genetic equipment gives us very little specification about any of this, it is fair to say that much of our behavior and social patterns are shaped by culture rather than biology. Exactly how much is debatable, and this issue has been at the core of an intense—but not very productive—debate between evolutionary biologists, anthropologists, and sociologists for about a decade.[2] People do not always conform to cultural norms, but we all experience powerful social pressures to conform and often face social sanctions if we don't.

But culture is hard to classify by this tripartite scheme because it has both symbolic and material dimensions. *Material* technology, for instance, includes the tools, factories, weapons, and computers, that relate to economic subsistence. Underlying these "things" are ideas, plans, recipes for doing things, and the innovative processes that are part of *symbolic* culture. To continue the computer analogy, if material culture is the hardware or mainframes, symbolic culture is the software programs of human systems. Thus, subsistence technologies really include all the ideas, formulas, tools, and gadgets that people use to convert raw biophysical resources into goods and services that humans find useful. Viewed as part of the material infrastructure, they relate "making a living" in the elemental sense of providing sufficient food, shelter, and clothing. But it also includes a lot of other "stuff" unrelated to basic subsistence like pet rocks, beanie babies, toenail clippers, computers, and sociology texts, which have economic utilities that would be quite baffling to most humans who ever lived.

Social Institutions

Social institutions are both left out *and* hard to classify by the foregoing scheme. They are nearly universal sociocultural formations, like families, economies, political systems, judicial systems, healthcare, and so on. Social institutions are both structural and cultural. That is, they include broadly established ideas, values, beliefs, technologies, and structural systems that address some enduring human concern related to collective survival. You can get a sense of the structural *and* cultural sides of institutions by thinking about families (groups organized around kinship). The operative structural units of American families, established by law and custom, are parents and their children (even though other relatives have important legal and cultural standing). On the cultural side, again established by both law and custom, married spouses are two (only two) people of the opposite sex. They ideally exhibit an interaction style shaped by the values of positive affection (love) and trust, rather than by economic utility or relations of domination-submission. Children, normatively now not more than two or three, are to be valued intrinsically, and not as utilities for family economic or sexual exploitation. Does this picture represent the empirical reality of all families in the United States? Of course not. But social institutions are imperative normative "shoulds" that we find hard to disagree with, supported as they are by powerful cultural customs and laws. Furthermore, this institutional template is very different from that of families in other cultures (as anthropologists have studied extensively). My point is that social institutions are as much cultural as structural.

Social Structure

Elementary structural units of human systems are statuses and roles. Your *status* is the position or "rank" you occupy in a social system. It is linked hierarchically with other statuses (like student and professors). Your social *role* is what you are "expected" to do while you are occupying a status. Professors, for instance, are "expected" to work hard preparing for their classes, do scholarly research, and take an interest in their students. But I'm sure you know that such role expectations vary a lot and are not always enacted anyway! *Status* is a structural term, and *role* is a behavioral or cultural one. (Again, things are not so neatly categorized.) Status roles exist in social systems of every size, and may coalesce into broad structures of social stratification in complex human systems.

The status-role concept is somewhat analogous to the way ecologists use the ideas of ecological habitats and niches—as the structural locations and functioning of organisms within an ecosystem. Furthermore, it is important for you to note that some other social animals, particularly primates, have almost humanlike status-role systems. As our evolutionary cousins, primates (and some other mammals) live not as unorganized mobs but in relatively structured *rank-dominance hierarchies*, usually with the older males in charge of things.

Population Size and Characteristics

The most obvious and elementary components of human systems are individual people and populations of various sizes and characteristics. Populations are *aggregates*, not systems, but their characteristics (such as size and age distribution) have a lot to do with what goes on in human systems and how they come to be structured. For instance, as human systems grow in size, they typically develop complex subsystems and experience problems of communication and coordination. The environmental implication of population size is that, *other things being equal*, larger populations make more demands on the biophysical environment that do smaller systems. But in the actual world things are rarely equal. Technology is a major force that makes the environmental impacts of populations *unequal*. Small populations with powerful subsistence technologies can impact environment far more that larger ones with less powerful technologies. I will have a lot more to say about the environmental implications of the interaction of population and technology in Chapters Five and Seven.

The Duality of Human Life

The cultural uniqueness of human beings has a profound implication. It results in what I take to be an existential dualism that underlies much of the debate about human-environment relationships, including the Cassandra-Pollyanna quarrels. This duality, inherent in the human condition, can be stated simply:

> *On the one hand*—humans and human systems are unarguably embedded in the broader webs of life in the biosphere. We are one species among many, both in terms of our biological makeup and our ultimate dependence for food and energy provided by the earth.
>
> *On the other hand*—humans are the unique creators of technologies and socio-cultural environments that have singular power to change, manipulate, destroy, and sometimes transcend natural environmental limits. (Buttel, 1986: 338, 343)

Biologists and ecologists usually emphasize the first part of this duality and social scientists typically place more emphasis on the second part. You probably recognize that *both* statements are true in some complicated and partial sense. Yet it makes a great deal of practical difference which assumption we use as a guide to action, choices, and policies. Since the industrial revolution, the second assumption—*humans as an exceptional species*—has been the dominant assumption and viewpoint. It is important to note that humans act on the basis of such viewpoints rather than on the basis of what world "really is." This is a subtle but important point that requires some elaboration.

Worldviews, Social Paradigms, and Cognized Environments

There is obviously a reality external to human beings that we live within. But human choices and policies are more directly related to our *definitions* of that reality than to what reality "really" is. In other words, human social behavior is more directly related to symbolic constructions and definitions of situations than by external environments per se. People *exist* in natural environments, but they *live and act* in worlds mediated and constructed by cultural symbols (Berger and Luckmann, 1976; Schutz, 1932/1967; Thomas, 1923).

Yes, there is an external biophysical environment independent of how people think about it, but they act on the basis of what they *think* the environment to be. To differentiate this environment from the "real environment," scholars have invented a rather awkward term, *cognized environment*, to mean their human definitions and interpretations of the biophysical environment. The very notion of nature itself is a way of *cognizing* the environment that didn't exist much before the eighteenth century. As a cultural conception and idea, nature was invented mainly by

English intellectuals, in the eighteenth century, particularly Romantic artists, writers, poets, and literati (such as Wordsworth and Ruskin). They sought a metaphor to contrast the "good" pristine natural state with the (presumed) evil artificiality of the cities, mines, and factories of the industrial world. Thus the notion of nature that has come down to us was originally part of the Romantic discourse and critique of the invasion and destruction of all that was "natural" by the barbaric machines of the industrial system (Harrison, 1993: 300; Fischer, 1976, chap. 2). "Mother Nature" is a more obviously gendered and anthropomorphized cognition of the biophysical environment (*anthropomorphized* means that something nonhuman is understood in human terms).

A cognized environment is part of the cultural *worldview*—the totality of cultural beliefs and belief systems about the world and reality that people share. The cognized environment is also an important component of the *paradigms* that people share. The notion of paradigm was originally used by philosopher of science Thomas Kuhn to describe the mental image that scientists had of their subject matter that guided their theory and research (1970). But there are paradigms not only in science, but in other forms of knowledge and the ways that people think about social life. Hence the concept of social paradigm. A *social paradigm* is an implicit model of "how the world works" that is broadly shared by people in society (Olsen, Lodewick, and Dunlap, 1992: 17–18). I will use the term *paradigm* (with appropriate adjectives) to describe the implicit mental models of the people in society and among more specialized communities of scholars.

A social paradigm is:

1. A narrower cultural element than the inclusive *worldview* concept, and pertains only to certain areas of life rather than—as worldviews—to the totality of existence.
2. Not the same as *ideology*, which is the parts of the worldviews that people *purposefully* use to justify action and political choices, such as nationalism, democracy, or individualism.
3. A "logic" or "mental model" that underlies the stated goals and ways organizing and governing the systems that make up social institutions. It is a taken-for-granted and commonsensical construction of reality that provides frameworks of meaning within which "facts" are defined and issues such as the environment are debated.

A *dominant social paradigm* (DSP) is the major one that operates in a given society. It may not be universally accepted, and often becomes "dominant" as part of the struggle for control in society, and may be challenged. Indeed, it may become understood only when it is challenged (Cotgrove, 1982: 26–27,33; Olsen, Lodwick, and Dunlap, 1992: 18–19).

HISTORICAL CHANGE, HUMAN-ENVIRONMENT RELATIONSHIPS, AND PARADIGMS

So far, this discussion has been pretty abstract. Now I to turn to a more concrete discussion of the historical *development of human societies* and their changing relations with the biophysical environment. This is what social scientists often refer to as *social evolution*, and the earliest evolutionary thinkers thought that there were "master" change processes at work in all societies through time. But recent developments in understanding social evolution emphasizes the accumulation of complex contingencies (such as the generation of novel forms and their transmission and selection over time), which is closer to the biological meaning of the term *evolution* (Burns and Dietz, 1992; Sztompka, 1993). My purpose here is more elementary: to provide a descriptive summary of the major historical and developmental changes in human societies, and their relevance for understanding human-environmental connections. What follows relies heavily on Lenski and Nolan (1999), Harper (1998: 71–85), Miller (1998), and Sanderson (1995).

Hunter-Gatherer Societies

Throughout about three fourths of our 40,000-year existence, humans existed as small bands of *hunter-gatherers*, who survived by gathering edible wild plants and killing animals (including seafood) from their immediate surroundings. They foraged from day to day and week to week with little accumulated economic or food surplus. They survived by the accumulation of a cultural stock of expert knowledge of their surroundings, such as knowing when the seasons changed and which roots and berries were edible, how to track game and follow migrations, and how to find water—even in the desert. Few true hunter-gatherers exist today, except in scattered places. Near-contemporary examples include the Native Americans of the Great Plains (in the 1700s), the polar Eskimos (Inuit), the Bushmen (!Kung) of the African Kalahari Desert, and the natives of the rainforests of Brazil and Venezuela (such as the Yanomamo).

Hunter-gatherers traveled in bands of about fifty persons, following game and the seasons, carrying virtually all their possessions with them. Though they were aware of other bands with whom they shared language, culture, and territory, each band was independent and had a simple division of labor and status-role system based on mainly on age and sex. Leadership tended to be informal and situational, and there were few nonsubsistence roles: Almost *everyone* helped with the search for food. Since there was little surplus to hoard or accumulate and everyone had rights to food, there was little social inequality among hunter-gatherers. While they had to endure periods of deprivation during drought or when game was scarce, usually they survived well; some anthropologists have estimated that many hunter-gatherers took only two to five hours a day for two or three days per week to fulfill their material needs. In sparse environments such as deserts or savannahs, it took a large territory to support small nomadic bands. Interestingly, anthropologist Marshall Sahlins describes hunter-gatherers as the "original affluent society." He argues that "an affluent society is one in which all people's material needs are easily satisfied" (1972). Hunter-gatherers eat well, work little, and have lots of leisure time, despite living on far less than modern people. Not only their basic needs but also their *wants* are satisfied because they don't want much (Bell, 1998: 40).

Human-Environment Connections and Worldviews of Hunter-Gatherers

Because of their small numbers, decentralized social patterns, use of natural materials for tools (such as pots, baskets, arrows and spears) and reliance on "muscle power," their impact on environments was typically small and localized. They were examples of *people in Nature*, who trod lightly on the earth because they were not capable of doing more. They survived by being keenly aware of their dependence on nature and each other. This embeddedness in nature was the central theme of the worldview of hunter-gatherers. Because they left no written record, what little we know about the worldviews or social paradigms of hunter-gatherers must be reconstructed from myths, folklore, and the oral records of near contemporaries. With exceptions, they did indeed think of themselves as people *in* nature. Their cognized environment was that of a *living natural world* (wilderness/jungle/forest/grassland) of things and beings governed by spiritual forces. Humans interacted with the benefits and constraints of this world. But they were capable of abusing their environments by, for example, driving great numbers of large game animals off cliffs and using only a few. And given more powerful technological means (such as guns), they were not necessarily better stewards of nature than people in industrial societies.

Agricultural Societies

About 10,000 years ago, people discovered how to cultivate crops and to domesticate and breed animals. The transition from hunter-gatherer bands to settled agricultural communities took place slowly in widely scattered places. By cultivating crops (yams, corn) and domesticating animals (sheep and goats), *horticulturalists* and *pastoralists* could produce a larger and more certain food supply. Examples of horticulturalists include the Eastern Woodland Native Americans (such as the Cherokee and the Iroquois confederation of tribes) and the Trobriand Islanders in New Guinea. Crops were planted and tended by hand, using tools such as digging sticks and hoes. Whereas pastoralists (who herded animals) continued a nomadic existence, horticulturalists began to live in larger settlements, and their villages could contain several hundred people. The material surpluses of horticulturalists produced population growth and an increase in social complexity and social stratification. Nonsubsistance roles and statuses evolved, such as leaders, craftworkers, warriors, and magicians. Complex and stable institutions separate from family and kinship began to emerge; probably the oldest of these was a separate political system in which villages came to be ruled by headmen and hereditary chiefs. The resulting increases in social inequality produced status distinctions among social elites, such as rulers, specialists, and magicians, and ordinary people who continued to produce food.

Pastoralists had to keep moving their herds and flocks, following water and fertile grasslands. When their herds overgrazed, they moved on. Often they moved "up altitude" in the summer and "down altitude" in the winter. Horticulturalists cleared small plots in tropical or temperate forests for their crops, practicing what was called *slash-and-burn*, or *shifting agriculture*. In two to five years, crops would decline because of overuse, erosion, leaching of soil nutrients, or the invasion of weeds. When this happened, they left the plot and moved on to clear a new plot. Ecological succession began, and they learned that it took between ten and thirty years to restore the soil productivity to the plot. With a relatively small population and a large area, these practices of moving herds or shifting agriculture could be sustained indefinitely.

Real *agricultural societies* emerged about 7,000 years ago with the practice of *intensive agriculture*, in which crops were grown year after year in the same fields, often using irrigation and fertilizers. These were the ancient civilizations you studied in history, including Mesopotamia, Egypt, India, China (and later Greece and Rome), and, in the New World, the Incas, Aztecs, and Mayans. The gradual shift to cereal grain agriculture (e.g., wheat, maize, rice) produced more food per hectare and could be more easily stored for long periods of time. In Old World agricultural centers, the invention of the metal plow, pulled by domesticated animals and steered by farmers, enabled people to cultivate much larger plots and to break up fertile grassland soils that previously couldn't be farmed because of their thick and widespread root systems. In contrast to horticulture, intensive agriculture enabled people to produce enormous amounts of food surpluses and sustain dense populations in large permanent settlements. The uneven but pervasive diffusion of intensive agriculture, as it is called, and agricultural technologies resulted in one of the great social transformations of humankind, which scholars have called the *agricultural revolution*. Besides farming and irrigation technologies, it stimulated the invention of metalworking, craft production of all sorts, mathematics, calendars, literacy, and—oh yes—military conquest, slavery, and empires.

The *scale* of human social life vastly increased. Populations grew much larger, and societies came to be dominated by urban centers of perhaps 20,000 people which controlled, coordinated, and extracted taxes from a much larger area of rural villages. Still, in agricultural societies, perhaps 90 percent of the people lived as farmers in rural villages, which were the real sources of material wealth. As the scale of life increased, so did social differentiation and complexity. Now there emerged a whole panoply of nonsubsistence status roles (traders, scribes, priests, potters, weaver, metalworkers, warriors, slaves, healers, and so forth). Gone also was the relative equality of hunter-gatherers as there emerged systemwide dominance hierarchies (classes) that included (1) kings and nobles, (2) priests and scribes, (3) merchants and warriors, (4) craft workers and artisans, and (5) the vast majority at the bottom who worked the land—peasants and slaves. There was an increase in trade and communication between city-state systems and usually an expansionary dynamic in which the stronger city-states tended to conquer the weaker ones, resulting in vast political empires ruled by hereditary dynasties (as in ancient Rome, China, and the Incan empire).

While complex "civilization" was born with agriculture, it is questionable whether the social or the physical quality of life improved for ordinary persons. Peasants in agricultural societies were controlled by landlords and rulers in ways inconceivable to free-roaming hunter-gatherers and pastoralists; the work day became long and monotonous and the patriarchal domination of women increased. In fact, contemporary anthropologists agree that gender inequality became far more pronounced among agriculturalists than among the more equalitarian hunter-gatherers and horticulturalists. Systems and ideologies of patriarchal domination really emerged with pastoralists and intensive agriculturalists. There is also evidence that life became shorter. The health status of people deteriorated because people were more sedentary, and their monotonous grain-based diets were less nutritious and less varied than those of hunter-gatherers (Goodman and Armelagos, 1985). They were also less healthy because large, dense, and settled populations of people and their livestock had to contend with their wastes, pollutants, and effluents in ways that mobile hunter-gatherers or pastoralists didn't. Urban centers in particular were places of highly contagious plagues and pestilence, carried by microorganisms in sewage and in impure water supplies. All this led one anthropologist to quip that for most people, with the possible exception of elites, the development of agriculture was the "worst mistake of the human race" (Diamond, 1987).

The Human-Environment Connection in Agricultural Societies

I remember my history texts (of previous decades!) emphasizing the "rise and fall" of civilizations and dynasties in terms of inept or corrupt leaders and military conquest. These reasons were important, but there was more to it. One factor is that in the *long term* many agricultural societies degraded the productivity of their soils and resource bases. To feed and fuel growing populations, food and fuel resources were overused and poorly managed: Grasslands were overgrazed, topsoil eroded, salt built up in irrigated soils, and canals and rivers became increasingly polluted and clogged with silt and effluents. As late as 7000 B.C.E., for instance, the Tigris and Euphrates valleys, home to Mesopotamian civilizations, were covered with productive forests and grasslands. But increased salt build-up of the soil from irrigation evaporation in a hot climate caused food production to decline an estimated 42 percent per hectare between 2400 and 2100 B.C.E. (Pointing, 1991). Much of this once productive land was turned into barren desert. Over the centuries, more and more human effort was required to sustain productivity. A combination of climate change, environmental degradation, and invading armies eventually put an end to the Mesopotamian civilizations. There are many other cases of complex agricultural societies weakened or destroyed by ecological collapse, including the western Roman Empire (Tainter, 1988).

Perhaps the clearest case of social collapse induced by environmental degradation is that of the Lower Mayan societies in Central America (in what is now Honduras and Guatemala). The Mayan population around the Copan urban center was about 5,000 in 550 C.E. Land was cleared for crops in the Copan river valley, and native hardwoods and pine trees in the surrounding highlands were cut for fuel and building construction. By 850 C.E., the Copan population had increased to about 20,000. But as deforestation continued, semi-tropical rains badly eroded soil cleared for agriculture, and productivity declined. By 1000 C.E. population had declined by 50 percent, and by 1250 C.E. the entire settlement was abandoned. Soil cores extracted from the times of the social peak and collapse found little tree pollen (indicating rather complete deforestation), and also evidence of dwellings covered by huge mud slides from eroding areas. Thus forest mismanagement was directly linked to accelerated erosion rates which were primary causes of the collapse of the lower Mayan states (Abrams and Rue, 1988: 337–393; Hammond, 1982; Sharer, 1983). The overshoot of the carrying capacity and eventual collapse of the Mayans resembles that of the reindeer on Matthew Island that I mentioned in Chapter One. It just took longer.

Not only intensive agriculturalists but pastoralists also degraded their environments by keeping too many animals on semi-arid rangeland. That has happened historically, but there are contemporary examples. Much of the Argentinean pampas is now badly overgrazed, and cattle production has been slowly declining in recent decades. But the most dramatic and well documented cases of overgrazing are in East Africa. In regions of Kenya, Tanzania, and Uganda cattle herds have grown enormously, badly devastating the dry grasslands and lowering water tables. By 1965, as many as 75,000 cattle depended upon just eight water holes in northeastern Uganda, which led to the dieback of herds and associated human deprivation (Netting, 1986: 51; Campbell, 1983: 141–150).

When agriculturalists endured total or near ecological collapse, people migrated and started afresh or sometimes collapsed into more rudimentary social forms (dispersed villagers eking out a meager existence where once there were powerful, prosperous, and complex agrarian states and empires). Since the 1970s, such ecological collapse has been a powerful indirect cause of the chronic political chaos and material deprivation common to much of sub-Saharan Africa. It is too simple, of course, to think that the ecological impact of agriculturalists was always one of despoliation. Historical evidence suggests that long-term processes of desertification and land degradation included times of accelerated rates of loss of arable land and times of land reclamation as well (Adams, 1981). Still, most scholars would agree that long-term pressures of population growth in agricultural societies produced strong pressures for environmental degradation.

The Dominant Social Paradigm in Agricultural Societies

With the evolution of agricultural societies, the dominant social paradigm (DSP) began to shift from *people-in-nature* to *people-controlling (or "against") nature*. The emerging theme of the DSP of agricultural people was the human domination of nature, as people learned to tame and control wild nature to accumulate wealth and material surpluses. Importantly, they got used to controlling not only nature, but each other as well in hierarchical and patriarchal structures of power. The concrete cultural origins of this paradigm shift have been debated by scholars. Historian Lynn White argued that for Western civilization, this view of human "dominion over nature" derives from biblical accounts in Genesis, which encouraged an *anthropocentric view* that nature is subservient to human needs and thus a lopsided idea of the interdependence of life on earth (White, 1967). This view of biblical themes has been challenged by other scholars, but they have been unable to demonstrate that the Bible emphasized ecologically minded stewardship any more than the dominion theme (Glacken, 1967: 157; Simkins, 1994). More recently, feminist scholars (ecofeminists) have argued that the domination theme in agrarian DSPs and connected ecological problems stem from the growth of patriarchy, which replaced the more gender-equalitarian status-role systems and ideologies of horticulturalists. According to this view, *patriarchy* is not only the source of the domination of women, but of nature as well (Eisler, 1988; Boulding, 1992: 330–331).

I don't believe, contrary to White's thesis, that Christianity was particularly responsible for separating the consciousness of humans from their environmental roots. Rather, all of the world religions of agricultural societies (Zoroastrianism, Buddhism, Christianity, Islam, Hinduism) were transcendental in two senses: by positing both a universal deity and another world after the present one, divorced from particular places or environments. The present world became a "transit hall" to the next, where humans were separated into the virtuous and the worthless and where God and soul were separated from matter, rather than being inherent in it (Harrison, 1993: 34). The agricultural revolution did much more, then, than transform human-environment relations. It provided DSPs that legitimated the domination of humans over nature, of humans over other humans, and of men over women.

The survival of wild plants and animals, viewed by hunter-gatherers as vital to humans, no longer seemed to matter as much. Wild animals, competing with livestock for grass and feeding on crops, were killed or driven from their habitats. Wild plants invading cropfields were a nuisance to be eliminated. Human monocultures and "built" environments began, slowly but inexorably, to expand and push back wild growths behind a cultural frontier that separated people from nature.

If the cognized environment of hunter-gatherers was a natural living *wilderness*, that of agriculturalists was more like that of a *garden*, still a natural system upon which people depended, but one that could be extensively cleared, plowed, weeded, tended, watered, mined, and dominated for human purposes.

Industrial Societies

Industrialization began about three hundred years ago in Europe. Like the invention of agriculture, industrialization depended upon some key discoveries and technologies—first in the textile industry in England—that substituted machine production for human and animal labor. Industrial production depended not only on new machines, but also on new energy sources to power them—water power, steam engines, hydroelectric power, petroleum, and so forth. Like the agricultural revolution, the *industrial revolution* eventually produced a quantum leap in the power to accumulate economic surpluses, and in the scale and complexity of human societies.

Since the new engines and machines were large and expensive, centralized production in factories began to supplant the decentralized "cottage" craft production of earlier times. People began to migrate to cities in unprecedented numbers, not only because the factory jobs were located there, but because the application of industrial techniques to agriculture—such as the introduction of farm machinery and new inorganic chemical fertilizers—reduced the demand for labor in rural areas. In industrial cities, wealth and power began to be associated not so much with control of land—as in agricultural societies—but with ownership and control of industrial enterprises. A new class system based on industrial wealth rather than the ownership of land began to emerge. Labor became increasingly a cash commodity rather than a subsistence activity with shares as taxes. Work became increasingly separated from family life and bound up with emerging bureaucratic systems of production. Modern complex organizations (bureaucracies) and nation-states were significant new social formations of industrialism.

Like the agricultural revolution before it, industrialism stimulated a whole basket of cultural and economic innovations: in transportation and communication, in medicine, sanitation, and disease control. Prominent among these innovations was the acceleration of the rate of scientific discovery and the application of science-based technologies to economic production. These developments, particularly improved disease control and the rapid accumulation of foodstocks, allowed unprecedented population growth and an extension of the human life span. Unlike agricultural societies, in which overpopulation, ecological collapse, and plagues kept global population rates modest (up to about the 1600s), in industrial societies rapid improvements in economic technology and disease control resulted in positive feedback between population growth and accumulating wealth. I will return to population-environment issues in Chapter Five.

However, as with the agricultural revolution, it is arguable whether industrialism improved the life of the ordinary person, at least until after the turn of the twentieth century. Early industrialism, as observed by both Charles Dickens and Karl Marx, was for the vast majority an uprooting from farm life into a bleak new life of misery, industrial hazards, and exploitation in early industrial sweatshops. Yet in the longer term, improvements in health and living standards diffused from social elites to ordinary people in the large middle and working classes of industrial societies, if not to those at the bottom. Some scholars argue that after the turn of the century industrial societies became more equalitarian than historic agricultural societies in terms of both political rights and the distribution of material well-being (Lenski and Nolan, 1991). Yet this is a slippery argument. Most people live longer, are materially better off, and have more individual freedoms. But have they traded overt forms of social domination and oppression for more subtle forms of control and pervasive alienation unique to the industrial world? Critics of urban industrial societies argue that they have separated humans from nature, destroyed or weakened the bonds of traditional communities (neighborhood, kin), weakened our sense of civic community, and made us dependent on vast international systems (market economies and the like) that elicit neither our loyalty or comprehension. Critics argue urban industrialism produces fragmented ("autonomous") individuals and families with little connection to community at several levels (Young, 1994).

For some time now, a *world system of nations* with its underlying *world market economy* has been evolving. These developments, along with shared cultural traits and aspirations among people in many parts of the world, constitute what is commonly called *globalization*. The important point is that because a "world system" of sorts is emerging, there are no hunter-gatherers or agricultural people anywhere on the earth who remain untouched by the expansion of the industrial societies. Although the diffusion of industrial technologies, consumer goods, and culture has been uneven, it is now found everywhere. For better or worse, Coca-Cola and Marlboro cigarettes are found in every Chinese village. The polar Eskimos (Inuit people)—those that weren't killed off by smallpox and measles—now zoom around the tundra hunting with snowmobiles and repeating rifles. Gone forever are igloos and dogsleds (except for sport), and their children are now plagued by dental caries from refined sugar in their diets, a problem virtually unknown when they were pristine hunter-gatherers.

Human-Environment Relations in Industrial Societies

Like agricultural societies, industrialism dramatically increased human use and withdrawals from the biophysical resource base. The key change in the human-environment relationship was the use of relatively cheap fossil fuels that supported industrialization, more intensive agriculture, and urbanization. This involved much more extensive exploitation of the physical and biotic resource base. It also produced more, and more difficult pollution, as production gradually shifted from natural materials (wood, paper, cotton), which are environmentally benign compared to synthetic materials that break down slowly in ecosystems and may be toxic to humans and wildlife (such as stainless steel, DDT, dioxin, and plastics—chemicals that Mother Nature never knew!).

No evidence yet exists of the weakening or total collapse of an industrial society—for ecological reasons (abundant such evidence exists for historic agricultural societies). This is because the industrial environmental degradation has so far been more than offset by increased investment and technological inputs. Whether this state of affairs will continue to be true in the future is arguable, as discussed in Chapter One. It is the big question I return to in Chapter Seven. Here you should note that it took the Copan Mayans more than 400 years to collapse, and much longer for Mesopotamians. By comparison, industrial societies have only been around for about 300 years and the growth of world population and technological prowess means that our biophysical impacts are on a much larger scale than in historic agricultural systems.

The Dominant Social Paradigm (DSP) of Industrial Societies

If the main cognized environment of agricultural societies was that of a garden to be tended, modified, and dominated by humans, that of industrial societies is a dramatic extension of this concept. It was amplified particularly by cultural developments of the European Enlightenment period (seventeenth and eighteenth centuries), which emphasized empirical reasoning, science, the world as a giant cosmic mechanism, and the ability of humans to rationally control nature through systematic innovation and experimentation. The earth and other species became cognized as a huge *resource base* and facility to be used, developed, and managed for human needs and desires. Unlike agriculturalists, industrial people not only tended the garden, they attempted to remake it.

Many scholars have attempted to describe the DSP of industrial societies. Although they differ about the details, they agree that industrial DSPs amplify the second part of the human duality already mentioned: that humans, by virtue of culture and technology, have a unique power to change, manipulate, and sometimes to transcend natural environmental limits. In one way or another, most scholars think that DSPs of industrial societies have the following themes:

1. Low evaluation of nature for its own sake.
2. Compassion mainly for those near and dear.
3. The assumption that maximizing wealth is important and risks are acceptable in doing so.
4. The assumption of no physical ("real") limits to growth that can't be overcome by technological inventiveness.
5. The assumption that modern society, culture, and politics are basically okay. Adapted from Milbrath, (1989: 119). For other versions see Cotgrove (1982), Dunlap (1983), Harman (1979), Olsen, Lodwick, and Dunlap (1992: 18), and Pirages (1977).

Some described this worldview as the DSP of free-market or capitalist industrial societies. But it is obvious that the former communist nations of Eastern Europe and the USSR damaged their environments to a much greater degree than did Western market economies. Most now believe that it applies generically to industrial societies. In view of the emergence of the world system of nations and world market economy, it is also fair to note that this DSP does not affect only the more developed countries (MDCs) of the northern hemisphere. Hardly anyone in the world today is immune from it. People in the less developed countries (LDCs) want the things of industrialism (TV, autos, vaccinations, Coca-Cola, and cigarettes). The industrial DSP is diffusing rapidly around the world, where the *desire* for progress is defined largely in terms of increasing material consumption, security, and well-being. This is true even in the poorest LDCs, where material and health standards are now very low and misery is widespread.

But please note diversity and change also. That is, the DSP does not control everything—there are competing social paradigms—and now even it is obviously in some kind of flux and transition. Since the beginning of the twentieth century, there *was* concern in the United States about maintaining natural environments (for both utilitarian and intrinsic reasons), which produced turn-of-the-century conservation movements. These movements led to the establishment of protected public lands, national forests and parks. Similarly, there was concern among agricultural agencies about soil preservation and erosion, which continues today. But increasing popular environmental and ecological awareness was stimulated most directly from environmental problems and *environmental social movements* beginning in the 1960s, in the United States as well as other nations. I will return in more depth to environmental social movements and some evidences of paradigm change in Chapter Nine.

Finally, it is also clear that industrial societies are in the midst of another profound transition, like the historic agricultural and industrial revolutions. People and scholars have labeled this large-scale transition variously: postmodernism, postindustrialism, the "global village," the information age, a New Age of integration and spirituality . . . or what? I will return to the nature of this transformation in the last chapter. For now, it is enough to say that whatever else, it involves a rise in ecological thinking and a change in the DSP just described (Olsen, Lodewick, and Dunlap, 1992). Part of this change involves the rise of environmental social sciences, to which I now turn.

EXPLAINING SOCIOCULTURAL EVOLUTION: HUMAN ECOLOGY AND POLITICAL ECONOMY

This chapter began by discussing the components of human systems, their broad patterns of historical evolution, and their changing human-environment relations, all sketched in a descriptive and synoptic way. Paralleling Chapter One about ecosystems, I turn to explaining change and evolution in human systems, from the small scattered human systems of prehistoric times to the larger, complex, and more inclusive ones of today. In doing so, I will use two kinds of interdisciplinary thinking shared by scholars from many backgrounds. One is *human ecology,* the study of how humans live and adapt in relation to both their biophysical and social environments. The second is *political economy,* the study of the complex interactions between economic and political systems (more informally, between money and power, if you wish). After long neglect, some scholars are reviving evolutionary thinking about human systems that has the potential to link large and small scale processes, explain the emergence of complexity and link social science to biology without misleading reductionism (Dietz et al., 1990: 155; Maryanski, 1998).

Ecosystem and Sociocultural Evolution

Earlier I noted parallels between ecological and human systems. Recall that ecological theorists, such as Lotka and Odum, argued that ecosystems evolve as different species competing for available energy in the physical environment selectively survive. If uninterrupted, the result, over time, is a larger, more complex, and inclusive structure of species connected in food chain niches and often in symbiotic relations that range from mutualistic to parasitic. In a parallel way, sociocultural evolution begins when humans compete for control over limited natural resources. As they do so, some persons and groups develop more efficient material subsistence infrastructures related to subsistence. Complex relationship systems of statuses and roles emerge that parallel niches in ecosystems. These relationships, and the exchanges of goods, labor, control, loyalty, and symbols on which they are based, broadly parallel the types of symbiotic relationships in the biological world. First, there are social *exchanges of reciprocity*, which produce egalitarian, mutual benefit relationships in nonhierarchical contexts. These exchanges are similar to the phenomenon of mutualism. Second, there are social *exchanges of redistribution*, wherein goods and services are shifted "upward" to persons or centers that reallocate them (like profits, plunder, and taxes). These exchanges result in relationships that are asymmetrical in terms of power and equity, and stratified relationships (Polanyi, cited in Rogers, 1994: 45). Reciprocal exchanges predominated among hunter-gatherers, whereas redistributive exchanges became more pronounced as human systems evolved. Redistributive exchange bears some resemblance to the asymmetry of parasitic and predator-prey relationships.

But you can't carry these parallels too far, because there are important differences between humans and other animal species. Consequently, ecological and sociocultural evolution are somewhat different. While all animals communicate—that is, transmit behaviorally relevant information—only humans do so extensively through the use of *cultural symbols. Homo sapiens* share this symbolic capacity with our evolutionary primate cousins, but that of humans is of such greater magnitude that it makes us, in effect, unique among animals. The communication mechanism of other species is largely genetically programmed and innate, unlike the meaning of human symbols, which are *arbitrary* and depend on a consensus of symbol language users (Sanderson, 1995: 32–33). In biological evolution, the units of transmission and selection are individuals and particular genes that survive (or do not) between generations. In sociocultural evolution, however, the units of transmission and selection may be individuals, a society, or its subsystems. But the generation of sociocultural novelty along with its intergenerational selection and transmission is Lamarckian rather than genetic (Jean Baptiste Larmarck, Darwin's most famous predecessor, argued that animals could inherit learned behaviors and characteristics). Moreover, "symbol systems can blend, and components can be added to or subtracted from culture, thereby making it difficult to predict what is being inherited or transformed" (Freese, cited in Maryanski, 1998: 29). In this sense, Freese argues that human systems do not evolve, but they do change and develop. Others retain the idea of sociocultural evolution but emphasize the accumulation of complex contingencies (such as the generation of novel forms, their transmission and selection over time), which is closer to the biological meaning of the term rather than fixed "stages" of development, common in the early history of the idea (Burns and Dietz, 1992).

These considerations led earlier scholars to abandon strongly deterministic approaches to the evolution of human systems in which environmental and material forces were thought to determine everything else. Earlier generations of anthropologists and geographers coined the notion of *environmental possibilism* for more flexible approaches. These models posit that material and biophysical factors are broad limiting factors for particular human systems, but the most immediate and particular causes of many social and cultural changes are *other* social and cultural factors. Anthropologist Julian Steward, who rekindled interest in sociocultural evolution, used the term *culture core* to describe a society's technology and subsistence economy (what I earlier referred to as material infrastructure). The biophysical environment has direct interacting effects solely on this culture core, but only indirectly with other elements of human systems. Relationships are two-way interactive ones with feedback, or *cybernetic* ones (Kormondy and Brown, 1998: 45–47). (See Fig 2.1.)

Having described causal connections between the biophysical environment and elements of human systems, I turn now to the question of how human systems evolve into larger, more complex and inclusive systems. I described that historical process earlier in this chapter, and it is briefly summarized in Fig 2.2.

Devolution as well evolution of more complex systems occurred periodically, when complex systems collapsed (like those of the Mesopotamians and Mayans) and resulted in smaller, simpler systems.

Many people think that the change from food foraging to horticulture and agriculture during the agricultural revolution happened because people traded a precarious and insecure way of life for one that was more secure and satisfying. Little evidence exists to support this view. Rather, climate changes that "shrank" livable environments, the growth of human populations, the gradual destruction of edible plant and large animal populations that supported life *and* the discoveries and technological innovations that made the transition possible all combined to cause the transition to agriculture. Furthermore, fossil record and archaeological evidence confirm that hunter-gatherers did not abandon their lifestyle until forced to do so by these prob-

Figure 2.1 Human Ecology Theory: Relationships between the Biophysical Environment and Sociocultural System Elements

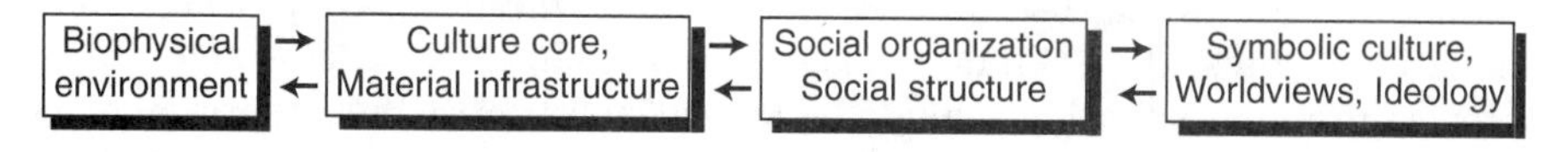

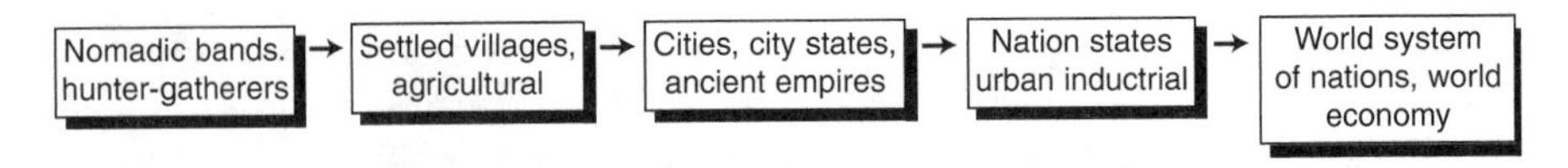

Figure 2.2 The Evolution of Complex Sociocultural Systems

lems, and did so at different times and in widely scattered areas around the world (Lenski and Nolan, 1999: 119; Sanderson, 1995). I discussed earlier the declines in the quality of health and social life associated with this transition. A similar combination of environmental problems, scarcities, and technological possibilities caused the rise of cities and ancient empires and the emergence of industrial societies (see pp. 61–65). Thus the growth of innovations and technologies produced more complex and inclusive human systems having ever-larger productive capacities to support human populations. These developments gave political and economic elites the ability to create large states, empires, and systems of control through the imposition and expansion of their domination by taxation and military means. Nonelites, however, often adopted the lifestyles associated with these transformations—not from positive attractions, but to survive when they had no other choices. In the nineteenth and twentieth centuries, for instance, established farmers rarely *willingly* gave up their farms and moved to cities seeking urban employment. Rather than city "bright lights," the story is of migration driven by progressive rural poverty, bankruptcy, and foreclosed farm mortgages.

Besides incredible productive capacities, industrial societies produced historically unprecedented sociocultural changes. Economic production for *use* became progressively eclipsed by production for *exchange* for other goods and services. To facilitate exchange, money increasingly replaced barter, as an abstract, portable, and more convenient medium for the exchange of goods and services. Even human labor became a "commodity for exchange" at a fixed monetary rate. Money came to symbolize the value of more concrete biophysical resources (e.g., land, minerals) and became, as finance capital, the premiere material resource, of industrial societies. These processes happened within exchange *markets*, a third form of exchange. Unlike the two mentioned earlier (reciprocity and redistribution), in markets social relationships become embedded in the economy instead of vice versa (Polanyi, cited in Rogers, 1994: 45).

The growing complexity of human systems produced another parallel with the evolution of ecosystems. Large-scale and complex market exchanges, particularly in industrial societies, dramatically increased occupation specialization (the "division of labor") and other kinds of social differentiation. That is analogous to *speciation*, the evolution of different biological species that use different niches of an environment. Social differentiation represents a kind of *quasispeciation*. In this process, we *Homo sapiens*, though remaining a single biological species, use the environment as if we were many species. Different institutions, industries, and occupations use the same biophysical environment in different ways for resources important to their specialized purposes. Thus in a highly complex social order equipped with modern technology, human beings become a *multiniche* species (Hutchinson, 1965; Stephan, 1970; Catton, 1993/94). Why is knowing this fact important? Because it should enable you to understand why people in modern societies have difficulty cooperating on problems of truly common interest without becoming sidetracked by their "special interests."

The Political Economy of the World System and Globalization

Earlier I mentioned that a sociocultural world system has been emerging in contemporary times. It is a rapidly growing world market economy, a loosely integrated but extremely volatile world system of nation states and international organizations, like the U.N. and NATO. There are important nongovernmental organizations (NGOs) such as the international Red Cross, scientific, and environmental organizations. People, ideas, money, technologies, products, and labor flow across borders, and the boundaries between "us" and "them" seem confused and often compromised. Since the eighteenth century, the growth of this world system has been associated with the following trends:

1. Dramatically growing human population
2. Growing average size of societies and communities
3. Urbanization
4. Increasing biophysical impacts of human systems on the biophysical environment
5. Increasing invention of new symbol systems and technologies
6. Increasing storage of technology and cultural information
7. Increasing inequality within and among societies
8. An accelerating rate of sociocultural change
   (Lenski and Nolan, 1999: 69)

What produced the emerging world system? There are two different kinds of explanations.

Political Economy I: Neoliberalism

The neoliberal perspective embodies the thinking of economists about markets and politics. It is rooted in a simple and dramatic assumption that will sound familiar to you—that the best human system results from individuals being free to pursue their own interests, under restraint of law. This may strike you as too obvious to mention, but it is a particularly modern view, not widely shared until recently, and not by everybody today. Until fairly recent times, people and governments assumed that it was right for governments to promote the society's well-being by special export subsidies to national industries and to protect them from foreign competition by import tariffs—just like armies protect from foreign invaders. That policy, called *mercantilism,* dated from the eighteenth century, when European kings controlled commerce and trade. Notwithstanding the emerging world system, it is by no means dead today, as when struggling businesses seek government protection from foreign competition.

But mercantilism (or economic nationalism) was questioned when international political, corporate, and banking leaders thought about the big traumatic events of the twentieth century. Two world wars with a world depression between them had produced prolonged devastating world-wide chaos and conflict. Leaders met toward the end of World War II at Bretton Woods, a resort in New Hampshire, to consider how to prevent similar events in the future. They thought that those events were born of the frustrations associated with excessive economic nationalism (undiluted mercantilism) that stifled trade, produced economic and political instability and unemployment, and raised consumer prices from what they hoped would be a free international market. The gathered political and financial leaders from all over the world agreed to embark on policies of international free trade. The *Bretton Woods system* was envisioned as a free system of international trade in open markets without barriers. But within this world trading system, individual nations would be able to conduct the kinds of policies for controlling inflation and unemployment and encouraging economic growth like those that influential economic theorist Maynard Keynes had advocated since the Great Depression. In other words, states should have important roles for economic regulation *within* nations, but free markets were intended to dominate relations between nations. The main theme to understand the political economy of the world system was no longer state versus market, but rather how much and what kind of state intervention was necessary within an overall system of open markets (Balaam and Veseth, 1996: 16, 42, 50).

The growth of the world market system and this understanding of it fit naturally with the goals of investors and corporations to expand their markets and profitability through international trade. This system was facilitated by the development of new technologies in agriculture, manufacturing, and transportation and particularly by the development of the new information technologies since the 1970s (computers and the Internet).These new technologies increased the cost-efficiency of doing business at widely scattered locations around the globe. In the context of the slow negotiating down of trade restrictions (the GATT "rounds"), these driving forces produced the dramatic growth of huge multi- or transnational corporations (TNCs) that came to dominate labor, capital flows, and production around the world. In 1970 there were some 7,000 TNCs, but by 1998 there were at least 54,000 (Brown, 1999: 24). Corporations, especially TNCs, prospered and the rate of aggregate output of goods and services mushroomed. *World aggregate economic output more than quadrupled since the 1950s, vastly exceeding population growth rates.*

Please understand that the *neoliberal* understanding of the political economy of the world system is not only an intellectual theory, but a major influence on international policy by governments, corporate leaders, and bankers around the world. It shaped negotiations between nations. It was, for example, the dominant thinking that resulted in the North American Treaty Agreement (NAFTA) and spawned international institutions like the World Bank, the International Monetary Fund, and the World Trade Organization. These organizations have become so powerful that they often control global political economic policy, and the fates of nations and regions. As you might guess, such broad ideas are difficult and contentious to practice in the real world.

This world system and market economy dramatically increased trade and production, but is certainly not free of deep problems and contradictions (look again at Lenski and Nolan's list of characteristics, p. 53). I discuss only a few, and social reactions to them. One of the most obvious is the size and net assets of TNCs, even compared with the gross domestic products of nations. By 1998, of the 100 largest global economies, more than half were corporations and not nations. Thus the net assets of Wal-Mart exceeded the gross domestic product of Greece, Philip Morris exceeded Chile, and the Nestlé corporation exceeded Hungary (Rauber, 1998: 17). The assets of the largest corporate players make it a very unlevel field for negotiation, even between many nations and TNCs. If policy based on neoliberal theory was compatible with corporate interests, it was less much compatible with the concrete interests of governments, since TNCs often evaded national taxes and regulations by producing in "tax and regulatory havens" and by shipping products and profits back to consumers and investors where they were headquartered.

The new world economic integration was also less compatible with the interests of people as workers, since they were often displaced by the new economic technologies and the search for cheap labor around the world. Human labor was often marginalized, and the resulting tensions of unemployment or substandard employment produced social burdens assumed by local communities and nations. Thus the neoliberal world market economy has a deep contradiction: Growing aggregate production alongside mushrooming inequality and poverty both within and among nations. Over 20 percent of the world's people are mired in absolute poverty, defined as a condition in which they cannot regularly meet their most basic subsistence needs (Giddens, 1995: 98). Neoliberal theorists have (recently) attempted to explain how this gaping inequality emerges, at least at national and corporate levels. An extensive literature exists about structures of domination (or *hegemonic structures*) in the world market economy (Balaam and Veseth,

1996: 51). I will return to this inequality in the world system in much greater depth in Chapter Seven, particularly as it relates to environmental issues.

Corporations, particularly TNCs, don't care much about place or, perhaps by implication, the geophysical environment—except as a resource base. But people, governments, and nongovernmental organizations (NGOs) *do*. So the tensions associated with the emerging world system have spawned a plethora of social movements and nongovernmental organizations (NGOs). These attempt to address the gaps left by corporations and governments in meeting social needs, including the adequate and equitable provision of health care, education, food, shelter, and environmental protection. By one estimate, the number of *international* NGOs grew from 1,000 in 1956 to over 20,000 in 1998 (Brown, 1999: 26). (I believe this to be an undercount!) Whether this emerging "civil society" layer of NGOs is adequate is an open question. I will return to the topic of environmental movements and organizations more in depth in Chapters Nine and Ten.

Political Economy II: World Systems Theory

A very different conception of the emerging global world system begins not with assumptions about states and markets, but with the political and economic history of the modern era. Most of the world has been in contact with modernizing European nations since about 1500, and by 1800 the scope of that contact had increase so that through colonial empires Europeans controlled most world trade. The global diffusion of Western technologies, culture, and values began during this period. Colonial nations imported cheap raw materials from their colonies and reexported more expensive manufactured goods in markets controlled by colonial administrations. But since 1900 the colonial empires (of the British, Dutch, French, and Germans) began to break up, and political control was replaced with economic control through a system of trade. The world market system mentioned earlier is, in this view, a global economic exchange network divided among competing national entities (corporations as well and governments). But it is a very unequal and stratified exchange system in which the industrial MDCs provided investment capital and technology, while the LDCs were the providers of raw material and, increasingly, of cheap labor. What characterized the evolving world system was a global hierarchy and division of labor as well as a highly unequal system of trade and profitability between the MDCs (largely in the northern hemisphere) and the LDCs (largely in the southern hemisphere). MDCs retain decisive control of the world systems because they control finance capital and the terms of trade. LDCs became increasingly enmeshed in the world system in dependent status as debtor nations; precisely how they became "less developed" to begin with. Thus MDCs and LDCs evolved together. The policies of the Bretton Woods institutions (the World Bank and the IMF) operate to amplify these inequalities and power-dependent relations. They encourage LDCs to borrow money for development, to open their economies to domination by TNCs, and to find money to pay external debts by cutting budgets that degrade items like education and health care.

This perspective was called *dependency theory* by economists (Frank, 1967), but it is increasingly known as *world systems theory*. Its theoretical reasoning extends Marxian thought about economic class, conflict, and inequality within societies to understand the world economic and political structure. Hence it is sometimes understood as a "new historical materialism." Wallerstein, probably its most articulate advocate, envisions the structure of emerging the world system in three tiers. *Core nations,* are powerful and affluent MDCs (like the United States, Germany, and Japan), have diversified industrial economies and exercise political, economic, and fiscal control over the world system. *Peripheral nations,* are most powerless, with a narrow economic bases of agricultural products or minerals, and often providing cheap labor for TNCs (e.g., Rwanda, Indonesia, Ecuador). Somewhere in between are *semiperipheral nations,* intermediate in terms of their wealth, political autonomy, and degree of economic diversification (e.g., Mexico, Malaysia, Brazil, Venezuela) (Wallerstein, 1980; see also Chase-Dunn, 1989).

It should be obvious to you why this theoretical perspective about the international political economy has been less influential in policies that shape the world market system. Core nations, understandably, resist attempts to equalize the terms of trade which work to their great benefit. But it is a thriving research perspective in the social sciences. It has been used, for instance, to explain the increasing poverty and environmental degradation in LDCs—such as in the Amazon Basin. This phenomenon results from the market power of MDC governments, banks, and corporations to control the world economy and to exploit weaker LDCs, human labor, and their land and natural resources (e.g., Ciccantell, 1999). The most obvious difficulties with world systems theory are that it depicts MDCs as acting too coherently in a complicated world, and it provides wealthy elites in LDCs with a ready set of ideas to blame the MDCs for their plight. Such leaders and elite classes in LDCs are very much involved in the problems of developing nations, from which they often benefit. For more information about and critiques of this perspective, see Harper (1998), Hecht and Cockburn (1989), Lenski and Nolan (1999), Sanderson (1995), and Wolf (1982). I will return in more depth to the *environmental implications* of globalization and international trade in Chapter Ten.

ENVIRONMENTAL SOCIAL SCIENCE

Ecology has been a part of biology since the 1930s, but environmental and ecological social sciences are relatively newer fields. They grew mainly since the 1960s and 70s as scholarly responses to the environmental problems, conflict, movements, and popular consciousness of those decades. But in fact the social sciences have a long and ambivalent intellectual history in how they conceptualize the environmental embededness of human systems. A more complete sketch of their development would include a wide range of specialties or disciplines, including psychology, anthropology, human geography, political science, and policy studies, but here I focus on economics and sociology.

Economic Thought

The founders of the field of economics all assumed that the earth's biophysical resources (land, minerals, living things) were the necessary basis for the economic production of useful goods and services. But, beginning with Adam Smith (1723–1790), they argued that *labor*, not nature, was the major source of economic value. Smith argued that the operation of private unregulated *markets* were the best natural mechanisms to determine the *economic value* of goods and services, and wages. Smith distinguished between market value and moral or social value, separating the latter from economics and thereby initiating the tendency of economic thought to treat the economy in abstraction from the rest of the sociocultural world.

Smith argued that the desire for profits and the "unseen hand" of unregulated markets would produce the best possible economic and social world. It would create a system that reflected "real" economic values and encourage the use of investment, labor, and technology in ways that increase production in response to consumer desires. Smith's view was buoyant and optimistic, reflecting a bustling and successful nation of English traders, shopkeepers, and merchants on the eve of the real industrial expansion that was to come. In the next decades, that optimism would fade. David Ricardo (1772–1823), for instance, argued that economic growth and the desire for profits would lead people to bring even marginal resources, such as poor and infertile land into production. As population grew, it would "become necessary to push the margin of cultivation further" (Heilbroner, 1980: 95). His message was ecological but also moral, for he argued that in the long term only the fortunate landlords stood to gain as their holdings rose in value—not workers struggling to make a living or enterprising capitalists laboring to maintain profits.

Thomas Malthus (1776–1834) argued that increasing production and improved living conditions would lead to population growth. But, he argued that population grows exponentially, while material resources such as food supplies increase in an arithmetic way.[3] Malthus predicted that after the bloom of initial growth would come the inexorable regression to scarcity (like the bacteria-in-the-petri dish mentioned earlier), bringing with it the "population checks" of misery, famine, pestilence, war, and social chaos.

Malthus not only was an influential figure in economics but also provided an early link among economic, demographic, and ecological thinking. Karl Marx (1818–1883), like others, argued that nature was an important factor in production but that social factors, in particular, the "ownership of the means of production" (land, capital, factories) was more important. Like Ricardo and Malthus, he saw chaos at the end of the capitalist era. But he argued that its sources were to be found not in the demographic-economic calculus of Malthus, but rather in inherent and eventually unmanageable conflicting material interests between economic classes of workers and the owners of the means of production. He believed that their struggle over wages and profits would eventually be resolved in the apocalyptic and revolutionary transition to socialism. The creative intellectual accomplishment of these classic thinkers was to move from anecdote to science, to comprehend economic markets as law-abiding systems whose dynamics could be understood and—perhaps someday—predicted. In this quest, they were only partly successful.

But as prophets, one has to admit that they were all a bust: We have not realized the capitalist paradise of Smith; Ricardo's landowners do not dominate the industrial world (certainly not at the expense of finance capital); capitalism has been able to politically contain the apocalyptic demise that Marx predicted (which ironically happened to state socialism in our times); Malthus certainly underestimated the amount of food that could be produced and the number of people who could be supported. But the greatest irony of all was that even though their views gave shape to modern economic thought, they all failed to comprehend the expansionary dynamic of industrial capitalism. They all thought that the growth they were witnessing would be short lived and that a "dull" steady state economy or systemwide collapse was only a few decades away (Heilbroner, 1980: 305–306). In that assumption they were dead wrong.

More than the classic thinkers, contemporary economists emphasize the second part of the human-environmental dualism I noted earlier. *Neoclassical theory*, the dominant perspective, views the economy as a circular flow of investment, production, distribution, and consumption, understood in abstraction from the rest of social life. To put it starkly, in the neoclassical view the economy contains the ecosystem (as resource bases and pollution sinks). Surely a natural scientist would put it the other way around—that the ecosystem contains the economy as well as other human institutions. Nonetheless, neoclassical theory implies that environmental and resource problems cannot be very important ones because the economy depends more on the boundless and continuous flow of money income. They maintain that the technological advance will outpace resource scarcity over the long run and that ecological services can also be replaced by new technologies. Economic markets work the same to make money and determine prices whether resources are plentiful or scarce. Nor do social values or questions of justice intrude: Markets work whether you are growing corn, producing health services, selling heroin, cleaning up toxic wastes, or selling slaves. Neoclassical economics deals extensively with "efficient" allocation, secondarily with distribution, and not at all with matters of scale. Though construing the world in narrow and abstract terms, neoclassical economics has become enormously influential in industrial societies in shaping debates about social, political, and environmental policy. This is true partly, I think, because the theory appears more objective by deliberately ignoring questions of human values, political, and ethical considerations. But these, I argue, are important human questions and considerations that really ought not to be "ruled out of court" (Costanza et al., 1995, 60, 80; Daley and Townsend, 1993: 3–6).

Even so, conventional neoclassical economic analysis is perfectly serviceable to shed light on any number of human-resource-environment problems. For example, it has been used to explain the collapse of complex agrarian societies through ecological degradation, such as that of the Mayans discussed earlier. After detailed investigation of the collapse of the Mayans, the Chaco Canyon societies of the American Southwest, and the western Roman Empire, Tainter concluded that decline begins when there are "declining marginal returns on investment." That is, at some point each society was investing more in maintaining essential institutions (e.g., temples, cities, armies) than it was able to benefit from them. Once the point of diminishing returns was reached, it required constantly increasing investment just to maintain the status quo. At some point the economy and social system began to collapse (Tainter, 1988: 187–195).

But there are important problems outside the scope of this useful insight. What, for example, led the Mayans to migrate to the Copan catchment area in large numbers rather than maintaining scattered plots of land and villages with (probably) sustainable agricultural practices? Why did they need cities, temples, and armies anyway? It is when you ask broader and longer-range questions that the limits of mainstream economic analysis appear. Many important issues are simply outside the framework of conventional economic analysis.

Emerging Ecological Economics

Some economists have been concerned with environmental resources issues for several decades without recasting the neoclassical theory significantly. But a small and growing band of economists are trying to recast economic theory by finding ways of incorporating both *nature* and *human values* into their economic calculus. They are producing, in other words, a real ecological economics. They begin with questions about how to assign economic values to nature and human values. Let me give you three examples of these dilemmas: (1) How can values ("prices") be assigned to goods that are held in common (the "commons") that are used by many and owned by none, such as the atmosphere, rivers, oceans, and public space? They cannot be privately owned in small pieces that can be meaningfully bought or sold. Hence, there *is* no "market" other than an invented or imagined one, and therefore no prices to limit use. The widely recognized problem is that we tend to overuse common as opposed to privately held goods (Hardin, 1968, 1993). (2) How can economic analysis incorporate and assign responsibility for the variety of environmental and social *externalities,* that is, the real overhead costs incurred in the production process that are born not by particular producers or consumers but by third parties, the larger social community, or the environment? (3) How can we calculate the value of using nonrenewable resources at the present time as opposed to reserving them for the future use or for future generations? We tend to value present consumption higher and discount future values. These are among the thorny conceptual problems you cannot really get to from the neoclassical model (Clark, 1991: 404).

> Conventional economics ignores all but humans . . . ecological economics tries to manage the whole system and acknowledges the interconnections between humans and the rest of nature. . . . The basic world view of conventional economics is one in which individual human consumers are the central figures. Their tastes and preferences are taken as given, and are the dominant determining force. The resource base is viewed as essentially limitless due to technical progress and infinite substitutability. . . . Ecological economics is prudently skeptical in this regard. Given our high level of uncertainty about this issue, it is irrational to bank on technology's ability to remove resource constraints. . . . If we guess wrong then the result is disastrous—irreversible destruction of our resource base and civilization itself. (Costanza, 1991: 6–7)

Thus, ecological economics attempts to deal with issues long ignored by the neoclassical model. Like neoclassical economists, they are concerned with *allocation*—that is, how much of investment and resources go into making which goods and services, and at what price. But ecological economists are also concerned with *distribution,* not only the initial distribution of money or resources before the market works, but "who gets how much of what products of the market?" They are also concerned with *scale,* referring to the physical volume of matter and energy throughput of the economy. This concern is related not only human consumption but to the natural capacities of the ecosystem to regenerate inputs and absorb waste outputs on a sustainable basis (Costanza et al., 1995: 80–81).

You need to understand that this view represents a frontal assault on the abstracted tidiness of the neoclassical model (not necessarily that of the classical founders!). By emphasizing scale and throughputs, ecological economics challenges the article of faith of mainstream economists that human ingenuity and technology will always overcome environmental limits and ecosystem capacities. It also recognizes important matters of value that can't be reduced to price-efficiency: "A good distribution is one that is *just* or *fair*, or at least one in which the degree of inequality is limited within some acceptable range" (Costanza et al., 1995: 80). How much inequality is just? As you can see, not only nature but human values and culture have been reintroduced center stage.

Ecological economists are busy tinkering with ways to "price" social and esthetic externalities, finding ways of circumventing "commons problems" and producing measures of human well-being broader than those that simply measure how much money a nation produces (e.g., gross national product figures). They are also rethinking tax and subsidy policy to reverse their historically damaging environmental impacts and ways of avoiding onerous regulation altogether by emissions trading schemes for toxic wastes and greenhouse gases that make being "good to nature" profitable. I will return to these strategies in Chapter Eight.

Sociological Thought

Classic economic thought affected early sociological thinking, which by the 1880s was taking shape across the English Channel in France and Germany. It was influenced not only by the English political economists but also by Darwin's theory of evolution, which by that time was in wide intellectual circulation. In vast oversimplification, the classic formulations in sociology can be understood to be largely the work of three paramount figures: Karl Marx, Émile Durkheim, and Max Weber.

Earlier you encountered Marx (1818–1883), whose ideas are claimed today—or *disclaimed,* as the case may be—by contemporary sociologists, economists, philosophers, historians, and political scientists. Among his multiple critiques of early capitalism, Marx noted how the growing concentration of ownership of land and productive resources (including capital) affected how economies worked and gave rise to classes of wealthy and classes of poor, to inequalities of all kinds, and to decisive political control by wealthy classes. In his view the dominant ideas and values, laws, philosophies, worldviews—in other words *culture*—simply represent the material interests of the dominant economic classes. His view is represented schematically in Figure 2.3.

This model should strike you as familiar. It shaped theoretical thinking in human ecology that I discussed earlier. *But note* that for Marx, the arrows indicating causal influence go only one way and are not feedback or cyber-

Figure 2.3 Marx's View

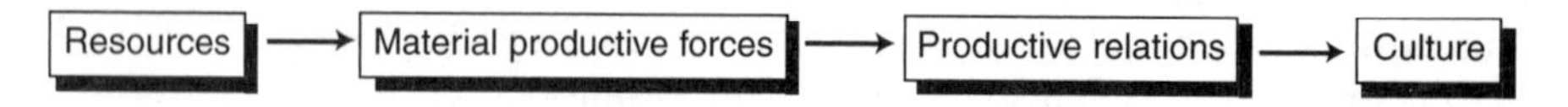

netic relationships. By resources, he meant such things as land and minerals, but also money. Material productive forces are factories, tools, and technologies that converted resources to commodities for exchange (e.g., farms, mines, factories). Material productive forces also include technologies. Productive relations are the *class relations* that reflect the ownership and control of these. Marx's famous illustration was that "the windmill gives you a society with the feudal lord, the steam-mill the society with the industrial capitalist" (1920: 119). The most controversial aspect of Marxian thought is that this is the "real" material substructure of society that gives rise to the "superstructure" (states, laws, philosophies, worldviews). Without denying the insights of this materialist conception of history, many critics and Neomarxians argue that these things (governments, legal systems, worldviews), once established, can exert powerful restraining and stimulative effects on the "material base." In other words, putting it more abstractly, elements of the superstructure, once established, can provide feedback loops that shape and change the material substructure. Others critics of standard Marxist thought add that important social conflicts of interest are not all based on economic class but can also be based on ethnicity, nationality, kinship, gender, and so forth.

Émile Durkheim (1858–1917) was greatly influenced by the evolutionary thinking of Darwin but he rejected the fashionable "biolologism" of the day (using biological analogies to understand society). He also rejected the notion that individuals were the important units for studying social evolution. For Durkheim, the operative units of evolutionary adaptation were sociocultural systems, and human evolutionary history was understood abstractly as a transition from simple and homogenous systems with powerfully binding cultural rules (*mechanical solidarity*) to complex and heterogeneous systems with weaker and less binding cultural rules (*organic solidarity*). You can get a concrete sense of this model by reconsidering my previous sketch of historical development (from hunter-gatherer to industrial societies).

Durkheim argued that increased population density and the intensification of the struggle over scarce resources were important antecedents to industrialism and the complex division of labor in industrial societies. This division of labor would, he thought, increase the adaptability of more populous and dense societies to their environments by decreasing direct competition over resources and causing cultural innovation—such as science—that would redefine and effectively expand resources (1893/1933). In contrast to Marx's view, he believed that industrialism would mitigate class conflict by reducing scarcity. In his view, the major problems of industrialism would stem from the weakening (cultural) bonds between groups in a complex division of labor, resulting in cultural confusion (anomie). Sociologist William Catton contends that Durkheim misreads both Darwin and contemporary ecology. The result of the growth of social complexity Durkheim could observe in his time was not a "mutualism of interdependent specialists," but rather a web of unequal power-dependent class relations more akin to the "parasitism" that Marx observed (Catton, 1997: 89–138). As with the critics of Marx, I'm not sure how devastating this critique is to Durkheimian thought. Though class relations in modern capitalist societies are vastly unequal, they are more equalitarian with regard to both resources and rights than preindustrial ones, as in the empires of the ancient world. Perhaps the point is moot: Predator-prey and host-parasite relations can be symbiotically stable, even if not equitable. Well-adapted predators do not decimate their populations of prey, and a well-adapted parasite doesn't quickly kill its host.

As you can see, Durkheim's thought is powerfully rooted in ecological thinking. But he attended to only one side of the agenda of environmental sociology: He paid a lot of attention to the mechanisms through which the biophysical environment affects society and hardly any attention to the processes by which social systems affect the biophysical environment (Buttel, 1986: 341). But by emphasizing the reality and critical role of "sentiments" and "collective conscience" (in other words, culture) in the provision of social cohesion, his ideas cannot be placed unequivocally in the category of a materialist conception of history and society, along with Marxian thought (Lenski and Nolan, 1999).

In contrast, Max Weber (1864–1920) is hardly ever regarded as an ecological thinker. But he is an important early sociological thinker because he was, even less than Durkheim, uncomfortable with the "materialist" interpretations of history of Marx and the English economists. In his view, the historical development of societies is shaped by a plurality of interacting causes and factors; ideas and values (culture) are given equal weight at potent "causes" of historical evolution and development. Weber argued this point by examining the critical role of the "ethic" of Calvinist Protestantism in the development of industrial capitalism (1905/1958). The main thrust of Western social development, he later argued, could be understood as the progressive elaboration and diffusion of the *cultural complex of "rationality"* in Western societies, which underlies the development of capitalism, bureaucracy, and empirical science. Similarly, although he did not disagree with most of what Marx said about the origins of economic *class* and inequality, Weber argued that social inequality (or *stratification*) could also be based on *status* (honor, prestige) or political *power* itself (which was often prior to, and controlled wealth) (1922/1968).Weber was an intellectual ancestor to sociological thinkers who give a significant role to *human agency* (vs. structural determinism) and the power of cultural ideas to shape the historical development process.

Emerging Environmental and Ecological Sociology

By the middle of the twentieth century, sociologists, like their economist cousins, had begun to distance themselves from seeing sociocultural systems as in an important relation with nature by similarly emphasizing the second part of the human-environment duality: that humans are an exceptional species. When you think about it, it is not really surprising that the dominant theories and intellectual paradigms of the social sciences were congruent with the DSP of industrial societies. They may have, in fact, contributed to its formation, or at least its legitimation. Catton and Dunlap called this anthropocentric socological paradigm the *human exceptionalism* or *exemptionalism* paradigm (1978: 42–43).To illustrate, one sociologist argued in the 1990s that "humans are not governed by the natural processes that govern plants, animals, and planets, *but by their own creations* [emphasis in the original] (Rossides, 1993: 31).

Nonetheless, like economics, sociology did not entirely loose touch with the importance of the biophysical bases of human social life, and environmental or ecological sociology gradually reemerged. By the 1920s, researchers were using ecological concepts (zones, niches, competition, ecological succession) to understand urban structure and change, and some, particularly rural sociologists, were studying the American conservation movement. Leonard Cottrell, almost singlehandedly among sociologists, was studying the social implications of different ways of producing energy (Cottrell, 1955; Duncan and Schnore, 1959; Hawley, 1950; Humphrey and Buttel, 1982: 49; Park, Burgess, and McKenzie, 1925).

As they did for economists, environmental problems and movements of the 1960s and 70s stimulated sociologists to coalesce and refocus scattered interest in human-environment relations. Some made attempts to define a coherent field from diverse antecedents. As some economists set about reshaping the neoclassical model, some environmental sociologists called for a social science undergirded by a *new ecological paradigm* (Catton and Dunlap, 1978: 34; Dunlap, 1980). But there was one critical difference. Unlike the virtual disciplinary dogfight between neoclassical and ecological economists, environmental sociologists have chosen, thus far, to work within the established theories of the discipline, and now occupy a rather comfortable subdisciplinary niche.

Both sociology and environmental sociology have operated from two contrasting paradigms that emphasize (1) how consumption, the economy, technology, development, population, and biophysicial resources shape us and our environmental situation, (like Marx and his heirs); and (2) how culture, ideology, moral values, and social experience influence how we think and act toward the environment (like Weber and his heirs). This realist-idealist paradigm difference underlies many of the debates among sociologists and other social scientists (Bell, 1998: 3). But many sociologists find virtues and partial truths in each view and strive to understand the interplay of material and ideal factors and a more inclusive theoretical integration of these historic paradigmatic differences. The latter is the position I take and try to develop in this book.[4]

Conventionally and more concretely understood (but with vast oversimplification) types of sociological theories compete that emphasize (1) the way systems and institutions have to work for social survival (*functionalist theories*), (2) the human struggles to control others, and the conflicts over scarce material resources (*conflict theories*), and (3) human agency, that is, the ability of people to creatively construct and negotiate "reality" through interaction, thereby to creating both society and culture (*interactionist, interpretive, or social constructionist theories*). The first two positions are taken by the intellectual descendants of Émile Durkheim and of Karl Marx, respectively. The third represents a confluence of ideas from various early thinkers, including Max Weber, George Herbert Mead (1934), and Alfred Schutz (1967). How has environmental sociology been "done" from each of these theoretical perspectives?

Functionalist theories assume that like other organisms, humans live in sociocultural *systems* that, like all systems, have parts or *subsystems* that work or *function* to keep the entire system going. Functionalists typically recognize social institutions as the relevant structures that carry out functional processes. To get a sense of this, try a mental experiment by asking a rhetorical question: What kinds of processes (functions) are critical to the viability and survival of *any* social system? Some are obvious: (1) producing enough individual people through reproduction or organizational recruitment, (2) socializing individuals well enough to be able to live in particular systems, (3) the production of sufficient goods and services to maintain individuals and organizations, (4) sufficient order and authority to resolve conflicts and allocate goods, and (5) the maintenance of enough shared culture to facilitate communication and consensus. Such functions were conventionally understood to be conducted by social institutions (e.g., the economy, politics, families, schools, churches) (see Parsons, 1951; Mack and Bradford, 1979). Note that I put adjectives in each, *enough* or *sufficient*. How much of each? Well, that is an open question about which observers disagree. But each is unquestionably necessary in *some* degree for a surviving social system. Further, note that consistent with the way functionalism was understood in the 1950s, a sustainable relation between humans and their biophysical environments was not a part of this list of functional processes. Nature is only implied as "out there" as a resource for economic functioning.

Dunlap and Catton expanded the framework to include the *three functions of the environment* for human society (as well as other species). The biophysical environment functions as a *supply depot* for human material sustenance. Environmental *sinks* function as *waste repositories* for wastes and pollution. In addition, the environment functions as *living space* for all activities, and overuse of this function produces crowding, congestion, and the destruction of habitats for other species. Further, they argue that overusing the environment for one function may impair the other functions (as when a waste site makes a neighborhood undesirable for living, or pollutes groundwater resources). Human impacts may become so large that they threaten to be *dysfunctional*, threatening human social viability on a global scale. This impairment may be of such magnitude to impair the environment's ability to fulfill all three functions, for humans or other species (Dunlap and Catton, 1983; Catton and Dunlap, 1986; Dunlap, 1992).

Conflict theories argue that the most important societal dynamics are the variety of processes by which the parts or subsystems of societies come into conflict about control of important but limited material resources *and* symbolic rewards of the society. In most societies, material resources still means biophysical resources like land, but more importantly, they become *money* as an abstract expression media of economic value. Through various forms of conflict and power struggles, society's parts or subsystems attempt to protect or enhance their share of resources and values. Relevant subsystems, in this view, are not only socioeconomic classes, but may include gender and age categories, ethnic groups, communities, and organizations, interest groups, and social movements of all sorts around which people can mobilize to defend and extend their "interests." These processes periodically exhibit visible tensions and conflict, resulting in inequalities of power and resources that biologists would call a *dominance hierarchy* and social scientists would call a *social stratification system*. Even so, the ability of one part to dominate the system is limited by the others with which they must contend, and society itself is likely to be controlled by a coalition of the most powerful subunits. Both social stability and change derive from such ongoing competition and conflict (Collins, 1975; Olsen, 1968: 151).

Noting large corporate organizations as important features of modern free market societies, Allan Schnaiberg and his colleagues developed a conflict theory of human-environment interaction.They argue that many social analyses of environmental problems have paid too much attention to consumption and too little to the dynamic of production. Competition makes higher profitability a key to corporate survival, and firms must continually grow to produce profits and attract investments. This imperative for continual growth becomes a *treadmill of production* in which each new level of growth requires future growth, and growth in production requires the stimulation of growth in consumption. The contradiction is that *economic expansion is socially desirable, but ecological disruption is its necessary consequence*. Environmental disruption limits further economic expansion. New technology may introduce efficiencies that reduce the environmental impacts per unit produced, but continued increase in total consumption offsets this effect. The deeper threat of the treadmill may not lie in technologies that pollute, but in the competitive logic of the maximization of market share values without limit (Schnaiberg and Gould, 1994: 53). Governments are in the ambivalent situation of being expected to encourage economic growth, pay the costs of environmental disruption, and regulate environmental abuse. The first of these outcomes is of overwhelming *political* importance.

Schnaiberg and his colleagues propose a *societal-environmental dialectic* as the most likely pattern of change:

1. *The economic synthesis*—the system of addressing the contradiction between economic expansion and environmental disruption in favor of maximizing growth without addressing ecological problems.
2. *The managed scarcity synthesis*—in which there is an attempt to control only the most pernicious environmental problems that threaten health or further production by regulation; governments appear to be doing more than they really are (the situation of U.S. environmental regulation policies since the 1970s).
3. *The ecological synthesis*—major efforts to reduce environmental degradation through specific controls over treadmill production and consumption institutions directed specifically to that end. Curtailment would produce an economy so that production and consumption would be sustainable from the use of renewable resources. This is a hypothetical case with no known examples; it would only emerge when the disruption of the environment is so severe that the political forces would emerge to support it (cited in Buttel, 1986: 346–347, see also Schnaiberg, 1980; Schnaiberg and Gould, 1994).

Conflict-based processes that result in such agreements or syntheses may result in different outcomes: (1) the most powerful entities perpetuate the status quo and enhance their domination, (2) a prolonged stalemate occurs between dominant and contending parts of the system, or (3) significant change takes place that redistributes power, wealth, and privilege. *In most historic moments, the first outcome is most likely.*

Interactionist and social constructionst theories, more that the other traditions, come from diverse sources that have some similarities in their understanding of social action and change. Beginning with Weber's emphasis on the importance of ideas in society and history, these theories focus on human action and the creation of meaning in various ways. To understand them, you can begin with an empirical observation. As humans interact, they constantly create, defend, rearrange, and negotiate social order and cultural meanings among themselves. The implication is that social and cultural reality is, in fact, a *social construction.* What we take as "things" like organizations, society, culture, or social institutions are really a shorthand way of describing the historical outcomes of interaction episodes between real human actors. Social construction is a form of social action in which competing groups seek to define issues in terms that support their material interest and thereby reshape underlying material and social processes Interactionist and social constructionist approaches also provide important insights about the environment and environmental problems. Social and cultural reality are not the only social constructions of reality; so are "nature" and "environment" (Berger and Luckmann, 1976; Ciccantell, 1999: 294–295; Hannigan, 1995; Rothman, 1991: 157).

This is a subtle and important point that I mentioned earlier when discussing culture and paradigms. To clarify again, there *is,* of course, a natural world that exists quite apart from human perceptions of it. Humans live within this world and its constraints, but they do so only in terms of how they understand and define it. Furthermore, different people *cognize* the natural world and environment in very different ways. This needs no extensive illustration here. Earlier in this chapter I noted that the idea of nature itself was not used before its invention by English artists and intellectuals in the eighteenth century, and that we have further anthropomorphized and "gendered" nature by coming to speak about Mother Nature. Furthermore, I described in some depth the ways that people have cognized and constructed the environment variously at different stages of human social evolution. It should be obvious that the *culture of nature*—that is, the ways we think, teach, talk about, and construct the natural world—is as important a terrain for action as nature itself (Wilson, 1992: 87). This culture of nature may differ between human systems and within complex ones by socioeconomic status, education, ethnicity, and so on. The important point here is that interactionist, and particularly social constructionist, approaches also provide important insights about the environment and environmental problems (Ciccantell, 1999: 294–295; Hannigan, 1995). After discussing global warming in Chapter Four, I will discuss that how we understand *risks* from such geophysical processes is very much a social construction. Please do not misunderstand: I don't mean that this makes it any more (or less) serious for us to deal with.

CONCLUSION: HUMAN SYSTEMS AND ENVIRONMENTAL SYSTEMS

The first chapter introduced some ideas about environments and ecosystems. This chapter provided you with overviews about sociocultural systems, their historical development, human-environment relationships, and social constructions of environments. It also discussed social science explanations of sociocultural evolution, and the development of environmental and ecological economics and sociology. I hope you can see connections between the two introductory chapters that set the agenda for topics yet to come. I conclude by summarizing how human systems impact environments and ecosystems, suggesting that every environmental problem is also a social issue.

The Human Driving Forces of Environmental and Ecological Change

Ecological theory now emphasizes that instead of a static equilibrium or "balance of nature" some change and flux is the normal state of affairs (Lewis, 1994: A56; Miller, 1998). Environmental and ecological changes today differ those of the past in at least two ways. The pace of global environmental change has dramatically accelerated, and the most significant environmental changes are now *anthropogenic,* caused by human impacts (Southwick, 1996: 345–348; Stern et al., 1992: 27). Indeed, everywhere you look there are signs of human modifications of the natural world: buildings, roads, farms, human-modified lakes, rivers, and oceans. Even the gaseous envelope surrounding the earth is becoming littered with human refuse—bits and pieces of satellite "junk" now in orbit. As nature recedes into the interstices of the planet, true wilderness is becoming so rare that there is concern with preserving the last natural refuges unmodified by human civilizations.

Four types of human variables are *proximate causes* or *driving forces* of environmental and ecosystem change: (1) population change; (2) institutions, particularly political economies that stimulate economic growth; (3) culture, attitudes and beliefs, including social constructions and paradigms about the environment; and (4) technological change (Stern et al., 1992: 75). Chapter Seven discusses another way of understanding them (the I = PAT model).

1. *Population change has obvious connections to environmental/ecosystem change.* Human population grew exponentially since the 1600s. It is now about six billion and will probably plateau somewhere between ten and fourteen billion in the next century. More people eat more, consume more, and pollute more. The straightforward population question is: If environmental degradation is obvious with a population of six billion people, how can larger numbers be accommodated without devastating the resource base of the planet? Simply providing food for that many people—even if food resources were much more evenly distributed than today—is a daunting prospect, and experts disagree about whether it can be done without devastating the agricultural and water resources of the planet.
2. *Political economies that stimulate growth cause environmental/ecosystem change.* Industrial market economies and governments have a central dynamic that promotes continuous economic expansion. This increasingly characterizes *all* economies and states because of their growing integration in a world market economy. Competition makes higher profitability a key to corporate survival, and firms must continually grow to produce profits and attract investors. Firms compete not only in sales and market shares, but in financial markets, and the ones that fail to attract capital investment will die. Corporations in market economies will inherently try to squeeze as much output as possible from a given level of inputs in order to continue to elicit investments. If a corporation is not to have a large "unsold inventory," growth in productivity requires parallel growth in consumption on the "other side" of the market (Schnaiberg and Gould, 1994: 52; Heilbroner, 1985: 252). Thus, economic growth is the "success indicator" for economies and corporations as they now exist. Nongrowth or low profits leads to disinvestment, contraction, which are taken as problems. The central economic function of the state is to provide a *political environment* for continuous economic expansion. Both political and economic leaders assume that unlimited economic expansion is desirable, possible, and necessary. Democratic politicians are preoccupied with providing for economic growth and are routinely turned

out of office if they can be blamed for not producing it (e.g., of "mismanaging" the economy). Authoritarian regimes may be tolerated if they can preside over a booming economy (such as in Singapore or the People's Republic of China), but the failure to do so figured large in the collapse of authoritarian regimes (such as the Soviet system in the late 1980s). *Development* in LDCs also requires self-sustaining economic growth, as it is understood.

However socially desirable, it also true that economic growth depletes the stock of nonrenewable resources, often overuses the stock of renewable resources (such as fresh water) by using them at a rate that far exceeds their replacement rate, and may overwhelm the capacity of the earth to absorb pollutants. To get some sense of this you need to understand the prodigious scale of economic growth since. In just three years—from 1995 to 1998—economic output exceeded that during the 10,000 years from the beginning of agriculture to 1900. Furthermore, the 1997 growth of the world economy exceeded that in the entire seventeenth century (Brown and Flavin, 1999: 10).

It is important to understand that growth in population and economic growth are related, but they are not the same thing. Environmental costs of population growth are significant, but they are independently compounded by affluent lifestyles that accompany economic expansion. In poorer LDCs, people spend more of their money on things like food and clothing, which do less per capita damage to the environment than does the per capita consumption of affluent people in MDCs on things like electricity, fuel, and transportation. In the 1990s, the typical person of an industrial country consumed three times as much fresh water, ten times as much energy, and nineteen times as much aluminum as did the average person in a developing country. Affluent industrial countries burned the fuels that released about two-thirds of the greenhouse gases" (Durning, 1992: 51–52).

3. *Cultural values and belief systems* are different from the expansionary dynamic of contemporary societies, but they are intimately connected with it. Values, beliefs, and ideologies legitimate and normalize the institutional arrangements of societies. Consumerism, the modern incarnation of materialist values, has elevated the consumption of goods to a defining feature of the "good life." Consumerist ideology has an obvious connection with supporting the demand side of the market economy, and in that sense, making the system work. The ability to consume more is taken as a measure of growth, progress, and social status. The notion that consuming more is always preferable to consuming less is so deeply embedded in American culture that the question "How much is enough?" appears unnatural. Materialism as a value (and temptation) is, of course, not new or unique to modern societies. Indeed, *all* of the world religions (e.g., Christianity, Islam, Buddhism, and Hinduism), whatever their disagreement about religious worldviews, agreed in denouncing the elevation of material consumption into a paramount principle of life.[5] But the culture of industrial societies became more centrally preoccupied with material and consumerist values than previous societies. How that came to pass was not entirely unintentional. After World War II, retailing and advertising analyst Victor Lebow declared: "Our enormously productive economy . . . demands that we make consumption our way of life, that we convert the buying and use of good into rituals, that we seek our spiritual satisfaction and ego satisfaction

in consumption. . . . We need things consumed, burned up, worn out, replaced, and discarded at an ever increasing rate" (cited in Durning, 1992: 21–22).

The "consumerist" cultural ethos is an important part of the *dominant social paradigm* in America and the MDCs, and it is spreading with the world market economy. It is historically new, yet contemporary and Western societies are not unique in degrading their environments. Rather, I think, the culture of consumerism combined with industrial technologies gave us more deliberate sanction for our ability to do so. While the relation between such values and behavior is complex, it is undoubtedly true that they, along with our social constructions of nature, are indirectly but pervasively related to environmental/ecosystem change.

4. *Technology is the cultural dimension that enables humans to transform the environment more than other species.* In market economies, the major role of technology has been to facilitate increasing productivity of capital (hence profits), mainly by removing human labor (Reich, 1991). Energy technology underlies all economic activity and growth (I will return to this point in Chapter Four). Resource depletion played a historic role in the invention of new technologies (Boserup, 1981; Simon, 1998); in turn, however, more productive technologies themselves meant new ways to exploit natural resources, resulting in resource depletion and pollution, a treadmill of new problems. Certainly, technology has great potential to address environmental/ecological problems, but that has not been its primary role to date.

System Connections

These human "causes" of environmental change are themselves a complex system that not only produce changes in global ecosystems, but cause changes in each other through complex feedback mechanisms. They are distinct but interdependent, and I am unwilling to argue that any is a "more basic" cause, as some scholars do.[6] It seems to me that given their interdependent character, doing so raises a number of chicken-and-egg arguments (which came first?) that are not very productive. I do think it is important to distinguish between more proximate causes (such as a particular technology or social arrangements that produce hunger or civil war) and more distant levels of causation (such as population pressure or global climate change). Which is more important depends on the time horizons and purposes of analysis.

Within the physical environment, ecosystems and human social systems are interconnected and interdependent, and the scope of human activity is so vast and powerful that hardly any ecosystem in the world is free from human impacts. But each ecosystem has its own internal dynamics of equilibrium and change quite apart from human systems. Similarly, each human system has its own sources of change apart from being embedded in ecosystems. The important thing is to understand the interfaces within which the dynamics of human societies become the proximate causes of ecosystem change, and the parallel interfaces between ecosystem change and the things that humans depend on and value. These relationships are summarized in Figure 2.4.

Intellectual Paradigms about Human-Environment Relations

Scholars from different disciplinary backgrounds have different assumptions about the "way the world works" and thus pose questions a bit differently. Here are three scholarly paradigms about human-environment issues.

- Natural scientists tend to view human-environment problems broadly in terms of the long range implications of continuing *growth in scale in a finite world.*
- Neoclassical economists "frame" the causes of human-environment problems in terms of more proximate causes of *market failure and resource allocation problems.*
- Other scholars, including some economists, sociologists, and political scientists, frame human-environmental problems in terms of other proximate causes, seen as *social inequality and maldistribution.* These include, for instance, national and global patterns involving the vastly unequal distribution of wealth, political power, information, technology and so forth.

Figure 2.4 Interaction between Ecosystems and Human Social Systems
Source: Adapted from Stern et al., 1992: 34.

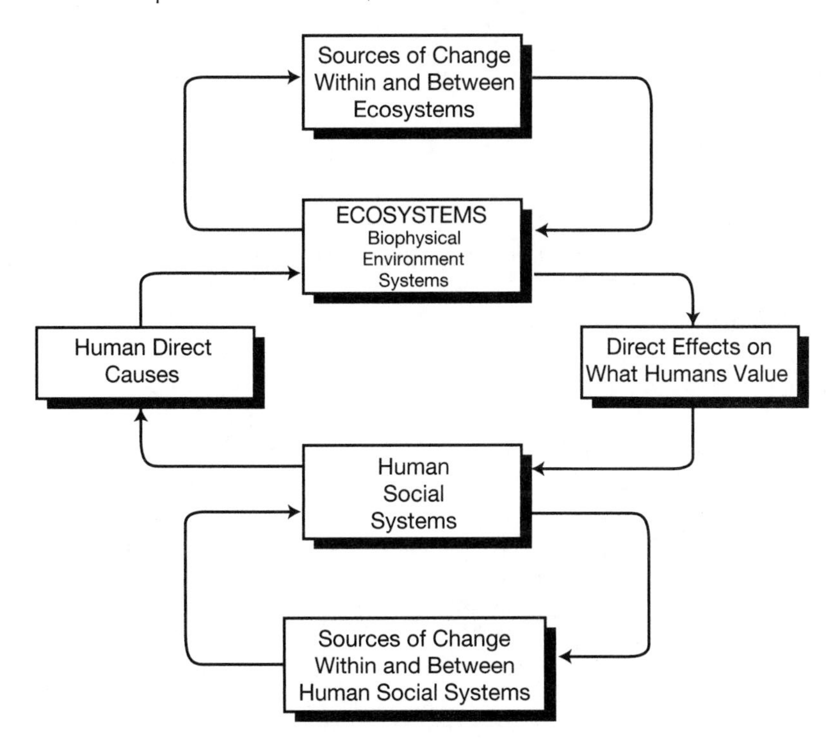

Illustratively, the problem of hunger in the world can be broadly framed as (1) too many people making demands on limited natural and agricultural resources, (2) the overregulation and failure of free markets that make producing food unprofitable compared to other investments, or (3) an adequate total food supply, but hungry people so poor that they cannot afford to buy food and so powerless that governments are unresponsive to their needs (I will return to this issue in Chapter Five). Such paradigmatic differences are keys to understanding many debates about the seriousness and causes of human-environment problems. Reconciling them as legitimate but different points of view is difficult, but I don't believe it is impossible. Subsequent chapters will return to these paradigmatic differences in various places. Stay tuned.

I end this chapter by reiterating a point that should be clear to you by now. Natural environmental/ecological phenomena and problems are social issues as well. For example, social questions, and often controversies, arise about natural resources like fertile land, mineral deposits, pristine forests, or fresh water. Who owns them? Will they be used or left alone? If used, for what and how fast? Who benefits and pays the cost of their uses? Which people or organizations have a stake in these questions, and whose preferences will prevail? If there is an environmental/ecological problem, such as pollution, species extinction, or climate change, who, if anyone, bears the costs of doing something about them? How are such costs fairly distributed? Putting it more abstractly, "What are called 'natural' resources are in fact social as well as natural; they are the products of historically contingent sociocultural definitions just as much as they are products of biochemical processes" (Freudenburg and Frickel, 1995: 8). Now most people are aware that environmental problems and change cannot be understood, much less dealt with, in the absence of substantial contributions from the social sciences (Stern et al., 1992: 24).

PERSONAL CONNECTIONS

Consequences and Questions

This chapter discussed the importance of the DSPs of people in society and the driving forces of ecosystem change. Those are very abstract notions, but if they are real, you should be able to find them illustrated in some ways in your own life or the lives of those around you. Try it. Here are some leading questions:

1. Look again at the description of the dominant social paradigm of industrial societies on p. 48. Can you see any connection between the DSP and your own perceptions about the "way the world works," or your values about what is good and bad? Is it reflected in your behavior, or the behavior of others around you? Is it reflected in your hopes about the future? How will you and others judge whether or not you are successful in life?

2. What kinds of personal inducements are there to keep you or your friends consuming a lot? Pressures from the expectations of others? Time? The media and advertising? What kinds of forces are there that inhibit you and others from adopting more environmentally frugal behavior? What's the connection among recreation, consumption, and waste in your life? How do you "have a good time"?
3. Are there ways of understanding personal growth, change, success, and development that have little to do with the consumption of "things"? If so, what?
4. How do attitudes about growth, consumption, and the environment vary among the people you know? How do you think they vary by age, education, or socioeconomic level among people?
5. Think about the times you have adopted a new technology of some kind. Did it solve problems? Did it create other kinds of problems? Did it make your life more or less complex? What are some technological developments that increase your impact on the environment? What are some that decrease your impact on the environment?
6. Do you think the costs and availability of different technologies will reinforce or increase differences between different kinds of people in society?

Real Goods

The telephone is one of the greatest innovations in human communications since Gutenberg's printing press. In spite of all the hype about computers, they are cheap and can save you a variety of trips to see people or do things. Telephones will never lose ground to newer and pizazzier communications technologies—though they may be combined with them. Unlike fax machines, personal computers and computer networks, televisions, VCRs and camcorders, CD-ROMs, and all the other flotsam and jetsam of the information age, telephones are a simple extension of the most fundamental means of human communication: speech (Durning, 1994: 97).

What You Can Do

The shopping malls are waiting for you to come. They know that you have friends and relatives who have birthdays, anniversaries, babies, and so forth. Of course, buying gifts to celebrate events in the lives of your friends is not evil. It is a way of symbolizing the relationships that mean a lot to you. But retailers in industrial societies have developed this need into an elaborate marketing and profit-making exercise. Ironically, with the telephone as a real good, you can now do a lot of shopping by catalogue and phone. By November the malls are stocked full of things for you to begin the holiday buying frenzy. The merchants depend on it, as do judgments about the retail "health" of the economy each year. To celebrate such occasions, consider a partial exit from this system.

Here is a list of more "environmentally frugal" gifts than overpackaged, plasticized "things" that often won't last long:

1. An experience (hot air balloon ride, picnic, day at the beach, use your imagination)
2. Season tickets to a sporting event
3. A house plant
4. Membership to a museum or organization
5. Photographs
6. A solar watch or calculator
7. For people with children, an evening of free babysitting
8. Donations to the needy in your friend's name
9. A tree to plant
10. A recording or compact disc
11. Food or confections that you cooked
12. A gift certificate for auto repairs, oil changes, etc. (repairs that would have to be done anyway)
13. A magazine subscription

ENDNOTES

1. There were, of course, kings and political empires throughout much of human history. But these were different from modern nation states—with their greatly expanded social functions (e.g., economy subsidy and regulation, public education and social welfare). Perhaps as important, modern nation states emphasize *sovereignty* as involving the right, not just the coercive power, to rule. Similarly, organizations in the bureaucratic sense are relatively new social inventions that arose at about the same time as nation states. The major difference between modern organizations and those of antiquity is that in modern bureaucratic organizations, accountability and authority are vested in organizational statuses and structures rather than in persons. The importance is that modern organizations have greatly enhanced stability and continuity. The army of Attila the Hun and the pyramid-building crews of the Egyptian pharaohs were both personal empires that did not long survive their founders (the classic formulation of the features of bureaucratic organizations can be found in Weber, 1922/1968.)
2. While you should not overdraw the similarities between *Homo sapiens* and other animal species, it would be an equal error dismiss human rootedness in the biotic world. The relative weights given to biological/genetic programming versus cultural learning as causes of the behavior of humans and other species is a perennial debate that surfaces about every decade in new guises. But this is surely a matter of degrees of difference rather than sharp differences of kind. It is, I think, a matter of "both-and" rather than "either-or." To say that, of course, only concedes an abstract principle and gives no help in knowing specifically how much of which to emphasize in what circumstances. New versions of this heredity-environment debate have been shaped in the subdiscipline of biology that has come to be called

sociobiology. For more about this see Barash (1979); Maryanski (1998),Van den Berghe (1977–78), and Wilson (1975).

3. Linear growth is additive (1,2,3,4,5,6,7 . . .), while exponential growth squares each new number (2,4,8,16,32,64 . . .). If Malthus was correct about this, then you can see his point about the inexorable tendency for population to outstrip supplies.
4. My friend and colleague Steve Sanderson vehemently disagrees with this strategy and has written creatively about human social evolution from a particular (materialist) position (1995). He sees the refusal to take a firm position in this debate as producing only eclectic "confusions," which violate "laws" of theoretical parsimony. But to some extent taking sides in this historic debate strikes me as "flag waving" that violates what we all know after years of research about the complexities of the real world. Parsimony is nice, but not if it impedes accounting for (real) complexities and the partial truths of different views. Indeed, much creative scholarship goes on at the boundaries or fault lines between fields and theories, as it does in Sanderson's work (I think in spite of what he says). I think we should strive for more inclusive and integrative theory that is admittedly a journey not an end state.
5. Here are some citations from sacred texts of world religions that illustrate the point:

   - Christian "It is easier for a camel to go through the eye of a needle than for a rich man to enter the kingdom of God." (Matthew 19:23–24)
   - Jewish "Give me neither poverty nor riches." (Proverbs 30:8)
   - Islamic "Poverty is my pride." (Muhammad)
   - Hindu "That person who lives completely free from desires, with longing . . . attains peace." (Bhagavad-Gita, 11.71)
   - Buddhist "Whoever in the world overcomes his selfish cravings, his sorrows fall away from him, like drops of water from a lotus flower." (Dhammapada, 336)
   - Confucian "Excess and deficiency are equally at fault." (Confucius, XI.15)

   Source: Durning (1992)

6. In what has come to be called the Ehrlich-Commoner debates among environmental scientists, biologist Paul Ehrlich has argued that population growth is the most important driving force for environmental change and problems (1974, 1992), while zoologist Barry Commoner has argued that advanced industrial technology is a more important and powerful cause (1971). Among many others, sociologist Alan Schnaiberg emphasizes the importance of the institutional arrangements, and in particular the political economy (1980). Other analysts, including social theorist Talcott Parsons, view culture, values, and paradigms as the most basic forces that sanction and limit the other variables.

PART II

READING THE EARTH'S VITAL SIGNS

CHAPTER THREE

The Resources of the Earth: Sources and Sinks

In the 1960s, when I was a young man, I took a canoe trip with a friend down the Current River in southeast Missouri. The water was clear and cold, and while the surrounding land was hilly, rocky, and not much good for agriculture except for grazing a few cattle, the river was lined with magnificent forests in the Ozark National Scenic Riverways and the Mark Twain National Forest. Tourism and outfitting canoeists was one of the major industries in the surrounding counties. My father told me that when he was a young man living near the area in the 1920s, the trees had been clear cut by lumber companies and soil erosion had turned the clear spring-fed river into a muddy mess. I marveled at the contrast between the merciless exploitation of resources that had taken place around the turn of the century and the restoration that I witnessed by the 1960s. Although the landscape was certainly not like it was before human settlement, the net effect of human activity over time had in some ways compensated for the damage done at an earlier time, at least in that particular area.

In Chapter One, I outlined a contemporary litany of environmental problems to frame the concerns of this book and the current debate and dialogue about them. In this chapter I return to some of these problems in greater depth. My agenda here is an assessment (reading) of the condition and change in the resources of the planet, mostly viewing the earth as a huge resource system that supplies the necessary materials which support human life. As noted in Chapter One, the earth is a huge recycling system of matter and energy, and from it humans withdraw resources, consume material, and contribute wastes that are absorbed, diffused, and recycled. As you can see from the chapter opening illustration, the important questions concern the net effects of human-environment interaction over time, when the current state of scientific evidence is adequate to know them. Often it is not, and this chapter will note ambiguity. So that you don't view nature as infinite and free, I will also note some important *ecosystem services* that such resources provide for human economies and ecosystems, and sometimes their approximate monetary value—if they were priced. The chapter will discuss some ideas for addressing resource/environmental problems and a few things about the demographic, social, economic, and political contexts that surround such problems, but most of that discussion will come later in the book.

In narrow anthropocentric terms, you can conceptualize the earth as a series of *sources* (from which resources are drawn) and *sinks* (into which human wastes and effluents go). I will discuss the current state and human use of *physical resources*: soil, water, biotic resources (forests and species diversity), and nonfuel mineral resources. Later chapters discuss climate and energy resource issues in greater depth. I will also discuss *pollution sinks* (of solid wastes and chemical pollutants). In simpler words, sources function as "supply depots" and sinks function as "waste repositories" (Catton and Dunlap, 1986). This is an abstract way of talking about the functions of the environment for people. A sink can refer to a trash dump, a river, or the atmosphere, which absorbs wastes of different kinds.

LAND AND SOIL

Soil is formed from the minerals derived from the breakup and weathering of rocks combined with deposits of organic material derived from wastes, and the dead and decaying remains of plants and animals. Since soil contains microbes and other detritovores, it is not only a variable mix of inorganic and organic compounds, but also a "living layer" of the biosphere. Topsoil layers are particularly rich in the nutrients necessary for primary producers to carry on photosynthesis. We are utterly dependent upon the land for food: 98 percent of human food is produced on the land. Worldwide, food and fiber crops are cultivated on 12 percent of the land surface, 24 percent is pasture used for grazing livestock that produces meat and milk, while forests cover another 31 percent, most of which is being exploited for fuel, lumber, paper, and other forest products. The remaining land, less than one-third, is desert, mountains, tundra, and other land unsuitable for agriculture (Buringh, 1989).

Land can be degraded and eroded so that it is less productive or even useless for human cultivation. In fact, land is always eroding naturally, topsoil is being dissolved, or carried away by water or wind, and the rate of this natural erosion varies with local geology, climate, and topography. The *critical question* is the rate of erosion and degradation in relation to the rate of soil formation, and in particular the impact of human activity on the relationship between these two processes. In their natural state, vegetation and decaying vegetation provide ground cover that replace lost minerals, "bind" soil particles, and slow soil erosion and nutrient degradation. After land is cleared for crops, it is relatively stripped of such ground cover and soil binders, and the recycling of minerals is significantly reduced. One consequence is that the arrival of humans practicing agriculture increases the volume of soil and silt being carried into the ocean by at least two and a half times the original rate (studies cited by Craig et al., 1988: 353). Human agriculture has without doubt led to the accelerating erosion of topsoil and soil nutrient loss, and subsequently to declining crop and livestock yields. This was often, as noted in Chapter Two, the fate of agricultural societies.

Soil and Food

But if human intervention produced a net degradation of soil, how can we explain the enormous increase in food production in recent times—which has until very recently grown faster than human population? From the beginning of agriculture until about 1950, nearly all the growth of food output came from expanding cultivated land area. Since 1950, at least four-fifths of the increase in food output has come from increasing *productivity* (Brown et al., 1992: 36). Part of this increase was from the growing irrigation of marginal and semi-arid land, from 95 to 235 million hectares between 1950 and 1989. This irrigation explosion took place in the MDCs and well as the LDCs. Equally important was the increasing capital and technical intensity of agriculture, which increased productivity through massive use of selectively bred livestock and seed hybrids, cheap inorganic fertilizers, herbicides, and pesticides. In addition, this intensive (or "industrial") agriculture required large amounts of fuel energy to produce agrochemical, deliver water, and power farm machinery (about 1,000 liters of oil equivalent energy per hectare in the industrial nations) (Pimentel, 1992b: 220).

While intensive agriculture dramatically increased productivity, it all but destroyed the traditional methods of preserving soil productivity that farmers everywhere had learned to practice, such as terracing, contour plowing, crop rotating, using fallow years, using organic fertilizer, and—in the tropics—shifting agriculture and herd migration. Intensive agriculture encouraged continuous cropping of monocultures without rotation or fallow periods, cropping on hilly and marginal land, and overgrazing in confined pasturelands. Since 1950, manufactured inorganic chemicals, capital, and energy inputs made it possible to expand food supply. For obvious economic reasons, this has been truer of MDCs than of many LDCs, which have been less able to afford such inputs. As a consequence, since 1950 the *net area* of cultivated land area has not grown much. Total agricultural productivity increased at a rate faster than population did, but the number of cultivated acres per capita declined steadily since 1950 (Brown, 1999b).

But copious food production had another less visible source: farming that overdraws and degrades natural resources (both water and soil) to maximize production.

Present and Future Status of the World's Soils

Experts have been concerned about the global soil situation for some time (e.g., Eckholm, 1976). Still, without very good global data, studies during the 1980s suggested that we could adequately feed the world's population for some time to come. That was the position of a study by the United Nations Food and Agriculture Organization (FAO) in 1982. Its authors argued that while some nations had little arable land left, other nations had uncultivated land that could be brought into production (particularly in Latin America), and that in LDCs traditional agriculture could become much more productive by using more intensive agricultural technologies (Hrabovszky, 1985). The study noted that hunger was more directly caused by poverty and price fluctuations in the world food market than by soil degradation. It was argued that an even larger population could be adequately fed by cultivating the remaining arable land, upgrading production, and by increasing world trade in food to those regions with food deficits (Crosson and Rosenberg, 1989).

This optimistic view weakened more recently. Experts now estimate that about one-third of the world's soil that ever existed has been lost (Southwick, 1996: 347). As you might guess, getting good data about the extent of soil erosion on a global basis is a formidable task. The best estimate of soil degradation to date is the 1990 Global Assessment of Soil Degradation (GLASOD), involving a team of 220 analysts from around the world who measured soil damage from overgrazing, damage from pollution, oversalinization, as well as water and wind erosion. GLASOD found that of the earth's present vegetated land area, 17 percent has undergone human-induced degradation since 1945 and 10 percent has been severely degraded. Land is degraded if the chemical, physiological, or biological characteristics of the land have been altered and fertility compromised. GLASOD found the causes of land degradation to be about equally divided between overgrazing of livestock, poor farming practices, and deforestation, with about 8 percent caused by urban and industrial pollution (Livernash and Rodenberg, 1998: 19; World Resources Institute, 1993). When the full ecological effects of soil erosion are considered—including reduced soil depth, reduced water availability, and reduction in organic matter and nutrients—agronomists report crop losses of between 15 percent and 30 percent in North America and between 38 percent and 52 percent in some tropical areas (Pimentel, 1992b: 331; Brown and Wolf, 1984: 27). These declines partly explain declining yields on some lands despite high use of costly fertilizers. "Because fertilizers are not a substitute for fertile soil, they can only be applied up to certain levels before crop yields begin to decline" (Pimentel, 1992b: 331).

There are not very many successful national programs to preserve topsoil on cropland. The United States is the only nation among the major food producers with programs to systematically reduce excessive soil erosion. The Conservation Reserve Program (CRP) encouraged the conversion of erodible land to grassland or woodland and penalized farmers who didn't manage soil responsibly by denying them the benefits of government farm programs (price supports, crop insurance, and low-interest loans). "In 1987 the CRP helped reduce U.S. soil losses by 460 million tons, the greatest year-to-rear reduction in U.S. history" (Brown, 1988: 49).The United States has made more progress than most nations in slowing soil degradation, but in spite of these efforts, in 1992 the U.S. Soil Conservation Service estimated that 25 percent of total croplands were eroding faster than they could be preserved (Stroh and Raloff, 1992: 215). In the early 1990s, the program was threatened by congressional budget cutters, but in 1996 it was reauthorized until 2002 (Miller, 1998: 560).

Addressing Soil Problems

A major reason to be concerned with soil problems is obvious. To feed a growing global population on increasingly degraded and expensive agricultural resources, we will need to increase the productive yield of agriculture while protecting the fertility of cropland soils. That's easy to state, but it is a *formidable* goal, particularly on a world basis. One possibility for increasing yield is in the new *biotechnology* (so-called genetic engineering) that some envision as a new agricultural revolution in progress. The prospects of genetic manipulation producing crops that are more drought resistant, salt tolerant, early maturing, or resistant to genetic pests are very attractive, indeed. By the 1990s, most American food consumers were probably unaware that most soy products and some vegetable oils (canola) were produced from bioengineered seeds. Benefits are undoubtedly real, but there are two very big questions about whether a mainly genetically engineered agricultural system would deliver on the world's soil and food problems.

First are biological difficulties: Even with genetic resistance to pests and weeds, there is no reason to think that they would not evolve rapidly into species of "superpests" that can thrive on the newly resistant crops. Such a *pesticide treadmill* accompanied the introduction of inorganic chemicals, and there is no reason to think that it would not happen again. *Second* are political and economic questions: Genetically engineered species are very expensive and are being developed and patented by private investment and multinational corporations. They may never be available at the "right price" to benefit the world's poor in LDCs, where increased yields are most needed (Gardner, 1998; Halweil, 1999). Aside from biotechnology, good fertile cropland should be protected both from conversion to nonfarm use and from erosion and nutrient loss. Erosion can be reduced by encouraging terracing, contour plowing, multiple cropping (planting ground cover crops in between rows of corn, for instance), and using low tillage methods (which leave crop residues on the land for soil binders and organic fertilizer). Nutrient recycling can be increased by using more organic fertilizer, which reduces the need for chemical fertilizers. One 1980s estimate found that if the use of inorganic fertilizers stopped suddenly, global agricultural production would decrease by more than 40 percent in a few years (Wolf, 1986). But the use of organic wastes for fertilizer is growing. Many Asian cities systematically recycle human wastes onto the immediately surrounding farmland. By the 1980s, Shanghai was virtually self-sufficient in producing vegetables grown in this manner (Brown, 1988: 50). Moreover, communities in the MDCs were returning organic material to soils at a growing rate in the 1990s (Gardner, 1998).

Contrary to global trends, some propose land reforms that encourage smaller privately owned farms because, compared with very large ones, these are more productive because they are more labor intensive. Small farmers with secure land ownership are also more likely, other things being equal, to care for the land more sustainably than are landlords or corporations operating remote large estates. But others suggest addressing problems of soil (and food problems) by the application of technology and the advantages of large-scale management. Many nations need food price policies that encourage the profitability of agriculture. But these ideas about land reform and price policy are *political dynamite* because they involve changing the rules about land ownership and often raising the price of food. You can understand why few governments have been willing to tackle them. Preserving the soil and increasing the world food supply will require the best efforts not only of agricultural scientists and geneticists, but also of energy planners, and economic and political policy makers of all sorts. This underscores the point the last chapter made in closing: All environmental problems are also social issues. Since it is so important, I will return to food issues in Chapter Six (about population) and other places.

Ecosystem Services: Pricing Soil Degradation

Lest you think that the costs of soil degradation are only abstract and long term, here are some recent efforts to price them. Direct costs of soil erosion, as measured by the costs of replacing lost water and nutrients on agricultural land, amount to about $250 billion per year globally. Additional costs, including damages to recreation, human health, private property, navigation, and so on, amount to about $150 billion globally, and $44 billion in the United States alone. Soil erosion is extremely costly in the short term, but the benefits of many prevention measures would greatly outweigh the costs. Even so, the political and economic barriers to the implementation of such measures are formidable (Daily et al., 1997: 127–128).

WATER RESOURCES

Even more clearly than soil, water is the lifeblood of the biosphere. Life is only possible because of the solar-driven circulation of water through the hydrological cycle from the ocean to the atmosphere, from the atmosphere to the land, and back to the ocean. Water is a renewable resource, but as you can see by reviewing Figure 1.2, most water circulates from the ocean to the atmosphere and back. A much smaller fraction falls as precipitation over land, and of that, much reevaporates or runs off back to the ocean so that an even smaller fraction is available for human agricultural, industrial, and household use. Usable water is *very* unevenly distributed over the earth's surface, so that getting enough water has often been a source of political conflict. Because water is renewed within the global cycle, we tend to treat it as a renewable, free, and unpriced common good. However, it is not the volume of water that determines how much is available for use over time, but its renewal or "recharge" rate for groundwater, lakes, and rivers (Falkenmark and Widstrand, 1992: 4–5). Worldwide, surface water and groundwater each supply about half of the needed fresh water, but the recharge rate for groundwater is *very slow*, about 1 percent per year (Miller, 1998: 491–493).

Unlike many resources, there are relatively fixed minimum requirements for water needs. To assure adequate health, people need a minimum of about 100 liters of water (about 26.5 gals.) per day for drinking, cooking, and washing. Many times this amount is necessary to support the economic base of a community, and persons in affluent and more water-rich societies use many times this personal daily minimum. Agriculture accounts for the most water use, about 70 percent worldwide, and is also the most inefficient use. It requires 4.2 million liters (about 1.1 million gallons) of water in a growing season to grow 1 hectare (2.4 acres) of corn. Accounting for runoff, evaporation, waste, and other factors, a hectare of corn requires on average 10 million liters of rainfall, spread evenly over the growing season. It is not uncommon for 70 percent to 80 percent of the water in irrigation systems to be lost by evaporation or to seep into the ground before reaching crops. Industry accounts for about 23 percent of global water use. It takes over 400,000 liters of water to produce an automobile, and industrial societies produce about 50 million cars every year. A nuclear reactor needs 1.9 cubic miles of water a year, and together all U.S. reactors use up the equivalent of one and a third Lake Erie's each year. Of all water uses, households consume only about 8 percent. Still, that adds up to a lot of water, and you probably use more water each day than you think you do, particularly in relation to the 26–7-gallon daily minimum need (Miller, 1998: 493; Pimentel, 1992b: 219; Sale, 1990; Falkenmark and Widstrand, 1992: 14).

Growing Water Use and Its Problems

Water usage has tripled since 1950, and in 1992 humans used about 4430 cubic kilometers per year—more than eight times the flow of the Mississippi River (Postel and Carpenter, 1997: 197). Planners have met this growing demand by so-called water development projects: dams, irrigation, and river diversion schemes. But limits to this ever-expanding consumption are swiftly coming to light. Around the world

> water tables are falling, lakes are shrinking, wetlands are disappearing . . . Perhaps the clearest signs of water scarcity is the increasing number of countries in which population has surpassed the level that can be sustained comfortably by the water available. As a rule of thumb, hydrologists designate water-scarce countries as those with annual supplies of less than 1000 cubic meters per person. Today, 26 countries, collectively home to 232 million people, fall into that category. (Postel, 1992a: 2332)

The worst problems are in the arid tropical and low-latitude nations, particularly in Africa and the Mideast, where demand is high and the ceiling of available water low. On the other hand, the high-latitude nations, particularly those whose climates are dominated by moist ocean systems (e.g., Scandinavia, the Pacific Northwest), are a long way from approaching the ceiling of available water. But water resource situations are critical in places as diverse as California, Mexico, Jordan, Libya, Pakistan, India, Australia, and most of the arid African Sahel (Falkenmark and Widstrand, 1992: 14).

Since irrigation has grown rapidly as a cornerstone of modern agriculture, groundwater supplies are particularly critical. Water is being pumped from wells much more rapidly than the recharge rates. In fact, irrigation overdrafts are so great that it is more like "mining" a finite source (such as coal) than using a renewable resource. Irrigation overdrafts lowered water tables by 20–30 meters in the Tamil Nadu region of India during the 1970s; around Beijing, China, water tables are declining at the rate of between 1 and 4.5 meters per year. Groundwater in the United States is being withdrawn at four times its replacement rate, and in some locations, like the Ogallala Aquifer of the Great Plains stretching from Nebraska to Texas, pumping is so great that the overdraft is 130 percent to 160 percent above replacement (Miller, 1998: 404–405; Pimentel, 1992b: 219). At present rates of consumption, much of the Ogallala Aquifer will be barren and production in the region that now supplies about 40 percent of the nation's beef and grain will drop sharply. As that happens, ripple effects will be felt in High Plains economies and communities as they begin to deal with slow depopulation and search for economic alternatives to their traditional agricultural bases (Postel, 1993). As you might imagine, since I am from Nebraska, this issue hits pretty close to home.

But similar water problems beset other areas, such as America's "salad bowl" in the San Joaquin Valley of California. Most of the water in the fertile but arid south comes via large aqueducts from reservoirs in water-rich Northern California. Illustrating once again our inability to price common goods realistically as well as the political power of California growers, agribusinesses have had to pay only about 5 percent of what it cost taxpayers to deliver water to them over the last forty years. To supply large amounts of such cheap water, major rivers, lakes, and large areas of prime wetlands have been drained, destroying thousands of kilometers of salmon habitat and contaminating rivers and groundwater with pesticides and fertilizers. A five-year drought (1986–1991) reduced irrigation water through the Central Valley project by 75 percent, causing thousands of hectares of cropland to be abandoned. Conflict has long been brewing between farmers (who use 82 percent of the state's water but produce only 2.5 percent of its economic wealth) and the huge cities of Southern California (Los Angeles, San Diego), which draw water not only from Northern California but also from the Colorado River (Miller, 1998: 501). Water resources are so chronically tight in California that some coastal cities (Monterey, Santa Barbara) have considered expensive desalinization projects to convert ocean water into fresh water for municipal use. The social consequences of these shortages are that litigation simmers between states, regions, and people between urban and rural water users, and innovative water management proposals proliferate.

Water and Political Conflict

The water conflicts in a wealthy nation like the United States will be mild compared to those in poorer, drier nations, which have neither the economic wealth nor technological resources to address water problems as do Kansans or Californians. Squabbles over water have been rapidly escalating, for instance, between Ethiopia and Sudan (which control the Nile headwaters and plan to retain most of it) and downstream Egypt (which is totally dependent upon the Nile flow).The decades-long conflict between India and Pakistan over Kashmir is partly about the water-rich region the foot of the Himalayas. India is damming up the river that ties the whole region together, and the Pakistanis fear that the Indians might turn off the floodgates in the middle of some hot summer and parch that part of Pakistan. On the eastern border of India, the poverty-stricken Bangladeshis have similar fears, because India is damming the Ganges only miles from the border. Bangladeshi officials viewed this act as a matter of life or death, and demanded that the matter be brought before the World Court in the Hague (*Der Spiegel*, 1992).

The political, military, religious, and ideological conflicts in the Middle East are well known. The proximate causes of these is certainly not water shortage, but a hydrological time bomb of sorts underlies and amplifies those intense conflicts—for instance, between the Israelis, Palestinians,

Iraqis, Syrians, Jordanians, Turks, and others. The Palestinian West Bank has a population of 12 million people with as much rainfall as Phoenix, Arizona. Water supplies are barely adequate to maintain a quality standard of living. They have overused an underlying aquifer to such a degree that much of the land is now below sea level. Meanwhile, Turkey has been building a vast complex of dams at the headwaters of the Euphrates River that will drastically reduce water flow to downstream Syria and Iraq. In 1999, a landmark study of water in Israel and the Palestinian lands found that ancient underground aquifers were being drained dry, scarce rains flow unused, and rivers diverted for water-intensive tropical agriculture. A groundbreaking comprehensive study involving collaboration among Israeli, Palestinian, and Jordanian experts urged their governments to jointly manage watersheds (Associated Press, 1999).

Addressing Water Problems

As with soil problems, solutions to water shortages involve improved efficiency and conservation. And the most obvious place to start is to replace today's wasteful irrigation systems with more efficient systems, such as drip irrigation. In spite of Israel's water problems (or perhaps because of them), Israeli agronomists pioneered irrigation technologies to reduce waste. With the tools and technologies now available, farmers could technically cut their water needs by 10–50 percent, industries by 40–90 percent and cities by one-third with no sacrifice of economic output or quality of life. Whether these efficiencies are politically feasible is another matter, because—as with soil—the substantial barriers are political and economic. Tax subsidies provide water to people and industries in the absence of "real cost pricing" that encourage wastefulness (Postel, 1992a: 2333). With water made so cheap by governments, there is no *real* incentive for growers to invest in more efficient systems. Moreover, as noted, the ability of nations to invest in technologies and conservation varies considerably.

> Rich societies with willing neighbors, such as in Southern California, can construct canals, pipelines, and pumps to import water. Rich societies with vast oil reserves, like Saudi Arabia, can use fossil energy to desalinate sea water. Rich societies with neither, like Israel, can come up with ingenious technologies to use every drop of water with maximum efficiency and can shift their economies toward the least water-intensive activities. Societies with none of those options must develop severe rationing and regulation schemes . . . [such] . . . poor societies experience famine and/or conflict over water. (Meadows et al., 1992: 56)

Even with vast differences in hydrological and technological resources, most human societies will be required to conserve or do with less water. Unlike petroleum and copper but like arable soil, water is, so far, nonsubstitutable for human economies and human well-being. You can get a sense of this by considering the most apparent technological alternative to conservation: *desalinization*, or the removal of salt from seawater. Desalinated water accounts for less than 1 percent of the world's water because it requires so much energy, costing four to eight times the average costs of municipal water, and ten to twenty times what most farmers pay for water. Most of the desalinated water is now produced around the energy-rich Persian Gulf, where nations are in essence trading oil for water. Similar strategies are unthinkable for the world as a whole, because desalinating the volume of water for world demand (at 1990s levels) has been estimated to cost 12 percent of the world gross economic product (Postel and Carpenter, 1997: 197).

Please note that as with other resource problems like soil, there is a *demographic trap* that limits the effectiveness of relying only on conservation and technological strategies. We can substantially conserve existing supplies and increase efficiency, but, given the nature of the hydrological cycle, we cannot expand the total supply much. In view of the relatively fixed human and agricultural needs for water use, how will we manage water sufficiency during the next decades, when human population undoubtedly will pass the six billion mark and the need for water increases as affluence spreads?

Freshwater Ecosystem Services

Rivers, lakes, aquifers, and wetlands provide a myriad of benefits to human economies. They provide water for drinking and hygiene, irrigation, manufacturing, and such goods as fish and waterfowl, as well as a host of "instream" nonextractive benefits including recreation, transportation, flood control, bird and wildlife habitats, and the dilution of pollutants. Such instream benefits are particularly difficult to quantify, since many are public goods that are not priced by the market economy. Thus the total global value of all services and benefits provided by freshwater systems is impossible to measure accurately but would certainly measure in the several trillions of dollars (Postel and Carpenter, 1997: 210).

BIODIVERSITY AND FORESTS

Every society has three kinds of wealth: (1) material, (2) social and cultural, and (3) biological. Of these, we understand the importance of the first two very well, because they are the substance of our everyday lives. Biological wealth, the diverse species of plants and animals, are probably underappreciated (Wilson, 1990: 49). Since that is true, I will discuss them in a bit more depth than I did soil and water issues.

Forest Resources

Two-thirds of the forests that historically existed around the world are gone now. Of the three major intact and unfragmented forest biomes that cover about 12 percent of the earth's surface, *boreal* forests that circle the northern latitudes (e.g., in Canada, Russia, and Scandinavia) are the largest (about 30 percent of the remaining forests), followed by *temperate zone* forests (in the United States and Europe), and *tropical* forests, which cover only about 6 percent of the earth's surface (about the size of the lower 48 states of the United States), and just four countries, Brazil, Indonesia, Zaire, and Peru, contain more than half of the world's total forests. Even with this small area, tropical forests receive more than 50 percent of the world's rainfall and provide habitat for the vast majority of the world's known species of other plants and animals, which gives them a unique and strategic importance on the earth as a global system (Myers, 1997: 215–216).

Both boreal and tropical forests are rapidly being destroyed by humans. In the north it is mainly caused by commercial logging, but in the tropics it is caused in various proportions by commercial loggers, farmers and ranchers (both peasants and corporate). Chances are that the next hamburger you eat or cup of coffee you drink—a cup is sitting by my computer right now—was produced on land that was formerly a tropical forest! Pollution and climate change also take their toll on forests, and the impacts of both will likely increase in the future.

In the temperate zone forests are now roughly stable in *area*, but in the United States much of the forests are regrown secondary forests after clear cutting the Northeast, Midwest, and Southeast before the turn of the twentieth century. They are much more fragmented and less biodiverse. Europe has virtually no primary forests left. In both the United States and Europe, a primary reason for reforestation, even more important than deliberate reforestation programs, has been urbanization, which left only a small fraction of the population on farms. As agriculture and livestock operations became concentrated on productive soils, the pressure on many previously forested lands decreased. This same pattern of reforestation occurred in Japan (Spears and Ayensu, 1985: 301). Even though temperate forests are now roughly stable in area and are often being "sustainably managed," some temperate zone forests exhibit declining growth rates, soil nutrients, and wood quality. Two decades ago, eminent biologist Norman Myers commented that "if we carry on business as usual, today's young people may eventually look out on a largely deforested world" (1989).

Tropical Deforestation

Will the history of temperate zone forests be repeated in tropical zones? Probably not, because they have different climates, soil types, and ecosystems. In general, tropical forests are richer in species, faster growing, more fragile, but more vulnerable. To a much greater degree than temperate forests, tropical jungle ecosystems depend on nutrient recycling within the forest itself rather than the (typically) nutrient poor tropical soil. Moreover, when cleared of tree cover, heavy tropical rains quickly leach and erode soil nutrients that exist, making agriculture unsustainable and forest regrowth long and difficult. Thus at their present rate of destruction, tropical forests are nonrenewable, not renewable resources (Meadows et al., 1992: 57–58).

Globally, half of the original tropical forests remain, but they are rapidly disappearing because they are being cut, logged, and degraded. The rate of their disappearance is well known—more than 150,000 square kilometers per year (Myers, 1997: 224). At present rates of exploitation, experts project that sometime between the years 2020 and 2090 virtually all of the world's tropical forests will be gone. As noted earlier, tropical forests are being cut for a variety of reasons by a mix of actors and agencies, including multinational lumber and paper companies seeking profits; LDC governments anxious to pay off international debts; landowners, ranchers, and farmers; and poor peasants scrambling for firewood (Meadows et al., 1992: 60–61). The last cause is especially significant in the drier regions of the tropics, such as Africa and India. Costa Rican history illustrates a pattern of tropical forest mismanagement.

> Since about 1940 much of the extensive rainforest was cleared to expand cattle ranching for beef export [much of it for North American hamburgers!]...Many of the new pastures proved unsustainable. Within a few years they were grazed down, eroded, and abandoned. On steep hillsides and in heavy rains there were landslides, which destroyed roads and villages. Silt from eroded lands filled up reservoirs behind hydropower dams or washed into the oceans, where it buried and killed coral reefs and destroyed fisheries. The land bears the scars from Costa Rica's few decades of intensive beef production. (Meadows et al., 1992: 58)

While less than one-third of Costa Rica's rainforests remain, that country has acted belatedly to stabilize the remaining areas in national parks and preserves. Although most LDCs have not acted as the Costa Ricans have, Brazil shows signs of reform. The South America Amazon (Brazil and Peru) has the greatest loss of historic forests, but by the 1990s that *rate* of deforestation was eclipsed by Southeast Asia (Indonesia, the Philippines, and Thailand). China, where forest cutting exceeds regrowth by 100 million cubic meters a year, is one of the single most deforested nations in the world.

Forest Ecosystem Services

Standing forests supply various ecosystem services. They stabilize landscapes and protects soils from erosion and help them retain moisture and store and cycle nutrients. They serve as buffers against pests and diseases.

By preserving watersheds, they regulate the quantity and quality of waterflows, and help prevent or moderate floods and store water against drought in downstream territories. They help keep rivers and seacoasts free from silt. They are critical to the energy balance of the earth, and modulate climate at local and regional levels by regulating rainfall. They shape they sunlight reflectivity of the earth (the "albedo" effect). At planetary levels, they help contain global warming because they store and sequester carbon as part of the earth's carbon cycle (see Figure 1.1). While all forests do these things, many of these functions are more prominent in tropical forests (Myers, 1997: 215–216).

Consider the costs of some particular forest ecosystem services, as their loss or modification does or might may effect humans. In Nepal between thirty and seventy-five tons of topsoil is washed away from each deforested hectare of land (about two and a half acres) each year, much of it unwittingly "exported" to India, where it winds up as turgid streams of mud in the Ganges and other rivers. The economic costs are substantial, particularly to Nepalese agriculture and to the Indians, where rivers have about 14 times the silt sediment as does the Mississippi, and rising riverbeds aggravate floods in densely populated regions. The on-site soil conservation benefits of India's tree cover has been estimated at between $5 billion and $12 billion per year, while the value of flood control has been assessed at $72 billion. Perhaps most mindboggling, what would the value of such forest services be in the year 2025, when at least 3 billion people in the LDCs are suffering from water shortages? If forests disappear, there would be a decline in nonwood products such as foodstuffs, wild game, fruit, Brazil nuts, and latex that are of great value to local populations and sometimes in trade. In the Mediterranean basin (Greece, Italy, Spain, France, Morocco, and others), trade in cork, resin, honey, mushrooms, wild fruit, and trees used in livestock production had an estimated value of between $1 billion and $5 billion in 1992. The old-growth forests of North America's Pacific Northwest protect habitats for 112 fish species—and the salmon industry alone is worth $1 billion per year (Myers, 1997).

Declining Biodiversity

We do, of course, appreciate the value of the species of things that provide our food, fiber, and wood products, but the value of the diversity of species itself in ecosystems is largely unappreciated by people (Wilson, 1990: 49). As noted, tropical forests and the world's wetlands (e.g., swamps, mangrove swamps, and saltwater marshes) are particularly rich repositories of species biodiversity, and they are now widely threatened. For example, the Eastern Madagascar forest houses 160,000 known species, at least 60 percent of which were endemic (found nowhere else). More than 90 percent of that forest has been eliminated, and scientists estimate that half the original species have gone with it (Miller, 1998: 343). Moreover, the problem is not just in tropical forests. In 1997, the World Conservation Union (IUCN) coordinated a study of 240,000 plant species around the world and found that one out of eight plant species surveyed is potentially at risk of extinction, and that more than 90 percent of these at-risk species were endemic (found only an a single country and nowhere else in the world). The United States, Australia, and South Africa had the greatest number of species at risk (Tuxill, 1999: 97). Many animal species are threatened also. You can see some illustrations of declines or threatened declines of animal diversity in Table 3.1, which summarizes findings from more than 25 studies.

To me it is particularly poignant that so many primates, our closest biological relatives, are threatened with extinction—sort of like deaths in the

Table 3.1 Declining Animal Diversity

• *Amphibians*	Worldwide decline observed in recent years. Wetland drainage and invading species have extinguished nearly half New Zealand's unique frog fauna. Biologists cite European demand for frogs' legs as a cause of the rapid nationwide decline of India's two most common bullfrogs. Of 497 species assessed by the IUCN, 30% were either threatened with, or in danger of immediate extinction.
• *Reptiles*	Of the world's 270 turtle species, 42% are rare or threatened with extinction. For the 1,277 species assessed by the IUCN, 26% were either threatened or in danger of immediate extinction.
• *Birds*	Three-fourths of the world's bird species are declining in population or threatened with extinction. Of 9,615 species assessed by the IUCN, 20% were either threatened or in danger of immediate extinction.
• *Fish*	One-third of North America's freshwater fish stocks are rare, threatened, or endangered; one-third of U.S. coastal fish have declined in population since 1975. Introduction of the Nile perch has helped drive half the 400 species of Lake Victoria, Africa's largest lake, to or near extinction. Of 2,158 species assessed by the IUCN, 39% were either threatened of in danger of immediate extinction.
• *Mammals*	Almost half of Australia's surviving mammals are threatened with extinction. France, western Germany, the Netherlands, and Portugal all report more than 40% of their mammals as threatened. Of 4,355 species assessed by the IUCN, 39% were either threatened or in danger of immediate extinction.
• *Carnivores*	Virtually all species of wild cats and most bears are declining seriously in numbers.
• *Primates*[a]	The IUCN considers primates the most imperiled order of mammals. 50% are threatened with extinction, and another 20% are "near-threatened." While many species are threatened, one (human beings) continue unprecedented expansion with a world population of more than 6 billion.

[a]An order of mammals that includes monkey, apes, lemurs, and humans.

Sources: Ryan, 1992: 13; Tuxill, 1997: 13; 1998: 128.

family tree. These particulars about species extinction add up to quite an impressive general picture. But compared to the data noted earlier about soil degradation and water problems, these estimates are even more uncertain. They are uncertain because nobody knows exactly how many species of living things there really are and therefore nobody can calculate with any precision what the *rate* of extinction actually is. Biologist Edward O. Wilson, working with other specialists, estimated that 1.4 million species have been formally identified and named, and they conservatively guess that at least four million species actually exist (Wilson, 1990: 49). Such uncertainty led some scholars to question whether a general decline in biodiversity is happening (Simon and Wildavsky, 1993). Yet most biologists believe that the direction of change is clear and that species are disappearing at an accelerating rate. A 1998 survey of biologists by the American Museum of Natural History found that a majority of biologists were convinced that a "mass extinction" is underway but that most Americans were only dimly aware of the problem (*Washington Post*, 1998). They think the present human-induced wave of extinction surpasses anything since the wave of species extinction that took place during the Cetaceous Age (65 million years ago) that ended the age of dinosaurs (Meadows et al., 1992: 64; Miller, 1998: 669–670; Tuxill, 1998: 128; Wilson, 1990: 58).

The Human Causes of Declining Biodiversity

Four kinds of causes drive declining biodiversity. *First*, the greatest threat to all kinds of wild species is the destruction and fragmentation of habitats as humans occupy and control more of the planet. According to conservation biologists, tropical deforestation is the greatest eliminator of species, followed by the destruction of coral reefs and wetlands. To reiterate: Tropical forests alone cover only 5 percent of the earth's surface, but contain more than 50 percent of all terrestrial species (and even higher proportions of arthropods and flowering plants). Wilson estimates that the current rate of species disappearance from tropical forests is about 4–6,000 species per year, which is about 10,000 times greater than the natural "background" rate of extinction before humans arrived (1990: 54). The two other great genetic storehouses of species—wetlands and coral reefs—are also under severe stress from human intrusion and both land and waterborne pollution (Miller, 1998: 675).

Second, modern agriculture is a powerful cause of declining biodiversity. People have historically used over 7,000 plant species for food, now reduced to largely 20 species around the world. These are mainly wheat, corn, millet, rye, and rice. Humans encountered these plants haphazardly at the dawn of the agricultural revolution, but they are now selectively bred into a few strains with greatly reduced genetic variability. In Sri Lanka, farmers cultivated some 2,000 varieties of rice as late as 1959. Today only five principal varieties are grown. India once had 30,000 varieties of rice; today most production comes from only ten. In a trip through your supermarket fruit section, you can purchase perhaps five or six varieties of apples; in North American alone more than a hundred varieties were grown and marketed in the late 1800s. The same sort of reduction in genetic variability has taken place in the herds of cattle, sheep, and horses that humans raise. The U.N. Food and Agriculture Organizations estimated that by the year 2000 two-thirds of all seeds planted in LDCs were of uniform strains (Wilson, 1990: 85; Miller, 1992: 374). In addition to destroying habitats and the impacts of agriculture, a variety of related human actions have reduced biodiversity. These include overfishing, commercial hunting and poaching, predator and pest control, the sale of exotic pets and plants, and deliberate or accidental introduction of alien or nonnative species into ecosystems. Nonnative species—usually highly adaptable plants and animals that spread outside their native ranges, usually with human help—do well in disturbed habitats (Miller, 1998: 673; Tuxill, 1998: 129).

Third, to the extent that the earth's climate warms (the *greenhouse effect*), it will cause reduction in biodiversity. Such a broad global trend would mean changes in seasons, rainfall patterns, ocean currents, and other parts of the earth's life-support systems. Global warming could cause an increased dieback and decomposition of forest biomass, triggering the release of more CO_2 and other greenhouse gases into the atmosphere. This release of gases would magnify warming trends. Since the largest forests are the boreal forests located in the northern high latitudes (e.g., Canada and Siberia), where temperatures would rise the most in a greenhouse-affected world, they could quickly start desiccation and dieoff, except in areas offset by increased precipitation. By one estimate, such warming could increase Canada's forest fires by 20 percent and their severity by 46 percent (Myers, 1997: 223). Although species responded to past climate changes by migrating or shifting their ranges, such adaptive responses will be more difficult in today's degraded habitats. How many species would be able to migrate or adapt to changing climate is not known (Tuxill, 1998: 129). Neither are successes in potentially moving much agriculture from the world's "breadbaskets" into regions with different soils, nutrient, water resources, and across political boundaries. I will have more to say about climate change in the next chapter. In sum, the earth's natural diversity of species is caught in a vise between declining agricultural diversity, habitation destruction, and the potential threats of global warming. Who cares?

Why Should We Care About Declining Biodiversity?

People should care about declining biodiversity for at least three reasons: (1) because the natural diversity of living things is has great actual and potential value as food, medicines, and other substances commercially important for humans; (2) because ecosystem services that a diversity of species in different niches in ecosystems upon which all life, including humans, ultimately depend; and (3) because as the earth's evolutionary and biological heritage, the diversity of species is irreplaceable and valuable in itself. Let me expand on each of these points a bit.

The first reason, in the most anthropocentric terms, is the great actual and potential economic value of natural species diversity. From tropical forests alone, we get essential oils, gums, latexes, resins, tannins, steroids, waxes, acids, phenols, alcohols, rattans, bamboo, flavorings, sweeteners, spices, balsam, pesticides, and dyes. Many wild plants bear oil-rich seeds with potential for the manufacture of fibers, detergents, starch and edibles. Plants called euphorbias contain hydrocarbons rather than carbohydrates; hydrocarbons make up petroleum. Of the species that are candidates for "petroleum plantations," some can grow in areas made useless by strip mining. Several tree species—including beech, elm, oak, sycamore, willow, and elder—can clean up urban pollution, particularly sulfur dioxide. They act as air coolants. A 20-meter shade tree can mitigate enough heat to offset three tons of air conditioning costing $20 a day in the United States.

This highly abbreviated list is just the beginning. Consider chemicals from "wild things" in medicine and pharmaceuticals. One in four medicines and pharmaceuticals has its origins in the tissues of plants, and another one in four are derived from animals and microorganisms. These include antibiotics, analgesics, diuretics, tranquilizers, and many others. The contraceptive pill derives from a tropical forest plant. Perhaps the most famous illustration is a substance call *taxol,* a compound in the bark of the Pacific yew tree. In the Pacific Northwest, loggers cleared it as a "trash tree." Researchers found that taxol damaged cancer cells that were unaffected by other drugs; it can help 100,000 Americans fight breast, lung, and ovarian cancer, not as a cure, but by enabling patients to live longer with less pain. Taxol may soon no longer require bark because by 1992 drug companies were trying to synthesize the active ingredient. A child with leukemia in 1960 had a one in ten chance of remission, but by 1997 such children had 19 chances out of 20, thanks to two powerful alkaloids derived from Madagascar's rosy periwinkle—found in Madagascar's threatened rainforests noted earlier. Other than the anticancer drugs, the commercial value of plant-derived products from all these sources topped $40 billion a year in the late 1980s. Plant-derived anticancer drugs now save about 30,000 lives annually in the United States alone and have a combined economic benefit for patients and society of at least $370 billion. Double that figure for all industrial nations. The potential for further discoveries is vast, particularly from tropical forests where the greatest majority of species extinctions are occurring (Myers, 1997: 263–267).

Medicinals and pharmaceuticals come from animals also. Amphibians have been a particularly good source, since they are beset by all kinds of predators and diseases. Medicine from an Australian tree frog protects against infections. An Ecuadorian rainforest frog secretes a painkiller with 200 times the potency of morphine. Insects secrete substances that promote wound healing and that fight viruses. An octopus extract relieves hypertension, seasnakes produce anticoagulants, and the menhaden (a fish) produces an oil that helps atheriosclerosis. A Caribbean sponge produces a chemical that acts against diseases caused by viruses, much as penicillin did for bacterial diseases. Even lowly barnacles, the bane of sailors, produce a chemical that could be adapted for cement for tooth fillings and could replace the binding pins now used to set broken bone fractures (Myers, 1997: 265).

Beyond its values for commercial goods and medicines, consider the value of biodiveristy for food and agriculture. Though farmers can now purchase and plant genetically engineered seeds, the productivity of our food supply still depends on the plant diversity maintained by wildlands and traditional agricultural practices. Wild relatives of crops continue to be used to maintain the resistance to disease, vigor, and other traits that produce billions of dollars in benefits to global agriculture (Tuxill, 1999: 100). The previous chapters noted the vulnerabilities inherent in such agricultural monocultures with reduced biodiversity. Remember the Irish potato blight famine and its social consequences? Here's a near-contemporary example. By 1970, when 70 percent of the seed corn grown in the United States owed its ancestry to six inbred lines, a leaf fungus infected ("blighted") cornfields from the Great Lakes to the Gulf of Mexico, and American's great corn belt was threatened. Fifteen percent of the entire crop (as much as half in parts of the South) were destroyed, increasing costs to consumers of more than $2 billion. The damage was halted with the aid of various kinds of blight-resistant germ plasm with a "wild" genetic ancestor that derived from Mexico. Wild species have great potential to help address the world's food problems. For example, a species of wild corn was discovered in the mid 1970s and preserved from extinction—just in the nick of time. Only a few thousand stalks were surviving in three tiny patches in south central Mexico that were about to be cleared by loggers and farmers. Remarkably, this was the only known *perennial* variety of corn, and crossbreeding it with commercial varieties could reduce the need for plowing that in turn would reduce soil erosion, water use, and energy use. More importantly, this strain was found to have built-in genetic resistance to four of the eight major viruses that affect corn, which heretofore breeders had not been able to breed into commercial varieties (Miller, 1998: 607).

Think about the value for humans of *pollinators*, including honey bees and many other species. Pollination services are provided to cultivated food crops both by wild (feral) and managed insects that nest in habitat adjacent to croplands and orchards. For example, the activities of honey bees and wild bees are essential in pollinating about $30 billion worth of U.S. crops in addition to pollinating natural plant species. One scientist estimated that in

New York state alone bees pollinated about *1 trillion* blossoms on a single summer day (Pimentel, 1992a: 219). The number of honey bee colonies in the United States has been declining since 1947 because of the use of organophosphate insecticides and diseases; they declined 20 percent between 1990 and 1994. If managed honey bee services were reduced for 62 U.S. crops (including 20 fruit species, 17 vegetable species, and 5 oilseed species) to relying mainly on wild pollinators, this would cost American consumers at least $1.6 billion per year. If native and wild pollinators were affected by the same conditions, the cost to the U.S. agricultural economy would be at least $4.1 to $6.7 billion a year (Nabhan and Buchmann, 1997)! As biologist David Pimentel commented sardonically, "Humans have found no technology to substitute for this natural service or for many others supplied by wild biota"(1992a: 219).

The important point of these illustrations is that humans clearly cannot survive by depending only on a few livestock and crop species. The diversity of wild species, whose role is not often appreciated, is also vital for humans and to maintain ecosystems themselves.

Beyond direct human benefits, a *second important reason* for valuing biodiversity is its *ecosystem services*, that is, how it influences the supply of ecosystem goods and services. Ecosystem services include the important roles in particular niches that a diversity of species play in maintaining the food chains, energy and matter cycles, and population balances of entire ecosystems. Diversity is fundamental to all ecosystems, and its decline has raised numerous concerns, including the possibility that the functioning and stability of the earth's ecosystems might be threatened by the loss of biological diversity (Schulze and Mooney, 1993).

Evidence suggests that diverse ecosystems are more *productive* because different species use resources in a more complimentary way, using resource limits more effectively. They are *more stable* and less susceptible to a variety of disturbances (e.g., drought or floods). Evidence for these ideas comes from a study of the severe drought in the midwestern United States in 1998. Less species-diverse Minnesota grassland plots fell to one-twelfth of their pre-drought productivity, while more diverse plots declined by only about half of that. More diverse ecosystems are *more sustainable* probably because they conserve and make more effective use of water and soil nutrients. The ecosystem services provided by biodiversity are not some magical direct effects, but rather increased functional roles that are possible in ecosystems that contain more species (Tilman, 1997).

The rapid expansion of human activities across the earth and the subsequent modification of natural ecosystems produced much lower diversity within managed ecosystems. As we destroy, alter, or appropriate more of these natural systems for ourselves, ecosystems and their services are compromised. At some point, the likely result is a chain reaction of environmental/ecological decline. When such thresholds will be reached, no one can say. ". . . Few would argue that every beetle or remaining patch of natural vegetation is crucial to planetary welfare. But the dismantling, piece by piece, of global life-support systems carries grave risks" (Ryan, 1992: 10).

A *third kind of reason* for valuing species diversity is very different than for human utility or ecosystem functions. Preserving biodiversity is important to many people for historical, esthetic, and spiritual reasons. The diversity of presently existing species is an irreplaceable product of an eons-long evolutionary process. Every living thing contains from one to ten billion bits of information in its genetic code, brought into existence by an astronomical number of mutations and episodes of natural selection over the course of thousands or millions of years. This process has enabled life to adapt to an incredible diversity of physical environmental circumstances. But as species diversity declines, natural speciation will not refill the gap left by extinction, at least not in any meaningful human time scale. After the last mass extinction (in the Cetaceous era, when the dinosaur age ended) five to ten million years passed before biodiversity returned to previous levels. Biodiversity—the world's available gene pool—is one of the earth's most valued and irreplaceable resources. Species diversity is also a meaningful source of mystery and great beauty to many people. These reasons obviously *transcend* viewing the world in narrowly anthropocentric terms as "sources and sinks" related to human uses of the earth and its creatures. I think that each nation should value the diversity of living things as a part of its planetary heritage as well as its resource base; it is the product of millions of years of evolution centered on that particular place. Hence there is as much national reason for preserving biodiversity as for concern and preservation of the nation's history, language, and culture (Wilson, 1990: 50, 58).

Addressing Deforestation and Declining Biodiversity

As with soil and water resources, there are many ways that the nations of the world could slow or halt unsustainable forest use. One of the most significant ways of reducing tree harvest rates is by greater efficiency in use, eliminating waste and recycling. The United States, for instance, has the world's highest per capita paper consumption (317 kilograms per person per year), of which fully half is quickly discarded packaging and only 29 percent is recycled. Japan, by comparison, recycles 50 percent of its paper but is the greatest consumer of valuable tropical hardwoods in many throwaway products. Half of U.S. wood consumption could be saved by increasing the efficiency of sawmills, plywood mills, and construction, by doubling paper recycling, and by reducing the use of disposable paper products (Postel and Ryan, 1992). Similar steps could be taken throughout the MDCs combined with the introduction of more efficient cooking stoves in the LDCs to reduce the world's demand for fuelwood. All of these would be ways of reducing the economic "throughputs" of forest products to reduce deforestation rates.

Let me mention some specific initiatives to preserve forests and biodiversity:

1. *Promoting sustainable use.* The sustainable exploitation of forests by local and indigenous people is often worth more than commercial exploitation (e.g., logging). From sub-Saharan Africa's relic forests, people derive almost 80 percent of their dietary protein. In Amazonia the harvest of mammals, birds, medicinals, fish, nuts, and the like can generate as much as $200 per hectare compared with commercial logging, which generates (unsustainably) less than $150 per hectare (Myers, 1997: 227). The main barrier is political: Big companies and national governments can turn lumber, cattle, or gold into hard currency (for debt payment) more easily than forest products, which offer more benefit to local people.
2. *Debt for nature swaps.* Participating nations act as custodians for protected forest reserves in return for foreign aid or debt relief. Typically a private organization pays a portion of the debt and supervises the swap. Conservation International purchased more than a half million dollars of Bolivia's $6 billion foreign debt from Citibank to preserve about 3.7 million acres of the Bolivian Amazon (Miller, 1998: 353). Several dozen similar projects exist.
3. *Preserving nature in place.* Conservationists have long sought to set aside parks and nature preserves. Such nature preserves now account for about 8 percent of the earth's surface. The problem with such wilderness preserves is when they conflict with cultural and economic uses. Some are protected from poachers in name only. To date, the best job of protecting tropical forests is in Costa Rica, which in the 1970s set aside 12 percent of its land (6 % for the exclusive use of indigenous people). In comparison, the United States has only 1.8 percent of its land as wilderness reserves, not used for any commercial purpose. Such conservation paid off for Costa Rica. Tourism (especially "ecotourism") provides most of its revenues from outside the country (Miller, 1998: 356).
4. *Gene banks and conservatories.* A major approach to preserving plants and animals has been to remove them from their habitats and protect them in specialized institutions such as zoos, botanical gardens, nurseries, and gene banks. By one estimate, nearly 25 percent of the world's flowering plants and ferns are now so protected. Gene banks focus almost exclusively on storing seed of crop variety and their wild relatives. They arose from plant breeders' need to have readily accessible stocks of breeding material, particularly after the near disaster of the American corn crop in 1970, noted earlier (Tuxill, 1999: 107).
5. *Bioprospecting.* In 1991, Merck & Company paid the Costa Rican Biodiversity Institute $1 million to search for and locate tropical organisms as sources of pharmaceuticals. In the event that marketable product result, the company will retain the patent and pay the institute an undisclosed royalty (rumored to be 1–3 percent of sales). Critics of such arrangements argue that most of the money should go to local or indigenous people, from whose land (and often from whose folk knowledge) such products were developed.

6. *International treaties.* The Convention on International Trade in Endangered Species of Wild Fauna and Flora (CITES) of 1973 provides a powerful legal tool for controlling international trade in threatened plants and animals. It requires that signatory nations issue permits for a limited number of species export and imports.

As you can see, there is a potentially powerful combination of strategies that could be used to stop the process of species extinction. Will it happen? Who knows? Could it happen? Yes.

NONFUEL MINERALS, MATERIALS, AND SOLID WASTES

Industrial economies depend on mining the earth for minerals and materials (such as metal ores and petroleum derivatives that makes plastics). Industries extract raw minerals from geological sources and transform them into usable "products" that you use, consume, and ultimately discard—sometimes quickly in a throw-away economy. But a broader and more integrated view of this process would include not only production and consumption, but prospecting for mineral sources, extraction from sources, discarding solid wastes in various sinks, and—sometimes—recycling materials. Capital and energy investments are required at every stage. The components of this process are illustrated in Figure 3.1.

Thus, rather than thinking only about producing and using products, you can envision the broader process as a sort of "industrial ecosystem" that would *ideally* function as an analogue to biological ecosystems (Frosch and Gallopoulos, 1990: 98). But unlike respiration in a biological ecosystem, or even the burning of materials for fuels, consuming materials such as metals, concrete, plastic, and glass does not turn them into gasses after use. They either accumulate somewhere as solid wastes, or they are vaporized, or otherwise dispersed into soils, water, or the air (Meadows et al., 1992: 79). An important difference between the way industrial economies work and real biological ecosystems is that—at present—a far greater proportion of the wastes are simply dumped somewhere without being recycled. But you can't really throw stuff away in the broader sense. Chapter One discussed the law of conservation of matter. You can, of course, spread it around in air or water or dump it in somebody else's neighborhood, state, or nation.

Reserves and Sources of Minerals

Some minerals are so plentiful in the earth's crust that they are practically inexhaustible (magnesium, aluminum, silicon, titanium, manganese, and iron), though their varying concentration or dispersion in natural formations

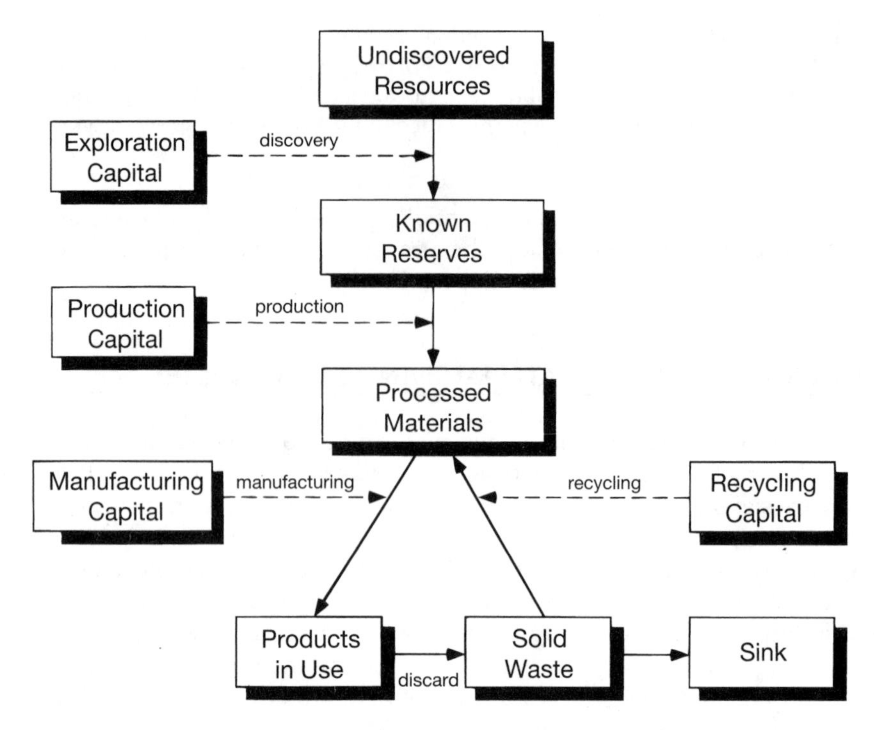

Figure 3.1 Components of the Mineral Production Process
Source: Adapted from D. H. Meadows et al., 1992: 79.

and ores means that they are often technically difficult and uneconomical to refine for human use. Of these minerals, the real backbone of industrial economies is still iron. Other important industrial minerals are scarce and rare in the earth's crust (e.g., chromium, cobalt, copper, gold, mercury, nickel, tin). These are used as chemical reagents to make alloys and different specialty products and are, in a sense, like the enzymes that keep your body healthy; they are necessary in smaller amounts in industrial economies. Even though they are needed in smaller amounts, they can be expensive because of their diffuse distribution in the earth's crust, the difficulty of finding new ore deposits, and the difficulty of extracting these metals from the host ore minerals (Craig et al., 1988: 167,198–199).

What is the available stock of raw minerals? You can see the best current estimates in Table 3.2. The numbers in this figure estimate the "years left till depletion" of various minerals at current consumption rates in the year 2030, on the assumption that a larger population (about 10 billion) will consume at current rates. The implication is that some essential raw materials will become in perilously short supply if the LDCs increase their consumption to match that of the industrial nations (Frosch and Galapoloulos, 1990: 98).

Table 3.2 Estimated Lifetimes of Some Global Nonfuel Mineral Reserves[a]

	Current Consumption Rates		*2030 Rates*	
	Reserves	*Resources*	*Reserves*	*Resources*
Aluminum	256	805	124	407
Copper	41	277	4	26
Cobalt	109	429	10	40
Molybdenum	67	256	8	33
Nickle	66	163	7	16
Platinum group	225	413	21	39

[a] Figures show reserves (quantities that can be profitably extracted with current technology) and resources (total quantities though to exist). Estimates of years left until depletion are based on current global consumption (left) or on the assumption that in 2030 a population of 10 billion will consume at current U.S. rates (right).

Sources: Adapted from R. A. Frosch and N. E. Gallopoulus (1990) "Strategies for Manufacture," in *Managing Planet Earth*, p. 98.

But wait. This is not necessarily a prediction of a grim future, but rather an indication of future costs and adjustments necessary to maintain industrial economies. How so? *First*, note that Table 3.2 has different years to exhaustion for reserves and for resources. *Reserves* are mineral sources currently identified and profitable or at least marginally profitable to extract by existing technologies. *Resources* are reserves *plus* known mineral sources that are "subeconomic"—that is, not profitable to extract with existing technologies—*plus* hypothetical resources in known districts *plus* "expert" speculation about what minerals might be available in undiscovered districts or forms. Mineral reserves and resources are calculated according to a standard formula developed by the U.S. Geological Survey and the Bureau of Mines.[1] As you might guess, as more minerals are discovered or as technology improves, the amounts of both reserves and resources can increase. *Second*, we will never actually run out of minerals, because the earth's crust is like a sponge saturated with minerals than a cup to be emptied of them. You can always (economically and technologically) squeeze harder and get more. The question is how much are you willing to pay for minerals when they become scarce (and often subeconomic, meaning that the cost of their extraction exceeds the value of their use)? *Third*, the *depletion curves* ("years to depletion") estimated by experts vary greatly because they are made with different assumptions. Some estimates are only of consumption (as those above), but some assume consumption with significant conservation efforts, and some assume significant conservation efforts *and* recycling (Keller, 1992: 354). *Fourth*, and more significant, substitutes for minerals can be discovered that effectively change consumer demands, cancel all such calculations, and require calculation of depletion curves for the new resource. Estimates of mineral resources mean something—they tell us what minerals are in short supply and are difficult and expensive to refine—but they are certainly squishy numbers and have historically been made moot because of the invention of substitutes.

Consider the case of copper. Between 1900 and 1950, the average grade of copper ore mined in the United States declined from that which yielded about 5 percent copper by weight to an average of just 0.4 percent. Yet the technology of mining, extraction, and refining improved so that the inflation-adjusted price of copper actually fell. At the same time, demand for copper fell because of the invention of substitutes (such as fiber-optic cables that began to replace copper in telecommunication wiring). The joint effects of technological improvements in extraction, substitute materials, and declining demand actually *increased* world copper reserves by 500 percent between 1950 and 1980 (Miller, 1992: 514). In fact, mineral reserves and inflation-adjusted prices have generally remained stable or fallen between 1950 and 1980 (Vogley, 1985: 462–463).

It is with nonfuel minerals that the arguments of neoclassical economic theory become most convincing. The economists are right—the operation of markets will prevent the complete exhaustion of mineral resources, because the last reserves of subeconomic minerals will always be too expensive to get. The operation of free markets do not, in themselves, increase supplies (in the absence of sunk investment in technology, exploration, or resource substitutes). They merely send you signals that stuff is getting scarce. That, of course, is little consolation to those individuals and nations that find themselves priced out of the market. Or for those nations or communities that are *extractive economies* dependent on production of a particular mineral that becomes technologically outmoded (Freudenburg, 1992a).[2] The consensus among environmental scholars is that such arguments about substitutability are more valid for nonfuel minerals than they are for soil, freshwater, and biological resources.

Problems with Mineral Production

Even so, there are reasons for concern. *For one thing*, even though the long-term costs of minerals have declined, there are undervalued and unrecognized costs in the production process. The more efficient extraction processes use, for instance, vastly greater amounts of freshwater and energy— commodities that are generally underpriced. Nor does the process include the environmental devastation caused by open pit mining or pollution from the extraction and refining process. As the amount of usable metal in metal ore falls, the amount of rock that must be mined, ground up, and treated per ton of product rises with rapidly. For example, "as the average grade of copper ore mined in Butte, Montana, fell from 30 percent to 0.5 percent the 'tailings' produced per ton of copper rose from 3 tons to 200 tons. This rising curve of waste is closely paralleled by a rising curve of energy required to produce each ton of final material" (Meadows et al., 1992: 84–85). The point is that the current measures of (falling) mineral prices do not include a full ecological accounting.

Other problems are more geopolitical than economic or technical, having to do with the global production and distribution patterns. As products become more complex, the search for economic high-grade ores has pushed exploration and extraction literally to the remote corners of the earth. Consider my (or your) personal computer.

- It still has about 2.5 pounds of copper that began as copper sulfide ore, much of it mined from the Chilean Andes for export to Asia. By law, 10 percent of Chile's copper revenues go to the Chilean military. Refining copper ore in the industrial nations produces about one-fourth of the sulfur dioxide emissions, which produce acid rain. Mining and producing metals account for about 7 percent of the global energy consumption.
- A glassworks in Kobe (Japan) made the glass for the front of the monitor, using local sand and electricity from a power plant burning Australian coal. The glass contains 5–10 percent each of strontium oxide (from Mexico), potassium oxide (from Russian ore,) and barium oxide (from Chinese ore).
- The shell was made of ABS (acrylonitrile-butadiene-styrene) plastic. It was mostly made from Saudi Arabian oil, refined using (Wyoming) coal.
- The silicon chip package was probably made in a factory in Kuala Lumpur, Malaysia, running around the clock by workers earning about $2 an hour and Japanese robots running on coal-fired electricity. Face-masked and gloved workers glued each chip to an etched copper frame and ran tiny wires of South African gold between the frame and the chip.
- Finally, these components were shipped to Texas or California (Silicon Valley), where these chip packages were installed in a motherboard and the case and disk drive were all assembled. Last, it was stamped with the logo of an American corporation. (Ryan and Durning, 1997: 48–51)

Far more than world trade in food and fiber, world trade in minerals has formed the skein of the emerging global economy and economic interdependency.

So what are the problems? Does this not simply illustrate the emerging world system of nations and world market economy I discussed in Chapter Two? The answer is yes, and it also illustrates one of the geopolitical problems of that world system. One of these problems has to do with the decreasing self-sufficiency of nations. When trading partners are reliable and friendly and the price is right, there are few problems. But where trading partners are unstable or potentially hostile economically disastrous supply interruptions are possible. Many high-grade ores in the industrial world are practically speaking, gone forever. Although the United States and Europe are still major producers of some minerals, they will never again be self-sufficient. Currently, no industrialized country is self-sufficient for minerals, although the former Soviet Union came close. The United States is critically deficient in at least 20 minerals and has essentially no reserves at all for manganese, colbalt, chromium, aluminum, tin, and fluorine. Japan is more dependent (Miller, 1998: 542–543). In addition to lacking coal and lumber, Japan has virtually no mineral resources whatsoever. A supply disruption or embargo would prove disastrous. But if the main problem for MDCs is the *potential* disruption of supplies, the problem for LDCs is more serious, *actual*, and ongoing. Simply put, the LDCs are becoming the mineral suppliers for the MDCs (as is elegantly illustrated above by my computer), that consume eight to ten times per capita more than do the LDCs. This means that as world mineral prices tend to fall, the MDCs are the main beneficiaries, while the LDCs are left holding the bag with declining export incomes. Each decline of a few cents per ton of Chile's copper or Zaire's cobalt means recession and economic hardship. Hence the world trade in minerals tends to amplify economic inequality and political tensions in the world system.

So even though the case made by neoclassical economists works better for minerals than other resources we have discussed, you can see that there are still environmental, economic, and geopolitical problems. Indeed, current environmental concern is less with source problems than with the enormous amounts of energy used in processing minerals and the land erosion, pollution, and ecosystem disruption that their extraction causes.

Sinks: Solid Waste Problems

In the 1970s, analysts often argued that we would run out of mineral supplies, but now most experts argue that more immediate problems with high consumption of minerals are the continuing damage that their extraction and processing imposes on the environment—at both ends of the production cycle. In other words, the worst problems with nonfuel minerals may not be with the sources but with the sinks. You probably identify the term *solid wastes* with municipal garbage, but that is only the small but highly visible portion of the solid wastes produced by industrial societies. In 1998, the U.S. Environmental Protection Agency and the Bureau of Mines found that 75 percent of solid wastes were produced by mining and oil and gas production, 13 percent by agriculture, 9.5 percent by industry, 1.5 percent by municipal garbage, and 1 percent by sewage sludge (Miller, 1998: 565–566). The solid waste produced by nonfuel mining alone amounts to more than a billion tons per year, at least six times that produced by municipalities. Yet even the 1.5 percent of total solid wastes from homes, businesses and municipalities represents a *significant* amount of discarded junk. Consider that Americans threw away

- Enough vehicle tires to encircle the planet almost three times
- Enough aluminum to rebuild the country's entire commercial airline fleet every three months

- About 2.5 million nonreturnable plastic bottles each hour
- About 18 billion disposable diapers per year, which if lined up end to end would reach to the moon and back seven times

About 60 percent of the total weight of typical household garbage is paper, cardboard, and yard waste from potentially renewable resources. Most of the rest is products made from glass, plastic, aluminum, iron, steel, tin, and other nonfuel minerals. In 1995, only 24 percent of the municipal solid wastes were recycled or composted. The other 76 percent was either hauled away to landfills or burned in incinerators (Miller, 1998: 566). Recycling rates for materials vary significantly among MDCs. The United States has the lowest rate; western Europe and Japan have higher rates. Americans have always generated a lot of trash. Remarkably, archaeologist William Rathje estimated that in the last 25 years Americans have consistently discarded 1.5 to 2.5 pounds of garbage per person a day (1990). But there are a lot more of us now, and in some ways it is more difficult to dispose of modern high-tech trash. The basic ways of dealing with solid wastes haven't change much, though. They are either dumped, burned, or—sometimes—recycled.

Addressing Solid Waste Problems

In rapidly growing cities of the LDCs, garbage disposal problems are nightmarish. Cities such as Mexico City, Manila, Caracas, Port-au-Prince, and Lagos are awash with enormous garbage dumps on their outskirts, untreated by any modern sanitary methods. Even though the sinks are full of infectious diseases and toxic chemicals of all sorts, poverty among the lower classes in such cities means that around the world there are garbage pickers who literally live by recycling usable materials from such dumps.

In the MDCs, modern sanitary landfills are state-of-the-art constructions vastly different from historic town dumps. They have impermeable plastic linings to prevent leachate (dangerous liquid waste residues from the degradation of solid wastes) from seeping into the surrounding soil, perforated drainage pipes, and methane-collection tanks. These tanks are capped with impermeable materials to keep out rainwater when they are filled to capacity. Still, there are problems: Most eventually *do* leak. Tree roots can perforate the barriers and caps. Leachate, which contains over 100 toxic chemicals, does seep out. Forty-six percent of America's landfills have wells close by, and 75 percent of municipalities do not test for leachate. Some estimates suggest that 2 percent of the entire world buildup of greenhouse gases derives from the methane generated by U.S. landfills. Measurably adverse health conditions correlate with proximity to landfills (Clyke, 1993: 53–54). And if the production and distribution of minerals relate to global social inequality, there is a similar national linkage here: Evidence shows that landfills are more likely to be found close to lower-income and minority populations (Bullard, 1990, 1993).

In addition to the environmental, health, and social justice issues that surround solid waste dumps, there are a host of other pressing concerns. One is that the dumps are rapidly filling up. Furthermore, people don't want them, and communities have become more effective in blocking the construction of new waste disposal facilities for both industrial and municipal wastes. NIMBY issues ("not in my back yard") have been prime organizing issues for grassroots environmental activists. Governments and industries responded by seeking to reduce the volume of the solid waste stream by either incinerating (burning) it, or recycling it. But there are problems with both.

Incineration does reduce the volume of solid wastes, by about 90 percent. But though it removes many harmful chemicals from wastes and has a potential to produce energy, the capital construction costs of incinerators are very high. Incineration produces ashes that are rich in other chemicals in significantly toxic amounts (dioxins, furnans, hydrocarbons, lead, cadmium, chromium, mercury, and zinc). It does little to discourage the production of such wastes and transfers many of them from one sink (the landfill) to another (the atmosphere).

Recycling has been the most publicized solution to reduce the solid waste stream. But for reasons involving practical difficulties, costs, and resistant American lifestyles, only about 11 percent of American trash was recycled in 1993. Despite such deep-rooted obstacles, recycling has grown steadily in the United States and other MDCs. Taken together, the proportion of glass recycled between 1980 and the mid-1990s rose from less than 20 percent to over 50 percent. The recycled proportion of metals Americans used rose from 33 percent to 50 percent between 1970 and 1998. Furthermore, around the world, the proportion of paper and cardboard recycled grew from 38 percent to 41 percent between 1975 and 1995 and is projected to reach 46 percent by 2010. Recycling is being transformed from a gesture to help the environment into a solid industry. The reasons are that the economic recovery in the 1990s increased demand for raw materials and that manufacturers built mills that can process recycled materials. It was often more profitable for them to do so (Gardner and Sampat, 1999: 45; Matoon, 1998: 144; *New York Times*, 1994). While most environmental scientists view recycling as a positive step, it developed in relation to rising landfill costs, not to environmental damage. Current recycling of materials like glass, aluminum, paper, and plastics is based on economic rather than environmental criteria, and other hazardous substances are not recycled. A product is considered recyclable only if it is economically profitable, and we have been less willing to bear the costs of materials that were not.

Reuse of materials is even more effective. It extends resource supplies and reduces energy use and pollution more than incineration or recycling. Many commercial and industrial materials could be reused, but think about two obvious examples: refillable beverage bottles and 1.5 liter refillable soft drink bottles made of plastic. Unlike throwaway or recycled cans or bottles, refillable beverage bottles create local jobs related to collection and refilling. In 1964, the majority of soft drinks and beer in the United States were sold in refillable bottles. But by 1995, only about 7 percent of all such drinks came in refillable bottles and only 10 states even had them. Some industries lobbied hard to prevent the return of refillable bottles. Other countries are more successful at reuse of such containers. Ecuador, a poor LDC, instituted a refundable beverage container deposit fee that is 50 percent of the cost of the drink. In Germany 95 percent of the soft drink, beer, wine, and spirits containers are refillable. But Denmark led the world in reuse of beverage containers, banning all that cannot be reused (Miller, 1998: 571–572).

The problem is that it is difficult to deal with solid waste (or any other pollutants) after it has been created. The most effective way of dealing with wastes is source reduction or, as it is sometimes called, the *dematerialization* of production and consumption. In other words, the most effective reduction of the solid waste stream could be realized by introducing efficiencies in extraction, production, or consumption so that the economic cycle simply doesn't generate as much solid waste. For consumers, this would mean manufacturing more durable, long-lasting goods (rather than disposable ones), reuse of things like glass bottles (rather than recycling them), and source reductions, such as reducing the layers of packaging (rather than recycling them). Roughly, a hierarchy of more effective ways to address solid waste problems would begin with source reduction, followed by reuse, recycling, incineration, and landfills, in that order. All of this would presume significant modifications in the *throw-away economies* of many industrial nations. In the current system, neither producers nor consumers bear the real costs of the solid wastes they generate. Governments (or aggregated taxpayers) do. This means that there are no real incentives that encourage the reduction of material throughput because there are no real market signals of the costs to particular producers or consumers. Again, a problem is to internalize and particularize the real costs of production. Or—to return to the initial analogy of this chapter—to make the industrial system function more like a real biological ecosystem.

CHEMICAL POLLUTION AND TOXIC WASTES

This chapter began by focusing on resources necessary for human well-being and ecosystem maintenance (soil, water, biotic and mineral resources). But in the preceding section, I began to consider the sink problems of dealing with the products—solid wastes—of human industrial and consumer activity. Besides solid wastes, an astounding variety of chemical pollutants and toxic substance resulting from human activity permeate the environment we live in. I referred to them briefly in several places, but here I discuss them at bit more directly.

Chemical Pollution from Agriculture

Agriculture, particularly intensive or industrialized agriculture, is an important source of pollution and toxic substances. Agriculture generates chemical pollutants in the residues from pesticides and herbicides, from the nitrates and phosphates remaining from the use of chemical fertilizers, and from the salt that accumulates in soil from irrigation water.

Among the most noxious are the wide variety of herbicides, fungicides, and pesticides (mainly insecticides) applied to croplands, which leave residues in the soil and water—as well as on the fruits and vegetables you buy in grocery stores. Though some of the most dangerous chemicals with long-lasting residues (the chlorinated hydrocarbons such as DDT and chlordane) have been banned from use in the United States, they have been replaced with others with equally high toxicity levels but shorter-lived residues (organophosphates such as parathion) (Miller, 1998: 621). But the production and use of the more dangerous agrochemicals has merely shifted overseas, particularly to the LDCs, from which the United States increasingly imports food. Ironically, the world market economy facilitates a circle of toxins as well as of goods and services.

Defenders of the use of agricultural chemicals argue that they increase food supply, lower costs by preventing crop loss (by as much as 35 percent to 40 percent), and increase the profitability of farming. How dangerous are they? Scientists associated with the agrochemical industries argue that their health risks are insignificant compared with their benefits, and that studies of pesticide hazards are based on worst-case scenarios.[3] Other scientists disagree. The problem is that agrochemicals exist in very low concentrations (measured in parts per million in water), and their effects are long-term ones that materialize statistically among populations after years of contact. Major health hazards of toxic agrochemicals are cancers; malignant lymphomas (tumors), which may take 20 years to develop; and DNA damage, which may result in birth defects in babies of the next generation. And the causes of these conditions are complex; it is difficult to link these conditions with a single set of causes.

Still, the World Health Organization estimated that each year 25 million agricultural workers in LDCs are seriously poisoned by pesticides and 220,000 die. In the United States, at least 300,000 farm workers suffer from pesticide-related illnesses each year and at least 25 die. The effects are particularly severe among Hispanic migrant farmworkers, who have very high levels of exposure. In fact, partly because of exposures to agrochemicals, farming was named the nation's most hazardous occupation, ahead of construction, mining, and manufacturing. Furthermore, every year about 20,000 Americans—mostly children—get sick from home misuse or unsafe storage of pesticides (Miller, 1998: 623;. National Institute for Occupational Safety and Health, 1992).

Pesticide and herbicide pollution in surface and groundwater is widespread. In Long Island, New York, 23 percent of the 330 wells tested were contaminated with Aldicarb at levels that exceeded heath guidelines. Aldicarb is a highly toxic pesticide once used on the region's potato fields (World Resources Institute, 1993: 40). In the 1990s, studies by the State Department of Health and the University of Nebraska found contaminated wells in 42 counties in my home state. One of seven wells tested was contaminated with the herbicide atrazine (linked to lymphatic tumors and cancers). In fact, the state of Kansas threatened to sue Nebraska because of the high atrazine levels in the Blue River flowing between the two states, which feeds into the municipal water supplies of Manhattan, Lawrence, Topeka, and Kansas City, Kansas. In 1987, the U.S. Environmental Protection Agency ranked pesticide residues in foods as the third most serious environmental health threat in the United States in terms of cancer risk (Citizen Action, 1992; Miller, 1998: 625).

As if this weren't bad enough, mounting evidence indicates that in the long run, pesticides are not effective in protecting crops from losses. The reason is that while insect pests may initially be suppressed by insecticides, they breed and mutate rapidly and tend to develop chemically resistant strains that then require *more* or different chemicals to suppress. Chemicals can also produce more insects by killing the predators (birds) that feed on them.

> In 1986, with rice fields plagued by pesticide-resistant brown plant hoppers, the Indonesian government withdrew $100 million in annual pesticide subsidies, banned 57 pesticides, and launched a national IPM program . . . [integrated pest management...which combines a diversity of nonpesticidal tactics with sparing pesticide use] . . . Since then, pesticide use has fallen by 60 percent, while rice harvest has increased some 25 percent. (Halweil, 1999)

Evidence of similar cases around the world is pervasive. Since the 1940s, the world has been on a *herbicide and pesticide treadmill* that has been very profitable for the agrochemical industry. But it is of dubious long-term value in increasing food security and is surrounded by evidence of pervasive long-term health hazards (Halweil, 1999).

Another category of agricultural pollutants consists of residues from the application of inorganic chemical fertilizers to croplands. These fertilizers unquestionably boost crop yields, but they leave large concentrations of nitrates and phosphates that wash into streams, rivers, lakes, and groundwater. During warm weather, this stepped-up nutrient level produces rapid growth of aquatic plants, such as algae, water hyacinths, and duckweed, which use (for their own respiration) most of the dissolved oxygen in the water. These plants then die and sink to the bottom to decay, along with most of the oxygen-consuming fish and aquatic animals. This process, called *cultural eutrophication*, may leave a body of water that is essentially dead except for the decomposers and the few scavenger species that can live in such an oxygen-depleted environment. A river ecosystem that undergoes cultural eutrophication from a specific point may recover miles downstream, but it is more damaging when it does not derive from a specific point, such as in the 221-square mile Florida Everglades, which is being degraded by broad water flows containing nitrates (and pesticides) from the sugar fields and orange groves to the north (Miller, 1998: 656). In the United States, many medium to large lakes, and more than half of the large lakes near major population centers, suffer some degree of cultural eutrophication. The long-term health risks for adults of exceptionally high nitrates in drinking water are not well understood, but there is some evidence that it is related to high rates of female breast cancer and miscarriages. There is clear evidence that in very young children nitrates can react with oxygen-carrying hemoglobin in blood producing a serious form of known as the "blue baby" syndrome (World Resources Institute, 1993: 40–41).

The *salinization* of land from long-term irrigation is a third important cause of chemical pollution. Fresh water contains between 200 and 500 parts per million (ppm) of salt. Crops take up the fresh water but leave salt in the soil, and daily irrigating a plot of land can add literally tons of salt to the soil each year, eventually exceeding the salt tolerance limits of crops. Unless soil is flushed with fresh water periodically—an expensive process in water-short irrigated areas—soil eventually becomes barren and useless. Remember that this was part of the plight of the ancient Mesopotamians and other agricultural civilizations I discussed in Chapter Two. Since World War II, growing irrigation increased agricultural productivity, but the long-term consequences of soil salinization may now be decreasing it (Postel, 1992b).

Chemical Pollutants from Industry

Other than those pollutants produced by agriculture, most toxic and hazardous chemicals are introduced into the environment by industries. These include heavy metals and synthetic organic chemicals that are not degraded for a long time by natural processes and are highly toxic or hazardous (as neurotoxins, carcinogens, or allergens, or mutagens).[4] There are two problems. One problem is that, as with agrochemicals, we are often uncertain about the cumulative effects of exposure to relatively low concentrations of such chemicals with known toxicity. The second problem is that we are introducing new technologies and synthetic chemicals at such exponential rates that our ability to estimate the risks and benefits of even a few of them has been overwhelmed. Of the approximately 72,000 chemicals in commercial use, only about 10 percent have been tested for toxicity, and we introduce about 1,000 new chemicals each year (National Academy of Sciences, cited in Miller, 1998: 266). Every year more than one billion tons of toxic substances are released into environmental sinks. In 1990, the EPA has identified over 3,000 contaminated sites (some experts thought there were as many as 10,000) and estimated that at 1990 prices it would take more than $500 billion and 50 years to clean them up. Even then, the government was uncertain about how to dispose of them to prevent future risks. (Environmental Protection Agency, 1990).

Urban and Municipal Pollution

Because industries are usually in human settlements and cities, they contribute directly to the water and air pollution around cities, but municipal pollution also includes pollution from municipal wastes and sewage as well as air pollution from the combustion of fuels in autos, factories and homes.

In LDCs, much of the sewage from human settlements is not treated and is highly contaminated with raw sewage and microorganisms that carry waterborne diseases such as dysentery, typhoid, and cholera. Poverty often means malnourishment and exposure to soil and water that carry disease. This high incidence of diarrheal diseases in LDCs reflects the lack of safe drinking water and the ingestion of food-and soil-borne microbes that causes disease. In 1995, for example, less than 25 percent of the populations of Afghanistan, Burkina Faso, Haiti, Liberia, and Vietnam were served by adequate sanitation systems for the disposal of human wastes. One of the most important things that could be done to improve health and nutrition in LDCs would be to provide clean drinking water to more people so that they could absorb the food they have. The problem, of course, is that most LDCs do not have the capital resources to build sewage treatment plants for towns and growing urban areas. Even in wealthier MDCs the poor often have greater exposure to such hazards. Partly for such reasons, disadvantaged populations everywhere have lower life expectancies (Olshansky et al., 1997: 14–15).

Most MDCs have long invested in sanitation and water treatment facilities that reduce the risk from these water-related diseases. Primary treatment involves filtration that removes the suspended junk, while secondary treatment uses settling basins were aerobic bacteria degrade organic pollutants. Sewage treatment leaves a toxic, gooey sludge that must be dumped or recycled as an organic fertilizer. About 54 percent of such municipal sludge is applied to farmland, forests, highway medians, or degraded land as fertilizer, and 9 percent is composted. The rest is dumped in conventional landfills (where it can contaminate groundwater) or incinerated (which can pollute the air with toxic chemicals) (Miller, 1998: 532). Yet conventional sewage treatment does not remove many toxic chemicals, nitrates, or phosphates. Special treatment can deal with many of these problems, but because of their costs (twice as much as conventional treatment facilities) they are rarely built. Half of U.S. drinking water is to some degree contaminated by leaking hazardous waste sites (Clyke, 1993: 58).

Chapter One described carbon dioxide as the by-product of all respiration and combustion of carbon-based fuels. But there are *many* other chemical byproducts of the combustion of fuels. One is suspended carbon particles (soot), which can remain in the air a long time and can contribute to respiratory disease. Another is carbon monoxide (CO), which is the result of incomplete combustion, particularly from cars and trucks. CO is an odorless gas that interferes with the body's ability to absorb oxygen and can exacerbate heart and respiratory disease or even cause death. Sulfur dioxide (SO_2) is produced from the burning of coal and oil. It can cause respiratory problems. More important from an ecosystem perspective is that SO_2 combines with water in the atmosphere to form acids (e.g., H_2SO_4) to form acid rain, which kills forests (miles away from the sources) and pollutes soil by making it too acidic for optimum plant growth. Wide areas of the United States and Central and Eastern Europe, as noted in Chapter One, have been affected by acid rain from urban and industrial sources. Nitrous oxides (NO_x) and other volatile organic compounds (VOCs) are also produced from the incomplete combustion of fuels and hydrocarbon compounds from autos and a wide variety of commercial and industrial sources. More important, in the presence of sunlight, SO_2, NO_x, ozone, and VOCs react to form *smog*, a hazy, dirty brown, toxic witch's brew of more than 100 exotic chemicals that hangs in a bubble over most cities when the weather is right. Smog is particularly a problem in cities such as London, Los Angeles, and Mexico City, where topography and inverted thermal layers of air (warmer aloft than on the surface) often hold smog to the surface in more concentrated forms. But look for it if you fly into any major large metropolitan center.

Pollution Trends

The awareness of environmental problems that developed in the 1960s began with an awareness of toxic wastes and water and air pollution, and since that time many MDC governments have instituted antipollution programs.[5] The good news is that there have been successes. The United States passed the Safe Drinking Water Act of 1974, and from 1972 to 1992 the organic material from sewage and industrial sources dumped into rivers declined significantly. Drinking water generally became safer during this period, except for nitrates and in private wells and very small water systems (World Resources Institute, 1993: 38,41).

But America's water quality is still very mixed. The measures of drinking water improvement rely mainly on measures of declining industrial heavy metals and fecal and choliform bacteria. But in 1994 a study for the American Waterworks Association found that the drinking water in the large central states regions of the Mississippi-Missouri-Ohio River basins contained the runoff from agricultural land. That study found that at least 87 million people in 119 cities were drinking water laced with 14 different insecticides and herbicides, and truly dangerous levels of two of them: cyanizine and metolachlor. The health risk was to place people of the region at 10 to 100 times greater risk of developing cancer. Municipal water departments reacted mostly with denial, protesting that removing these chemicals would triple the costs of water purification. The Environmental Protection Agency (EPA) stuck to its claim that most drinking water was perfectly safe, to which the study authors retorted that EPA safety standards, developed a decade previously, were woefully too lenient, given what is now known about the carcinogenic properties of many agricultural chemicals. Even EPA officials acknowledged that they permit up to 100 times the established "safety benchmark" for certain chemical contaminants (Flannery, 1994: 10; Thomas, 1994: 7). This controversy became a classic scientific-technical-political controversy about appropriate safety standards and the criteria for such standards.[6] Across the nation, people reacted to fears about water quality by the proliferating sales of bottled water and home water purification systems (both of questionable health value). By 2000, Congress was being pressured by water-polluting industries to weaken the Safe Drinking Water Act in various ways, while environmentalists or were calling for it to be strengthened (Miller, 1998: 536–537).

In the United States, Congress passed Clean Air Acts in 1970, 1977, and 1990 that required the EPA to set national standards for ambient air quality and emission standards for toxic air pollutants. The legislation was successful because between 1970 and 1995 levels of air pollutants decreased nationally by almost 30 percent even though both population growth and economic growth continued in those years. Lead, in particular, has virtually gone out of air pollution because it was removed from gasoline. The result was that 50 million people now breathe cleaner air, and the economic benefits greatly exceeded the costs. The United States spent about $346 billion between 1970 and 1990 to comply with the Clean Air Acts, whereas the human health and ecological benefits in that same period were estimated at between $2.7 and $14.6 *trillion* (in 1990 dollar values). Even so, in 1995 and

1996 Congress was under intense pressure from officials of polluting industries to weaken the 1990 Clean Air Acts. Most such efforts were vetoed by the President.

Even with such obvious progress in the United States, air is probably not clean enough and problems remain. Nitrogen dioxide levels have not dropped much since 1980 because of a combination of inadequate automobile emission standards, and more vehicles traveling longer distances. Urban smog remains a problem in many areas. In 1994, 100 million people in 43 metropolitan regions continued to live under ground level ozone and smog conditions that violated federal air safety standards. The worst offender for unhealthy air was—surprise—the Los Angeles basin (Associated Press, 1994a: 19; Miller, 1998: 482–483).[7] Similar trends exist in other MDCs. But air pollution continues to exceed health guidelines in many cities in LDCs, including Beijing, Calcutta, Teheran, and Cairo (O'Meara, 1999: 128).

CONCLUSION: THE RESOURCES OF THE EARTH

In the short span of a single chapter, I have tried to provide you with a fairly rigorous yet synoptic reading of the state of the earth's important resources for humans (land, water, biotic resources, nonfuel minerals). The chapter described the physical and environmental nature of each kind of resource and the magnitude, impacts, and consequences of human resource use, both for humans and ecosystems. After that it described how the use of the earth's resources accumulate as wastes in various sinks as solid wastes, soil, water, and air pollution. Throughout, I discussed some of the current technical suggestions for addressing various environmental problems and touched lightly on the demographic, sociopolitical, and economic dimensions connected with them.

So what is a fair summary of this reading of the earth's "vital signs"? Here is how I would summarize, going in reverse order from the last issues discussed back to those at the beginning of the chapter. Evidence exists for both progress and many problems in dealing with specific types of pollution in the MDCs. Although some things are improved in LDCs, much of the world still lives in unsanitary conditions that threaten life and health. Because of changes in technologies, the supply of nonfuel minerals is not an immediate problem, although the environmental costs of their extraction and use often create problems as well as geopolitical tensions in the world system of nations and world market economy. Wild biological resources and diversity are threatened globally by human activities and are underappreciated for their usefulness to humans, their role in maintaining ecosystems, and their part in the world's heritage of species and genetic diversity. A severe and rapidly emerging general problem with water supply is emerging that has to do with (1) the increasing human demand for water, and (2) the limited or uneven water supply in regions and the hydrological cycle. Though there are many technical efficiencies to be gained in water use, these problems seem intractable and politically explosive. Similarly, problems of land and food security suggest an upper limit to many of the previously successful techniques of intensive "industrial" agriculture. Human material security requires maintaining soil quality, conserving water, and preventing further decay of forests and biodiversity. Addressing these problems will require a great deal of technical innovation *and* institutional changes in the way that the nations of the world now operate. I summarized this material in reverse order to make a point: It vaguely retraces changes in popular and scholarly ways of understanding environmental problems over the last few decades, moving from concern with specific environmental problems (such as pollution), which we have had some success, to a more integrated and holistic view of ecological problems. In that view, they can't really be considered in isolation (Dunlap, 1992; White, 1980).

I need to make a final important point to end this chapter. Obviously, the human presence or "footprint" on nature is so large and intrusive that there is very little untouched "pristine nature" left anywhere in the world. What humans mostly deal with is managed or "socialized" nature. In terms of human-environment interaction, it has become an *ecosocial system*—a humanly organized environment. All landscapes, from Los Angeles to Amazonia, are such ecosocial arenas. This doesn't mean that human management of nature has been wholly successful; the boundaries of such control are exposed by the very failures of attempts to extend it indefinitely. Nor does this mean that there are no circumstances from which human beings should attempt to withdraw interventions that affect the environment or try to eliminate side effects. This chapter discussed many such circumstances related to use of the earth's resources. But since all landscapes are ecosocial and we can no longer disentangle what is "really" natural from what is social, we must now deal with environmental/ecological problems not only with appeals to "pristine nature." Chapter Two closed by arguing that environmental problems are also social issues. Likewise, how far we defer to or try to restore natural processes does not depend only on how complex they are, or the fact that many are too large for us to encompass—though they undoubtedly are. Trying to improve environmental resource problems depends on how far we agree that some natural process we have influenced could best be stabilized or reinstated. That depends on a consensus of human values that form the parameters of protection. In other words, the criteria addressing such problems are not given in nature itself, but with the values that guide management, no matter whether we speak of urban areas or of wildernesses (Giddens, 1985: 210–211).

PERSONAL CONNECTIONS

Personal Consequences and Questions

This chapter had a lot of information about sources and sinks; about resource use and the accumulation of wastes in sinks. Think about these large-scale issues in terms of the material flows in your daily living patterns.

1. You might be interested in calculating about how much water you or your family use during a day. Here are some typical amounts:

Use	*Gallons*
washing a car with hose running	180
watering lawn for 10 minutes	75
washing machine at top level	60
10-minute shower	25–50
average bath	36
handwashing dishes with water running	30
automatic dishwasher	10
toilet flush	5–7
brushing teeth with water running	2

Source: American Water Works Association, cited in Miller, 1992: 356.

How much water do you think you use in a day? A month? If you live in a college dormitory or apartment setting, how much do you think is used?

2. What is the source of water in the community where you live? What conflicts are there about water use? About its costs, about allocating it to industry, agriculture, or consumer use? Has there ever been news about impending water shortages or rationing for particular purposes? What local issues are there about water purity? You might call the local utility company or government regulatory agency about this.
3. As I noted in this chapter, most of the solid waste produced in the United States is from mining, industrial, and agricultural sources. But municipal trash from households is still a prodigious amount. About 10 percent of your shopping bills go for packaging costs alone (for highly packaged convenience foods, the proportion is a lot more). The typical family creates two or three large cans of trash each week. Do a two-part mental experiment: (1) Keep track of the trash you create for several days. What is it, mainly? Food wrappers? Newspapers? Pop cans? (2) Suppose that instead of having it hauled off, you just let it pile up in your yard. How long do you think it would take to fill your yard?
4. If you can, ask about the material flows where you work. What kinds of raw material are required? What kinds of behavior on the job amplify or reduce material flows through your workplace? Ask what proportion of the expenses of your workplace are accounted for by the disposal of materials. By the water bill? (You

may have a hard time getting answers to these questions, but somebody knows.). The results may astound you.

5. Do you have or use a computer? If the answer is yes, do you think it has caused you to use more or less paper?

What You Can Do

Quite a lot, actually, if you want to. Its hard to know how to limit the possibilities of altering your lifestyle to be more environmentally frugal. There are whole books written on this subject, and you can find them in any library or bookstore. Here are a few tips:

About Water:

- Don't leave the tap running while you are doing other things.
- If you buy a new dishwasher, get an energy-efficient one.
- Take showers rather than tub baths (it uses about a third of the water that a bath does, unless you use the shower for relaxation therapy rather than cleanliness). Take shorter showers (sing shorter songs).
- Put some bricks or a quart bottle full of water in your toilet tank to use less water per flush. When you buy a new toilet, buy a water-saving one.
- Install a flow constrictor device on faucets, particularly your showerhead (I have to tell you that my family hollered a lot when I did this. But they got used to it and now accept it as normal).
- For your garden, rig a hose for drip irrigation rather than spraying with a hose or sprinkler so that much of the water evaporates. Better yet, connect your gutter downspouts to barrels of some sort and use stored rainwater to water your plants.

About Trash:

- Separate trash into recyclables (aluminum glass, metal, and plastic) and organic wastes (leftover vegetable peelings and so forth).
- If your city or workplace does not have a recycling program, try to find out why not.
- Compost organic wastes: It's easy; just a frame of some sort outside into which go things like vegetable wastes, egg shells, and leaves. (*Don't* put leftover meat or bones in your compost in the city. You'll attract varmits and maybe the city health department!) After several months, take a shovel and turn the compost, or add water or preparations that make it biodegrade faster. After some months, you will have nice rich organic humus and fertilizer for somebody's flowers or garden. Some communities have community compost sites.
- When you separate your trash like this, you'll find that most of what you have left is packaging material. That suggests another dimension more important than recycling: Buy things that have less packaging, and if you can, carry them home without a bag from the store.

Real Goods:

- Buy durable or repairable goods rather than disposable goods, including disposable cups, pens, razors, and so forth. Where available, buy beverages in refillable bottles.
- Avoid red or yellow packaging; they are likely to have toxic cadmium or lead.
- Choose items that have less packaging. Store things in your refrigerator in reusable containers rather than wrapping them in plastic.
- Reduce the amount of junk mail you get by writing to Mail Preference Service, Direct Marketing Association, 11 West 42nd Street, P.O. Box 3681, New York, NY, 10163-3861.
- Share, trade, barter, or donate things you no longer need rather than throwing them away.
- Conversely, if you can find used things and they work, buy them. It used to be that thrift stores were only for the destitute. But they now attract a more diverse clientle. And a used car will cost much less than a new one. I guess it's cultural heresy among the fashionable affluent to buy it used. But if you can wait a year, you're likely to find it.

As you probably realize, such a list has no logical ending. But you get the idea.

ENDNOTES

1. Here is a graphic representation of how these categories fit together.

	Identified	*Undiscovered*	
		In known districts	In undiscovered districts or forms
Economic	Reserves	Hypothetical resources	Speculative resources
Marginally economic	Marginal reserves		
Subeconomic	Subeconomic resources		

← Increasing degree of geologic assurance

Reserves | Resources

Source: U.S. Geological Survey, 1980.

2. If you want to see some neat ghost towns, drive through the American West and visit old deserted mining towns. You can also buy a house pretty cheaply in Anaconda, Montana, if you can find work there, after the big copper smelter shut down and moved to Chile. In fact, if you want to talk with people who have some unkind things to say about the collapse of the world market for copper ore, talk with Chileans!

3. Whelan (1992), cited in Miller (1998: 621). Elizabeth Whelan was director of the American Council on Science and Health, which represents the position of the pesticide industry.
4. *Toxic* substances are fatal to humans in low doses or fatal to at least 50 percent of test animals at stated concentrations. *Hazardous* substances are harmful because they are flammable or explosive, or cause strong allergic reactions or damage to human immune systems. *Carcinogens* are cancer-causing substances, *neurotoxins* affect the nervous system, a mutagen causes inheritable damage in the DNA molecules in the genes. *Allergins* cause strong allergic reactions.
5. In the United States, for example, the Clean Air Act (1963), the Solid Waste Disposal Act (1965), the Water Quality Act (1965), and the National Environmental Protection Act (1960), which established the Federal Environmental Protection Agency.
6. Different criteria include (1) *no unreasonable risk,* as in the regulations in the Food, Drug, and Cosmetic Act; (2) *no risk,* such as the Delaney clause in that legislation, which prohibits the deliberate use of any food additive shown to cause cancer in test animals or the zero discharge goals of the Clean Water Act; (3) *risk-benefit-balancing,* such as the regulations that govern the use of pesticides; (4) *standards based on best available technology,* such as those embodied in the Clean Air Act; and (5) *cost-benefit balancing,* such Executive Order 1229, which gives the Office of Management and Budget the power to delay indefinitely, and in some cases veto, any federal regulation that is not proved to have the least costs to society. All of these criteria have been strongly criticized, both by politicians and the scientific "risk" community.
7. According to the EPA, the worst cities in terms of air pollution in 1994, were:

Extreme Pollution:
Los Angeles Basin

Severe Pollution:
San Diego
Southeast desert region, California
Ventura County, California
Chicago
Baltimore
New York–Long Island
Philadelphia
Houston
Milwaukee

Serious Pollution:
Sacramento, California
San Joaquin Valley, California
Greater Connecticut
Washington, D.C.
Atlanta
Baton Rouge, Louisiana
Boston
Springfield, Massachusetts
Portsmouth–Rochester, New Hampshire
Providence, Rhode Island
Beaumont–Port Arthur, Texas
El Paso, Texas

CHAPTER FOUR

Global Climate Change, Scientific Uncertainty, and Risk

Unlike the weather, the world's climate rarely sends clear signals. Climate is determined by the large-scale and long-term interaction of hundreds of variables—sunlight, ocean currents, precipitation, fires, volcanic eruptions, topography, human industrial emissions, and the respiration of living things—that produce a complex system that scientists are just beginning to understand, and which defies precise forecasts. Indeed, feedback relationships between the biosphere and global climate suggest that life and climate coevolved, a process in which the close interaction influenced the evolutionary paths of both systems in ways that would not have happened had they not been in each other's presence (Schneider and Londer, 1984). But the weather, in any given year, is so variable that some regions are warmer than normal, some cooler, some wetter, some drier, and many riddled with "severe weather events" like floods, droughts, and hurricanes. Almost all of these phenomena can be understood as falling within the enormous range of climatic variability. Thus trends of global climate change have been notoriously hard to detect and measure. *Until recently, that is.*

The 1990s, followed closely by the 1980s, were the hottest decades since recordkeeping began in 1866, and Antarctic ice core analysis suggests that the late twentieth century is part of a warming trend, with warmer global mean temperatures than at any time since at least 1400 C.E. Geophysical signs of this trend abound. Ice caps in the Andes Mountains are melting more quickly since the 1970s, glaciers atop European Alps have lost half their volume since 1850, and satellite radar shows that Greenland's northern ice cap is noticeably thinning. The sea ice around Antarctica has virtually disappeared since the 1950s, and average sea level has risen 10–25 centimeters in the last century as water has expanded and ice melted. So what?

To get some idea of what a generally warmer climate might mean, consider the consequences of a warm summer—the summer of 1988—that represented a dramatic spike in generally increasing global mean temperatures.

The North American corn crop was stunted by drought in the grain belt, and corn production fell below consumption (probably for the first time in U.S. history). No grain was added to the nation's reserves. Electricity use sky-rocketed as people ran air conditioners around the clock, and public agencies distributed electric fans to the elderly, for whom heat exhaustion was a significant health threat. Water levels in thoroughfares like the Mississippi River dropped so low that barges and their cargoes were stranded for weeks. Forest fires burned uncontrollably in America's great natural parks, a super-hurricane threatened the Gulf Coast, and in Asia floods in Bangladesh killed 2,000 and drove millions from their homes. That illustrates a statistically unusual clustering of severe weather that has racked the world.

In 1998, China was swept by its worst flood in three decades, displacing 56 million people from the Yangtze basin alone, and the $36 billion in estimated weather damages matches or exceeds the total weather-related losses in China since 1995—enough to put a serious dent in the nation's buoyant economic growth. Meanwhile, torrential monsoon rains cascaded down from the (deforested) Himalaya slopes and combined with ocean storm surges to keep Bangladesh under water for most of the summer, destroying the crowded country's rice crop. That same year, 54 other countries were stricken by droughts, many of which led to runaway wildfires. By early summer, scores of fires were sweeping the subtropical forests of Florida, leading to the evacuation of an entire county (Flavin, 1998b: 12). Heat radiating from warmed tropic waters created two of the most powerful and destructive storms ever to come out of the Atlantic, hurricanes George and Mitch, each with sustained winds of 180 miles per hour. Mitch, the deadliest Atlantic storm in 200 years, whacked struggling central American nations, devastating their "coffee and banana economies," and killing 11,000 people (Brown, 1999a: 18).

As I write, the records are not yet in about 1999, but newsweeklies described its late summer as saunalike conditions that gripped much of the country, from the East Coast to the Rockies (Thompson, 1999: 56). The sweltering heat left 300 people dead since mid-July, and drought emergencies were declared in scores of counties, from Montana to Maryland. The nation, and particularly the Atlantic coast from Boston to Charleston, experienced rolling power outages as utilities strained with peak loads. Crops, particularly in the East, shriveled and were written off as total losses after weeks without rain. Of course, you *cannot infer* from a specific summer that a general global warming has begun, but a generally warmer climate would increase the probabilities of increasingly severe weather events, and for widely disruptive changes in ecosystems and human societies (Silver and DeFries, 1990: 63–64). Evidence consistent with a warming pattern continues to mount.

As an environmental problem, climate change shows an important difference from most problems discussed in Chapter Three. Problems with soil, water supplies, deforestation, biodiversity, mineral resources, solid wastes, and water and air pollution do have global ramifications, but they are mainly visible as *ecosystem problems*. There are differences as well as similarities in the type and severity of these problems among ecosystems, but these are still problems that are visible *within* particular ecosystems. By contrast, atmospheric and climate problems are *biospheric problems*. As energy and matter circulate in atmosphere around the globe, their consequences effect all individuals, societies, and ecosystems, though certainly not in the same way or with the same intensity. In the first chapter I noted that the environment has complex sets of *limiting factors* that determine the success and distribution of living things on earth. The physical and chemical nature of the envelope of gases surrounding the earth—the atmosphere—is among the most important but also the most taken-for-granted of these.

Problems like climate change also have a unique phenomenology in how they are experienced, understood, and studied. Such *megaproblems* are unique in their vast scope, abstract nature, and long-time horizon over which they develop. Furthermore, they present very high-order risks in terms of their consequences. No one is really exempt from their effects, and they exemplify a negative side to the rapidly burgeoning human interdependence in the modern world. Conventional scientific inference related to such megaproblems is always contentious, since it cannot be based on experimental methodology. A pattern of climate change, for instance, cannot be conclusively demonstrated from any particular measured weather data at a particular time and place. Moreover, such megaproblems are typically remote from the concrete experience of individuals and seemingly unaffected by anything that individuals do. The very existence of such problems and their remedies are so abstract and complex that people are dependent on cadres of experts and their particular scientific (social) constructions of the problem. That means that such problems and their remedies have a peculiar counterfactual nature: If the remedies work, we will never know whether the original diagnostic claims were right. With or without remedies, the experts who make diagnostic claims are likely to find themselves branded as doomsday merchants (Giddens, 1984: 219). In the case of climate change, that social and political dynamic is obvious, even though a significant scientific consensus exists that global warming is a real phenomenon. A few scientists, however, in conjunction with some influential and powerful segments of American society maintain that global warming is a pseudo-problem, just a "lot of hot air."

This chapter is about climate change as a geophysical problem, but it is also about the problems of scientifically studying such problems, their associated risks and consequences, and alternative strategies to deal with them. Concretely, the chapter will focus on (1) *ozone depletion* in the upper atmosphere and its relationship to increasing levels of solar ultraviolet radiation, (2) *global warming* predicted to occur within the next century, and (3) a general discussion of *uncertainty and risks* in scientific analysis and policy. This seemed to me a perfect place for such a discussion of some of the complexities of science and policy issues as well as a discussion of risk in modern societies. In conclusion, (4) I will raise some thorny *ethical questions* about when and whether we know enough to act in the face of considerable uncertainty.

OZONE DEPLETION AND ULTRAVIOLET RADIATION

The destruction of significant portions of the stratospheric ozone layer graphically illustrates the unintended long-term consequences of a remarkable human technological achievement. It also illustrates how the nations of the world recognized the overshoot of a particular environmental limit, decided to back off, and gave up a profitable and useful industrial product before there was significant human or ecological damage. In that process the scientific community and the United Nations effectively communicated to governments evidence of an undeniable international problem and negotiated with them to conclude treaties about the problem. In fact, the resolution of the ozone depletion crisis shows nations, international organizations, and scientific communities at their collective best. We may have resolved the problem in time to prevent drastic damage (O'Meara, 1998).

High up in the stratosphere, twice as high as Mount Everest or as jet planes fly, is a gossamer veil of ozone with a crucial function. Ozone is made of three oxygen atoms stuck together (O_3) compared with ordinary atmospheric oxygen, which has two (O_2), and ozone is so unstable and reactive that it attacks and oxidizes almost anything it contacts. Low in the atmosphere, where it has a lot of things it can react with (including plant tissues and human lungs), ozone is a destructive but short-lived pollutant. High in the atmosphere, where ozone is created by the action of sunlight on ordinary oxygen molecules, there isn't much to react with, so the ozone layer lasts a long time. But there is enough ozone to absorb much of the most harmful ultraviolet wavelength from incoming sunlight (UV-B), which tears apart organic molecules that make up all living things. In humans it can produce cornea damage, reproductive mutations, and skin cancer while suppressing the immune system's ability to fight cancer. It damages single-celled organisms and could damage floating microorganisms (plankton) that are at the base of ocean food chains. Exposure to UV-B light stunts the growth and photosynthesis of green plants; in two-thirds of the crop plants that have been studied, crop yields go down as UV-B goes up. The ozone layer is in fact a stratospheric sunscreen that protects humans and ecosystems from damage in ways that are difficult to predict (Meadows et al., 1992: 141–147).

Destroying the Ozone Layer

In 1974, two scientific papers published independently stated that chlorine atoms in the stratosphere could be powerful ozone destroyers and that chlorine atoms could be increasing as chlorofluorocarbon molecules (CFCs) reach the stratosphere and break up to release them. Their hypothesis was controversial but treated seriously enough by nine countries, which banned the use of CFCs in spray cans in the late 1970s. The first unmistakable sign of the destruction of stratospheric ozone arrived in 1985, when a team of British scientists published findings that stunned the world community of atmospheric scientists. They presented evidence that between 1977 and 1984 the concentration of ozone above Antarctica had plunged more than 40 percent below the 1960 baseline measurements of the southern hemispheric spring season. Ground-level ozone measurements had not hinted at the decline, but the stratospheric depletion was confirmed by analyzing data from NASA satellites and a 1986–87 Antarctic scientific expedition of the U.S. National Oceanic and Atmospheric Administration (NOAA).

CFCs, widely used as solvents, refrigerant chemicals, and in the production of plastic "foam," were manufactured mainly in Europe and North America, but they were mixed throughout the lower atmosphere so that there are as much CFCs over Antarctica as over Colorado or Washington, D.C. Researchers surmised that upon reaching the stratosphere, CFCs encounter high-energy ultraviolet light, which breaks them down, releasing their chlorine atoms. These then engage with ozone in a catalytic reaction in which each chlorine fragment converts ozone to ordinary oxygen. But through a series of reactions, each chlorine atom can cycle through this process many times, destroying one ozone molecule each time and becoming like the "Pac-Man of the higher atmosphere, gobbling one ozone molecule after another and then being regenerated to gobble again" (Meadows et al., 1992: 148). Each chorine atom can destroy up to 100,000 ozone molecules before it is finally removed from the atmosphere. Chemicals thought most dangerous (CFC-11, CFC-12, and CFC-113) were increasing in the atmosphere by between 5 percent and 11 percent annually.

By the late 1980s, there was virtual agreement among the scientific community that CFCs were responsible for Antarctic ozone depletion. The most severe ozone depletion was limited to the Antarctic because the reaction requires the cold temperatures, stratospheric ice crystals, and sunlight characteristic of the early Antarctic spring and also because the circulation of winds (the *polar vortex*) tends to trap the depleted ozone over the Antarctic for several months. Less severe but record ozone losses have also occurred over the populous and agriculturally abundant mid to high latitudes of both hemispheres. Scientists speculate that increases in sulphurous particles, water vapor, and various pollutants in the stratosphere may provide material surfaces for the ozone-depleting reactions to take place much as ice crystals do in the Arctic and Antarctic (O'Meara, 1998: 70; Silver and DeFries, 1990: 103–112; Stern et al., 1992: 57–59).

A Cautionary Tale: Technology, Progress, and Environmental Damage

Here's a brief detour from the physical facts of the problem into its social and historical contexts. In the first chapter, I argued that underlying modern environmental problems were the economic, social, cultural, and technological issues. Following is a dramatic example related to ozone depletion. It is also a classic illustration about how undeniable progress can result in unanticipated long-run problems. To really understand the causes of ozone depletion, you need to reach back through a century's history, long before CFCs were invented (the following discussion relies heavily on Stern et al., 1992: 54–59).

Until almost the end of the nineteenth century, refrigerating food and drink depended on ice from natural sources that was chopped from local ponds and stored in warehouses or pits for use in the summer. Households used this ice, but breweries and restaurants were the heaviest users, and stored winter ice was sometimes shipped hundreds of miles to provide refrigeration (Boston ice merchants shipped ice as far as South Carolina and the Caribbean). Because this system of using stored winter ice was difficult and expensive, most food was preserved by chemical additives (most commonly salt, sodium chloride). Pork became the most popular meat because its decay could be easily arrested by salt. Preserved beef was much less popular, and those who ate beef preferred to buy it freshly slaughtered from local butchers. To increase their profits, meatpackers began experimenting in the 1870s with ice-refrigerated railway cars to ship dressed beef, slaughtered and chilled in Chicago to consumers hundreds of miles away. Soon this new ice storage and delivery technology was used to ship fruits and vegetables from California and Florida and dairy products from urban hinterlands to remote customers. This technology drastically lowered the rate at which food spoiled and made perishable crops available to consumers through much of the year. Eventually refrigeration changed the whole nature of the American diet. But natural ice was unreliable, and in two warm winters (1889 and 1890) the failure of the natural ice crop encouraged the packers to seek more reliable forms of refrigeration.

The principle of mechanical refrigeration—by which compressed gas was allowed to expand rapidly and lower temperature—had been known since the mid-eighteenth century. But mainly urban brewers used the first commercial adaptation of this process in the late nineteenth century. These early refrigerant systems used ammonia, sulfur dioxide, or methyl chloride as refrigerant gases, but they had serious problems. For efficiency, they required high pressures and powerful compressors, which increased the risk of equipment failures and explosions. They were toxic gases that caused a number of deaths. Toxicity and the need for expensive compressors kept mechanical refrigeration from making headway with retail customers, who represented a huge potential market. Which led Thomas Midgely, working for General Motors Frigidaire division, to develop in 1931 a new chlorinated fluorocarbon (CFC), patented as Freon 12, as a perfect alternative to existing refrigerant gases. Freon was chemically stable, nonflammable, nonexplosive, nontoxic, and required less pressure to produce the cooling effect.

Because smaller compressors were required, American consumers could soon own their own "refrigerators," making it possible to sell chilled foods in retail-sized packages. Frozen foods were marketed in the 1950s, as were the fresh vegetables and dairy products that became rapidly accepted as ordinary parts of the American diet. Europeans followed Americans in adopting these technologies.

Equally important, the properties of Freon made it possible for the refrigeration technology to be applied to space cooling in buildings, thus creating another important market for it. Air conditioning became common to offices and finally to residences. This development had an enormous impact on the American social pattern. Air conditioning promoted urban growth in the American sunbelt—from Florida to California—and in tropical regions around the globe. For many Americans, it would be difficult to envision life in the summer months or warm climates without air conditioning in their homes, autos, stores, and offices. It shifted the peak use of electricity from the winter (when its use for lighting and space heating peaked) to the summer, when air-conditioning systems use electricity at unprecedented rates. From the 1950s, the sales of CFCs were increased by other uses: as nontoxic propellants in aerosol sprays and as solvents for the manufacture of integrated electrical circuits. Taken together, these technologies had an enormous impact on improving the nutrition, comfort, and physical quality of life for many people. But the very *stability* of CFCs that made them so useful ultimately proved to be their greatest environmental hazard. As they leaked from refrigerators, air conditioners, and spray cans at an ever-increasing rate, they eventually found their way to the stratosphere, where they encountered ozone. The problem with ozone depletion was a direct but long-term consequence of a social pattern—the technical innovations, search for profitable markets, the residential, consumption, and lifestyle patterns and expectations of people—that evolved in the MDCs.

A Happy Ending?

Even with the scientific consensus about the relationship between CFCs and ozone deterioration, little would have happened without the United Nations Environmental Program, which hosted and prodded the international political process. Its staff assembled and interpreted evidence, created a neutral forum for high level discussions, and patiently reminded all nations that no short-term selfish consideration was as important as the integrity of the ozone layer. In consequence, the DuPont company, which produced 25 percent of the world's CFCs, declared its intent to phase out CFC production and search for more environmentally benign refrigerant chemicals. In 1987, 49 signatory nations to the *Montreal Protocol* announced their intention to cut CFC production and consumption by 50 percent by the year 2000. An even more stringent protocol was subsequently signed in London. Electronics firms are coming up with substitute solvents for cleaning circuit boards, insulating plastic foam is being blown with other gases, hamburgers are being wrapped in paper or cardboard, and many consumers are returning to washable ceramic coffee cups instead of foam ones. The costs of complying with the protocol, estimated to be $235 billion, are significantly less because many of the industries that relied on CFCs saved money by redesigning manufacturing processes or using simpler, cheaper, substances. The Montreal Protocol will save the world some 19.1 million cases of skin cancer through 2060 and at least $459 billion in damages to fisheries, agriculture, and plastics, according to a study by Environment Canada, a government agency (O'Meara, 1998: 70; Meadows et al., 1992: 155–58).

Stratospheric ozone will continue to deteriorate for some time because CFCs remain in the atmosphere a long time, and the new chemicals with be phased in over time as the existing stock of auto, home, and office air conditioners continue to leak. Ozone layer destruction hopefully peaked in 2000 and should slowly start getting thicker and better able to block UV-B radiation. In sum, dealing with the ozone problem represents a model for addressing environmental problems involving scientific consensus and its interpretation for policy, international mediation, responsible political and corporate behavior, and public education. Yet there are still reasons for concern about the solution and its broader applicability to climate problems.

First, it epitomizes a "technological fix": a solution to human-environmental problems that merely substitute one category of chemicals for others, supposedly more environmentally benign. Ironically, you need to remember that for over fifty years CFCs seemed like the perfect benign technical and engineering solution with no negative side effects whatsoever. *Second,* as a technological fix, it requires little change in our social pattern or behavior and may remove a barrier to the spread of current MDC energy-intensive architectural, food production, and transportation systems around the planet. Why is this of concern? Because many scientists think that the spread of industrial society lifestyles and consumption patterns is unsustainable on a global basis and will certainly amplify other environmental problems. *Third,* and most probably most important, the encouraging resolution of the ozone depletion problem may depend upon special circumstances not applicable to many other environmental problems. There were only about two dozen CFC producers worldwide, and banning production threatened few existing firms or long developed technical infrastructures. So the Montreal Protocol is a risky predictor for how quickly and expeditiously other international negotiations may turn out. Even if there is scientific consensus, such changes will be much more difficult (1) if the need for change requires greater alterations in social behavior and lifestyle expectations, (2) when there are many millions of responsible actors, or (3) when the costs and benefits of change are less evenly distributed around the planet (Stern et al., 1992: 59). By these criteria, the predicted problem of global warming will be more difficult to address.

TURNING UP THE HEAT: GLOBAL WARMING?

Gases in the atmosphere play a critical role in trapping enough infrared solar radiation (heat) to keep the mean temperature of the earth fluctuating within relatively narrow limits that make life possible. The most important of such gases, present in trace amounts, are water vapor, carbon dioxide (CO_2), tropospheric (low altitude) ozone, methane, CFCs, and nitrogen oxides (NO_x). Water vapor and CO_2, the most important of these, account for probably 90 percent of the heat-trapping capacity (Miller, 1998: 365–367). The role of such gases in maintaining the temperature of the earth has been known for more than 150 years. Fourier was probably the first to discuss the heat-trapping role of CO_2 in 1827, from which it was dubbed the *greenhouse effect* because he compared it to the warming of air isolated under a glass plate. Remember returning to your car on a sunny day with the windows all rolled up? In principle, the greenhouse effect explains the very cold climate of Mars, where water vapor, a highly efficient greenhouse gas, is virtually absent, as well as the hot climate of Venus, where the atmosphere is so thick with CO_2 and conditions are so hot that life—as we know it—could not exist (Silver and DeFries, 1990: 64).

After water vapor, CO_2 is the most plentiful and effective greenhouse gas. It occurs naturally as a consequence of the respiration of living things (remember the carbon cycle discussed in Chapter One?). But CO_2 is also produced in great quantities by the burning of fossil fuels—natural gas, petroleum, and particularly coal. Other greenhouse gases are much rarer, but molecule for molecule they trap more heat than CO_2. Methane, also known as *natural gas*, is produced through bacterial activity in bogs and rice paddies and in the digestive tracts of ruminant animals (cows, sheep). Most atmospheric methane is from biological sources, although some is produced from decaying human garbage dumps. CFCs, discussed earlier, not only destroy stratospheric ozone, but are effective greenhouse gases at lower levels, trapping 17–20,000 times as much heat per molecule as CO_2. Nitrous oxide (N_2O) is produced naturally through microbes in soil and in the burning of timber, the decay of crop residues, and the combustion of fossil fuels.

Speculations about the implications of anthropogenic increases in greenhouse gases are not really new. At the turn of the twentieth century, Swedish naturalist Arrhenius argued that increasing concentrations of CO_2 would raise the global mean temperature. In 1941 Flohn noted that anthropogenic CO_2 perturbs the carbon cycle, leading to a continual CO_2 accumulation in the atmosphere, and in 1957 Revelle and Suess concluded that "human activities were initiating a global geophysical experiment that would lead to detectable climatic changes in a few decades" (cited in Krause et al., 1992: 11). Also in 1957 the systematic measurement of CO_2 began at the Mauna Loa (Hawaii) observatory and at the South Pole. In 1979, the World Meteorological Organization convened a World Climate Conference in Geneva to discuss the issue. Following this conference were a host of national meetings about climate change issues that led to the first meeting in 1988 of the Intergovernmental Panel on Climate Change (IPCC), sponsored by the United National Environmental Program and the World Meteorological Organization.

In sum, the greenhouse effect and the possibility of global warming has been known for more than a century, but only in the last two decades has this threat begun to be taken seriously, and only in the 1990s did questions about "preventative" policy measures enter the international political arena (Krause et al., 1992: 15). There are some things about the threat of global warming for which a fairly strong consensus exists among climatologists as well as many loose ends and missing pieces of the puzzle of climate change. In addition to scientific uncertainties, there are *many* uncertainties about the costs, benefits, and wisdom of trying to do anything about the problem. I will return to these critically important issues of scientific and policy consensus and dissensus, but first I think you need a bit of background about how climatologists study these issues.

General Climate Models

All of our understanding about greenhouse gases and climate change are based on *general climate models* (GCMs) by which climatologists try to construct mathematical models to represent or "simulate" the complex workings of the earth-atmosphere interactions. As you might guess, these global interactions are very complex and involve many feedback loops that are only imperfectly understood. So like all models, GCMs represent only a vastly simplified version of the real world. Despite this fact, these computerized mathematical models that predict the ways in which temperature, humidity, wind speed and direction, soil moisture, sea ice, and other climate variables evolve through three dimensions and over time are the only tools available for understanding global climate change. There are five GCMs, and they are not alike: Some include scenarios of gradual addition of greenhouse gases into the atmosphere, whereas others assume massive, one-time doubling of the gases; some attempt to include the effects of the circulation of ocean currents, which others ignore.[1]

In addition to the mindboggling complexity of the global climate system, such GCMs are limited because the power of computers is not sufficient to handle the calculations required to understand all significant climatological variables. For example, they do not have the power to depict particular areas. In 1993, models could handle measurements for "grids" at about 300 miles (or 5° latitude) intervals around the globe. At that spatial resolution, particular regions—Japan, for example—do not exist, and the model cannot with certainty account for the role of forests or clouds, which typically occur in smaller areas. If the resolution were reduced to grids of 60 miles on a side, climatologists could get a much clearer picture of the consequences of global warming for the temperature, water supplies, agricultural implications, and so forth, in particular ecosystems. But this would require much more powerful computers and great expense, and it would not be feasible for at least two decades (Silver and DeFries, 1990: 73–75).

With their limitations, how can climatologists assess how well the GCMs simulate the dynamics of the climate system? In three ways: (1) by starting with existing data, the model can be "run forward" to simulate "today's climate," especially the large temperature swings of the seasonal cycle, (2) by determining whether the model can realistically simulate an individual physical component of the climate system, such as cloudiness, and (3) by running the model backward in time to see whether it can reproduce the long-term changes climate of the ancient earth (about which, surprisingly, a great deal is known.[2])

The performance of the five models by these methods has been constantly appraised. With what result? According to Stephen Schneider, a climatologist at the National Center for Atmospheric Research, the success of different models to have similar results show that the models are getting better at predicting climate change (1990a: 74). By now, there is strong scientific agreement that in the present century as well as throughout the earth's geological history there has been a strong *positive correlation* between atmospheric concentrations of greenhouse gases and fluctuations in the earth's mean temperature (see Figs. 4.1 and 4.2). Furthermore, by 1992 there was "virtually no debate in the scientific community that continuing rises in the atmospheric concentrations of CO_2 and other greenhouse gases will lead to global warming (Kraus et al., 1992: 4, 28; see also CDAC, 1983). Neither is there any real debate that global mean temperature has risen sharply in the twentieth century. According to the Intergovernmental Panel on Climate Change (IPCC), the official scientific body representing the world's climatologists, "Global mean surface air temperature has increased by between

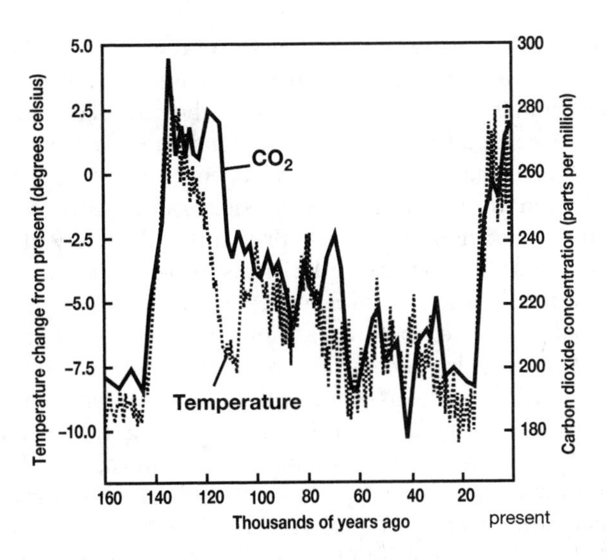

Figure 4.1 Carbon Dioxide and Temperature, Long-Term Record

Source: S. H. Schneider, "The Changing Climate," in *Managing Planet Earth*, p. 29. Copyright © 1990, W. H. Freeman and Company. Used with permission.

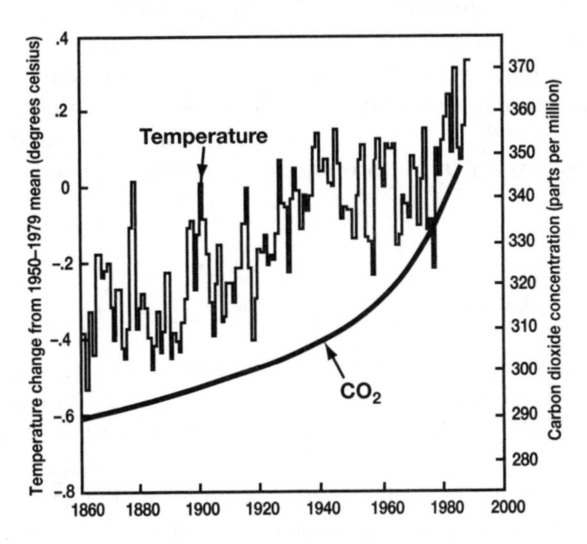

Figure 4.2 Carbon Dioxide and Temperature, Industrial Era

Source: S. H. Schneider, "The Changing Climate," in *Managing Planet Earth*, p. 29. Copyright © 1990, W. H. Freeman and Company. Used with permission.

about 0.3 and 0.6° C since the late nineteenth century. Since 1990, re-analyses have not changed this range of estimated increase . . . [and] . . . global sea level has risen by between 10 and 25 cm over the past 100 years, and much of the rise may be related to the increase in global mean temperature" (IPCC, 1999). Similar consensual statements have been published by the American Geophysical Union (representing geophysicists and earth scientists), the National Academy of Sciences, and the British Royal Society of London (the world's oldest scientific organization). For instance, in 2000, an independent panel of the National Academy of Sciences National Research Council said that "with global surface temperatures rising at an accelerated rate in the last decades, climate change is *undoubtedly real*" (*Ecology USA*, 2000: 26). There is also strong scientific consensus—but still some disagreement—that while there are natural "cycles" of warming and cooling that occur in about 100,000-year intervals, this century's rapid increase in gases and rise in mean global temperatures is substantially anthropogenic and *not* primarily caused by geophysical "long cycles" of cool glacial and interglacial warm periods (Krause et al., 1992: 27). According to the IPCC, "the observed increase over the last century is unlikely to be entirely due to natural causes," and "the balance of evidence suggests a discernable human influence on global climate" (IPCC, 1999). Studies published in 1998 and 1999 in prestigious journals *Science* and *Nature* both concluded that human activity has, in fact, been a significant cause of the warming that has taken place in the twentieth century, and particularly since the mid-1970s, even when accounting for the natural influence of sunspots and volcanic activity (Tett et al., 1999: 569–572; Wrigley et al., 1998). Putting it in terms of probabilities, Klaus Hasselmann, director of the Max Planck Institute for Meteorology in Hamburg, Germany, says that there is a 95 percent chance that the rise in temperature over the past century is caused by anthropogenic greenhouse gases (Miller, 1998: 369).

More and Less Uncertainty

What do these findings really mean for the human future? How likely are greenhouse gases to continue to rise? How much? One, 5 or 8 degrees? And how soon? Fifty, 100, or 200 years? And what would a temperature rise of this magnitude mean for humans and ecosystems? Beyond global averages, how would the effects be experienced in *different* regions and ecosystems? These are contentious questions, but most fractious of all are the policy questions: Do we know enough to act, and if so, what should be done? And when? Scientists can't answer many of these questions with certainty because there are still too many missing pieces of the puzzle. Here are some of the more important ones.

- While the correlation between greenhouse gas concentration and temperature fluctuation works well for geological history, it works less well for shorter time spans, particularly in the current period for which we have the best data. Greenhouse gases have increased rather steadily since the turn of the century, but temperature increases have not: Most warming occurred in the 1920s and the 1970s and, in fact, during the 1940s and 1950s the world cooled slightly! Given the current theory, this should not have happened and has been compared to a "murder mystery in which the whereabouts of principal suspects are unknown"(Schneider, 1990b: 34–35).
- There are other important missing suspects. Among these is the missing *carbon sink*. About 45 percent of the total anthropogenic CO_2 emissions since preindustrial times are unaccounted for ("whereabouts unknown"). Scientists believe that the ocean helps moderate tropospheric temperature by removing about 29 percent of the excess CO_2 we pump into the atmosphere, but they don't know if they can absorb more. If the oceans warm significantly, they may no longer be able to act as great buffer systems (Miller, 1998: 371).
- Beyond their role as carbon sinks, the role of the oceans in the warming process is largely unknown with current methodologies, but it is likely to be large. The oceans store most of the planet's heat and CO_2 and have deep circulation that is not well known or modeled. Their enormous mass will act as a thermal sponge slowing any initial increase in global warming while the oceans themselves heat up, but the magnitude of this increase in temperature will depend on ocean circulation, which may *itself* change as the earth warms (Schneider, 1990c: 31).
- Similarly, the inability of GCMs to factor in effect of vegetation and forests means ignoring their effect on ground surface *reflectiveness* (or *albedo*), their function as carbon sinks, or the significance of their release of water vapor and cloud formation.
- Likewise, the interactions between temperature change and cloud formation and the resulting feedbacks are unpredictable. Will heating of the atmosphere create more or fewer clouds? And would more clouds trap more heat at the earth's surface or reflect more solar radiation into space?
- Most scientists believe that warming the atmosphere would melt polar ice caps, causing sea levels to rise. But that effect is not beyond question. It has also been hypothesized that warming the air would accelerate air circulation and polar precipitation, snowfall, and the formation of polar icepacks. In fact, in 1991 one research team documented a significant accumulation of ice in eastern Antarctica since 1960 (Sullivan, 1991).
- The interaction between temperature change and the photosynthetic processes of plants and the resulting feedback mechanisms is unclear. Warmer temperatures are known to accelerate plant growth and hence the absorption of CO_2 *from* the atmosphere. Would that be sufficient to dampen global warming at some point? Or would greater cloud cover block enough sunlight to retard plant growth even in warmer conditions? Flip a coin!
- Perversely, the human production of pollution, smog, and soot may act to absorb some of the radiation that would warm the atmosphere. Whether these effects would be large enough to be significant no one knows.

- In sum, *human-environment interactions are complex and include many nonlinear relationships and feedback mechanisms.* Given the current imperfect state of knowledge of these complex system connections, our ability to predict the timing and magnitude of global warming is impaired, and in particular the more concrete changes in windpatterns, rainfall, and humidity that would differentially effect regions and ecosystems (Schneider, 1990a; National Academy of Sciences, 1991: 88–94).

With all of these uncertainties and unknowns, you may be wondering how one can have any confidence in the threat of future global warming. If so, you have some respectable scientific company. But just enumerating the unknowns and anomalies understates the degree of consensus in the climatological scientific community about the global warming and its probable consequences. That has have been summarized by the National Research Council and other policy groups (National Research Council, 1987; Silver and DeFries, 1990; Krause et al., 1992: 28–29; National Academy of Sciences, 1991: 94). Here are some conclusions, arranged from the virtually certain to the uncertain. *Virtually certain* means that there is nearly unanimous agreement within the scientific community that a given climatic effect will occur. *Very probable* means greater than about a 90 percent (9 out of 10) chance, and *probable* implies more than about a 67 percent (2 out of 3) chance. *Uncertain* refers to hypothesized effects but for which there is a lack of appropriate modeling or observational evidence.

- *Large stratospheric cooling* (virtually certain). Upper atmosphere destruction of ozone by chlorine and other gases will markedly increase the loss of infrared radiative heat in the upper stratosphere.
- *Global mean surface warming* (very probable). For a doubling of atmospheric CO_2 (or its equivalent from all greenhouse gases), which is expected sometime during the next century, the long-term global mean surface warming is predicted to be in the range of 1.5–5.0 (centigrade)°. Currently the most widely used models predict a narrower range of 3–5.5°, but when a broader range of possible feedback effects is considered the average warming from doubled CO_2 could be as high as 6.3–8° or more (Dickinson, 1986; Lashof, 1989). The most important uncertainty arises from the difficulties in modeling the feedback effects of clouds, and the actual rate of warming over the next century will be governed by the slowly responding parts of the climate system, such as the oceans and glacial ice.
- *Global mean precipitation increase* (very probable). Increased heating of the earth's surface will lead to increased evaporation and subsequently to greater global mean precipitation. Nonetheless, precipitation may well decrease in many individual regions, which would be hotter and dryer.
- *Northern polar surface warming* (very probable). Winter surface temperatures in polar regions would be much warmer than they are now (three times the global mean warming), with a greater fraction of open water and thinner sea ice as well as a probable reduction in sea ice.
- *Northern high-latitude precipitation increase* (probable). The increased poleward penetration of warm, moist air may increase the annual average precipitation

and river runoff in high latitudes (e.g., such as in northern Canada and Scandinavia).

- *Summer continental dryness/warming* (probable). Soil moistures in the mid-latitude continental interiors may decrease during summer, caused mainly by an earlier termination of snowmelt and rainy periods and thus an earlier onset of the normal spring-to-summer reduction of soil moisture.
- *Rise in global mean sea level* (probable). Sea level is likely to rise as seawater expands in response to the warmer future climate. Far less certain is how much this will be affected by possible melting of glaciers.
- *Regional vegetation changes* (uncertain). Climactic changes in temperature and precipitation must inevitably lead to long-terms change in surface vegetation, but the exact nature of these and how they in turn might affect climate are uncertain.
- *Tropical storm increases* (uncertain). A warmer, wetter atmosphere has been hypothesized to lead to more frequent and more intense tropical storms, such as hurricanes. But this effect has not been satisfactorily addressed in the coarse-resolution climate models because tropical disturbances are relatively small.

To grasp the enormity of these probable changes, you need to compare them to the climate history of the earth. A global average warming of 1.5° would represent a climate not experienced since the beginning of agricultural civilization some 6,000 years ago; 3–5° would represent a climate not experienced since human beings appeared on the earth some 2 million years ago. The last time the earth was this warm was in the Pliocene period (some 3–5 million years ago), and more than 5° warming would mean a climate not experienced since the Eocene period (40 million years ago), before the evolution of birds, flowering plants, and mammals, when there were no glaciers in the Antarctic, Iceland, and Greenland (Krause et al., 1992: 28). Furthermore, the projected rate of warming is 15 to 40 times faster than the "natural" warmings after the major ice ages and much faster than what most species living on the earth today have ever had to face. Warming could far outstrip the ability of ecosystems to adapt or migrate (Silver and DeFries, 1990: 71). A several-degree warming over a 100-year period would greatly exceed that natural rates of change, pushing forests poleward by 2.5 km per year, compared with the less than 1 km per year migration of even fast migrating tree species (CEC, 1986). The result would be a rapid dieback while new species take root much more slowly.

Impacts on Society

A growing body of empirical research is focused on how climate change might affect human societies, although with the uncertainties just mentioned, much of it is highly speculative. Let me just mention a few issues being addressed in this research:

1. *Food security*: Many crop yields are delicately dependent on a particular mix of temperatures, soil conditions, and rainfall patterns that could be disrupted by global warming. High latitude regions that could—in principle—become available for agriculture may not provide such favorable conditions. If the mid-latitude continental dryness materializes, the world's "breadbasket" (such as the U.S. Midwest, and the Ukraine and Kazakstan regions in Asia) would suffer a 50 percent drop in grain productivity. Heat stress could also severely reduce the productivity of Asian "rice bowl" regions. Reduced yields and less-than-needed yield improvements, combined with growing population and higher food prices, could seriously jeopardize the world's food security.
2. *Regional impacts*: The amount of warming is expected to be greatest in the northern latitudes, harming some growing areas and expanding others, but computer models suggest that the effect on crops is likely to be more uniformly severe in the southern latitudes. In short, the impact is likely to increase the price of grain on a worldwide basis, but the impact is likely to be especially severe in the LDCs (largely in the southern hemisphere) than in the MDCs (largely in the northern hemisphere). Similarly, increases in greenhouse gases will have the least damaging effect on social well-being in the MDCs, which have far greater technical and economic resources to respond and whose living standards are less dependent on fossil fuels than people of the LDCs (which include 70 percent of the world's people).
3. *Land use and human settlements*: A modest rise in sea level would threaten the coastal settlements in which *half* of humanity lives. They include Boston, New York, Miami, New Orleans, Los Angeles, Seattle, and Vancouver as well as Tokyo, Osaka, Manila, Shanghai, Guangshou, Calcutta, Lagos, London, Copenhagen, and Amsterdam. The entire Maldives Republic and much of Bangladesh, Indonesia, the Netherlands, and Denmark would be under water. Rich farmland in river deltas would be lost, salinity would move upstream, and high tides and storm surges would penetrate further inland. The economic costs of adapting to this change—of population relocation and protecting coastal infrastructures—would be enormous, with cumulative costs of billions for MDCs and probably prohibitive costs for LDCs. In the United States, for instance, economic analyses estimate that a 50-cm rise in sea level by the year 2100 would cost between $20.4 and $138 billion in lost property and damage to economic infrastructures (Alexander et al., 1997: 86). A one-meter rise would flood most of New York City, including the entire subway system and all three major airports. According to the OECD, a one-meter rise could cost the global economy $910 billion by 2100 (McGinn, 1999: 88).
4. *Freshwater supplies*: Global warming would reduce stream flows and increase pressure on groundwater while worsening the pollution discharge into smaller flows. This effect could exacerbate the world's existing water problems, which, as previously noted, are substantial.
5. *Planning uncertainty*: In the planning of human resettlement, flood control, revamping agriculture for changing growing seasons, society might find itself in a constant treadmill, trying to catch up with perpetual change in an environment that is changing rapidly, unpredictably, and differently in different regions.
6. *Other impacts*: Global warming could involve increased human health risks as a result of heat stress and more vigorous transmission of tropical diseases over

larger areas; increased in energy consumption for air conditioning, losses in hydropower availability; and losses in revenue from tourism and fisheries. (For more detailed reviews of the social consequences of warming, see Alexander et al., 1997; Cairncross, 1991; Kates et al., 1985; Parry, 1988; Rosenzweig and Parry, 1993; Rosa and Krebill-Prather, 1993; and Smith and Tirpak, 1988, which specifically focuses on the United States.)

As you can see, there is a broad consensus among the scientific community that global warming *is* a real threat but also that the many concrete questions about how much, when, and with what effects on which regions are very much open to question.

Policy Options: What Can Be Done about Global Warming?

Faced with a significant but still "iffy" array of threats, human responses could fall into three types: *adaptation, mitigation,* and *geoengineering*.

Those who urge *adaptation* believe that the large uncertainties in climate projections make it unwise to spend large sums trying to avert outcomes that may never materialize (Schneider, 1990a: 34). They believe that human systems can adapt to climate change much faster than they occur. Those advocating adaptation do not eschew all active policies (such as anticipating flood control or water supply problems), but they generally argue that human individuals, organizations, and communities will quickly adjust to such changes so that much organized governmental responses will be superfluous and unnecessary. This is a favorite argument of neoclassical economists. They maintain that while the projected doubling of CO_2 will take place over the next century, financial markets adapt in minutes, labor markets in several years, and the planning horizon for significant economic and technological change is at most two or three decades (Stern et al., 1992: 110). So there is plenty of time to adapt to whatever happens. A panel of the National Academy of Sciences has ranked the ease and costs of adaptation in various areas. They determined that in the areas of industry, energy supply, and human health, adaptation to global warming could take place quickly and affordably and that, for example, machinery, buildings, and energy systems can be renewed and modified much faster than the pace of global warming. The same, they argue, is true of medical technology and disease control. Managing agriculture, forestry, and water resources could be done, but at considerably higher costs. Managing the relocation of coastal settlements, flood control systems, and human migration flows would also be feasible but at great costs. Providing for political stability would also have considerable costs. All costs of adapting to global warming would be more easily born by the MDCs than by LDCs (National Academy of Sciences, 1991: 43–44).

The real issue here is not climate change per se, but how rapidly it will occur. If moderate change takes place gradually over several hundred years, adaptation is a perfectly feasible and adequate response. But if the projected warming takes place over several decades, or all within the next century, the costs will be substantial and accompanied by considerable social, economic, and political turmoil. Such relatively rapid changes in climate would also reduce the earth's biodiversity because many species couldn't adapt (Miller, 1998: 374–375).

Mitigation means curtailing the greenhouse gas buildup to prevent, minimize, or at least slow global warming. Advocates of taking action now to mitigate warming argue that because of the time lags in the global environmental system, it may be too late to prevent catastrophe by the time it becomes clear that a response is needed. Even if catastrophe is unlikely, mitigation that slows the rate of change means that successful adaptation would be easier and less costly. They argue that mitigation actions begun now allow for more modifications in process, and even blunders, than if begun at a later time when the situation may be more critical. Mitigation is like insuring against disaster: The costs of the "premiums" are onerous though bearable, but the costs of a disaster (fire, flood) may not be. It seeks to avoid the high-risk uncontrolled experiment now taking place with the global environment (and we only have one global environment on which to experiment). Furthermore, the advocates of mitigation believe that the economic arguments (which underpin the rationale for adaptation) are specious in the general case. This is because the costs and benefits of postponing action are not always comparable. If current economic activity destroys the life support systems on which humans depend, what future market adjustments or investments could ever recoup this cost? Neither do economic arguments include some environmental goods (such as biodiversity), which have both economic and intrinsic or spiritual benefits that people value. Furthermore, as I mentioned in earlier chapters, economic accounting undervalues "common property," which cannot be privately owned (the vast atmosphere is a case in point), and for which prices and property rights are fictitious and only potential (Stern et al., 1992: 111–113).

Mitigation strategies would curtail greenhouse gas buildup by various energy conservation measures, alternative energy sources such as a switch from coal and petroleum fuels to natural gas and other fuels with a lower CO_2 content and by stopping the production of CFCs altogether. All of these would reduce or slow the atmospheric accumulation of CO_2, as would reforestation programs. A wide variety of mitigation techniques have been suggested. As with adaptation, a panel of the National Academy of Sciences has collected and categorized them. Table 4.1 lists a sample of such proposals, and Table 4.2 compares the net implementation costs and potential for reducing greenhouse gas emissions for the United States.

The third type of policy options are *geoengineering strategies*, which use technical measures to counteract climate change. Proposals have included several ways of reducing temperature increases by screening sunlight (e.g.,

Table 4.1 Sample Mitigation Options

Residential and Commercial Energy Management	
White surfaces and vegetation	Reduce air condition use and the "urban heat island effect" by 25% by planting vegetation and painting half of the residence roofs white.
Residential water heating	Improve efficiency by 40–70% by efficient tanks, increased insulation, low-flow devices, and alternative water heating systems.
Residential appliances	Improve efficiency of refrigeration and dishwashers by 10–30% by implementation of new appliance standards for refrigeration and no-heat drying cycles in dishwashers.
Residential space heating	Reduce energy consumption by 40–60% by improved/increased insulation, window glazing, and weather stripping along with increased use of heat pumps and solar heating.
Residential and commercial lighting	Reduce lighting energy consumption by 30–60% by replacing and incandescent lighting with compact fluorescent bulbs; use reflectors, occupancy sensors, and daylighting.
Industrial Energy Management	
Co-generation	Replace existing industrial energy systems with an additional 25,000 MW of cogeneration plants that produce heat and power simultaneously.
Fuel efficiency	Reduce fuel consumption up to 30% by improving energy management, waste heat recovery, boiler modification, and other industrial enhancements.
New process technology	Increase recycling and reduce energy consumption primarily in the primary metals, pulp and paper, chemical, and petroleum refining industries by new, less energy-intensive process innovations.
Transportation Energy Management	
Vehicle efficiency (autos)	Use technology to improve fuel economy to 25 mpg with no changes in existing fleet, and to 36 mpg by gradually downsizing existing fleet (to 33 and 47 mpg, respectively, in CAFE terms).[a]
Alternative fuels	Over time, replace gasoline vehicles with those that use methanol produced from biomass, hydrogen created from solar electricity, or fuel cells.
Transportation demand management	Reduce solo commuting by eliminating 25% of employer-provided parking spaces and placing a tax on the remaining spaces.
Electricity and fuel supply	Replace fossil fuel-fired plants with ones powered by either hydroelectric or energy alternative sources (geothermal, biomass, solar photovoltaic, or solar thermal sources). Collectively could account for 13 quads of energy, or about half of the energy used by U.S. electric utilities.[b]
Nonenergy Emission Reduction	
CFCs	Find benign substitutes, alter production, and gradually retrofit existing stock of refrigerators, air conditioners, etc.
Agriculture	Eliminate all paddy rice production; reduce ruminant animal production by 25%; reduce nitrogenous fertilizer use by 5%.
Landfill gas collection	Reduce landfill gas generation by 60–65% by collecting and burning in a flare or energy recovery system.
Reforestation	Reforest 28.7 million Ha of economically or environmentally marginal crop and pasture lands to sequester 10% of U.S. CO_2 emissions.

[a]CAFE, or "corporate average fuel economy."
[b]1 quad-1 quadrillion BTUs (British Thermal Units)-10^{15} BTUs. Data about U.S. electric utility energy consumption from U.S. Energy Information Administration, cited in Craig et al., 1988: 75.

Source: National Academy of Sciences, *Policy Implications of Greenhouse Warming*, pp. 54–57. National Academy Press. Copyright © 1991. Used with permission.

Table 4.2 Comparison of Selected Mitigation Options in the United States

Mitigation Option	*Net Implementation Cost*[a]	*Potential Emission*[b] *Reduction ($t\ CO_2$ equivalent per year)*
Building energy efficiency	Net benefit	900 million[c]
Vehicle efficiency (no fleet change)	Net benefit	300 million
Inductrial energy management	Net benefit to low cost	500 million
Transportation system management	Net benefit to low cost	50 million
Power plant heart rate improvements	Net benefit to low cost	50 million
Landfill gas collection	Low cost	200 million
Halocarbon-CFC usage reduction	Low cost	1400 million
Agriculture	Low cost	200 million
Reforestation	Low to moderate cost[d]	200 million
Electricity supply	Low to moderate cost[d]	1000 million[e]

[a]Net benefit = cost less than or equal to zero
Low cost = cost between $1 and $9 per ton of CO_2 equivalent
Moderate cost = cost between $10 and $99 per ton of CO_2 equivalent
High cost = cost of $100 or more per ton of CO_2 equivalent
[b]This "maximum feasible" potential emission reduction assumes 100 percent implementation of each option in reasonable applications and is an optimistic "upper bound" on emission reductions.
[c]This depends on the actual implementation level and is controversial. This represents a middle value of possible rates.
[d]Some portions do fall in low cost, but it is not possible to determine the amount of reductions obtainable at the cost.
[e]The potential emission reduction for electricity supply options is actually 1700 Mt CO_2 equivalent per year, but 1000 Mt is shown here to remove the double-counting effect.
Note: Tons are metric.

Source: National Academy of Sciences, *Policy Implications of Greenhouse Warming,* p. 59. National Academy Press. Copyright © 1991. Used with permission.

space mirrors, stratospheric dust or soot, reflective stratospheric balloons, stimulating cloud condensation) as well as stimulation plankton growth to increase the uptake of CO_2 by the oceans. Reforestation, already mentioned, is really a *sort* of geoengineering. You can see several proposed geoengineering technologies in Table 4.3.

The synthesis panel of the National Academy of Sciences estimated that geoengineering options are technically feasible, their net implementation costs are low to moderate, and they could have an effect in mitigating global warming that far exceeds the effects of the mitigation strategies mentioned above (National Academy of Sciences, 1991: 60).

Yet if natural climate changes cannot be predicted with much certainty, the effects of such countermeasures are still more unpredictable. Such technical fixes run a real risk of misfiring because of unpredictable side effects (for example, chemical reactions that particles introduced into the atmo-

Table 4.3 Some Geoengineering Options

Sunlight screening	Place 50,000 100 km^2 space mirrors in the earth's orbit to reflect incoming sunlight.
Stratospheric dust	Use guns or balloons to maintain a dust cloud in the stratosphere to increase sunlight reflection.
Stratospheric bubbles	Place billions of aluminized, hydrogen-filled balloons in the stratosphere to provide a reflective screen.
Low stratospheric dust	Use aircraft to maintain a cloud of dust in the low stratosphere to reflect sunlight.
Ocean biomass stimulation	Place iron in the oceans to stimulate the production of CO_2 absorbing plankton.

Source: Adapted from National Academy of Sciences, *Policy Implications of Greenhouse Warming,* p. 58 National Academy Press. Copyright 1991. Used with permission.

sphere might alter the ocean chemistry—and food chains—in effort to stimulate plankton growth). Geoengineering options have a potential to affect global warming on a substantial scale, and some are relatively inexpensive, but all have large unknowns concerning possible environmental side effects. If we don't understand planetary dynamics very well, do we really know enough to re-engineer the earth on such a scale? Yet the National Academy of Sciences panel argued that we need to know more about these options, because they may be crucial if global warming occurs, particularly at the upper range of temperature projections. If adaptive efforts fail, and efforts to restrain greenhouse gas production on a global basis fail—for either technical or political reasons—such geoengineering options might be the only effective ones available (National Academy of Sciences, 1991: 62–63).

Strategies and Change

Adaptation and mitigation strategies would both involve significant change. In terms of mitigation, the IPCC, estimated that a reduction in greenhouse gases of between 60 percent and 80 percent below 1990 levels would be required to stabilize global mean temperature (Flavin, 1998a: 14; Miller, 1998: 377). Achieving such results *would not be easy*. Some measures might involve largely technical changes, as in designing more energy-efficient industrial and consumer equipment and in utilizing alternative (vs. carbon based) energy sources (for example, the old stock of refrigerators, autos, and dishwashers could be gradually replaced by more efficient ones). But other changes—such as car pooling, shifting to cycling or mass transit, and evolving away from the energy-wasteful low-density residential patterns characteristic of American cities—would involve significant changes of social behavior, consumption patterns, and established preferences and values. Producing the political consensus and mechanisms to develop, market, coordinate, monitor, and control such changes among the multitude of diverse communities, corporations, and households on a societal scale is a *daunting* prospect. This is particularly so when the worst of the effects—however serious—are still hypothetical and the initial costs would not be evenly shared.

A broad "carbon tax" may be one elegantly simple solution to this difficult problem. Advocates argue that it would provide the incentives to impel us to adopt energy conservation and efficiency measures, both industrially and at the household level (Amano, 1990; Goldemberg, 1990). But there is compelling evidence that the burdens would fall unevenly on different socioeconomic classes. Higher income households tend to purchase energy-efficient technology and to make building changes, while lower-income households tend to curtail consumption through behavior changes (Dillman et al., 1983). I noted in earlier chapters that environmental problems and their remedies are often related to inequality and social stratification so that their burdens and benefits are not equitably shared. In the absence of collateral policies that more evenly distribute the burden sharing (in housing and transportation, for example), such policies would have highly inequitable and regressive effects that would burden low-income households much more than more affluent ones (Lutzenhiser and Hackett, 1993).

There is a similar difficulty regarding international inequality between nations. With a small proportion of the world's population, the MDCs produce 45 percent of global carbon emissions. The United States alone produces about 23 percent. Russia and other Eastern European nations produce about 14 percent, and LDCs, with vast and growing populations, produce 41 percent of global carbon emissions. The richest fifth of the world contributes 63 percent of global emissions, while the poorest fifth contributes just 2 percent. Putting it in individual terms, the average emissions of one American equal those of seven Chinese, 24 Nigerians, 31 Pakistanis, or hundreds of Somalis (Dunn, 1999: 60).

Climate Treaties and Greenhouse Diplomacy: Kyoto and Beyond

The 1992 U.N. World Conference on Environment and Development in Rio de Janeiro intended to initiate a global greenhouse treaty, much like the successful Montreal ozone treaty. But this agenda proved much more contentious and difficult. In spite of the urgings of the National Academy of Sciences, the U.S. government refused to sign, sabotaging an initial agreement because it had quantitative national targets for emission reduction. What emerged from the Rio meeting was a statement that nations signed pledging "voluntary reductions" in greenhouse emissions, with no quantitative targets or sanctions for noncompliance. As you might guess, not much happened. As evidence of the reality of the problem continued to mount, political mobilization around the issue continued in the 1990s. Scientific and environmental organizations, along with insurance liability carriers, urged action. Many industry groups, sensing greater restrictions, regulations, and lost profits to come, campaigned against doing anything about global warming. Some labor groups, fearing loss of jobs, joined them. Even though greenhouse mitigation policies would benefit some industries and create new jobs, like most large-scale changes, there would also be big losers. Although the new business created by moves to reduce greenhouse emissions is likely to roughly equal the business lost, the potential losers are far better organized and well funded. Especially among these are the mining, petroleum, and lumber industries, which stand to lose from the reduction of carbon emissions and protection of forest carbon sinks. Industry associations started spending millions on "sponsored" research, lobbying, and advertising to derail further mitigation treaties and policies. While corporate opposition grew, evidence of the reality of global warming continued to accumulate. Around the world, scientists and environmental advocacy organizations urged action, as did a majority of the American people, according to public opinion polls.

Responding to global pressures, in December 1997 some 10,000 government officials, lobbyists, representatives of advocacy organizations and the media gathered for a high-profile world climate conference at the ancient Japanese city of Kyoto to negotiate a better treaty than the Rio accord. After 10 days of chaotic, complex, complicated, and contentious negotiations, 160 nations formally adopted a Kyoto Protocol, legally committing industrial countries to reduce their emissions of greenhouse gases early in the twenty-first century. Why only the industrial nations? Because LDCs objected, noting that their greenhouse emissions are far lower than those of MDCs, even though they are rapidly growing. Led by India and China, they argued that meeting targeted reductions early in the twenty-first century would destroy their fragile developing economies—without considerable financial and technical help from the MDCs. The Kyoto centerpiece was an agreement by all "Annex I" nations" (MDCs and former East bloc countries) to cut their output of climate-altering gases collectively by 5.2 percent below their 1990 levels between 2008 and 2012. Most contentious was the target and timetable negotiated for each nation's contribution to the collective goal, which was resolved after many concessions and protracted debate (Dunn, 1998b: 33). You need to understand that even in the most ideal terms, Kyoto only started the global political process, and the protocol is riddled with contradictions and exemptions. It has four major kinds of weaknesses:

1. *Weak commitments*: Its goal of 5.2 percent reduction in emissions is anemic compared to between 60 percent and 80 percent reductions below 1990 levels that the IPCC says are necessary to stop global warming. Little noticed outside climate policy circles was the curious fact that total CO_2 emissions by MDCs was—and is—already below 1990 levels, due to steep declines in the former Soviet Union. Furthermore, when emissions by the LDCs are added, the global emission total is projected to increase some 30 percent *above* the 1990 level by 2010. The most hopeful thing that can be said about the Kyoto Protocol echoes

Lao Tse's comment that a journey of a thousand miles begins with a single step. Though it will not affect global mean temperatures, it moves the world political process.

2. *Searching for "flexibility"*: Some countries, particularly the United States, were anxious to find provisions—critics call them loopholes—that would make it less expensive to meet the protocol's goals and avoid the need to take a big bite out of domestic CO_2 emissions. They targeted a "basket" of emissions rather than focusing on each individually, particularly CO_2, which would arouse a hornet's nest of industry outrage. At the insistence of the United States, Canada, and New Zealand, countries could count—and subtract—carbon absorption by forests and peat bog sinks.
3. *Hot air trading*: Another form of flexibility is the concept of emissions trading, supported by the U.S. regulators, private companies, and some environmental organizations. It is modeled on the U.S. Clean Air Act that allows power companies to "trade" their sulfur dioxide reduction obligations for cash on the theory that this will encourage cuts to be made wherever it is least expensive to do so. Related to global climate change, nations would have the option of buying greenhouse gas emission allowances from other countries that have more than met their own requirements. The idea is nice, but it opens the door for possible loopholes. For example, under the protocol signed, Russia and Ukraine must only hold their emissions to 1990 levels, which, given their depressed economies, would allow them to increase emissions 50 percent and 120 percent respectively. Even if their economies rebound robustly, experts do not expect either country to come close to such increases, which would allow the United States and Russia "make a trade," letting the United States take credit for emissions reductions—without reducing emissions by one molecule.
4. *The ratification trap*: These are thorny but (in theory) surmountable problems. Complicating the process further is a political ratification trap. The treaty would go into force if ratified by enough MDCs to represent at least 55 percent of the industrial country emissions. So in theory the protocol could go into force without U.S. ratification. Understandably concerned with both competitiveness and fairness, neither the Europeans nor the Japanese want to move forward with an agreement that excludes the world's largest greenhouse emitter. Meanwhile, leading U.S. senators said they would not ratify the treaty unless it had "new specific scheduled commitments by developing countries to reduce their emissions." As noted, the LDCs are wary of being asked to reduce their emissions, which average less than one-tenth of the U.S. per capita level. Prolonged arm twisting yielded little progress, giving the United States an effective veto over the protocol (Flavin, 1998: 14–16).

As you can see, progress toward emission reductions have bogged down in "fiddling with sinks," trading, and seeking other forms of "flexibility" that derail the main business of carbon reductions. There is an urgent need to pin down actual reductions in emissions, in order to give the treaty some teeth, with penalties for nations that don't meet their goals, and to fairly phase in LDC commitments that enable developing nations to leapfrog the industrial world in decarbonizing efforts.

Ironically, while national and international political processes have stagnated, opportunities for economically cutting emissions have blossomed. In 1997, the president of British Petroleum Oil Company announced that climate change is real and serious and that BP will reduce its emissions by 10 percent and step up investment in solar energy. While the American Petroleum Institute denounced BP for "leaving the church," Enron, Dutch Royal Shell, and several others followed BP's initiative. In 1998, the Pew Center on Global Climate Change announced that Weyerhaeuser lumber company, one of the world's biggest, formally joined the center's effort to combat the problems of climate change. During the Kyoto conference, Toyota stunned the auto world with the delivery to its showrooms of a "green" sedan (Prius), the world's first hybrid electric car, with twice the fuel economy and half the CO_2 emissions of conventional cars. The Prius sold so quickly in Japan that Toyota had to open a second assembly plant. By 1998, the Detroit Big Three and most European automakers had announced plans for new generations of hybrid and fuel cell cars. While national governments dither, a surprising number of cities are moving forward with active efforts to reduce their emissions. Over 100 cities representing 10 percent of global emissions have joined the Cities for Climate Protection program by investing in public transportation, tightening up public buildings, planting trees, and installing solar collectors. Since 1995, there has been a doubling of the world's wind-generating capacity, noticeably in the United States but particularly strong in Germany, Denmark, and China (Flavin, 1999: 54). By 1998, public opinion polls made it clear that the U.S. Senate was out of step with the American public. The majority of Americans feel so strongly about the need for a global warming treaty that they are willing to go forward even if the developing countries do not join in. That was the opinion of 80 percent of Americans, including 79 percent of the self-identified political independents, 84 percent of the Democrats, and 73 percent of the Republicans in the sample. The director of the poll taken for the World Wildlife Fund said that "these large and growing numbers show that despite a well-financed misinformation campaign by the fossil fuel industry, most Americans know that global warming is a serious problem" (*Ecology USA*, 1998b).

Taken together, these efforts suggest that it will be easier and less expensive to reduce CO_2 emissions than it seemed to be a few years ago. As with most environmental problems, once the world's nations get serious, they are likely to find a host of innovative and inexpensive ways to address them. I think the question is how to accelerate the process (Flavin, 1998: 16–17). In November 1998, more than 160 nations meeting in Buenos Aires adopted a plan of action with a two-year deadline to tighten and clarify rules of the Kyoto accord. Several nations announced voluntary limits on emissions, and the United States became the sixtieth signatory of the accord. Yet as scientists observed in *Nature* magazine, carbon-cutting strategies have yet to address the need for major mitigation and adaptation measures (Dunn, 1999: 60;

Ecology USA, 1998a; *Ecology USA*, 1998b). Having said all this about the scientific complexity and policy debates related to global warming, I thought this was a good place to discuss scientific uncertainty and risk more broadly.

UNDERSTANDING UNCERTAINTY AND RISK

While there is scientific consensus that global warming is a significant threat, there is much uncertainty about some basic facts, about the magnitude and timing of the threat, and about the environmental and societal impacts and costs. Given this, there is compounded uncertainty about policy options and, as you can see from the foregoing, about the feasibility of creating national and international political arrangements to address the problem. In fact, in the wake of publicity about the greenhouse effect, public understanding is (justifiably) confused and policymaking is paralyzed. Public understanding has been blurred by the complexity of the issues, by media miscommunication about them, by political controversies surrounding them, and by the debate among battling scientists themselves (Schneider, 1990b: 30).

Uncertainty and risk are not the same thing, but they are connected. *Uncertainty* is related to our knowledge of how true, real, or factual something is (e.g., global warming), while *risk* is a situation or event where something of human value (including humans themselves) has been put at stake and where the outcome is uncertain (Rosa, 1998). In addition to risks from some existing circumstance, both "doing something" and "doing nothing" about it have risks, and it is important for us to try to understand which are greater (but as you might guess, this is often very complex). Let me try to unravel some of the sources of uncertainty about scientific understanding and assessing risks as they relate to a variety of environmental issues.

Sources of Scientific Uncertainty

I think there are four main sources of uncertainty in scientific understanding. The most obvious is uncertainty that flows from existing state of theory and data. But, there are also paradigmatic, semantic, and social contexts of scientific uncertainly. The following explores each.

Uncertainty from Theory and Data

Scientists are often uncertain about things because they don't have enough or the right kind of facts ("data") or the right theory to explain them. Sometimes there is uncertainty because competing theories are equally plausible. They can't construct a good working model (a vastly simplified theory) of the earth's climate dynamics because the computer capacity does not now exist to handle all the necessary measurements at the same time, and they don't know how some important variables (e.g., clouds, oceans) interact with the atmosphere. Toxicologists don't know the exact exposure levels at which many pesticides or pollutants cause illness or death, because they have never done well-controlled experiments on human beings to find out (for some pretty obvious ethical reasons!). Scientists know how to generate electricity from solar cells, but not on the right scale and at the right price for widespread commercial and household use. Some scientists believe that genetically engineered species of plants or bacteria could have dangerous and risky consequences when introduced into existing ecosystems in which they have not evolved. Other scientists don't think so. Good evidence about this issue is nonexistent because it has never been done before. This list could go on and on, but you get the picture.

But there is another source of uncertainty built into science that may be less obvious to you. Contrary to common belief, the core of science is not a body of well-ordered facts and findings with clear meaning, but rather institutionalized skepticism and uncertainty. The deepest impulse of science is to question accepted truth and the certainty of things that are taken for granted. Scientific inquiry creates consensus and reduces (but does not eliminate) uncertainty. In doing so, it normally finds new areas of uncertainty. Some things become understood beyond much doubt, but areas of *current* scientific interest are *always* uncertain. (If they weren't, why would anyone want to study them?) Scientists may challenge not only popular assumptions but also the science of other scientists. My point is that dispute and contention are normal and routine processes that give science life. Scientists, appropriately, spend a lot of their time arguing about what they *don't* know, and that is true for both natural and social sciences. When dispute and contention spills into public visibility and the media you have the case of "battling scientists." Alas, the "experts" often don't agree. And scientific conclusions about important things are tentative, conditional, and hedged with ifs, ands, and buts. The intrinsic toleration of uncertainty and ambiguity is normal for science but terribly frustrating for people who just want answers, and particularly for those trying to make policy out of science.

Uncertainty from Different Paradigms

In Chapter Two, I introduced the notion of social paradigm as a belief structure that organizes perceptions about how the world works that people use—often implicitly. A scientific paradigm is an intellectual image containing explicit and implicit assumptions—broader than particular theories—that guide theorizing and research. Scholars and scientists in different scholarly communities (or disciplines), such as biologists and economists, or sociologists and psychologists, operate from significantly different paradigms, which often makes communication between them difficult.

Ecologists and environmental scientists tend to see environmental problems through paradigm lenses that are very different from those of neo-classical economists, and this paradigmatic difference is at the heart of the much of the disagreement and controversy about seriousness of environmental problems and dangers. Here are some examples discussed in this and earlier chapters.

Mineral resource issues (addressed in Chapter Three) have long been, and continue to be, a point of contention between economists and environmental and resource scientists. Neoclassical economists tend to view the supply of minerals as essentially infinite because of our ability to develop improved technologies to find and process minerals, and to find substitutes. Environmental and resource scientists argue that because of the ultimate finite supply and uneven concentrations of key minerals, there are limits on the grades of ore that can be processed without spending more money than they are worth and causing unacceptable environmental damage.

Economists argue that if global warming occurs sometime during the next century, well-functioning markets—without the distortions of excessive regulations or subsidies that make things unrealistically cheap—will be important. Markets will work to stimulate investments in increased efficiency, resource substitution, conservation, innovation, and promote behavior changes by which firms, families, and communities adapt to changing climatic conditions. Economists typically recognize the need to price *externalities* (e.g., environmental degradation) not considered in present cost-accounting systems; to figure out property rights and prices for common goods (of which the atmosphere is a case in point); and to consider the value of resources to future generations when pricing present consumption.

Natural and environmental scientists generally argue that effective markets and real-cost pricing will be helpful but are not sufficient, particularly when the full brunt of global warming may soon be upon us. We may not have the options or resources that we possess today. They have proposed a wide variety of mitigation measures to reduce greenhouse gas emissions that I discussed above. Responding to the notion that the magic of the market can save us from ecological devastation, ecologist Lester Brown of the Worldwatch Institute, in a pointed exchange with Harvard economist Theodore Panayotou, asked whether

> an economic system that is destroying 17 million hectares of forest each year [can] sustain progress[.] What about an economic system that's adding 90 million people a year—or even 50 million people a year in countries where demand on basic biological systems—whether grasslands, forest, seas, or soils—is exceeding the sustainable yield? Will an economic system that's pumping six billion [1 billion: = 1,000 million] tons of carbon into the atmosphere each year from the burning of fossil fuels sustain progress? Will an economic system that's converting six million hectares of productive land into desert each year sustain progress? (Brown and Panayotou, 1992: 354)

I will return to this paradigm disagreement when discussing population issues in Chapter Five and problems with growth and sustainability in later chapters. But these examples should be enough to make the broader point. Much of the disagreement and controversy about environmental issues occurs because different sorts of scholars use different paradigms about the way the world works, and therefore they frame problems in different ways. Physical and environmental scientists tend to see the problem in terms of the longer-term implications of the growth in scale in a finite world. Neoclassical economists tend to frame resource problems in terms of more immediate market failures and resource allocation problems, such as overusing resources by pricing them too cheaply. Others, as noted in earlier chapters, tend to frame environmental problems, and particularly starting points for policies to address them, in terms of dealing with the consequences of social inequalities related to environmental use and impacts. Communication across such paradigmatic divides is difficult but, I think, not impossible.

Uncertainty from Semantics

Given that scientists spend much of their time arguing about uncertainties, it is especially important they clearly communicate to the public and policy makers how much confidence they have in scientific findings. But the translation of complex scientific findings and "probabilities" into ordinary and clear language is fraught with difficulty and potential for confusion. To illustrate, consider an exchange between noted climatologist William Schneider of the National Committee for Atmospheric Research (whom I have cited earlier) and Andrew Solow, a statistician for the Woods Hole Oceanographic Institution. In testimony before a congressional subcommittee, Schneider argued that global warming represents a substantial threat of unprecedented climate change during the next century. Solow considered the probability low and Schneider's argument to be irresponsibly alarmist. Here, in some length, is Schneider's account of the conversation after bantering the issue back and forth.

> "I just can't see how you don't think there's at least a 50 percent chance that the next century will see 2 degrees C or more warming" [said Schneider].
> "I never said that," Solow responded.
> "But I thought you said it was a low probability," I replied.
> "Well, that *is* a low probability," said the statistician.
> "What would you consider a moderate probability?" I asked.
> "Oh, 95 percent," he said.
> "And a high probability?"
> "99 percent," he said.
> "Eureka!" I thought to myself, I now understand our "debate." Solow, a statistician, inhabits a culture in which the traditional standard of evidence for

accepting any hypothesis is the 95 percent or 99 percent confidence level. (Schneider, 1990b: 33)

Schneider argued that such criteria for evidence might be appropriate for accepting or rejecting scientific hypotheses but are very unrealistic guides for understanding risks or policy making.

> What business, individual, or government leader ever has the luxury of having 95 percent certainty about the facts underlying any major decision? Do [for instance] Pentagon officials know what the probability is than any particular terrorist incident will occur? Of course not; yet they run hundreds, probably thousands of alternative conflict scenarios defining how they can be strategically prepared for a range of outcomes . . . whose probabilities . . . are probably by a factor of 10 less than that for unprecedented global warming over the next 50 years. (Schneider, 1990b: 33–34)

In other words, the disagreement between Schneider and Solow was largely semantic. One expert said a probability was high while another said it was low, even though scientific consensus put probability at 50 percent or better. This is a slippery issue as experts attempt to communicate complex findings to the media and policymakers. They are, at minimum, required to clearly interpret what words like *high*, *medium*, and *low* mean.

Uncertainty from Social Contexts

Uncertainty deriving from inadequate theory and data, from paradigm differences, and from semantics is difficult but is still straightforward intellectual disagreement that is—potentially—resolvable by argumentation and debate. But some uncertainty and dispute also derives from social settings—the organizational, economic, and political settings in which scientists live and work—as well as from purely intellectual disagreements. About environmental issues, scientists are likely to differ in relation to their sector of employment. Academic institutions, government agencies, industry, and private environmental organizations employ scientists, and they provide different political climates and research agendas. Although the media and policy makers draw on information provided by all four sectors, they do not do so with equal frequency. The media are likely to give more attention to those scientists whose work suggests serious problems and has implications that affect large numbers of people. This means that those affiliated with academic and private institutions are likely to have high media profiles.

Certainly, those affiliated with environmental movement organizations (e.g., the Sierra Club or the Environmental Defense Fund) are most likely to dramatize environmental threats. Indeed, advocacy organization of scientists themselves, such as the Union of Concerned Scientists, have become important organizations within the American environmental movement.

Somewhat less likely to emphasize environmental threats are academic and government scientists, although this is highly variable (e.g., as seen earlier, some climatologists affiliated with government meteorological organizations have been most insistent about the threats of global warming). At the very least, government scientists work under pressure to represent the interests of the agencies in which they work. Scientists working for the Corps of Engineers or the Environmental Protection Agency (EPA) routinely experience pressure, for instance, not to emphasize the regulatory "failures" of those agencies to monitor and manage the environment.[3] Least likely emphasize environmental threats are industry scientists, whose work is shaped by corporate needs for profits, as well as by the more restricted range of issues that concern them. Indeed, scientists ranging from the chemical and pesticide industries to the tobacco industry have normally been busy defending the benign character of their products. Scientists working for the fossil fuel industries have been less likely to emphasize the dangers of global warming. Industry scientists are likely to have a relatively low media and policy profile except during times of intense political controversy surrounding issues related to a specific chemical or industrial practice (Clyke, 1993: 33–34; Chomsky, 1989; Schnaiberg and Gould, 1994: 146).

I don't mean to suggest that scientists are nothing more than hired guns, dressed up in white coats and armed with charts and statistics. In fact, scientists are prominent among organizational *whistleblowers*, who report organizational malfeasance to the press and public. They do so often at considerable personal and career costs, a fact that has made whistleblowers an interesting subject of study (Tsoukalas, 1994; Nixon, 1993). The point I am making is that scientists work in agencies that are rarely fully objective about scientific research outcomes, and they experience varying degrees of organizational and political pressures to represent the interests of those organizations. In addition to organizational pressures, other social factors limit the objectivity of science, including (1) methodological specialization between scientific disciplines, (2) the dominance within scientific disciplines of established theories—sometimes independent of a compelling evidential basis, and (3) research driven by the goals of a policy system for "correct answers" in the face of uncertainty.

Such considerations led many scholars to argue that scientific knowledge claims, likely other knowledge claims, are ultimately *socially constructed and therefore subjective* (Knorr-Cetina, 1981). Some sociologists have adopted social constructionist approaches to the study of risk (Dietz et al., 1989; Buttel and Taylor, 1992). In truth, social interaction, organizational interests, and cultural transmission shape all knowledge claims. They are therefore socially constructed and fallible. *But not all claims are equally fallible,* and many facts of science, including environmental risks, have a basis in reality that is more than a simple social construction. For example, the health effects of lead exposure have many subjective features, which is why they were ignored until the 1950s. But sufficient doses of lead are toxic in all societies, whatever the shared beliefs of people or scientists. It is just as clear that science, particularly "policy science," is deeply intertwined with vested interests and politics (Dietz et al., 1992). Having discussed some sources of uncertainty in science, I turn the coin: How do science and society assess risks, dangers, and hazards?

Assessing Risks, Analytic Modeling, and the Risk Establishment

Risk analysis can be traced back to ancient Babylon, where it is mentioned in the code of Hammurabi. But using risk as an analytic and management tool is really a twentieth-century phenomenon (Rosa, 1998). Early research about risk assessment focused on the individual perceptions of risk. These studies, conducted mainly by psychologists, assume that individual members of the public are the final arbiters of acceptable risk (Starr, 1969; Kahneman and Tversky, 1972; and Kahneman et al., 1982). Such studies concluded that (1) people take mental shortcuts in making sense of risk information, (2) perceptions of risk vary with the social location of the person, and (3) people are more tolerant of risks where exposure is thought to be voluntary. An important conclusion was that the general public's sources of information (primarily TV and newspapers) systematically distort data about risk, because people need cognitive heuristics (simplified pictures of complex realities), shortcuts, and stereotypes to make sense of complex data about hazards (Clarke, 1988: 23). Because the public is often ignorant of the details and environmental and technological hazards and does not think in terms of complex probabilities, many believe that only experts are best qualified to make decisions about the risks society faces. This is a best a half-truth to which I will return.

Analytic Risk Models

The environmental movement, increased public concern, and growing pressures for the state to rationalize and regulate technological and environmental risks stimulated concern with measuring risks like the hazards of pesticide exposure, toxic waste dumps, ocean oil spills, space exploration, and the commercial production of nuclear energy (Rosa, 1998). This led to attempts to quantify risks and produced a *risk establishment*, by which I mean networks of scholar-specialists who conduct risk analyses, translate them for policy makers, and weigh them for policy making. Many formal models were developed to measure and quantify risks. It is beyond the scope of this book to discuss these in much depth, but I will briefly explore three of them to give you a sense of what they are like.

Fault Tree Analysis. Modern formal risk modeling began when the nuclear power industry attempted to add objectivity—or at least its appearance—to traditional engineering judgements about the safety of nuclear reactors. It was a way of estimating the probability of a nuclear core meltdown, and since—at that time—this had never happened, such estimates were necessarily hypothetical (Mazur, 1981). The basic logic of this method is to specify the combinations of separate breakdowns that would have to occur to produce a failure of the total system. Then, if the probabilities of the separate component failures are known, they are combined (multiplied) to get the probability of a total system disaster. If, for example, the chance of breakdowns in components A, B, C, and D is one chance in ten (0.1) each and they are truly independent of each other, then the chances that they will simultaneously malfunction is .0001, or one 1 in 1,000 ($0.1 \times 0.1 \times 0.1 \times 0.1 = .0001$). When the probabilities of a total system breakdown are calculated for complex technical systems with many fail-safe procedures are calculated (e.g., the likelihood of the crash of a jet liner, an ocean tanker oil spill, or a nuclear meltdown), they are small. The likelihood of a nuclear power reactor disaster was calculated as being minuscule.

Social Impact Assessment. Social impact assessment (SIA) developed out of the impact assessment provisions of the National Environmental Policy Act (NEPA). The goal of SIA is to predict the social effects of policies and programs on communities and to use these predictions to aid public decision making and debate. Early SIAs used computer simulations, beginning, for example, with predictions of labor required for new projects (a dam, prison, mine, or nuclear power plant). It then estimated the demand for secondary jobs (e.g., all the teachers, cooks, barbers, police, and doctors required to serve the primary project workers). It continued to estimate the extent to which the demand could be locally met or would require in-migration, and finally to assess the stresses on government services and budgets (Freudenburg, 1988; Finsterbusch et al., 1983; Finsterbusch and Freudenburg, 1993; Dietz, 1987: and Stern et al., 1992: 187–188). The risks and hazards assessed in this case are not technological ones, or necessarily even monetary ones, but the risks and costs associated with rapid change, growth, and potential for social disruption of community life.

Cost-Benefit Analysis. Cost-benefit analysis (CBA) has been the premiere technique among economic thinkers for *valuing*, or attempting to assess the values of risks (costs) and benefits of various impacts and options. SIA might tell us, for example, about the probability of impacts from a project—say, increased jobs, lower energy costs, loss of wildlife habitat, increased cultural diversity or traffic congestion—but how do we *value* these impacts as costs or benefits? CBA assigns market values to impacts, either directly or by imputation. Future values are discounted. Then, having assigned a value to all costs and benefits in current dollars, the ratio of benefits to costs, net present value, internal rate of return, and other measures of the efficiency of some project or option can be calculated. While the practitioners of CBA typically caution against taking the final calculations too seriously as decision criteria, it does provide an explicit framework for valuing otherwise incommensurable options or impacts and may provide a basis for preferring one option over others (Stern et al., 1992: 193).

The reason for discussing formal models and techniques of risk assessment is that attempts are being made to adapt them to understand the impacts of the global environmental and climate change, and policy options suggested to respond to them. I mentioned some of this work earlier (see pp. 139 and following, especially Table 4.2). According the National Academy of Sciences, for instance: "The demand for SIAs of the anthropogenic effects of global warming is already large and will certainly grow" (Stern et al., 1992: 189).

Limitations and Criticisms of Risk Research and Modeling

The appeal of such research and models will be very great, but you need to know the limitations of these techniques and the strong criticisms that have been made of them.

Consider the fate of fault tree risk analysis when employed by the nuclear power industry. In the 1960s, they concluded that core damaging accidents could be expected only once every 10,000 "reactor years."[4] However, a number of observers had a difficult time reconciling these low probabilities with a number of near misses, and particularly the reactor meltdown at Three Mile Island in Pennsylvania (during which a string of six low-probability errors happened simultaneously). A new study raised the risk to once every 4,000-reactor years. Yet the uncontrolled disaster at Chernobyl happened only 1,900 reactor years after Three Mile Island. At this rate, one could have expected three more nuclear disasters by the year 2000 (Milbrath, 1989: 238). Fortunately, they did not occur. In the mid-1980s, Swedish and German scientists estimated that there was a 70 percent chance of another accident in the next five to six years (Flavin, 1988). How can such much-revised values have any objective meaning?

According to evaluation by a panel of the National Research Council, the ability of SIA computerized projections to make accurate forecasts is limited because many key parameters assumed to be constant (such as the ratio between primary and secondary employment, the generation of crime or of traffic problems) are themselves highly variable over time. Despite the use of such models for 20 years, there have been few attempts to validate them by comparing projections with actual events. The Research Council argued that *post hoc* SIAs, which study the actual social impacts of particular programs, may be more useful (Stern et al., 1992: 190).

The largest body of these address the effects of energy development boomtowns on the communities and their members (Freudenburg, 1984; Finsterbusch et al., 1983). But it is not clear how findings from such studies could be used to understand historically unprecedented problems such as declining biodiversity, water shortages, and climate change on a global scale.

CBA has been criticized on similar grounds. Many of the costs and benefits are not real market-price values at all, particularly future use values and social costs (such as the costs of community disruption, which people care about very much, but which is hard to assign a numerical price tag to). Many of these costs are speculative "guesstimates" that may reflect ideological biases of the analysts. Or—in light of the discussion of structural contexts—the agendas and interests of the sponsoring organizations. Therefore, it is fair to ask about the *distribution* of costs and benefits in such analyses. Whose costs? Whose benefits? In an outrageous case, Ford Motor Company executives once decided to continue producing a car they knew to be dangerously defective (the Ford Pinto). Their CBA suggested that the costs of legal defenses and compensation to the aggrieved families of victims would be less than retooling assembly lines to correct a simple but potentially lethal design defect. Don't misunderstand. I'm not arguing that this is a typical consequence of the use of CBA. But it did happen.

A problem with all of these formal models for risk-cost assessment is that such concrete-looking quantitative analyses often come, in the heat of policy debates, to be taken as objectively factual when in reality they are heuristic ("as if") models. While the research community understands the limitations of such models, they are often abused in the policy-making process by being interpreted as what *will* happen or given an undue aura of scientific objectivity. At best, they may stimulate or clarify assumptions about decisions related to risks. But the lesson of the last two decades is the limited uses of such formal models in aiding scientific knowledge or serious discourse on policy (Stern et al., 1992: 190; see also Freudenburg, 1988; Finsterbusch and Freudenburg, 1993: 28–37).

Social Contexts of Risk: Public Opinion, Organizations, and Society

Beyond methodological problems in the work of experts, deeper questions exist about the role of experts in such questions relative to the broader social process and tradeoffs within a political system. As noted earlier more than once, scientific and technical analysis is effected by the social context in which it takes place. While scientific methods can never replace the political process in assignment values to impacts and making decisions about risks and hazards, it is equally naive, I think, to believe that science cannot inform the public and political decision process.

Studies of opinion formation about environmental and technological risks have mainly been conducted at the individual level. Experts generally bemoan public ignorance. But while the public tends to be weak on many details, research suggests that the public is surprisingly strong on the bigger picture and may be more rational than they appear to experts (Slovic et al., 1979; Einhorn and Hogarth, 1981, cited in Freudenburg, 1988: 49; Dietz et al., 1989: 50). According to Freudenburg, himself part of the scholarly risk establishment, "just as scientists' risk estimates need to be treated with something less than reverence, the views of the public may need to be treated with something better than contempt" (1988: 47). But individuals are not very good at assessing the probability that they will *personally* be affected by a particular health, technological, or environmental threat. Most public perceptions come through the media, as shorthand heuristics, that dramatize some risks while ignoring less dramatic risks and events. The public systematically diverges from experts because they overestimate low-probability, high-consequence events (such as a nuclear accident) and underestimate high-probability, low-consequence events (such as exposure to medical X-rays). Such differences probably derive from media dramatizations that provide the shortcut heuristics through which people perceive complex risks (Dietz et al., 1992: 23–25; Slovic, 1987). This suggests that deliberate efforts to improve public understanding of environmental risks could be effective, and many scholars argue that laypersons should play a more central role in assessing, evaluating, and managing environmental and technological risks (Fischoff, 1990; Freudenburg, 1988; National Research Council, 1989a).

Organizations, Risks, and Bounded Rationality. Many of the problems investigated by psychologists about risks have a large role for individual choice. Examples include using seatbelts or dietary choices, but the role of organizations in shaping policy about risks is often neglected (for instance, in shaping laws about seatbelt use, diet and health, and labeling the contents of foods) (Freudenburg, 1992b: 2). I noted earlier the powerful role of big organizations in shaping the outcomes of global greenhouse treaty negotiations and the contexts within which scientific research is conducted. In fact, what drives democratic political systems more clearly and directly than individual preferences are the interactions between community organizations, interest group organizations, corporations, regulatory agencies, and scientific communities—in other words, interest-group politics at its finest. As many have observed, sociologist Lee Clarke argued, with considerable force, that organizations, not the public, are the crucial actors in dramas of risk assessment, and that what shapes policy outcomes is *organizational power* as much as *objective risk* (1988). Furthermore, there is evidence that organizations often misperceive risks or define an environmental threat in terms of the threat of any action to their organizational interests (which is a kind of rationality about the organization, but not pure scientific rationality about the wider problem). Regulatory agencies and organizations are often "captured" by the industries they are supposed to regulate and often do not effectively enforce existing regulatory laws for a variety of reasons (Clarke, 1993; Freudenburg, 1993).

In short, the real process of assessing risks is a complex interaction among big organizations, political and institutional actors, scientific communities, environmental movement organizations, the media, and the public. In this complex social process, various actors strive to frame controversies and to legitimate different definitions of environmental risks. Participants enter this process with different values, interests (what they stand to gain or lose), and capacities to mobilize "expert knowledge" on their own behalf (Freudenburg and Pastor, 1992). One study of conflict and controversy about environmental risk found that leaders of environmental movement organizations attributed conflict largely to different organizational interests and value differences. It found that the risk experts attributed conflicting opinions to public ignorance, while congressional staffs were more distrustful of expert opinion, particularly when connected with private industry (Dietz et al., 1989).

Having said this, I think it is still true that a broader or *bounded rationality* sometimes works to shape risk assessments and policy, even for big issues. In the United States, the components of that bounded rationality involve, I think, (1) a rough scientific consensus about a problem (not empirical certainty), (2) a generally knowledgeable public, (3) effective advocacy and social movement organizations, (4) receptive media and political leaders, (5) corporations willing to find profits in more benign ways, and (6) sometimes pressure from international organizations. Those were at least present in the social process that led to the mitigation of the ozone hole problem and some other environmental problems (water and air pollution) as well as some public health problems like cigarette smoking. Even with the strong limitations on rational policy noted earlier, I argue that under *some* conditions the democratic interest group jockeying and politics can be bounded by a broader problem-solving rationality. Will these elements be sufficiently present to produce a rational international policy about climate change? Who knows?

Contemporary Societies and Risk. Looking at the big societal picture, European sociologists added new dimensions about understanding risk (Giddens, 1991; Beck, 1986; Luhmann, 1993). They observed that people in earlier societies were mainly threatened by external and objective dangers, like floods, disease, droughts, storms, and famines. Life was hazardous and often dangerous. Premodern societies had some experts, but it was possible to carry on one's life, if one so wished, in terms of folk knowledge. By contrast, life in modern technological societies is often safer because hosts of special experts have constructed defenses and plans for dealing with dangers of all kinds. But "to be an expert in one or two small corners of modern knowledge systems is all that anyone can achieve means that such abstract knowledge systems are opaque to the majority" (Giddens, 1991). Thus in the vastly expanded scope and scale of modern life, with greater awareness of problems of many kinds, we do not feel safer. The reflexivity of modern life means that by our very powers to control ourselves and our environments, we wind up with extremely complex sets of contingencies and must think in terms of probabilities and risks rather than certainties. German theorist Ulrich Beck termed modern societies *risk societies* (1986). This important development connects risk to the fundamental nature of modern societies. It is also a challenge because their profound metaphor for defining the spirit of our age has (yet) to be related to the analytic literature about risks mentioned earlier (Rosa, 1998).

CONCLUSION: CRITERIA FOR POLICY AND ACTION

Turning from problems of the complexities of social and political process, I close by examining more fundamental kinds of assumptions, logic, and attitudes. There are two fundamental attitudes about making decisions in the face of considerable scientific uncertainty:

- Don't act until you are certain, or wait and see. Those holding this view believe that existing scientific uncertainties are still too large to warrant costly preventative action. Instead, more research should be pursued to reduce scientific uncertainties.
- Act now to minimize risks. Those with this view believe that uncertainty cuts both ways: If major climate change, for example, should occur, inaction could have catastrophic consequences. Society should therefore pursue investments and policies now to minimize such risks (Krause et al., 1992: 3). This is widely known as a "no-regrets" strategy.

This is a way of laying bare the differing rationales for those advocating wait-and-see *adaptation strategies* and those advocating proactive *mitigation strategies.*

Do We Know Enough to Act?

In a word, I think so. One reason is that the necessary data-gathering and computational capabilities are so enormous that it would probably take at least two decades before results would significantly improve climate modeling accuracy and improved scientific understanding (Schneider, 1989). In other words, by the time scientific knowledge improves, the full brunt of climatic change may be upon us. There are enormous risks, in other words, and a huge gamble, in the wait-and-see-while-collecting-more-data option (Krause et al., 1992: 5).

> We take huge risks if we continue to increase concentrations of greenhouse gases at the present rate and, in the face of uncertainty, the prudent course is to take some action early in the hope of cutting off the worst possible outcomes. The imprudent course is to do nothing, awaiting a complete confirmation of the models. As a recent World Bank paper states: "When confronted with risks which could be menacing, cumulative and irreversible, uncertainty argues strongly in favor of prudent action against complacency." (MacNeill et al., 1991: 17–18)

If we do something about global warming and the threat is real, we win. If we do something about global warming and the threat is not real, we lose something, but only the investments and insurance premiums. If, on the other hand the threat is real and we don't do anything about it until it is too late, we risk losing on a catastrophic scale.

Can We Afford the Costs?

Cost estimates vary widely. As a caution, I warn you that—at best—they are informed guesstimates with large speculative elements. In 1990, a federal government report said that controlling greenhouse gases would cost the United States about $10 billion a year (Lovins, 1998). The National Academy of Sciences "most likely" forecast put the costs at about $1 trillion over the next century (Schneider, 1990b: 38). Shortly before the Kyoto treaty, global warming skeptics, such as the Center for the Study of American Business at Washington University in St. Louis, Missouri, warned that even cutting emissions to 1990 levels would cost the United States $7 trillion and a million jobs. Accumulating evidence about global warming, and particularly the IPCC report, caused some skeptics to modify their thinking. In 1997, more than 2,000 economists, including prominent skeptic William Nordhaus of Yale University, signed a statement that the United States should take "preventative steps" to avoid global warming. A Reagan administration economist who coordinated the statement said that businesses have exaggerated the costs of other antipollution programs, such as acid rain controls and the phaseout of CFCs. (New York Times News Service, 1997). In any case, doing something meaningful would cost a lot of money. For instance, the $10 billion that the 1990 government report estimated *is* a lot of money. But put that amount in context: In 1992, the United States spent $341.1 billion on military outlays alone. Furthermore, physicist Amory Lovins argued that cutting oil imports by 20 percent and introducing energy efficiencies would almost cover $10 billion a year (1998). Even with substantial costs, action about global warming would not be all cost—like most insurance premiums. Look again at the number of items in Table 4.2 that are calculated to be net benefits. In other words, some of the possible mitigative options, such as measures that increase efficiencies, have long-term financial benefits that outweigh their costs. But note that these are long-term benefits, and changes of such magnitude produces both winners and losers on the short-time horizon.

To summarize, I think that climate policy should be developed in the context of four main assumptions. *First*, the very nature of the threat obviously requires a cooperative arrangement among nations that recognizes the different resources, circumstances, and capacities of each nation to address the problem. *Second*, many mitigative carbon-cutting measures, at levels beyond those specified by the existing Kyoto treaty, do pass accepted cost-benefit tests, and governments ought to be persuaded to take them quickly. Some adaptive measures (e.g., protecting ocean flood plains, reconstructing human habitats, and adapting agriculture to new climate realities) have substantial costs that will be necessary to phase in, as circumstances require. *Third*, investments in efforts to check global warming in the longer term need to be balanced with those related to population stabilization and preserving forests and biodiversity, both of which would yield immediate returns (Cairncross, 1991: 165; Myers, 1997: 221–225). *Fourth*, there are good reasons for addressing the problem aside from the threat of global warming itself. Such efforts would have side benefits for human welfare and other environmental threats, which make them worth doing on other grounds. They could, for example, improve problems of urban air pollution, acid rain, wasteful consumption of nonrenewable resources and energy inefficiency. They could promote international global cooperation in programs about reforestation, sustainable agriculture, soil conservation, and probably land reform and the alleviation of the most wretched global poverty. In other words, addressing global warming could be an important reason to synergistically unify a basket of separate measures addressing human-environment problems and preserving the "global commons."

PERSONAL CONNECTIONS

Consequences and Questions

Assume that the intermediate projections about global warming are right, that a general and significant rise in mean global temperatures will happen during the next 40 to 50 years. You would probably experience some consequences during your lifetime. Assume that such change has real but unpredictable consequences. They may develop slowly or in fact suddenly as the various parameters of global change begin to interact. We cannot predict, but we can make informed guesses about probable consequences. What would be your share of these consequences? Think about the following questions. Publicly or privately, you will bear a share of the costs of coping with these changes. How will your interests as an individual be similar and different from the interests of society?

1. In terms of mitigation, think about the 1990 federal government estimate of $10 billion a year to control greenhouse gases in the United States. That is probably a conservative estimate. But even that again is not a trivial tax or economic surcharge. Think about some ways the costs or taxes might show up for you as a typical consumer or taxpayer.
2. An enormous number of the earth's people, including citizens of the United States, live in coastal regions. Look at the location of the really big American cities and metropolitan areas in this light. Do you live close to the ocean? Chances are that if the ocean level rises because of climate warming, you will be directly affected. If the ocean level rises modestly, dense human settlements will have to be protected, people relocated or evacuated. Increasing sea levels would flood scores of estuaries, freshwater aquifers, and other resources on which societies depend. In Galveston (TX) a 1-meter sea rise would place the whole city in a flood danger zone, and in Charleston (SC) 60 percent of the city would be flooded on average every decade. With much larger populations, New York, Los Angeles, Seattle, and Miami would be severely impacted. A 50-cm rise in sea level by the year 2100 would cost the United States somewhere between $20.4 and $138 billion in lost property and damage to economic infrastructures. A 1-meter rise over that time would escalate costs and impacts dramatically. Either privately or publicly, you will bear a share of those costs, and some of the costs will be shifted to you even if you don't live on a coastal region, either through private insurance costs or public disaster relief programs.
3. Climate zones and vegetation may shift unpredictably. Some regions will become drier and produce less food, others wetter and warmer and produce more. Climate warming would increase both evaporation and precipitation, and atmospheric models suggest that regional effects would be extremely uneven. The most likely outcome is that water and food availability would change, and for the poor everywhere basic food security would become more difficult.

   For all of these observations, what do you think some of the implications would be for you in terms of prices? Taxes? Pressures to reduce consumption? For your general expectations and outlooks about "how things are these days"? Maybe we are talking about changes that would be so gradual that you wouldn't notice them. But maybe not, either. What are some ways that you think you would need to change your lifestyle to these changes?
4. You may already *be* beginning to absorb such costs. One likely consequence of global warming is rapid weather changes and more frequent severe weather patterns. A warmer atmosphere and warmer seas may result in greater exchange of energy and momentum to the vertical change processes important to the development of cyclones, tornadoes, thunderstorms, and hailstorms. In fact, between 1990 and 1998 there was a remarkable clustering of highest floods, longest droughts, most severe wild fires, and worst heat waves ever. For Americans, the standout examples of the early 1990s were Hurricane Andrew, which devastated South Florida in 1992, and the 1993 floods in the Midwest. (If you have friends

in Des Moines or St Louis, ask them about that one.) For much poorer Central Americans, it was Hurricane Mitch in 1998, and for China and Bangladesh, devastating floods. In China, only 1.5 percent of the damage had insurance of any kind.

The insurance industry has not exhibited the same scientific reserve as have some others when paying losses for flooded farmland or crop losses, or hurricane, tornado, or other disaster losses. For instance, after South Florida's Hurricane Andrew, major insurance companies (including the Prudential Insurance Company, Allstate, and State Farm) paid out huge claims. If the storm had directly hit either Miami or New Orleans, much of the U.S. property insurance industry would have been flat on its back. Globally, the insurance industry paid out a whopping $22 billion in that year. The industry paid out $9 billion, the fourth highest ever, in 1996. In that year, for the first time ever, a large delegation of insurers attended the Conference of the Parties, the official climate convention in Geneva, Switzerland. The insurers signed a statement calling on governments to reduce substantially the emissions of climate-altering greenhouse gases (Flavin, 1999: 70).[5]

Again I ask you: How will such costs percolate through the economy, and what might be your share of absorbing such costs? As a future homeowner, investor, or business owner, what problems do you foresee with maintaining adequate insurance? If you think that's spooky, put yourself in the place of a Bangladeshi, in a country where millions live in typhoon-vulnerable areas and can't afford any kind of insurance.

5. These are tough questions, and you can, of course, believe that none of these concerns are worth worrying about. But you need to recognize this is not the consensus of *most* of the world's scientific community. They could all be wrong. Who do you trust? What's your best bet?

Real Goods

1. *Ceiling fans*. Appropriate technology's answer to air conditioning, ceiling fans cool tens of millions of people in LDCs. Air conditioning, as found in about two-thirds of U.S. homes, is a real electrical juice hog and the bane of the stratospheric ozone layer because of its CFC coolants. Ceiling fans, on the other hand, are simple, durable, repairable, and take little energy to run. Ceiling fans run at very low speeds (summer and winter) and help even out the "layers" of room temperature. A fan over your bed circulates enough air that you may not have to run your air conditioner as much.
2. *The reel-type push lawnmower (without a gas engine)*. They're back! For about one-fourth of the cost of a self-propelled power mower (probably a sixth of the cost of a riding mower). Made with metal alloys that are much lighter and easier to push than historic versions. Easy to push, even up a 45-degree incline on my front yard (where my previous power mower would stall because it drained oil into the engine, and where I always feared that it would tip over and slice my foot off). No gas, tuneups, smoke, pollution, or noise. There is only a quiet *clik, clik, clik* as it moves that brings back nostalgic childhood memories for me. The kids in the neighborhood have never seen one and come over to ask me what the thing is.

3. *Compact fluorescent light bulbs.* They are three or four times as efficient as regular incandescent bulbs. One 18-watt compact fluorescent light bulb provides the light of a 75-watt incandescent bulb and lasts ten times as long. Currently they are pretty pricey but should get cheaper as more people use them. Even so, over the life of its use, an 18-watt compact bulb can keep more than 80 pounds of coal in the ground and about 250 pounds of CO_2 out of the atmosphere.

ENDNOTES

1. The five models are the NASA/Goddard Institute for Space Studies (GISS) model, the National Center for Atmospheric Research (NCAR) model, the NOAA Geophysical Fluid Dynamics Laboratory model, the model developed at Oregon State University (OSU), and the model developed by the United Kingdom Meteorological Office (UKMO).
2. Throughout the earth's long geological history, the expansion and contraction of glaciers, for which there are many physical markers, is a fairly good measure of past temperature fluctuations. Other evidence comes from studies of fossilized pollen grains, annual growth rings of trees, and the changing sea levels measured by the presence of coral reefs. Cores of sediment extracted from the floor of the deep oceans are particularly informative because their chemical composition and the presence of warm- or cold-water fossils provide clues to change in ocean temperature and the volume of polar ice caps. The most useful information about the presence of greenhouse gases throughout geological history comes from analyzing changes in the concentrations of gas bubbles through time (of, for example, CO_2 or methane) from ice cores extracted from ancient glaciers in Greenland and Antarctica (Silver and DeFries, 1992: 25).
3. For example, one of my neighbors, a freshwater biologist working for the Army Corps of Engineers, felt compelled to seek congressional "protection" from his senator before publicizing a shoddy and deceptive job done by the Corps in compiling data for an environmental impact statement about one of their proposed reservoir projects.
4. A reactor year is one reactor operating for one year. In 1986 the world had 366 operating reactors, so in that year there were 366 reactor years (Milbrath, 1989: 238).
5. As I write this in mid-September 1999, Hurricane Floyd is poised between Florida and the Bahamas in the South Atlantic. It is a large and terribly powerful storm with 155-mph winds, just below category 5, the highest rating for tornados (only two category 5 tornadoes have hit the United States since 1935). Floyd will hit land somewhere along the 350-mile Atlantic coast between Miami and middle Virginia. More than a million people have been evacuated from the Atlantic coast, and national guards are on alert in all affected states. This represents the most extensive and expensive peacetime evacuation in American history. By the time you read this, you will be able to find out exactly where Floyd made landfall.

PART III

THE HUMAN CAUSES OF ENVIRONMENTAL PROBLEMS

CHAPTER FIVE

Population, Environment, and Food

Imagine a human community with 100 people, 50 women and 50 men. Imagine further that during the next 25 years each of the women had four children (two boys and two girls) and that each of the girls grew up and also had four children. Thus, the original 50 mothers had 200 children (50 × 4 = 200). Of these, 100 became mothers, giving birth to 400 grandchildren (100 × 4). Our hypothetical community has now grown from 100 to 700 (100 + 200 + 400), a sevenfold increase. This imaginary scenario illustrates *exponential growth,* and like all living populations human populations have the capacity to grow at exponential rates. In fact, the human population of the world has grown at a dramatically exponential rate.

For thousands of years, the human population grew at a snail's pace. It took over a million years to reach about one billion people by the beginning of the nineteenth century. But then the pace of population growth quickened: A second billion was added in the next 130 years, a third in the next 30 years, and the fourth billion in just 15 years (McNamara, 1992). By the 1990s, there were more than five billion people on the planet, and the United Nations estimated that in early October 1999 human baby number six billion was born. The overwhelming odds are that baby six billion was born to a poor family in a poor nation (Gelbard et al., 1999). (See Figure 5.1.)

Another way of expressing the rate of exponential growth is by computing the *doubling time*—the number of years it takes for population size to double. From 1750 to about 1950, the doubling time for the world population was about 122 years. But by the 1990s, the doubling time was only about 35 years.[1] World average growth rates mask lots of variation between nations: For the MDCs, the doubling times are 60–70 years and for LDCs they may be as low as 15 years (Weeks, 1994: 34). Think of that: Every 15 years the poorest nations of the world (such as Haiti, Bangladesh, and Rwanda) must double their supplies of food, water, housing, and social services just to maintain current dismal living standards. The global mean growth rate has declined somewhat in recent

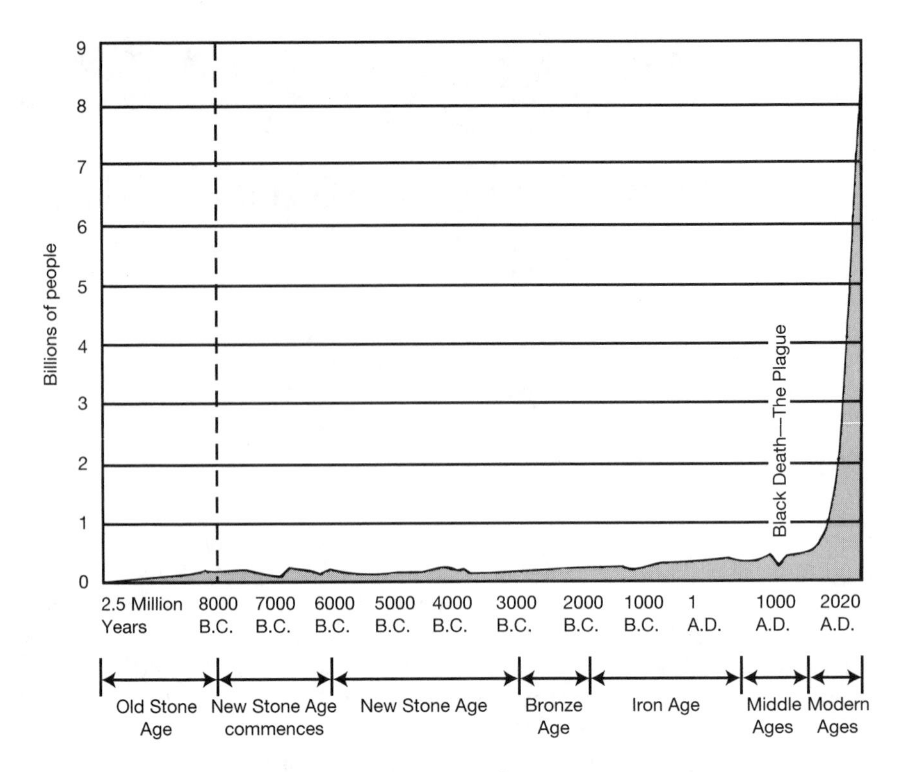

Figure 5.1 World Population Growth throughout History
Source: Adapted from M. Kent (1984), *World Population: Fundamentals of Growth.* Population Reference Bureau.

decades, and in 1998 world population was growing at a rate of 1.4 percent per year. Even so, a world population of seven billion could come very quickly with such a large base of absolute numbers, and many women in their prime childbearing years. The U.N. Population Division projects standardized world future growth outcomes using different scenarios for fertility and mortality. Although no single scenario forecasts the future, three of them—the low, medium, and high scenarios—span the range of plausible outcomes. In 1998, projections for the year 2050 were 7.3 billion (low scenario), 9 billion (medium scenario), and 11 billion (high scenario) (United Nations, 2000).

These numbers are truly staggering, and the popular term "population explosion" is indeed a proper description for the *demographic history* of recent times.[2] If the present six billion humans have visibly stressed the environmental carrying systems (as demonstrated in Chapter Three), what impact will eight to twelve billion have? Larger populations will eat more, consume more, and pollute more.

In this chapter I will discuss (1) the dynamics of human population change, (2) the controversy about role of population growth related to environmental and human problems, (3) the relationship among population growth, food supply, and the prospects of feeding a much larger population, and (4) some contentious policy questions about stabilizing the growth and size of the world's population.

THE DYNAMICS OF POPULATION CHANGE

Concern with exponential population growth is not new. Contemporary concerns about population growth are still framed by questions raised by Thomas Malthus (1766–1834) in his *Essay on Population,* first published in 1798.[3] His book went through seven editions and has undoubtedly been the world's single most influential work on the social consequences of population growth. Malthus and other classical economic thinkers wrote at the start of the nineteenth century, when accelerating population and industrial growth were raising demands for food faster than English agriculture could respond. They saw real wages falling and food imports rising. Most classical economic thought emphasized the limits that scarce farmland imposed on agricultural expansion, arguing that applying ever more labor and other inputs to a fixed land base would inevitably encounter diminishing returns (you might want to review the discussion of the classical economic theorists in Chapter Two). Their argument was that limited productive land as well as limits of the supply of capital and labor would determine how many people could be supported by a nation.

Malthus turned these arguments upside down. He argued that since "sexual passion was a constant," human population would increase exponentially (in his words, "geometrically"), while the supply of land, food, and material resources would increase arithmetically. Thus instead of limited natural resources (land) and labor causing limits to population growth, Malthus believed that population growth caused resources to be overused and the market value of labor to decline. Population growth rather than lack of resources and labor produced poverty and human misery. "Overpopulation" (as measured by the level of unemployment) would force wages down to the point where people could not afford to marry and raise a family. With such low wages, landowners and business owners would employ more labor, thus increasing the "means of subsistence." But this would only allow more people to live and reproduce, living in poverty. Malthus argued that this cycle was a "natural law" of population: Each increase in the food supply only meant that eventually more people could live in poverty.

Malthus was aware that starvation rarely operates directly to kill people, and he thought that war, disease, and poverty were *positive checks* on population growth (the term "positive" in this context these has always puzzled me!). Although he held out the possibility of deliberate population control (*preventative checks)* on population growth, he was not very optimistic about their effectiveness. Rejecting both contraception and abortion as morally unacceptable, he believed that only moral restraint (such as sexual abstinence and late marriage) was acceptable.

In sum, Malthus argued that poverty is an eventual consequence of population growth. Such poverty, he argued, is a stimulus that *could* lift people out of misery if they tried to do something about it. So, he argued, if people remain poor, it is their own fault. He opposed the English Poor Laws (that provided benefits to the poor) because he felt they would actually serve to perpetuate misery by enabling poor people to be supported by others (Weeks, 1994: 61–65). Interestingly, many in our day criticized the governmental welfare system on just such grounds. Malthus's ideas were attacked from all sides in his day. I will save these criticisms for later, because they foreshadow many contemporary objections to demographic explanations of environmental problems. Certainly, in the short run, events have not supported the Malthusian view. He did not foresee

> [the] expansion of world cropland to more than double its 1850 acreage; development of agricultural technologies capable of quadrupling yields achieved by traditional farming methods . . . the diffusion of health services and improved hygiene, lowering death rates and then birth rates. He would never have predicted, for instance, farmers being paid not to plant, in order to cut surpluses and to reverse erosion. . . . And he would be amazed at the growth in world population. (Hendry, 1988: 3)

Whether Malthus will continue to be seen in error during the next century is another matter, as world population and related problems continue to grow dramatically. As you can see from the questions I raised here and in earlier chapters, there are plenty of grounds for concern, and indeed, *neo-Malthusians* today are alarmed about population growth as a cause of environmental and human social problems. But before returning to this issue, I'll examine the general outlines of population dynamics and change, as it is understood by demographers.

The Demographic Transition Model

One of the most universally observed but still not clearly explained patterns of population growth is termed the *demographic transition*. By the 1960s, George Stolnitz reported that "demographic transitions rank among the most sweeping and best documented trends of modern times . . . based upon hundreds of investigations, covering a host of specific places, periods, and events" (1964: 20). This model of population change has three stages:

> Primitive social organization (stage I) where mortality [is] . . . relatively high . . . and fertility is correspondingly high; transitional social organization (stage II) where mortality is declining, fertility remains high, and the population exhibits

high rates of natural increase; and modern social organization (stage III) . . . where mortality has stabilized at a relatively low level, fertility is approaching the level of mortality, and a stationary population size is possible in the near future. (Humphrey and Buttel, 1982: 65)

You can see this process schematically in Figure 5.2.

Explanations of this transition vary and are pasted together from somewhat disparate elements, but in general they flow from assumptions about the demographic consequences of modernization and industrialization. First, industrialization upgraded both manufacturing and agricultural productivity so that the economic base could support much larger populations. Second, medical advances in the control of epidemic disease and improvements in public services like urban sewerage, water systems, and garbage disposal contributed to improved health and reduced mortality rates. Third, as populations became increasingly urbanized, family changes occurred. The children of rural peasants are generally an economic asset: They eat little and from an early age contribute substantially the family farm and household. But urban children—their education and rearing—become more of an economic burden than an asset (Harper, 1998: 261).

Industrialization was also coupled with opportunities for women to work outside the family and eventually improved the status of women. Birth rates are high where the status of women remains low and they are economically dependent on men (Keyfitz, 1990: 66). Industrialization also produced societies that established national social security programs apart from kinship, which meant that parents were less dependent on the support of their children in old age. Industrial modernization had, in other words, a variety of incentives that promoted smaller families. As social and economic

Figure 5.2 Demographic Transition Model

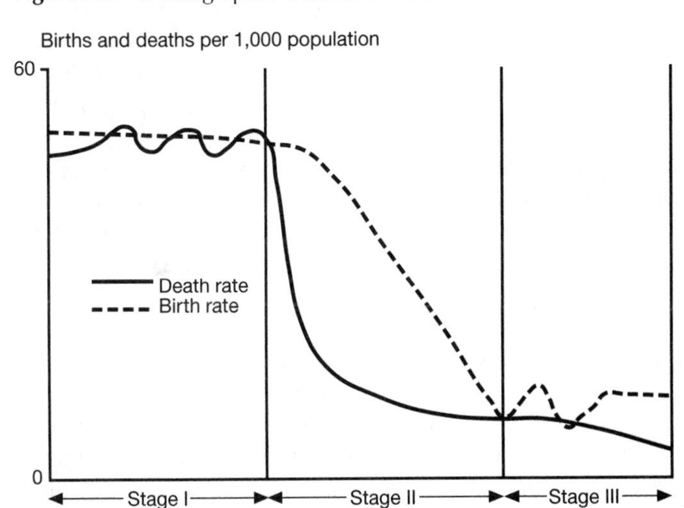

incentives changed, cultural norms promoting large families began to weaken. Finally, research demonstrated that while industrialization was inversely related to fertility, it also changed the level of economic equality. In the European nations "the demographic and economic transitions led to a general improvement in living standards for all persons and a gradual reduction in income inequalities" (Birdsall, 1980). There is good reason to doubt the unique impact of family planning programs as a cause of fertility decline apart from deeper socioeconomic causes, but abundant evidence exists that information about birth control and access to contraceptives have been important factors in fertility declines in all countries (Keyfitz, 1990: 66).

However it happened, the demographic transition process has meant that beginning with social and economic modernization, death rates declined, followed after a time interval by declining birth rates. But between these events was a period of *transitional growth* when birth rates remained high but death rates rapidly declined. That transitional growth period is what the population explosion since the beginning of the industrial era is all about. As you can see, when applied at a global level, the demographic transition model provides grounds for expecting the world population growth to eventually stabilize. It is a broad abstraction that fits the facts of long-term population change in the MDCs, but the variety of causal factors suggested do not form a very coherent theory about this change.

There are at least two other limitations of the demographic transition model. It is ethnocentric in assuming that historic processes of demographic change in MDCs are being repeated in the LDCs, when in fact the historical, political, and economic, circumstances in which they entered the modern world differ importantly. Related to this criticism is another—that the model has not been capable of precisely predicting levels of mortality or fertility or the timing of fertility declines at national, much less at global, levels. This is both because the causes of demographic transition are not well understood, and also because historical events (such as wars or economic collapse) cause unpredictable changes in the stability of demographic projections. Small differences in projected numbers stretched over long periods of time can add up to big differences. That is why agencies that make population projections typically make high, medium, and low ones, letting the user decide which is most reasonable. This means that some really important questions such as "How rapidly will global stabilization occur?" and "At what equilibrium number?" cannot be answered with much certainty. The uncertainties here much like those discussed about climate change in the Chapter Four.

Demographic Divergence: MDCs and LDCs

As MDC populations went through the period of transitional growth, they expanded into less densely populated frontier areas, rich with land and resources to be developed. This process of European expansion and colonization began in the 1500s, before the industrial revolution. Until 1930, European and North American countries grew more rapidly than the rest of the world. But since then, population growth slowed and geographic outward expansion virtually ceased (Weeks, 1994: 37). Today most MDCs are far along the path toward population stabilization, well into stage III of the demographic transition. They exhibit declining birth rates and slow rates of growth. Many are coming close to the equilibrium or replacement rate of fertility, which would result in zero population growth (2.1 children per female). By 1998 in Western Europe, population growth was almost zero or even declining, even with the impact of immigrants from other parts of the world. In Austria, France, the United Kingdom, and Norway the populations grew between 1 percent and 3 percent a year. Germany was declining by 1 percent each year. In much of postcommunist Eastern Europe, year including Russia, Romania, Lithuania, and Ukraine, economic and social conditions were so bad that birth rates were below replacement levels and population size declined slightly each year (Population Reference Bureau, 1998: 8).

But in LDCs, the story is very different. Their rapid transitional growth came later in the twentieth century without the benefit of territorial expansion—that is, without the relatively unpopulated land or colonies to absorb the pressure of population growth. In addition, they have birth rates and levels of mortality much higher than European MDCs, at a comparable stage. As a result, LDC populations are growing rapidly, especially in the poorest of the poor nations. In the MDCs, demographic transition proceeded apace with internal economic development. But the decline of death rates in LDCs was more related to the rapid introduction of effective techniques of disease by outsider agencies like the World Health Organization. Babies born in the poor nations today have a historically unprecedented chance of surviving to adulthood, and the average life spans of nations have converged. The vast majority of babies born in the world today live in the LDCs. At the turn of the year 2000, the world added about 86 million people per year, and at least 90 percent of this growth was happening in the LDCs.

Even so, economic development—with its widespread improvement in living standards, improved education and opportunities for women, incentives for smaller families, and the establishment of national social security systems—has not kept pace in the poorest LDCs. Cultural and religious norms favoring large families are still powerful. Even when the world economy was growing, people in the poorest nations experienced little economic growth, while population growth continued vigorously. Often economic growth has been literally "eaten up" by exploding populations. The continuation of this demographic divergence between MDCs and LDCs into the next century may increase geopolitical tensions, pressure on migration and refugee flows, and a corresponding social and environmental duality among rich and poor nations. Even though they consume more per capita, MDCs with more economic and technological resources will find it easier to maintain environmental quality than poorer LDCs. Among them, both rural and urban populations are rapidly growing, pressures on natural resources are increasing, and economic and technical resources are often overwhelmed as local and national governments try to provide employment for increasing labor forces and infrastructure for expanding cities, like electricity, clean water, and waste disposal (Livernash and Rodenburg, 1998).

Population Redistribution: Urbanization and Migration

So far, I have focused on population growth in terms of the dynamics of demographic transition. But another type of population change is *population redistribution*, meaning the net spatial changes in population as individuals and families move from place to place. The two most important forms of population redistribution are urbanization and migration. Both are related to the pressures of population growth.

Urbanization

Most North Americans now live in—and were born in—cities. While we may be attracted to the attractions of cities or curse their problems, we recognize that urban life is the cultural, economic, and political center of modern society. Urbanization, or the redistribution of people from the countryside, is not new but has dramatically accelerated with the explosive transitional growth just described. Compared to rural dwellers, urban dwellers made up only about 11 percent of the world's population in 1850, but 26 percent in 1900, and 50 percent in 2000. Among the MDCs, at least 75 percent did so by the turn of the twentieth century (Gelbard et al., 1999: 17).

Cities are, of course, nothing new. They emerged with the agricultural revolution, but those cities were not very large by today's standards. Ancient Babylon might have had 50,000 people, Athens maybe 80,000, and Rome as many as 500,000 (Weeks, 1994: 354). To put this in perspective, Rome, the premiere imperial capital of much of the Mediterranean world and hinterlands beyond, was at its peak a bit smaller than my hometown of Omaha, Nebraska. Ancient cities were unusually dense settlements that were the political, ceremonial, and administrative centers in a diffuse "sea" of rural villagers. Villagers made up perhaps 95 percent of the total population of such societies, and their crops and livestock were the real sources of wealth, on which urban elites lived by imposing taxes. Ancient (and medieval) cities were neither economically or demographically self-sustaining. Poor sanitation and the rapid spread of epidemic disease (the plagues of ancient and medieval worlds) meant that they had higher death rates and lower birth rates than the countryside. They often had an annual excess of deaths over births, which meant that they had to be replenished by migrants from the countryside.

The self-sustaining demographic and economic character of modern cities began with transforming the basis economic production and wealth from agriculture (produced in the country) to manufactured and traded goods (produced in cities). Industrialization interacted with improvements in sanitation and epidemic disease control to dramatically accelerate urbanization.

Urbanization of the MDCs. Industrial era urbanization was fueled not only by expanding urban opportunities, but also by the push of rural overpopulation, poverty, consolidation of land holdings and declining farm labor markets resulting from the industrializing of agriculture (noted in Chapter Three). As economic development proceeded in Europe and North America, cities grew because they were more efficient. They brought more raw materials, workers and factories, financiers, and buyers and sellers together in one location than did dispersed rural production. Furthermore, as industrial societies developed, evolving modes of production continually reshaped the economic base of cities from the commerce and trading centers of the 1600s and 1700s (e.g., Amsterdam, London, Boston), to those centered on factories and industrial production in the late 1800s (e.g., Birmingham, Pittsburgh, Chicago). Since World War II, improvements in technology and the growth of an economy based on "services and information" has meant that the economic base of many cities is no longer manufacturing but, more often, the corporate headquarter locations of far-flung multidivisional and multinational firms and banks (e.g., Minneapolis, Dallas–Fort Worth). Now the largest MDC cities, such as Tokyo, New York, and Los Angeles, are really "world cities" that produce wealth by organizing and controlling international trade, commerce, and finance.

After the year 2000, the world passed something of a milestone when over half of its population was classified as urban. Fifteen years later (in 2015), the LDCs will be more than 50 percent urban (in 1950 only one fourth were).[4]

Urbanization of the LDCs. If you look at Table 5.1 closely, you can see that LDC urbanization is now taking place more rapidly than in the MDCs. Another way you can get a feel for this is to consider the world's ten largest cities. In 1950, only two of the ten largest urban conglomerations in the

Table 5.1 Percent of Population that Is Urban, 1950–2015

	1950	*1975*	*1995*	*2015*
World	29.7	37.8	45.3	54.4
MDCs	54.9	69.9	74.9	80.0
LDCs	23.8	36.2	47.1	57.3

Source: Adapted from United Nations, 1998b, *World Urbanization Prospects: The 1996 Revision.* United Nations Population Division.

world (Shanghai and Calcutta) were located in the LDCs. But by 2025 United Nations demographers project that 9 of the top 10 will be in the LDCs. In order, they are Mexico City, Shanghai and Beijing (China), Sao Paulo (Brazil), Greater Bombay and Calcutta (India), Jakarta (Indonesia), Dacca (Bangladesh), and Madras (India). New York, Chicago, London, and Paris, all on the 1950 list, are nowhere in sight. While Tokyo-Yokohama will still be the largest urban area in the world, it will be followed in 2025 by the demographic giants of the third world, Mexico City and Sao Paulo (Brazil) (United Nations, 1982, 1998).[5]

As in the MDCs in an earlier era, the explosive urbanization in the contemporary LDCs is fueled by the poverty, hunger, and destitution of peasants pushed off the land and also by the less visible forces but powerful force of high birth rates and population pressure. But there is a fundamental difference between the two eras. MDC urbanization was also accompanied by the pull of exploding economic opportunities in the industrializing cities. Urbanization in the LDCs today is largely a matter of the push of rural poverty without the simultaneous pull of dynamic urban economic growth. In other words, the LDCs have developed very rapidly in the post–World War II period, but they have skipped the prolonged period of industrial and manufacturing economic growth the MDCs experienced. Although less developed, many LDC cities have come to represent service economies without passing through the transitional stage of industrial growth (Walton, 1993: 289–302).

A service economy, as we have discovered in the United States, often produces less employment and comparatively lower wages for many people than do industrial and manufacturing economies. Thus, cities such as Calcutta, Cairo, Dakar, Jakarta, and Rio de Janeiro are becoming awash with displaced peasants with grim prospects for fruitful urban employment.

> To escape deepening rural poverty . . . [millions] of "environmental refugees" are on the move in Latin America, Africa, and parts of Asia, mostly from rural to urban areas. City services are collapsing under the weight of urban population growth, and unmanageable levels of pollution are creating a variety of threats to human health...solid waste could quadruple . . . [many] rivers are virtual open sewers, and many waterways flowing through metropolitan areas are biologically dead. (Camp, 1993: 130–131)

Such urban masses live in shantytowns and typically scrape out a meager existence as street vendors of petty goods and services.

Migration to these cities is fueled not only by rural misery, but also by political policies that give preferential treatment to city dwellers. In cities national governments concentrate schools, receive investments from multinational firms, and are most concerned with regulating the price of foodstuffs. By subsidizing the price of food (a policy practiced among most LDCs), life is made easier for urbanites while farm incomes are depressed (Harper, 1998:

263). It is also true that urban migrants are typically younger and usually more energetic and talented than their kin left behind in the countryside. This combination of demographic characteristics and political policies mean that, as bad off as they are, urban migrants are often better off than their relatives and friends who remain in the countryside and often send cash remittances back to them. Urbanites have fewer children and higher incomes.

The urban-to-rural diffusion of new consumption patterns and diets also exacerbate rural deprivation. Rural dwellers quickly learn to desire and emulate consumption patterns of the MDCs, and there is an increasing demand for goods (such as rice, hybrid grains, beef, tea, bread, biscuits, and soft drinks) that cannot be produced by the average rural farmer. Consequently the demand for the traditional cereals and foodstuffs of the countryside decreases while the most successful and "modern" farmers produce for export markets (Hendry, 1988: 22). Understand what is going on here: In a bizarre and perverse urban development process, production of the traditional food available to the poor in both cities and the countryside declines as products (including foodstuffs) are increasingly manufactured for export markets in a world market economy. Government investment and price policies, intended to benefit urban dwellers, depress the income of small traditional farmers (who would produce cheap food). This situation has been replicated in dozens of LDCs in recent years. The most striking example was Brazil during the 1960s. The Brazilian economy boomed while the production of black beans—the staple diet for low-income Brazilians—plummeted. And poverty became more general in both the cities and the countryside (Barnet and Muller, 1974).

In spite of the fact that the newly urban dwellers of the LDCs are somewhat better off than their village cousins, such rapid urbanization has overwhelmed the ability of cities to provide jobs, water, sanitation and food, and the resulting misery and degradation among recent migrants is historically unprecedented. Desperate peasants left behind in declining economic circumstances are most likely to survive by overfarming marginal land.

Migration

Urbanization is a really a special form of *migration*, which means the relatively long term movement of an individual, household, or group to a new location outside their community of origin (de Blij, 1993: 114–115). Being cultural foreigners and new claimants for existing jobs and services, their presence in new host communities is usually contentious and difficult. They may send money and information to their nonmigrant kinfolk back somewhere. Indeed, you need to understand migration as not only the numerical redistribution of people, but also as a slow but pervasive *social interaction process* which diffuses and reshapes human cultures—and the distributions of power and wealth.

Migration may be *forced*, as in the case of prisoners that the British shipped to penal colonies in Georgia and Australia. It was also the case of the African slaves brought to the New World, and the 50,000 Asians forcibly expelled from the African nation of Uganda in the 1970s—with only the belongings that they could carry on their backs. But migration may also be *voluntary*, as in the case of most Europeans who came to North America in the late nineteenth and early twentieth centuries seeking material improvement and greater opportunities. While they were attracted by better opportunities, they were also often fleeing from rotten conditions in their homelands. Some, such as the Irish immigrants to Boston and New York, came fleeing from famine, poverty, and unemployment in their homelands (remember the Irish potato blight and subsequent famine mentioned in Chapter One?). Others fled wars or political and sometimes religious oppression.

High-volume waves of internal migration weaken but do not destroy extended kinship networks. The phenomenon requires that host institutions adjust to shifts in the numbers and characteristics of people served. It alters, for example, the availability of labor, the demands for geriatric medicine, and the numbers and characteristics of students to be served by educational systems. Since migrants always insert themselves into or remove themselves from community status hierarchies, they always change the stratification system of communities: In-migrants tend to improve their status by moving into communities, while out-migrants improve it by moving out. In sum, adjustments, often difficult ones, are required in both the communities that migrants leave as well as in their new host communities. *Internal migration* is usually "free" in the sense that people are choosing to move in relation to their perception of better living conditions elsewhere. *International migration* is sometimes free, but it usually means that the migrant has met fairly stringent entrance requirements, is entering illegally, or is being granted refugee status (Weeks, 1994: 194).

Explaining Migration. The most common theory about the causes of migration is what demographers and geographers have called the *push-pull theory*, which says that some people move because they are pushed out of their homelands, while others move because they have been pulled or attracted to a new place. In reality, a complicated mix of both push and pull factors operates jointly to impel migratory behavior. Pushes can include poverty and lack of economic opportunity; fears for personal safety; political, cultural, or ethnic oppression; war, including civil war; and natural disasters such as droughts, floods, and so forth. Often underlying the push of these concrete factors is population pressure from rapid growth. The pulls are the mirror image of these and are likewise complex: the perception of better economic opportunities, greater social stability, and affiliation (desire to join relatives and friends). At any rate, social science conjures up the migrant as a rational decision-maker who calculates the costs and benefits of either pulling up stakes and moving or staying put. This thesis was posed a long ago as 1885 by British demographer Ernest Ravenstein, who studied internal migration in the British Isles (1889).

Ravenstein found, as have many investigators since, that migrants have some common characteristics: They are younger than nonmigrants; they are less likely to have families, or if they do, they have fewer and younger children; and they are likely to be better educated (Weeks, 1994: 197–203). In fact, voluntary migrants are a select population, usually more talented, capable, adaptable, and ambitious than nonmigrants. In addition to personal characteristics such as these, the push-pull causes of migratory behavior are also conditioned by intervening factors or barriers. These include the costs of moving, lack of knowledge about migration options or managing complicated moves, broad themes of the sociocultural environment like established values about the importance of geographic "roots," risk taking, and openness to change. As you can see, in spite of the simple attractiveness of the push-pull thesis, the actual situation is quite complicated and not simple to predict. (See De Jong and Fawcett [1981] for an ambitious effort to conceptualize the complex causes of migratory behavior.)

Old and New Patterns of Global Migration. We do not know exactly how many persons have migrated around the world at any given time, but beginning with the modern era (in the 1500s) there were discrete waves of immigration involving particular locations that accounted for the greatest volume of immigrants. One such stream, as I'm sure you are aware, virtually *constituted* the nations of North America. Except for Native Americans, the citizens of the United States, Canada, and Mexico are *all* descendants of immigrants from somewhere else. Immigrants from Europe (and particularly Britain) were always more welcome, and by the 1920s the United States was so concerned about the flow of "unsuitable" non-Anglo-Saxon immigrants that it passed laws establishing quotas by nations that severely restricted non-European immigration.

Before World War II, the main currents of migration were out of the more densely settled regions in Europe and Asia and into North and South America and Oceania. Since the 1950s, that changed so that the net migration flows were back into Europe, out of South America, and (still) into North America, but increasingly from non-European nations. About half of all international migration is from one LDC to another, but the net flow of international migration is now from the LDCs to the MDCs (Gelbhard et al., 1998: 16). The pressure of rapidly growing LDC populations since World War II enormously increased the pressure on natural resources and the demands for employment and social services, while in the MDCs a slowly growing population and buoyant economic growth often created a demand for lower-cost workers from the LDCs. Thus "guest" workers flowed into northern

Europe from nearby Algeria, Egypt, Turkey, and the Middle East as well as from comparatively less developed southern and Mediterranean Europe and—more recently, as the communist world collapsed—from Eastern Europe. In the short time between 1985 and 1992, immigrants to Western Europe from North Africa and Eastern Europe *tripled,* rising from 1 to 3 million persons. War, including civil war, often creates the social chaos that stimulates a flood of immigrants and refugees. Immigrants are considered refugees or asylum seekers if they can demonstrate that they left their home countries to avoid persecution. The number of asylum seekers around the world peaked at 17.6 million in 1992. In 1998, an estimated 5.7 million refugees lived in the Middle East, 2.9 million in Africa, and 2.0 million in Europe (Gelbhard et al., 1998: 16; Pomfret, 1993: 7).

Between 1970 and 1997 18 million people immigrated to the United States. These "new immigrants" come from a variety of places: 42 percent from Latin American and the Caribbean, 33 percent from Asia. They particularly migrate across the 1,500-mile open border across the Rio Grande that divides the United States and Mexico.[6] As Mexico's population has grown, so has the difficulty of finding adequate employment for the burgeoning number of young adults. That push, in combination with the pull of higher wages in the United States, stimulated an enormous migration stream. Between 1820 and 1992, more than 5 million Mexicans immigrated *legally* into the United States with more than half of those arriving since 1980, and census data suggest that at least another million were undocumented (or illegal) migrants from Mexico alone (Warren and Passel, 1987). In addition, immigration from Asia significantly increased after the 1960s, made much easier by the reform of the Immigration Act in 1965, which replaced the old system of national quotas with new criteria based on skills, refugee status, and family ties (meaning that they could come because they had relatives in the United States). Immigrants came from Taiwan, China, Korea, the Philippines, and Indochina (collectively more than 1.6 million between 1971 and 1980). War created special circumstances and obligations in the 1970s, when many Indochinese (such as Vietnamese, Laotians, Cambodians, and the Hmong) were granted special entry as refugees after the Vietnam War. There is also a significant outflow (emigration) from the United States, but between 1960 and 1990 the ratio was about six immigrants to every one emigrant (Weeks, 1996: 236–245).

Refugees and asylum seekers to Europe and America are both legal and illegal. They arrive on foot and by rail, air, and sea from diverse origins. Some are smuggled in trucks and ships jammed together under terrible conditions. As always, floods of immigrants create conflict and controversy as they seek employment and raise questions about the political and cultural coherence of nations. By 1993, immigration was a significant and volatile political controversy in the United States, and many European nations. Germany in particular was busy revising its generous asylum laws. As LDC populations rapidly grow, it becomes harder and harder to find a niche in the domestic labor force, and people are often compelled to move. And as the United States and Europe are learning, it is very difficult and costly to stem the tide of immigrants who want to move and find ways. Whether we like it or not, a significant portion of people from the LDCs are coming to be our neighbors. And they will change the demography, culture, and eventually the politics of the nation. The U.S. Census Bureau, for instance, predicts radical shifts in the racial and ethnic composition of the nation, fueled by both immigration and the higher birth rate of ethnic minorities. The proportion of whites is expected to diminish from about 74 percent in 2000 to a tenuous majority of 53 percent by 2050, and Hispanics may well replace African Americans as the largest minority group (Martin and Midgeley, 1999: 23). The prime immigrant entry ports, such as Los Angeles, Miami, and New York, may in fact become more Third World than American cities.

Population, Environment, and Social Stability. I have discussed types of population change—growth, urbanization, and migration—in some detail. Now I would like to summarize their relevance as hypothetical causes of environmental problems. It has been argued since the time of Malthus that the tremendous population growth of modern times has damaged the environment. It did so by increasing demands for food, water, energy, and natural resources, and most think that this problem will become increasingly acute as the world population increases to nine or ten billion in the next century. Recall the discussion of soil erosion and water problems in Chapter Three. Population pressure contributes to both migration and urbanization so that the environmental impact of population growth is not evenly distributed. Problems are particularly acute in urban areas where the air, water, and land cannot absorb the wastes and toxic by-products of industry and dense populations. Other than problems of population density, the very *location* of cities causes environmental hazards. Because urban populations and industries need lots of water, they tend to be located along lakes, rivers, and bays. As a consequence, rivers like the Missouri, Mississippi, and Ohio; lakes like Erie and Michigan; and bays like the Chesapeake and New York become badly polluted (Eitzen and Baca Zinn, 1992: 101). Finally, by creating chaos and hardship in the LDCs, population growth will further accelerate the streams of internal and international immigration. However enriching immigration is in the long term, at a given time host nations and communities will find it a socially and politically disruptive burden. Evidence suggests that large flows of refugees are associated with social disruption and civil violence (Homer-Dixon, 1994). This is particularly so when the world economy is sluggish, as it was following the collapse of Asian economies in 1998. It is a fantasy to think that because of the demographic divergence just noted, the problems associated with population growth will be "contained" in the LDCs. Like it or not, much of the Third World is coming to live with us!

In sum, many demographers and ecologists argue that population growth threatens global social stability, human material well-being, and environmental integrity. In the next century, population growth may effectively overwhelm the carrying capacity of the planet. That, at least, is the *demographic* and *neo-Malthusian* interpretation of things. But, as I noted earlier, it has been a controversial and contentious point of view since the time of Malthus. Many scholars, then and now, have found it fundamentally flawed. How so?

HOW SERIOUS IS THE PROBLEM OF WORLD POPULATION GROWTH?

Most contemporary objections to Malthusian theory were raised 150 years ago. One of his contemporaries, French political economist Condorcet, foreshadowed contemporary technological optimists by arguing that scientific advance would offset diminishing returns. Condorcet said: "New instruments, machines, and looms can add to man's strength . . . [and] improve at once the quality and accuracy of man's productions, and can diminish the time and labor that has to be expended on them. . . . A very small amount of ground will be able to produce a great quantity of supplies . . . with less wastage of raw materials" (Condorcet, 1795). Fifty years later, Marx in particular fulminated against Malthus's theory. He dismissed it as nothing more than a rationale for class exploitation and argued that the real cause of human misery and deprivation was the increasing concentration of resources in the hands of capitalist owners. It was they, who exploited workers to the point of misery and exhaustion, rather than population pressure, who were the cause of human poverty and misery. Then as now, the dominant currents of economic thought discounted natural resource constraints (including population size) to emphasize the adaptability of market-induced substitution and innovation. In another classical objection to Malthusian views that foreshadowed modern objections, Nassau Senior asserted that improved living standards for the poor would not lead them blindly to expand their numbers but rather to restrict their fertility in order to preserve the gains they had realized (Hutchinson, 1967). So you can see that even though his book was a bestseller for decades, then as now Malthus got it from all sides (Poor Tom!). Even so, scholars have been unable to dismiss completely his haunting forecast of an impending demographic apocalypse.

Few debates in the social and natural sciences have been so heated or protracted as this one about the consequences of population growth. In the contemporary discourse, there are three broad positions (paradigms again!). *One position* argues that population growth is a severe threat, perhaps *the* most significant underlying cause of environmental degradation and human misery. A *second position* argues that population growth is not an important threat because markets will allocate scarce resources and stimulate efficient innovations. A more recent variant of this position, termed *supply-side demography*, argues that population growth may in fact be a benefit because the historical record demonstrates that as world population has grown, human welfare has improved: The more people, the better. *A third position* argues that human misery and environmental problems are caused by maldistribution that results from the operation of social institutions and economic arrangements (global or national inequality, poverty, trade policies, high prices, wars) rather than population growth per se. This argument, in effect, turns the table on Malthus, arguing that structurally induced misery causes both population growth and environmental deterioration, rather than the other way around. Let me elaborate each of these perspectives.

Neo-Malthusian Arguments

The standard neo-Malthusian perspective is that population growth causes human misery and environmental degradation. This has been the position of many demographers, but particularly of biologists, ecologists, and natural scientists (Ehrlich and Holdren, 1974; Ehrlich and Erlich, 1992). Some predictions of global demise have been concrete and dramatic. In 1968, Stanford University zoologist Paul Ehrlich wrote, "The battle to feed humanity is over. In the 1970s the world will undergo famines—hundreds of millions are going to starve to death" (cited in Stark, 1994: 558). There were indeed famines and widespread malnourishment in the 1970s, in particular parts of the world, such as sub-Saharan Africa. But nothing on the magnitude predicted, and global food production continued to outstrip population growth.

Modern history has not been kind to the neo-Malthusians, who have been arguing that "the wolf is at the door" routinely since the 1940s. But the wolf has—so far—failed to materialize. *Or has he?* Neo-Malthusians don't believe that one actually dies from overpopulation, but from other, more concrete causes (disease, war, malnutrition or famine). They argue that the doubling of the world's population in about one generation is the broad underlying cause of the stress placed on the global environment and human well-being, even though it is manifest in more concrete causes. For example, population growth helps widen income disparities among nations. In the past 20 years, the LDCs as a group have actually raised total economic output more rapidly than have the MDCs. But many of these gains have been offset by higher population growth rates. In per capita terms, the relative gap has narrowed negligibly while the absolute gap has widened substantially. Compare India and the United States from 1965 to the mid-1980s. Total GNP grew significantly faster in India, but because population grew twice as fast, India's average annual per capital income growth was 1.6

percent, slightly less than that of the United States, 1.7 percent (Repetto, 1987: 13). As population has mushroomed, so have wars. The number of armed conflicts around the world has grown from 12 in 1950 to 31 in 1998, with an all-time high of 50 in 1991 (Renner, 1999: 112). These were intrastate conflicts, but often having international dimensions and involvement, such as in Somalia, Rwanda, Serbia's Kosovo province, and East Timor.

Neo-Malthusians do not think that other factors (drought, poverty, wars) are unimportant sources of environmental or social stress, only that population growth must be considered primary. If, they think, all other factors could be made environmentally neutral, population growth of this magnitude would still spur resource social stress and environmental degradation (Stern et al., 1992: 76–77). Indeed, they argue that once population has reached a level in excess of the earth's long-term capacity to sustain it, even stability and zero growth at that level will lead to future environmental degradation (Ehrlich and Ehrlich, 1992). These scholars believe that, indeed, there is a carrying capacity and that in the long run, it applies to humans as it does to the bacteria in a petri dish. At some point there are limits to the physical capacity of the planet to sustain growth.

Economistic Arguments

Neoclassical economic theory maintained that population growth is not a problem, and may be a source of progress (Boserup, 1981; Simon, 1990; Simon and Kahn, 1984). It argues that population growth—and other resource problems—stimulates investment in increased efficiency, resource substitution, conservation, and innovation. When resources become scarce, well-functioning markets encourage people to allocate them in the most efficient ways and protect them by raising the price. It is a fact that in the long sweep of human history, population growth has been correlated with growing, rather than declining, resources—as well as with improvements in human health, longevity, and well-being. Today more people live longer and better than when the human population was much smaller. Even in the rapid post–World War II population explosion, global food production always outstripped population growth. Contrary to neo-Malthusian expectations, shortages—whether the result of population growth, increased consumption, or environmental problems—have left us better off than if shortages had not arisen.

The reason is that the accumulating benefit of intellectual inventiveness (human capital) met and overcame the challenge of shortages. We have found manmade substitutes for natural resources and more abundant natural resources for scarce ones, and we have invented technologies that allow more efficient use of the resources available. Neoclassical economists argue that finding substitutes for scarce natural resources is likely, and they rely on the ability of markets to respond effectively to resource scarcities (Jolly, 1993: 13). In this view, the cause of problems is not growth, but policies and market failures that do not price things realistically and that subsidize waste, inefficiency, and resource depletion. You get what you pay for, and you lose what you don't pay for (Panayotou, cited in Brown and Panayotou, 1992). Neoclassical economists argue that the neo-Malthusians ignore the role of markets in generating adjustments that bring population, resources and the environment back into balance (Simon, 1998).

A newer variety of this argument, termed *supply-side demography*, maintains that population growth is not a problem, but a positive benefit (Camp, 1993). In contrast to the Malthusian view of diminishing per capita resources over time, the holders of this view argue that the ultimate resource is human inventiveness, which itself accumulates over time as populations grow, and has multiplied resources as they are available to people. A wide range of illustrations can support this view. When a shortage of elephant tusks for ivory billiard balls threatened in the last century and a prize was offered for a substitute, celluloid was invented, followed by the rest of our plastics. When whales were almost hunted to extinction in the nineteenth century to produce oil for lamps, petroleum distillates such kerosene was substituted to fuel lamps and thus created the first petroleum industry. Englishmen learned to use coal when trees became scarce in the sixteenth century. Satellites and fiber optics (derived from sand) replaced expensive copper for telephone transmission. Importantly, the new resources wind up cheaper and more plentiful than the old ones were. Such, it is argued, has been the entire course of civilization (Simon, 1990). Since people create wealth, population growth can never long be a problem in a properly organized free-market economy. To neoclassical economists, the notion of a human "carrying capacity" is a static population-resource equation that conceals more than it reveals and has no empirical validity. It ignores technical inventiveness and market allocation. Counterintuitive as it may seem, as populations grow, resources multiply rather than become scarce. Rather than stressing a finite resource bases, it is more correct to recognize that 10,000 years ago only 4 million humans could keep themselves alive, but by the nineteenth century the earth could support 1 billion people and today it can support 6 billion (Simon, 1998). This view is a recent and radical articulation of the notion that the unique potentials of humans make them almost exempt from the physical limits of the earth.

Inequality Arguments

The inequality (or stratification) argument maintains that human misery and environmental degradation, as well as population growth, are caused by social structural arrangements that produce inequality. This is a more complex and nuanced argument. It is favored by neo-Marxians, but also by a wide variety of other social scientists, economists, agronomists, and some biologists. Unlike the neoclassical economists, they argue that population size *is* a problem. It's just that Malthusians have always got the causation wrong. The operation of global political and economic structures and inequality cause population growth, human misery, and environmental problems rather than the other way around. They argue, for example, that instead of rapid population growth stalling economic development, economic stagnation in the LDCs is caused by poverty, inequitable trade policies, and ongoing dependencies. (You might want to review the discussion of dependency and world systems theory in Chapter Two.) In other words, continued LDC poverty is maintained by the operation of the global economy, and in a condition of deep poverty and stalled development there are few incentives to have smaller families.[7] The final act of the world demographic transition, so the argument runs, is delayed by stalled economic development in the LDCs, not overpopulation.

Strongly objecting to the neo-Malthusian arguments of Paul Ehrlich and others, biologist Barry Commoner argued that plans to limit population that focus on birth control, abortion, or sterilization of people in LDCs ignore the principle cause of rapid population growth—poverty. Furthermore, Commoner argued that on the whole, advanced technology and affluent lifestyles are more environmentally damaging than growing numbers of people (1992). He and many others argue that the reality of global environmental deterioration is that large multinational corporations, not the growing masses of the poor in the LDCs, are responsible for most environmental destruction. It is not, for instance, the indigenous people and subsistence farmers who are destroying the world's rainforests. It is the lumber companies, large cash crop estates, and mining companies.

In a similar vein, others argue that neither the malnutrition that now routinely afflicts at least one-fifth of humanity nor the periodic famines in which people actually starve are produced by population growth. The most direct cause of hunger is not too many people, but lack of money and high food prices. At the *system level* of analysis, hunger and malnutrition are most directly caused by the political economy of agriculture: here meaning patterns of investment and land-holding, and the structure of trade in the world economy (Norse, 1992; Sadik, 1992). Consider, for instance, the following:

- The 22 most food-deficient African countries could meet their food needs with just 11 percent of the food *surplus* held by neighboring countries.
- China has only half the cropland per capita as India, yet Indians suffer widespread and severe malnutrition while the Chinese do not.
- In Thailand, rice production increased 30 percent, but with exports of rice increasing nine times faster, per-person availability of rice has fallen.

- In Chile, farm exports have increased over 30 percent since the early 1970s. However, 40 percent of Chileans consume only 75 percent of the calories necessary for survival.
- In the 1970s, when India had more than 300 million malnourished people, the Indian government, working with large corporations, ensured that India ranked as one of the biggest exporters of food among the LDCs. (Lappé et al., 1998)

Globally, the LDCs now export more agricultural products *to* the MDCs than they receive back in food aid or agricultural subsidies. Consequently, the majority of the world's population remains poor and often hungry. The problem, then, is not with the lack of food but its global distribution patterns. Frances Moore Lappé, one of the most outspoken critics of the neo-Malthusian arguments noted earlier, suggests that to argue that overpopulation causes world hunger is akin to arguing that a person whose throat has been slashed is killed by loss of blood!

Another variety of the inequality argument finds causes of human misery and environmental degradation not in the operation of world markets or structures of inequality, but in authoritarianism and the absence of responsive governments or free markets (Sen, 1981). Nations with democratic regimes, free markets, and a free press can deal with droughts and fluctuations in prices and food supplies to prevent famine, whereas authoritarian regimes do not. It is no accident that the worst starvation happened in one-party states, dictatorships, or colonies: Maoist China, British India, or Stalin's USSR. The last great Chinese famine, in which perhaps 30 million starved, was in the 1960s during Mao's Great Leap Forward, which forcibly confiscated and collectivized the landholdings of villagers. Famine vanished when the Chinese reprivatized agriculture during the reforms of the 1970s. And while there were food shortages in India in 1967, 1973, 1979, and 1987, and western India had half the food per capita of sub-Saharan Africa, democracy, relief, and public works programs averted widespread starvation. Not so in Somalia, Ethiopia, and the Sudan, where wars, corruption, the absence of democracy, and government reluctance to admit problems let droughts grow into mass starvation. Meanwhile, more democratic Zimbabwe, which also experienced decreased production of grain during the 1979–1984 African drought, not only averted famine, but posted significant declines in mortality during that period (Sen, 1993).

Perhaps more familiar to Americans were the gruesome pictures of starving Somali children that dominated the media in 1992–93. But that starvation was not caused—most directly, anyway—by too many people or even too little food, but rather by civil war, chaos, and the looting of the nation's food supplies by warring clan factions.[8] Even the historic Irish famine (1846–1851), which I have mentioned several times before, was not caused directly by population pressure or ecosystem collapse. While it is true that monocrop cultivation (potatoes) failed because of blight, the English

Parliament would not raise the taxes to provide famine relief though it raised over $60 million for the Crimean War campaign. While millions starved (or emigrated), Ireland continued to export large quantities of wheat to England, some 800 boatloads in all (Kinealy, 1996). A more democratically constituted regime might have been compelled to do otherwise.

The inequality perspective maintains that poverty is not only the more direct cause of high fertility and human misery, but is also connected to environmental destruction. Notwithstanding the larger role of multinational mining, agribusiness, and lumber companies on the environment, it is still true that poverty adds considerably to the resource pressures in the LDCs. Poor households are often virtually forced to overuse natural resources daily for subsistence. Thus, desperate farmers grow cassava and maize on highly erodible hillside. Rural households in fuelwood-deficit countries strip foliage and burn crop and animal residues for fuel rather than using them for fertilizer. This practice also contributes to desertification, since land stripped of trees and plant residues is less likely to hold moisture. Underemployed men in coastal villages overexploit already depleted inshore fisheries (Repetto, 1987: 13).

Finally, the inequality perspective would maintain that the economists' emphasis on market and technological solutions provides little help in understanding the structural barriers that impede the access of poor LDC farmers to the technologies or capital necessary to use their resources efficiently. People's income affects their ability to adopt new technology. As demographer Nathan Keyfitz has argued, it is the people who are the best off who have the greatest potential to innovate (cited in Jolly, 1993: 15). Thus the "magic of the market" works well in New York or Tokyo but is of little help to the folks in Mogidishu (Somalia) or Khartoum (Sudan)!

Population Policy and Politics

I'm sure by now you are aware that the controversy about the significance of population growth is not, and never has been, just an academic question. It has officially engaged the world's policy makers at least since the 1950s. And in fact, the three positions that I outlined have been thrashed out not only within scholarly communities but have enlivened (putting it mildly!) the world population policy conferences sponsored by the United Nations.

By 1984 the World Bank had concluded that "for the poorest countries, development may not be possible at all, unless slower population growth can be achieved soon (Merrick, 1984; World Bank, 1991: 185). Yet not everyone agreed. In 1974, at the first United Nations Conference on population Policy held in Bucharest, seven countries refused to send delegates even to discuss the problems, and the nations that did attend quickly separated into contending factions reflecting the theoretical positions described here.

The major cleavage at the 1974 conference was between the representatives of the Western MDCs and the LDCs and socialist bloc nations. In general, the Western capitalist nations argued the familiar neo-Malthusian position: that the world is poor because it is becoming overpopulated. They argued that the pressure of exponential population growth undercut economic development efforts in the LDCs and were quick to point out that since the growth rates in the MDCs were declining, the major responsibility for curbing world population growth lay with the LDCs themselves. LDC representatives argued, to the contrary, that the world is becoming overpopulated because it remained poor. They held that to expect population growth to level off before economic development and redistribution of wealth takes place is to put the cart before the horse. Furthermore, LDC representatives pointed out that if the concern was overconsumption of global resources and degradation of the environment, the real locus of the problem was with the MDCs, which consumed 60 to 70 percent of the earth's annual production of natural resources to maintain affluent "northern lifestyles." Meanwhile, obtaining raw materials to satisfy minimum basic human needs was often difficult in the poor countries. LDCs maintained, consistent with the inequality perspective noted earlier, that underdevelopment, poverty, and environmental problems were maintained by structural barriers and global inequity rather than by population growth. Less polite LDC representatives held that neo-Malthusian theory enabled the world's white people to hog the planet's wealth while justifying racial genocide against the world's people of color. Marx would have loved it!

In a heated rejoinder, Western representatives argued—with considerable supportive evidence—that under contemporary conditions it was simply not possible to support the world's population at a level consistent with Western living standards without devastating the resource base of the planet.[9] And so it went. As you might guess, the Bucharest meeting ended in a rancorous stalemate between those favoring policies of *birth control* and those favoring policies of *wealth control*.

In spite of the Bucharest conference's inconclusiveness, during the following decade LDC leaders became more supportive of population control policies. Some countries that had been vocal critics in Bucharest, most notably China, were leading supporters of family planning activities by the time of the next world population conference, which took place in Mexico City in 1984. Ironically, while the weight of opinion among the LDCs had shifted to recognize that population growth was a significant problem, opposition to population and family planning programs at Mexico City came from the nation that had promoted neo-Malthusianism in 1974: the United States! This opposition reflected both intellectual developments and American domestic politics. The intellectual development, already noted, was the emergence of supply-side demographic theory. This was received with caution by scholars (including many economists), but with enthusiasm by the conservative thinkers of the Reagan administration, who wanted to demonstrate that the real problem was the absence of free markets, not too many people. Supply-side demography provided the intellectual rationale to dismiss population problems as a hoax, and, more important, to solidify the administration's relationship with religious conservatives by opposing family planning programs both at home and around the world. Thus the second U.N. Population meeting ended in disarray, but this time it was because of the objections to the whole notion of a population problem by the U.S. representatives.

The U.S. policy reversal at Mexico City was followed by a decision to end seventeen years of support for the London-based International Planned Parenthood Federation, and then in 1986 by the withdrawal of all support from the United Nations Population Fund. Twenty years of U.S. leadership on global population issues came to an end. The consequence of this action—since the United States had provided much of the aid money—was that funds for family planning programs, which had grown during the 1970s, remained essentially flat during the 1980s (Merrick, 1986: 20; Camp, 1993: 127, 132–133).

MAKING SENSE OUT OF THIS CONTROVERSY

Are you now thoroughly confused? If so, don't worry, because you're in pretty good company. As you can see, these are complex issues, but also terribly important ones. As a beginning, you need to recognize that this, like some of the controversies I have noted earlier, is not only a debate about facts but also about different paradigms.

Physical scientists and ecologists—and many demographers—see the world in terms of problems of growing scale in a world with ultimately physical limits. Neoclassical economists, in contrast, see the world as a largely mutable system of possibilities because of human technical inventiveness and the capacity of market allocation to adjust to scarcities and stimulate investment in resource substitution. They argue that ecologists simply fail to appreciate the magic of the market.

Ecologists retort that the reason that economists believe this is that they miss entirely the environmental "debts" that growth incurs, which results in a delayed form of deficit financing. Those who fail to recognize the ultimate physical limits of the planet, says environmental economist Herman Daly, are "treating the earth as if it were a business in liquidation" (cited in Brown, 1991: 9).

Inequality and stratification arguments are similar to economic ones because they emphasize the importance of human social factors rather than natural limits as causes. But proponents of this view are like the ecologists in seeing both exponential population growth and environmental degradation as real problems. Briefly, in understanding the relationships between population growth and human and environmental problems, neo-Malthusian arguments emphasize *scale issues,* neoclassical economic arguments emphasize *market allocation issues,* and inequality arguments emphasize *distribution issues.* Although these paradigms have very different views of the way the world works, they are each partial—and not necessarily mutually exclusive (Jolly, 1993: 21). I think it is possible to reconcile some of their differences.

Considering the broad sweep of human history, the neoclassical economists and technological optimists have a better factual argument. There were, to be sure, particular times and cases where population growth contributed to environmental and social disasters, particularly in the preindustrial world. (Remember the cases of the Copan Mayans and the Mesopotamians discussed in Chapter Two?) But in the industrial world as a whole, technological progress has always outrun the pressure of population growth. In sum, the neo-Malthusians have always been wrong about a global demographic disaster: The wolf never *was* really at the door.

In its own way, however, the neoclassical economic paradigm is as static and ahistorical as the physical science notion of fixed limits. It posits an unchanging linear relationship between population size and the ability of technological innovation and markets to overcome problems. It fails to recognize that the enormous *growth in scale* of the human population since World War II has put us much closer to absolute physical planetary limits than ever before in human history. To put it in economic terms, the "elasticities of substitution" between natural and humanmade resources are historically quite variable and are now declining.[10] Furthermore, there are physical limits beyond which *no* substitution is viable. Wheat, for example, cannot be grown with only labor, or without water (Jolly, 1993: 15). I think that the enormously large world population—which will may reach 10 or 11 billion in the next 50 years—means we will have fewer options, less maneuvering room, a more degraded resource base, and less ability to absorb and recover from environmental damage than ever before in history. We may face an "ingenuity gap." I believe that the dependability of economic and technological capabilities diminish relative to the threats of scale posed by the present and future population size. Neo-Malthusian theory should be taken more seriously because the population-environment equation is historically dynamic. The wolf is not yet at the door, but he's certainly in the neighborhood, and a lot closer than he was as recently as 100 years ago!

Finally, I think that the conflict between neo-Malthusian and inequality arguments is more apparent than real. Neo-Malthusian arguments are more persuasive in the abstract and on the long-term horizon. But stratification arguments are more convincing explanations of human misery and environmental degradation in the concrete here and now. In other words, things like hunger, poverty, and water pollution are more directly and concretely caused by social, political, and economic arrangements than by the underlying specter of overpopulation. Whether you prefer a demographic or a stratification argument depends upon whether you prefer more direct and concrete or more distant and underlying causes. It also depends on whether you emphasize short- or long-term time spans. But as you can see from the foregoing, they do have very different policy implications for how human and environmental problems are addressed.

These are my own thoughts about the different perspectives about population growth and change. They are consistent with, but not identical to, a large middle-ground consensus emerging within the world scholarly and policy communities.

An Emerging Consensus?

The emerging consensus argues that population growth is one among many causes of human misery and environmental deterioration, but not the only one. It emphasizes that other factors, such as economic policies and political institutions, influence human well-being and environmental integrity at least as much as variations in population growth rates, and often more directly (King and Kelly, 1985; Repetto, 1987: 12). While exponential population growth is not to blame for all of our problems, present and future, neither is it insignificant, and we ignore it at our peril. Furthermore, market solutions to population-environment problems are important but by themselves not sufficient, and the most pressing problems are institutional rather than technical (Crosson and Rosenberg, 1990: 83).

Here is how one scholar described this consensus as it applies to problems in the LDCs:

> The most important lesson learned from continued study of the relationships between population and development is the key role of institutions in mediating these relationships. . . . Institutional obstacles . . . [include] the unequal distribution of wealth and political power, poor management and organization, and waste of resources on military activities. Rapid population growth exacerbates many of the resulting problems but slowing population growth will not remedy the situation without positive steps toward change. Some . . . characterize rapid population growth as the "accomplice" rather than the "villain" in this story. (Merrick, 1986: 29, citing King and Kelly, 1985)

This is a more complex and nuanced consensus than the three paradigms described earlier.

One more thing: Contrary to the arguments of the new "supply-side demography," the vast majority of responsible scholars now believe that in general more people is not necessarily better and quite probably worse. The most damaging evidence comes from a review of existing evidence from a panel of experts of National Research Council (within the National Academy of Sciences), who found little evidence that "lower population densities lead to lower per capita incomes via a reduced stimulus to technological innovation, efficiency, and economies of scale." Regarding the LDCs, the panel concluded that "slower population growth would be beneficial to economic development for most of the developing world, but . . . a rigorous quantitative assessment of these benefits is context-dependent and difficult" (National Research Council, 1986: 90). In sum, there is a large consensus that virtually all current and future problems with resource supplies, human material security, and environmental integrity would be easier to deal with if world population growth slowed more rapidly and stabilized at a lower "equilibrium number" (Reppetto, 1987; Brookfield, 1992; National Research Council, 1986).

POPULATION AND FOOD

In the 1990s, the world's farmers produced enough cereals, meat, and other food products to adequately feed the world's population. Taken all together, there was enough to provide 3,800 calories per day per person, well over the minimum daily calorie requirement, even for those whose jobs involve hard physical labor.[11] But the world's food supply is not distributed evenly throughout the world, within nations, or sometimes even within households (Bender and Smith, 1997: 5). Many low-income countries do not grow enough food to feed their citizens and cannot afford to import enough food to make up the difference. Thus the most concrete cause of hunger is not overpopulation or a degraded environment but *poverty*, often exacerbated by wars, oppression, authoritarian governments, and the like. Occasionally people do starve, but the broader and more chronic problem is not starvation but chronic malnutrition. Most people in the world are able to get enough food to survive, but many have dietary deficiencies. You can think of global food consumers as being on three levels or tiers.

At the bottom are more than 900 million people (about 20 percent of the world's people) who are unable to provide themselves with a healthy diet. These people are classified as *food-energy deficient*, and at least 60 percent of them are children. Such chronic malnutrition may not be as grotesquely visible as massive famine, but its consequences are nonetheless devastating. In children it delays physical maturity, impairs brain development, and reduces intelligence, even if replaced by an adequate diet later on. Malnourished adults are unable to work hard or long and have lower resistance to diseases. The danger of epidemics is always high in overpopulated and underfed areas. *On the middle level* are about 4 billion grain eaters, who get enough calories and plenty of plant-based protein, giving them the healthiest basic diet among the world's people. They typically receive less than 20 percent of their calories from fat, a level low enough to protect them from the consequences of excessive dietary fat. *At the top* are the world's billion meat eaters, mainly in Europe and North America, who obtain close to 40 percent of their calories from fat (three times that of the rest of the world's people). As people in the middle level (in China, for instance) become more affluent, they tend to "move up the food chain" to emulate people at the top (Brown, 1994b). The high meat diet of those at the top is not only unhealthy, but creates a demand for meat production that causes a substantial share of the global inequity of food resources and environmental abuse. To illustrate, ignore the high inputs of fuel and chemicals it takes to produce meat and consider only how many liters of water it takes to produce 1 kilogram of various foods:

Potatoes	500
Wheat	900
Maize (corn)	1400
Rice	1910
Soy beans	2000
Chicken	3500
Beef	100,000

(Baylis, 1997)

At least a third of the world's grain is fed to animals to produce meat. Hence the simple act of eating less meat could "stretch" the world's grain supplies, making it possible to feed a much larger population and to significantly reduce the current global food inequity.

In spite of these inequalities, increasing food production and rising incomes in recent decades have reduced the incidence of malnutrition, from a third to around a fifth of the world's population. Progress was particularly dramatic in the LDCs. One-third of the world's population was classified as "food-energy deficient" in 1970. By 1990, the percentage of malnourished people had fallen by 8 percent in the MDCs and by 43 percent in LDCs. Moreover, if everyone around the world adopted a vegetarian diet and little food was wasted, current production would be enough to feed 10 billion people in 2050. So *if* the diets of some changed and less food was wasted and became more evenly distributed, there would be enough for some time. Those are very big *ifs*! These observations are the good-news picture. The bad news is that, in spite of the declining percentage of malnourished people, in a large and still growing world population the absolute number of malnourished people continues to grow. In parts of the world, such as sub-Saharan Africa, both the absolute numbers and the percentage of people suffering from malnutrition continues to grow (Bender and Smith, 1997: 6–7). Please understand that chronic hunger doesn't exist only in poor nations. Hunger agencies estimate that about 30 million Americans are malnourished, and the U.S. Department of Agriculture reported in 1999 that at least 10% of all American households do not have access to enough food for a healthy diet (Charles, 1999). In spite of the progress in recent decades, some evidence suggests that in the future our capacity to produce enough food for all without severely stressing the life-support capacity of the biophysical environment is uncertain. How so?

Pressing the Earth's Limits?

Chapter Three discussed evidence that you can understand as related to this question about whether or not we are approaching carrying capacity to adequately feed the growing world population over the next century. Consider again in that context the findings of the U.N. study (GLASOD) that found much degradation of the world's soil, representing 17 percent of all the land that grows vegetation (Livernash and Rodenberg, 1998). Recall also that water requirements for human well-being are relatively fixed, and that the total water supply through the hydrological cycle can't be increased much. If the "population per water flow unit" (1 million cubic meters per year per person) remains low, water supply is generally not a problem. As population pressure increases, better water management is required, and signs of water stress appear. When there are over 1,000 persons per flow unit, a country invariably experiences chronic water shortage; under current water technology, extreme scarcity occurs if the ratio between population and flow units exceeds 2,000. As overuse and irrigation shrinks river flows and lowers water tables, many nations and regions now experience growing political and legal conflicts about water. By the year 2025, population growth will push many countries into higher stress zones and chronic water shortages (Falkenmark and Widstrand, 1992: 19–20). Nations will respond to this situation differently. Rich societies can draw down groundwater more efficiently, import water from other regions, and experiment with expensive desalinization. These are strategies that can support locally—for a while—a society that has grown beyond its water limit. But none of them can work globally, in poorer nations, or for very long anywhere. At some point, nations and regions that have grown beyond their water limits will have to deal with continuing growth and consumption of all kinds in relation to the renewable but fixed water budget of the earth (Postel, 1992a, 1992b, 1993).

There is another measure of how close humans are now pressing on the biophysical capacity of the planet that Chapter One mentioned. Ecologists define the *net primary production* (NPP) of the biosphere as the amount of energy captured from sunlight by green plants and fixed into living tissues, which is at the base of all food chains. The NPP is the energy flow that powers all of nature. Several years ago, Stanford University's Peter Vitousek and his colleagues calculated how much of the biological product (NPP) of the planet is appropriated for the use of human beings. Their results were astounding: Humans used 25 percent of the photosynthetic product of the earth as a whole and 40 percent of the photosynthetic product on land!

Humans directly consumed only about 3 percent of the land-based NPP (through food, animal feed, and firewood). But indirectly another 36 percent went into crop wastes, forest burning and clearing, desert creation, and the conversion of natural areas into settlements. And the impact of pollution on reducing the NPP was not even considered (Vitosek et al., 1986). If that 40 percent figure was even approximately correct, it raises an interesting question about the consequences of having significantly larger populations in the future. What would the world be like if humans coopted 80 percent of the NPP? Or 100 percent? No one is really sure. At best it might look like the Netherlands or England, totally manicured and under human control: livable but with no wilderness and no room for expansion or mistakes. *But wait.* The Netherlands and England import food, feed, wood, and fiber and therefore depend on far more than 100 percent of the NPP of their national areas.[12] Some countries can do that, but the world as a whole cannot (Meadows et al., 1992: 49–50). Importantly, the more the NPP is appropriated for humans and their chosen life forms (corn and cows), the less is left for other species, producing a drastic decline in biodiversity. Is a world totally appropriated and managed by *homo sapiens* a viable biosphere? Again, no one knows for sure. But there is reason for thinking it is not (as discussed in Chapter Three).

There's more. In 1994, the Worldwatch Institute published evidence that growth in world food supplies are now no longer keeping up with population growth (see Table 5.2). The strategies and technologies that served to keep food production growing more rapidly than population (adding more land, new genetic hybrids, increasing fertilizers, pesticides, herbicides, and irrigation) apparently no longer do so. Though total food production still grows, *per capita* production has been declining for some time. From 1984 to 1990, the annual growth in grain production was 1 percent while that of population was nearly 2 percent.[13] Two of the four kinds of meat produced (seafood and beef, but not pork or poultry) showed per capita declines. These trends in world food are indeed worrisome trends—particularly grain, which is the food staple of the world. As noted earlier, there is still

TABLE 5.2 Production Trends Per Person of Grain, Seafood, and Beef and Mutton, 1950–1993

		Trends Per Person		
Foodstuff	*Growth Period*	*% Growth*	*Decline Period*	*% Decline*
Grain	1950–84	+ 40	1994–93	−12
Seafood	1950–88	+126	1988–93	− 9
Beef (and Mutton)	1950–72	+ 36	1972–93	−13

Source: L. Brown, "Overview: Charting a Sustainable Future" in L. Brown, H. Kane, and D. M. Roodman (Eds.), *Vital Signs 1994: The Trends That Are Shaping the Future*, p. 20. Copyright 1994 W. W. Norton. Used with permission.

more than enough food to adequately feed everybody, but this evidence suggests that per capita production has declined and the world's "margin of safety" regarding food has declined.[14]

These aggregate statistics conceal important regional variations. Until the mid-1970s, there were per capita increases in most regions of the world, but since then regions diverged significantly, with some (sub-Saharan Africa, the Middle East, and parts of South Asia) often posting absolute (not only per capita) declines in food production. You can now count the major food exporters on the fingers of one hand (Argentina, Australia, Canada, France, and the United States).

Other equally pessimistic evidence exists that we are pressing critical thresholds in the per capita production of food. Consider the natural resource bases that produce food (good grain-producing land, irrigated land, and grazing land). They are not expected to increase much in the near future, but population growth will continue, which means that the per capita availability of these key resources will shrink significantly (see Table 5.3).

HOW CAN WE FEED NINE BILLION PEOPLE IN THE NEXT FIFTY YEARS?

Food experts find the Worldwatch Institute data controversial, but most take it seriously. Since we were having difficulty feeding six billion people in 2000, I think that adequately feeding a much larger population in the future

Table 5.3 Population Size and Availability of Renewable Resources, c. 1990, with Projections for 2010

	Circa 1990	*2010*	*Total Change*	*Per Capita Change*
	Million		*Percent*	
Population	5,290	7,030	+33	—
Fish catch (tons)[a]	85	102	+20	−10
Irrigated land (hectares)	237	277	+17	−12
Cropland (hectares)	1,444	1,516	+5	−21
Range and pasture land (hectares)	3,402	3,540	+4	−22
Forests (hectares)[b]	3,413	3,165	−7	−30

[a]Excludes aquaculture.
[b]Includes plantations; excludes woodlands and shrublands.

Source: S. Postel, "Carrying Capacity: Earth's Bottom Line" in L. Brown et al. (Eds.), *State of the World 1994: A Worldwatch Institute Report on Progress toward a Sustainable Future,* p. 11. Copyright 1994, W. W. Norton. Used with permission.

will be an enormously difficult task in building a sustainable and equitable world. "The world's total demand for food is likely to nearly double its present level by 2030" (Brown, 1997: 9). Furthermore, accommodating the much larger population that will appear by the end of the next century will require a much greater increase of current food output levels on stressed global food resource bases. These feats will challenge the ingenuity of the world's policy makers and farmers under *any* circumstances, and particularly if they are done in a sustainable way. We must simultaneously produce more food and halt the destruction of the agricultural resource base. How?

Increasing Food Security: Technical Options

The most obvious way of increasing food supplies is to extend the technologies that served us so well since the 1950s: bring more land into cultivation; use more fertilizer, pesticides and herbicides; irrigate more, and so on. Continuing these techniques produces little significant increases in crop yields. The J-shaped curve of early rapid growth slows down, reaches its limits, and levels off, becoming an S-shaped curve. Grain yields per hectare still increase in most nations, but at a slower rate (see endnote 13). Since 1985, yields for major grains in the three countries with the highest yields—the United States (corn), Great Britain (wheat), and Japan (rice) have leveled off (Miller, 1998: 607). Not only do the intensive agricultural techniques from the 1950s no longer produce increasing per capita yields, they measurably degrade the resource bases for agriculture (Bender and Smith, 1997: 25–40). It is doubtful whether even current yields of such intensive agriculture as practiced in Europe and the United States are environmentally sustainable in the next century, without considerable modification. It is even more doubtful that temperate zone monoculture agriculture could be successfully exported wholesale to the tropic and subtropics—even if companies and governments were willing to *give it away* or the LDCs had the money to *buy it*. On the scale required, we won't and they don't.

Biotechnology?

Some view new biotechnology (or genetic engineering) as a technological panacea of the coming decades that will give an enormous boost to agricultural productivity, becoming a gene revolution like the green revolution seed hybrids of the 1960s. The *green revolution* refers to a massive global effort to cross-breed species producing crop seeds that were much more productive per unit of cultivated land, thereby increasing total food production. As noted in Chapter Three, the global diffusion of the new green revolution hybrids significantly decreased the genetic diversity of crops around the world. By gene splicing and injection, the new genetic engineering techniques can produce new varieties that "Mother Nature never knew"; more pest resistant, earlier maturing, drought resistant, salt resistant, and more efficient users of solar energy during photosynthesis. Because of such potential benefits and their profitability, bioengineered crops were rapidly entering the American farming/food system by the year 2000. For instance, about two-thirds of soybeans were grown from engineered seeds species.

But there are ecological reasons for caution. Without huge amounts of fertilizer and water, most green revolution crop varieties produced yields that were no higher (and sometimes lower) than traditional varieties. Similarly, if genetically engineered crops increase productivity by accelerating photosynthesis, they could also accelerate the loss of soil nutrients, requiring more fertilizer and water. Without ample water, good soil, and favorable weather, new genetically engineered crops could fail. Furthermore, new species would be inserted into natural food chains, predator systems, and mineral cycles with unpredictable results. Some scientists think they could, over time by cross pollination, result in *superweeds,* or pests. Weeds might acquire the special defenses or enhanced photosynthetic capacity of a genetically engineered crop plant, and crop plants with built-in pesticides might harm many insects other than target pests.[15] Furthermore, new organisms introduced into an environment can themselves become pests. Please don't consider this an unimportant issue: In the United States, nonnative plant invaders cause an estimated $138 billion in damage, including the costs of controlling them (Pimentel, 1999). Historically, more that 120 intentionally introduced crop plants *have* become such weed pests in the United States. Unlike people in the United States, Europeans have demonstrated strong skepticism about the biotechnology industry's claims that no adverse health effects are associated with consuming bioengineered food. Europeans are also wary of the unintentional—and damaging—introduction of genes or substances into the environment. Europeans protest the importation of U.S. bioengineered foods. At the turn of the twenty-first century, a serious food trade war between the United States and Europe was brewing about this issue (Anderson, 1993: 144; Halweil, 1998, 1999; Union of Concerned Scientists, 1996).

Other reasons why biotechnology is a questionable panacea for malnutrition around the world have to do with economics and institutional contexts. Genetic engineering requires heavy capital and technical investments and is being conducted by large private companies that will hold patents on "their organisms," available to buyers at the right price—rather than cheaply to those most in need of food. So far, biotechnology research has been more driven by the desire for agribusiness sales and profits rather than for food for the hungry or agricultural sustainability. Priorities have been, for example, to develop herbicide-resistant crops producing higher sales and profits for herbicide companies, and foods that retained freshness while being shipped long distances. In the most widely known illustration, the Monsanto company was developing a high-yield seed with a *terminator* gene; meaning that after the crop was grown, harvested seeds could not be regrown. Rather than being saved by farmers, each year's seed had to be purchased anew from the company. Reactions were so negative that the company has abandoned the project, but in corporate circles the race is on. Because of risky but extraordinarily high profit potentials, agribusiness firms now compete vigorously to develop and patent engineered species. The prospects of producing more food cheaply for the world's poor and hungry has so far eluded researchers, and—more important—attracted little interest by investors.[16]

None of this means that genetically engineered crop species should be rejected out of hand, particularly if the research agenda could be redirected toward more food and fewer ecological impacts rather than more profits. Doing this would mean shifting some control of research and development agendas to the world's food consumers and farmers (Anderson, 1993: 146–148). But lest you think it is only environmental scientists and industry critics who doubt that biotechnology is a solution to the world's food problems, listen to Donald Duvick, for many years director of research at Pioneer HiBred International (one of the world's largest seed producers). "No breakthroughs are in sight. Biotechnology, while essential to progress, will not produce sharp upward swings in yield potential except for isolated crops in certain situations" (cited in Miller, 1998: 607). Like many new scientific technologies, genetic engineering has impressive promises mixed with serious and sometimes sinister possibilities—environmental but also economic and political.

Sustainable Agriculture: Agroecology and Low-input Farming?

As the limitations of modern intensive agriculture and the hazards of biotechnology became apparent, agronomists and ecologists rediscovered some of the virtues of more labor-intensive traditional agricultural practices. These are most obvious for increasing food in tropical LDCs, where rural labor is plentiful but capital and technology are scarce. Though often less profitable in the world market economy, many traditional methods were superior in productivity per hectare when energy inputs and long-term sustainability were considered (Rappaport, 1971; Armillas, 1971). Now a newer agricultural paradigm of *agroecology* recognizes that a farm is also an ecosystem and uses the ecological principles of diversity, interdependence, and synergy to improve productivity as well as sustainability (Altieri, 1995). The tools of industrial intensive agriculture are powerful and simple and mean using products like insecticides bought off the shelf. By contrast, agroecology is complex and its tools are subtle. They involve intercropping (growing several crops simultaneously in the same field), multiple cropping (planting more than one crop a year on the same land), crop rotation, and the mixing of plant and animal production—all time honored practices of farmers around the world (Lappé et al., 1998: 77–78). Agroecology can be combined with *organic* and *low-input* techniques. Farmers can, for instance, recycle animal manures and "green manure" (plant residues) for fertilizer, and they can practice low-tillage plowing that leaves plant residues to prevent erosion and improve soil productivity.

Consider an example. In 1999 on a 300-acre farm near Boone Iowa, farmer Dick Thompson rotated corn, soybeans, oats, and wheat interplanted with clover and a hay combination that includes an assortment of grasses and legumes. The pests that plagued neighboring monoculture farms were less of a problem because insect pests usually "specialize" in one particular crop. In a diverse setting, no single pest is likely to get the upper hand. Diversity tends to reduce weed problems because complex cropping uses nutrient resources more efficiently than monocultures, so there is less left over for weeds to consume. Thompson also keeps weeds in check by grazing a herd of cattle, a rarity on Midwestern corn farms. Most cattle are now raised in feedlots. Cattle, hogs, and nitrogen-fixing legumes maintain nutrient-healthy soil. Moreover, Thompson is making money. He profits from his healthy soil and crops and the fact that his "input" costs—for chemical fertilizer, pesticides, and the like—are almost nothing (Halweil, 1999: 29).

Such techniques can be highly productive, *but only when human labor is carefully and patiently applied*. Evidence from developing nations is impressive. The agriculture of China, Taiwan, Korea, Sri Lanka, and Egypt is now close to this mode—with high yields to show for it. And in tropical rainforests such as in Brazil there is evidence that small-hold agroforestry systems can combine the production of crops, trees, forest plants and/or animals to "mimic" tropical ecosystems. They protect soil from leaching and eroding while replicating the natural succession of plant growth. There is evidence that over five years such systems can produce yields 150–200 times higher than more modern farming or livestock production (Hecht, 1989). But it was in Cuba that such alternative agriculture was put to its greatest test. Before at the collapse of the communist world, Cuba was a model green revolution–style farm economy, based on enormous production units using vast quantities of imported chemicals and machinery to produce export crops while over half the island's food was imported. When, around 1990, Cuba lost trade and subsidies from socialist bloc nations, Cuba was plunged into the worst food crisis in history, with per capita calories dropping by as much as 30 percent. Faced with the impossibility of importing either food or agrochemical inputs, Cuba turned inward to create more self-reliant agriculture based on higher crop prices to farmers, smaller production units, and urban agriculture. By 1997, Cubans were eating almost as well as they did before 1990 (Rosset, 1997).

Urban agriculture is based on the idea of getting urban dwellers to grow vegetable crops in empty lots, back yards, and other spaces in and around cities. Such gardeners in Havana now supply 5–20 percent of the city's food. Urban gardening is not a new idea. For instance, during World War II such "victory gardens" produced 40–50 percent of the fresh vegetable in the United States. Urban gardening is now a major source of food in the large cities of the Third World, such as Shanghai and Calcutta, where food security is often a matter of survival. In the United States, organizations formed in many American cities to support urban gardeners, who meet regularly to sell and swap their produce. Advocates see urban agriculture as one means of helping urbanites to reclaim neighborhoods from crime and pollution, training low-income residents business skills, and teaching young people about nutritional, environmental, and food security issues. Thus, a movement toward community-supported agriculture that started in the 1970s now includes 600 programs around the world (Nelson, 1996).

Is more sustainable agriculture economically viable? A landmark study by the prestigious National Research Council found that in the United States "alternative farmers often produce high per acre yields with significant reductions in costs per unit of crop harvested" (National Research Council, 1989b: 8). Regarding the use of agrochemicals, an eleven-year study found low-input farming to outcompete high-input farming in terms of profit per acre (Strange, 1994; for similar findings see Kleinschmit, Ralston, & Thompson, 1994). In the early 1990s, low-input and low-tillage agriculture were practiced by less than 10 percent of American farmers; more recently it grew rapidly because of obvious economic and environmental advantages. In South India, a 1993 study that compared ecological farms with conventional chemical-intensive farms found that ecological farms were just as productive and profitable as the chemical ones (Lappé et al., 1998: 81).

Systems of alternative are also preferable in to industrial farming terms of the human ecology and communities they create. Just as the modern industrial system values managers and technical experts and treats labor as just another production commodity, industrial agriculture no longer values workers who are adaptive, enterprising, self-steering decision makers. Seen in ecological terms, industrial agriculture tends to remove an essential niche from the system: the owner-operator who adapts farming to uniquely local conditions has a long-term interest in protecting the land's resources and continued productivity. Compared with absentee or corporate contract farmers, owner-operator farmers have been found more concerned with environmental problems (Williams and Moore, 1994). The farmer-owner farms to maintain a quality life for his family rather than profits for remote landlords, corporate investors, bankers, or government bureaucrats. A system of small holders sustains rural communities in ways that large-estate production does not. Industrial agriculture, to the extent that its technology takes away autonomy from individual farmers removes valuable custodians of natural and social resources, and no one takes their place (Anderson, 1993: 146).

After noting all this evidence, it should be obvious that American agriculture is *not* presently evolving toward such smaller alternative farming systems, but rather toward larger, chemically intensive monoculture farms owned or controlled by large agribusiness firms. This is true for both grain crops and animals, as illustrated by the huge cattle feedlots, and "factory farms" that raise hogs and chickens. Agricultural research, economic subsidies, and pricing policies have favored such operations. As they have for 100 years, American private farmers often feel themselves like an endangered species. The study quoted earlier that reported the high productivity and economic advantages of alternative farming also concluded that "many federal policies discourage the adoption of alternative practices" (National Research Council, 1989b: 10). Even so, Altieri, the agricultural scientist and advocate who coined the term *agroecology,* recently observed that "it is clear that the future of agriculture will be determined by power relations, and there is no reason why farmers and the public in general, if sufficiently empowered, could not influence the direction of agriculture toward goals of sustainability" (1998: 71).

This is another way of underlining a point noted earlier, that the problems of agriculture and food are as much political-economic as they are technical.

Increasing Food Security: Political-Economic Options

Scholars and policy makers agree that part of the answer to increasing food security and protecting the environment has to do with changing the political economy of agriculture. But as with the technical options we have explored, there is little agreement about how to do this or precisely what institution factors are the culprits. There are four kinds of competing perspectives about desirable transformations in the political economy of food production: (1) monetarist, (2) structuralist, (3) reformist, and (4) world systems perspectives.

Monetarist Perspectives

Monetarist perspectives derive from neoclassical economic theory and maintain that government interventions in food markets are the most important cause of the maldistribution of the world's food. Food may be underpriced, discouraging investment in agriculture and depressing productivity. In many LDCs, for instance, farmers must sell their crops to monopolistic public marketing firms at well below world market prices. The government firm then resells the crops at higher prices on the world market, profiting from the difference. At the same time, food production in the MDCs is typically subsidized, encouraging overproduction and consequently the necessity to export (or dump) cheap, subsidized, food on the world market—which further depresses the true market price for agricultural goods. In both cases, the theoretical price of maximum efficiency is distorted by government intervention, resulting in price/market distortions, reduced competition, and misallocation of resources (Johnson, 1973). In the monetarist view of things, government intervention causes overproduction in some countries (MDCs) and depresses production in others (LDCs), resulting in the badly misallocated global availability of food. Shipping excess wheat and corn to LDCs is a desirable form of charity to relive a food emergency, but not when the foreign grains discourage local agriculture. In a world with less government intrusion into world agricultural markets, the LDC small-hold farmers would not be required to sell their crop to a government ministry for an artificially low prices nor compete from a disadvantaged position with subsidized American, Canadian, or French farmers.

Structuralist Perspectives

Structuralist perspectives emphasize that low food productivity is related to the land tenure system. They note that there is an almost universal tendency for small- and medium-sized farms to be more productive than very large farms, both in the LDCs and the MDCs (Dorner, 1972). Large farms tend to produce for export markets rather than domestic food markets, thus reducing domestic food supplies (e.g., both Mexico and Brazil became less food sufficient while food exports have grown). Large farms tend to replace labor with technology. In the LDCs, the consolidation of small landholdings has created a class of dislocated and landless peasants, depressing both rural and urban labor markets and exacerbating poverty. Absentee ownership, common on large farms, typically leads to inefficiency and resource mismanagement. *Land reform* has been the principal policy prescription of this perspective. Giving land to former landless peasants would increase food productivity and particularly production for domestic markets, encourage more intensive use of natural resources, and make food available to those who suffer the greatest hunger and malnutrition—the rural poor. China, Taiwan, and Korea had the most successful land reforms, with a substantial impact on agricultural productivity, but land reform was so minimal in other parts of the world (e.g., Latin America and the Philippines) that its effect on raising productivity was minimal (Humphrey and Buttel, 1982: 208).

Reformist Perspectives

Reformist perspectives see the need for change across the entire society, not just in the agricultural subsector. Unequal land tenure systems are, in this view, merely one aspect of more general *societal social inequality*, which is held to be the root cause of agricultural stagnation and hunger. Widespread poverty in both rural and urban settings inhibits the growth of effective demand for agricultural commodities, keeps prices low, and discourages the expansion of production. Lipton (1974) argued that the fundamental cleavage in most LDCs is an urban-rural one, and that politics in the developing world has a systematic urban bias. Governments keep food prices low to benefit urban areas, where their most powerful political constituencies live and which are also the source of the most threatening riots and political challenges to the regime. At the same time, large landowners are given subsidies to produce for cash crop export markets, which increases national earnings from world markets but diminishes production for domestic food markets. Thus a powerful coalition of interests exists against most of the rural population. Food production tends to stagnate and rural smallholders are politically marginalized. In fact, the urban bias disadvantages all but the rural and urban elites. Reformist theorists such as Lipton place their main hope on political coalitions between the rural and urban poor, and particularly on rural small holder political movements to challenge the existing structure of policy making (1974: 328–329).

World Systems (or Dependency) Perspectives

World systems (or *dependency*) thinking was described in Chapter Two as a perspective about the emerging global political economy. It views agricultural stagnation and hunger as a consequence of the asymmetrical relationships among nations in the world market and political systems. The penetration of capitalism into traditional societies produces dependency and the progressive marginalization of the rural poor. It encourages the consolidation of land holdings and the introduction of technology and capital-intensive crop production. LDCs become "specialized" in an international division of labor where they are the producers of primary goods (minerals and food) at relatively low prices. LDC economies become more oriented to producing for external markets than for domestic needs (Johnson, 1973; Frank, 1967). The world systems perspective also emphasizes the damaging consequences of investments by multinational firms on LDC agricultural productivity. Dole and United Brands profit from cheap labor, produce for export markets, and do little to manage land in a sustainable way. The world systems advocates see the same problems as do the structuralists and reformists, but they see them as manifest within an overarching global structure of dependency and inequality. And in this view, breaking the chains of poverty and agricultural stagnation is a more formidable task, not only generating the political momentum for redistributing land and redressing rural-urban inequalities, but also disengaging these nations from the world food market system (Humphrey and Buttel, 1982: 210; Payer, 1974).

As you can see, this analysis came full circle. It moved in stages from perspectives viewing the solution to the food problem as removing subsidies, market barriers, and opening up a "freer" world trade system in food-stuffs as the answer (monetarist). It resulted in a perspective viewing the structure of the world market economy itself as a source of the problem (world systems). Concrete policies about food and malnutrition by government and planners of multinational organizations are likely to mix useful ideas from a variety of these abstract perspectives. In a broader sense, two views exist about political-economic remedies to world food problems from which concrete policies are constructed. On one hand, some view the solution to world food problems to be creating a more dynamic and open world market and trading system for food with fewer political barriers. That would not encourage national food self-sufficiency, but each nation would try to specialize, producing and selling thóse products for which it had a natural "comparative advantage" (Avery, cited in Jordon, 1993). On the other hand, others advocate that each nation should strive for food self-sufficiency, producing food for domestic consumption and becoming less dependent on the vast, and remote international trade of foodstuffs (Lappé et al., 1998). The first is a rationale for industrial agriculture and a global trade system managed by multinational companies. The second emphasizes small-hold labor intensive production that is at least compatible with low-input, sustainable production. You can probably guess by now that, other things being equal, I favor the second view and the reasons why, but the case is certainly arguable.

Food Options and Policy Questions

Uncertainty exists about the concrete numbers associated with certain future population growth, and it compounds questions about the future food security for the world's people. No disagreement exists, however, about the difficulty of *equitably* feeding a growing world population with stable or shrinking per capita agricultural resource base will be a daunting challenge.[17] You can think of these difficulties in terms of questions raised by the previous discussion. Can the extension of industrial intensive agriculture produce more food in the future that is affordable by most of the world's people (particularly those currently malnourished)? Can this be done without further degrading the world's agricultural and biophysical resource bases? Can sustainable and low-input farming produce enough food along with self-sufficiency and environmental sustainability? Will biotechnology research result in crops more productive and adaptive to changing conditions while being environmentally safe? Will people who desperately need more food be able to afford them? On one level, these questions have to do with the possibilities of different production strategies and technology questions. On another level, producing enough food for the 9 billion who will most likely appear in the next century depends partly on the efforts of farmers and partly on the success of governments and international development agencies in dealing with population stabilization and environmental threats. Efforts to stabilize population may be more important for the future than anything agricultural policy makers themselves can do.

STABILIZING WORLD POPULATION: POLICY QUESTIONS

The rate of population increase has been falling around the world now for about a decade as fertility rates fall around the world. Several causes contribute to the world decline in the rate of growth, which are enormously variable between nations and regions: (1) the socioeconomic development and falling birth rates that complete the demographic transition in some LDCs, (2) the successes of family planning programs, (3) the global diffusion of feminism and women's rights movements, and (4) the increasing malnutrition, misery, and AIDS that increase the death rates.

The postwar population explosion really began to decline in the early 1970s, and not primarily as the result of Malthusian catastrophes such as diseases, famine, or war. Most important, family planning programs started by many LDCs in populous Brazil, Egypt, Indonesia, and Mexico, and most particularly the world's two largest nations (India and China), drove this turnaround. Policy makers would like to know how much of the fertility declines of recent decades can be attributed to these programs and how much of it would have occurred in any case because of general socioeconomic improvement. Using data from 19 LDCs, Timothy King of the World Bank estimated that family-planning programs accounted for 39 percent of the decline and overall socioeconomic improvement accounted for 54 percent. Other studies using different data report family planning accounts for between 10 percent and 40 percent of observed fertility declines, and no study has failed to find some effect (Keyfitz, 1990: 66–67). Research on the determinants of contraception and fertility shows that most couples want to plan smaller families for reasons of health and personal aspirations, whether or not they perceive the effect that their action may be having at the societal level (Merrick, 1986: 48).

Curiously, in the 1980s the decline in fertility rates stalled, contrary to the projections of the United Nations, the World Bank, and other trend watchers. It stalled for several reasons, in part because some LDCs (particularly in sub-Saharan Africa and South Asia) were expected to start their demographic transitions and failed to do so, probably because lagging socioeconomic development caused them to remain very poor, with an average of five to six children per family. In part it stalled because of the changing world *age structure*. In the LDCs, and particularly in India and China, a large number of persons born in the 1960s reached childbearing age in the 1980s. It lagged in part for political reasons, because the Chinese relaxed their "one child" policy and allowed local interpretations of it and because in India the Congress Party, which had promoted birth control programs, was defeated in the 1977 elections. Perhaps most important, it stalled partly because aid for population programs from donor nations remained essentially flat, and particularly from the United States, as the Reagan administration enforced the Mexico City Policy (discussed earlier) withdrew support from such efforts in 1984. As you can see, world population policy and programs are often overwhelmed by political events.

Promoting Fertility Decline

During the 1970s and 1980s, women around the world began forming small nongovernmental organizations (NGOs) to lobby for improvements in their social, economic, and political circumstances. By the 1990s, women in LDCs were advocating improvements in family planning programs in order to improve information, access, and encouraging service providers to treat clients with greater respect. Opposition by women's groups to existing family planning programs as well as ethical, scientific, and religious debates about population growth formed the backdrop for the fifth U.N. conference on population. When the International Conference on Population and Development (ICPD) met in Cairo Egypt in 1994, the level of participation by NGOs was unprecedented, as over 1,200 NGOs participated as observers or delegates and worked with government officials to craft the ICPD program of action (Gelbard et al., 1999: 34).

By an overwhelming consensus, delegates of the ICPD argued that population growth *is* a serious problem that exacerbates core social and environmental problems, while they rejected the notion that population growth is *the* cause of all human problems. They emphasized the necessity of creating conditions under which couples willingly lower the number of children they have. Like previous conferences, they affirmed (1) making the traditional strategies of family planning/contraception available to all people, and (2) addressing poverty and destitution that amplify population growth. Powerful evidence suggests that everywhere these strategies have made a difference. But the ICPD emphasized something quite new: (3) empowering women. Many women—particularly in LDCs, where 90 percent of the world's population growth will happen—have large families simply because they have no other way to achieve social and economy security for themselves. Women in strongly patriarchal (male-dominated) societies are often forced to marry young. They get paid much less than men when they are allowed to work, have little access to land or bank credit, and have few opportunities to participate in political life. A pervasive consensus, among women's organizations as well as scholars about development and population policy, maintains the policies designed to improve the well-being of and expand the social choices available to women would go far to limit population growth, address environmental problems, and promote human development. Where women have low status and are financially dependent on their husbands, fertility remains high. There are no known exceptions to this generalization (Camp, 1993: 134–135; Sachs, 1995: 94). But you can understand why those in power in patriarchal societies may strongly resist such changes.

What would be the report card grade a half decade after the ICPD? "Fair." The rapid global decline in fertility rates and resumed after the 1980s pause (the lost decade). Progress in improving the status and social choices of women is measurable in most nations. LDCs' initiatives and spending is largely on track, but the MDCs are not keeping their part of the bargain. While the Clinton administration succeeded in overturning the Reagan-era Mexico City policies, congressional opponents succeeded in reducing the U.S. family planning aid from $547 million in 1995 to $385 million in 1997. If the Cairo program continues to be underfunded, 96 million fewer couples had access to modern contraceptives in 2000 than if commitments had been met (Mitchell, 1998: 26–27).

CONCLUSION

While the signs that the demographic transition is working in some fashion on a global basis provide the basis for some optimism, world population is an enormous problem because of the built-in momentum of absolute growth. Using a metaphor of a semi truck speeding toward us for population growth, the optimist would note that it has slowed from eighty to sixty miles an hour. The pessimist would note that while we looking the other way, someone just doubled the weight of the cargo!

PERSONAL CONNECTIONS

Implications and Questions

You can intellectually comprehend large-scale population change, but my guess is that it is so abstract and pervasive that you rarely think about your everyday life circumstances, problems, and opportunities as related to population change. Here are a few leading questions to help explore the demographic contexts of your life:

1. In the absence of lots of room in which to expand, population growth must mean an increase in population density. That means that people live more closely together, interact more frequently, and complete with each other more intensely for living space and all resources for which supplies are limited. Think of the times when you have lived in a smaller dense environment with others (in a

shared apartment, a college dormitory, boarding school, or military base, for instance). How do you describe the experience? What kinds of problems did you and others experience? What kinds of things became important that weren't important in a less densely populated living environment? What kinds of special rules or regulations evolved to deal with problems of increased population density? You might think of all the special rules that college dorm systems and military bases need to deal with problems of living in such facilities. Not all such rules deal with crowding and density problems, but many do.

2. Unlike other species, attitudes and cultural definitions about "how many are enough" mediate human reactions to different population densities. Humans can live comfortably in communities of various sizes and densities. They can be comfortable in very dense populations, particularly if they are not poor, are culturally homogenous, and are connected by a "culture of civility" (Amsterdam may be a case in point, with one of the highest population densities in the world but with a minimum of urban problems). But if people are crowded and poor, culturally dissimilar, and have only a weak culture of civility about how to deal with each other, the experience is a problem. Given the expectations for population growth around the world and the fact the vast majority of Americans will live in or close to large heterogeneous cities, what kind of experience do you think that will be for people? For you personally, what will determine whether is (or would be) a good or a bad experience?
3. High population density is one thing in Amsterdam, but quite another in the dense, rapidly growing, and poverty-ridden cities in many poorer LDCs. There, growing population density produces such competition for scarce resources that likely result in political turmoil and pressures to immigrate. Indeed, problems with political instability and increased immigration and refugee flows are a widespread problem in the contemporary world. Consequently, citizens of many nations international organizations will share in the costs of maintaining an orderly world—if not a humane one. Thinking more concretely, what might your "share" of these problems entail? How will you and your friends react to the pressures of immigrants seeking entry, or the necessity intervention to address demographically driven economic and political disasters in poor, densely populated LDCs? How will you feel when your children may be required to leave on military "peacekeeping" missions to places like Rwanda, Haiti, or East Timor?
4. The stabilization of population growth has been on the world's political agenda for some years, and most notably from the ICPD conference at Cairo. That conference defined strategies for slowing population growth that involved the continuation of established family planning programs, social development in LDCs, with assistance from international agencies, and enhancements in the status of women around the world. How much of a priority do you think this should be, compared to other issues? How urgent should it be for the politicians who collect your tax money? How do your age, family status, education, political attitudes, or religious background shape answers to these questions?

What You Can Do

This chapter's twin concerns were population and food. Food security may be an alien concern to you, unless you're among the minority of Americans whose food supply is chronically in jeopardy. But food security is a problem for an estimated thirty million Americans, in addition to people in many other nations. In the midst of a seeming surfeit of food in America, what contribution could you make to increase the food security in the world?

1. First, you need to recognize that concern with food problems isn't only about hunger or food security. Agriculture and food processing have a greater environmental impact than any other human activity. In the predominant food systems in MDCs, capital investment in agriculture, application of agrochemical, packaging of foodstuffs, and its distribution over vast areas positively or negative impacts what happens to land, water, ecosystem diversity, human health, and what kinds of food-related economic markets thrive or falter.
2. To wit: Buying food in bulk, uncooked, with fewer layers of packaging makes food cheaper per unit of production, is likely to be healthier, involves less energy to produce, and creates less trash. More of your food costs go directly to producers and to corporate intermediaries who process it. And by selective buying, you can support traditional or organic farming, or local, regional, or large-scale markets. These ideas may be very difficult to practice among busy dual-income families, and in food systems increasingly dominated by fast foods, supermarkets, and prepared meals. They are for me!
3. As to hunger and food security itself: The most obvious way of helping is to give generously to food banks and international food relief agencies. That does help feed people who are desperate, but it does not contribute in any way to increase their ongoing food self-sufficiency. Most food relief agencies, such as Oxfam International, now emphasize contributing to the development of food producing capacity. You can contribute to both public and private food development programs. If you or your friends want a real challenging but important project, try to organize on behalf of the world's hungry people. Try to get food agricultural development programs to those who directly produce food rather than state ministries or firms. While you're at it, you might try to redefine domestic political priorities at any level—city, state, federal—more toward enhancing the food for the hungry.

   As you can see, addressing food security issues is not easy, and can be as much political as personal.
4. Among the important personal things you can do is to grow some of your own food in a backyard plot, a window planter, a rooftop garden, or a cooperative community plot. Spending $31 to plant a living room–size garden can give you vegetables worth about $250. Try getting a return like that in the stock market! In the 1990s, about 30 percent of the U.S. population grew fruit, vegetables, and herbs at home, and about 57 percent were grown organically (Miller, 1992: 386).
5. Even more important is *eating lower on the food chain*. This means eating less meat and more grains, fruits, and vegetables. If this lifestyle change became common, the benefits for environmental problems, dietary health, and food security would be enormous. It would save money and energy and reduce your intake of fats that contribute to heart disease and other disorders. It also would reduce air and water pollution, water use, reforestation, soil erosion, overgrazing, species extinction, and emissions of greenhouse gases (methane) produced by cattle. In the United States, animal agriculture pollutes more fresh water than all municipal and industrial uses combined. If Americans reduced their meat intake by only 10 percent, the savings in grain and soybeans could adequately feed 60

million people. More than half of U.S. cropland is devoted to growing livestock feed. Livestock also consume more than half of the water used in the United States, either by direct consumption or irrigating to grow their feed or processing their manure. Each time a single American becomes a vegetarian, 1 acre of trees and 1.1 million gallons of water are saved each year, and that individual pollutes half as much water. Currently only about 3 percent of Americans are vegetarian (Miller, 1992: 368).

6. The beef about beef: I hate to mention this. Particularly since I live in Omaha, which comes close to being the beef capital of the nation. Its hinterlands are loaded with cattle ranches, feedlots, and packinghouses, and the beef industry is terribly important to the local economy (Have you seen those ads for "luscious" Omaha steaks that could be shipped to you?). In fact, in Nebraska nothing comes closer to sacrilege than encouraging people to eat less beef. But you should. Why? It is not as healthy for you as fowl or fish because it has more saturated fat. It requires more inputs of feed and other agricultural inputs per pound than any other livestock. It takes about 9 calories of energy input to get 1 calorie of food output from beef. So you can see that in energy terms, it's a net loss. Most rangeland degradation in the United States is from cattle, not hogs or chickens. A large proportion of rangeland in the western United States is owned by the federal Bureau of Land Management and is leased to ranchers who use it for small fees. In fact, when you figure it out per hamburger, American cattlemen pay an average of 1 cent per hamburger for the use of government land (Worldwatch Institute, 1994: 39). Not all the beef we eat comes from the United States. The most ecologically expensive beef is from cattle raised on tropical soils of Latin America.

   After all this, I have to be honest and tell you that my family and I still eat meat, including beef. But we eat a lot less of it in relation to vegetables and carbohydrates than we used to, and turkey has replaced much of the beef (particularly ground beef) in our diet. I'd feel a lot better about eating beef if more of it were raised on grass fed on ecologically managed rangeland rather than in crowded feedlots where cattle are fattened up with processed food, pumped full of growth hormones and antibiotics, and produce concentrated waste disposal problems. But little beef is currently produced on open range land.

Real Goods

The Chinese Diet. It consists overwhelmingly of rice, other grains, soybeans, and locally produced beef, with bits of pork and fish for variety. The Chinese eat one-fifth as much meat as Americans, making them paragons of low-on-the-food-chain ecological correctness. It also reduces their saturated fat and cholesterol consumption to levels the National Cancer Institute and the American Heart Association don't let themselves dream of in America. Consequently the Chinese suffer fewer heart attacks, strokes, and cases of breast cancer. They also have lower levels of anemia and osteoporosis in spite of their lower calcium intake (Durning, 1994: 98). As China develops, many Chinese are giving up their traditionally healthy diet and learning to eat more like affluent Americans.

My family likes Chinese food and we cook a lot of it, both with meat and without. I've discovered that I can buy a 25-pound cloth bag of American rice from Arkansas for about $13.00. That is more than 20 times cheaper than a box of name-brand rice, and you can fix a *lot* of stir fry pretty inexpensively for that price. Chinese foods in American restaurants tastes about the same as we found food in China to taste, but here they use a lot more meat in proportion to vegetables and rice than they do in China. And watch it, Chinese-American restaurant food often comes with fried rice, which loads the fat and cholesterol back in. Most restaurants are happy to substitute ordinary steamed rice.

ENDNOTES

1. The doubling time can be computed by the *rule of 70*—that is, 70 divided by the growth rate per year (expressed in per centage). So at the growth rate in the 1990s of about 2 percent per year, the doubling time was 35 years. Exponential growth is expressed in logarithms. So to find the doubling time, you must find the natural logarithm (or $\log_e$) of 2, which turns out to be 0.70, which is multiplied by 100 to get rid of the decimal point. If we wanted to find the tripling time of an exponentially growing population, we would find the natural log of 3, which is 1.10, or 110 when multiplied by 100.
2. *Demography* is the study of the characteristics and change processes of human populations. It focuses on the social implications of aggregate biological characteristics of populations, such as the age structure, sex ratios, or population size.
3. The full title of the substantially revised 1803 edition was *An Essay on the Principle of Population; or a View of Its Past and Present Effects on Human Happiness; with an Inquiry into Our Prospects Respecting the Future Removal or Mitigation of the Evils which It Occasions.* Titles were longer in those days!
4. You should treat these numbers with caution, because what gets classified as "urban" varies between nations. In Italy, towns of 10,000 or more are counted as urban; in the United States, places of only 2,500; and in Iceland, 200. But in Japan the cutoff for urban population is a lofty 50,000 (Haub, 1993: 2)
5. As early as 1992, Mexico City was projected to the word's largest city. But because of a Japanese reclassification of what constitutes an urban area, it now follows Tokyo-Yokohama (Population Reference Bureau, 1993b).
6. According to the 1990 census between 1981 and 1990 there were 974,000 *legal* immigrants from Mexico, 267,000 from Puerto Rico, 136,000 from El Salvador, 120,000 from Haiti, 149,000 from Cuba, 56,000 from Guatemala, 32,000 from Nicaragua, 100,000 from Columbia, and 25,000 from Panama (U.S. Bureau of the Census 1991: 10)
7. Although I have noted this several times before, you may still be wondering just how this works. Indian scholar M. Mamdani (1972) has provided what I think is the clearest explanation why the poor in developing nations have large families: (1) Children provide a form of old-age support in nations that provide no public retirement security; (2) children provide economic support through their labor on the farm or the sale of their labor to others; and (3) children add little to house-hold

expenditures in a condition of deep poverty. Living in chronic poverty, he argued, does not provide the incentives for reduced fertility and population control policies are likely to fail.

8. Consider one of the worst famines in recent times—Bangladesh's in 1974. It took place in a year of unusually high global rice production. According to Martin Ravallion (cited in Nasar, 1993), a World Bank economist who studies poverty in Asia, severe flooding disrupted rice planting and threw landless rural laborers out of work. Then false fears of food shortages doubled rice prices in a few weeks. For the poor, who spend more than three-quarters of their wages on food, the blow was catastrophic. But the famine, which was largely over even before the rice crop was harvested, was hardly inevitable. "Almost everything the government did made things worse," said Ravallion. Bangladesh's authoritarian ruler sent the army out to "bash hoarders," convincing the people that it had lost control and further fueling the price surge. And in the middle of all this the United States contributed to the panic price hike by announcing that it would withhold food aid to punish Bangladesh for, of all things, selling jute to Cuba!
9. Consider the consequences of miraculously and suddenly raising the Chinese alone, with a population of over 1 billion, to American consumption standards. The average American was supported by 11 tons of coal-equivalent energy in 1974, while the average Chinese consumed only two-thirds of a ton. Sustaining the Chinese at an American level would "require 9 million metric tons of coal-equivalent in additional energy each year, which would increase present world energy consumption by more than 100% . . . and these figures do not even consider . . . [the high levels] of other minerals that would be required to develop the industrial infrastructure needed to raise China to these high levels of development" (Pirages, 1978: 62).
10. *Elasticities of substitution* simply asks how much human technical capacities can stretch enough (are elastic enough) to surmount natural limits. If elasticity is high, there is no problem, but if elasticity is low, then beyond a certain point, human inventiveness is insufficient to overcome resource limits. I have argued that they are much higher in industrial than in preindustrial societies but are now declining in the industrial era because of absolute population growth and accumulated environmental damage.
11. Calorie requirements range from 1,000 per day for infants, to 3,900 per day for adults under extreme physical stress (Bender and Smith, 1997: 11).
12. The Netherlands is said to "occupy" somewhere between five and seven times its own territory, largely because of imports of animal fodder from the LDCs (Rijksinstituut voor Volksgesondheit en Milieuhygiene, 1991).
13. Grainland has declined in per capita terms, and since 1981 in absolute terms (from 735 to 693 million ha in 1991). Growth in irrigated areas has slowed dramatically since 1978. Response of crops to the additional use of inorganic fertilizer is diminishing, and in many countries (such as the U.S.) using more fertilizer now does little to boost production. Global fertilizer use has declined by about 15% since the 1980s (Brown et al., 1993: 19, 42). And in 1993 research by the International Rice Institute (IRI) in the Philippines, an agronomy research organization that helped disseminate the "green revolution" in Southeast Asia, found that the use of pesticides and herbicides was no longer producing increasing rice

yields, and had produced health problems among exposed farmers that should have been "subtracted" from past increases. Significantly, IRI recommended drastic cuts in the use of herbicides and pesticides among Southeast Asia's rice farmers (Zwerdling, 1993).

14. A key measure of food security is the "carryover stocks" measured in days of consumption. Carryover stock is the amount of grain left in the bins when the new yearly harvest begins. In 1993 the carryover stock would feed the world for 73 days, down substantially from the 1960s (when it was typically in the 80s) and from the all time high of 104 days in 1987. When stocks drop below 60 days grain prices become highly volatile, sometimes doubling, as they did between 1972 and 1973 (U.S. Department of Agriculture, 1993).
15. For example, in 1999 researchers found that bioengineered corn with genes from soil bacteria (bt) produced pollen that killed the caterpillars of monarch butterflies, compared with no deaths among those exposed to traditional pollen varieties. Scientists and government regulators should have known it would be lethal to caterpillars, since the engineering was undertaken to kill corn pest caterpillars like the European corn borer and the corn ear worm (Rissler, 1999).
16. But consider that only example I know of for a biotech food that would manifestly address the needs of the world's hungry. In 2000, a Swiss research institue was developing a strain of rice that would supply vitamin A (beta carotene) and not bloc the absorption of iron, both a problem among rice-eating populations. Moreover, the new rice strain was not patented or sold by a multinational corporation, but given to the International Rice Research Institute for distribution in the third world (NPR, 2000).
17. I put the word *equitably* in italics for a reason. People and organizations who do not care whether the food products of the world are distributed in a minimally equitable fashion to prevent malnutrition when it is technically possible to do so bear a heavy ethical responsibility for that choice.

CHAPTER SIX

Energy and Society

Population growth, increasing the food supply, and general increases in human living standards have only been possible because of substantial increases in the amount of energy consumed. By 1990, the total energy consumption by humans around the world was fourteen times larger than it was in 1890, early in the industrial era. Growth in energy consumption vastly outstripped population growth, which doubled during the same time period. But the human use of energy—its mining, refining, transportation, consumption, and polluting by-products—accounts for much of the human impact on the environment (Holdren, 1990: 159). Earlier chapters argued that human societies are "embedded" in the biophysical environment. Most fundamentally, in fact, they are embedded in systems of energy production and consumption. In other words, energy mediates between ecosystems and social systems and is a key to understanding much about the interaction between humans and environmental systems.

Energy is fundamentally a physical variable—measured variously as calories, kilowatt-hours, horsepower, British Thermal Units, joules, and so forth. But energy is also a social variable, because it permeates and conditions almost all facets of our lives. Driving a car, buying a hamburger, turning on your computer, or going to a movie could all be described in terms of the amount of energy it took to make it possible for you to do those things. A kilowatt-hour of electricity, for instance, can light your 100-watt lamp for ten hours, smelt enough aluminum for your six-pack of soda, or heat enough water for your shower for a few minutes (Fickett et al., 1990: 65). All of social life, from the broad and profound things to the minutia of everyday life, can be described in energetic terms.

This chapter discusses (1) social and environmental problems associated with our present energy systems; (2) studies about the relationship between energy and social life, or what some scholars have termed the *energetics* of human societies; (3) the variety of current energy resources and some possibilities for alternative methods of producing energy; and (4) some policy issues about transforming existing human energy systems.

It may well be that energy mediates between ecosystems and human systems, but that's a very abstract way of putting the human-energy-environment relationship, and its implications may not be clear to you. So before I address this agenda, let me provide a concrete but limited illustration of this statement by taking you on a historical detour, back to . . .

THE WINTER OF 1973

In most of the industrial world, it was an awful one, and not because of the weather. The reason was a sudden change in the availability and prices of energy supplies. The world market for oil, which had become the industrial world's premiere source of commercial energy, was very tight, meaning that in previous decades the global consumption of petroleum products had almost outgrown the world's capacity to produce, refine, and distribute them. U.S. domestic oil production was declining. The MDCs were increasingly dependent on the oil reserves of the LDCs such as Nigeria, Venezuela, and particularly the nations around the Persian Gulf, which possessed most of the world's known reserves. In September 1973, Japan's prime minister predicted that an oil crisis would come within 10 years. It came in more like 10 days, with the surprise attack that launched a war between Israel and her Arab neighbors that was later called the Yom Kippur War. In retaliation for the Western support of Israel, the cartel of oil-producing nations (OPEC), led by the Arab nations, declared an embargo on the export of oil to the MDCs. Nations and oil companies scrambled to buy, control, and ration existing supplies in storage and in the pipelines around the world. Oil prices zoomed from $2.50 to $10.00 a barrel, and the world economy went into rapid downturn—with price increases of almost everything, rapid inflation, plant closings, and layoffs. Rationing of energy supplies meant sudden uncertainty about the supplies of industrial, heating, and transportation fuels that Westerners had taken for granted as cheap and plentiful (Stanislaw and Yergin, 1993: 82–83). American President Richard Nixon left it to the energy departments of each state government to figure out how to allocate existing fuel. As increased costs of energy percolated through the whole economy, every facet of the American economy and lifestyle seemed threatened.

In the winter of 1973, Christmas displays were turned off. In my city (Omaha) the Salvation Army "Tree of Lights," a holiday tradition at the county courthouse, burned for only one hour a day. Lights in urban office buildings were turned off. Everyone worried about keeping enough gas in their cars as gas stations periodically ran out of gas. Nebraska gas stations were closed on Sundays, and every Saturday night there were long lines. The days of supercharged V-8 muscle cars were numbered, as was the 75-mph interstate speed limit. Thermostat settings in offices and homes were turned down. In the state of Iowa, individual coffeepots were banned in the statehouse, and all high school basketball games were banned after December 22 (Kotok, 1993: 1). The latter was *serious* business, if you were a high school student in the rural Midwest! The crisis continued in 1979, when a revolution in Iran disrupted world supplies and created a panic that drove oil prices from $13 to $33 a barrel. All this seemed to foretell permanent shortage and continued turmoil. Adding to the mood of crisis, a prestigious group of scholars and computer modelers (the Club of Rome) produced studies to show that among other things, the world would be visibly "running out of gas" in the future (Meadows et al., 1972).

But none of these fears caused by the "oil shocks" of the 1970s really came true. The ability of the OPEC nations to control the world's oil supply declined as non-OPEC production increased at a rapid pace. OPEC's share of the world oil market fell from 63 percent in 1972 to 38 percent by 1985 (Stanislaw and Yergin, 1993: 82–83). People responded by changing the way they lived and worked. They insulated homes and bought more fuel-efficient autos and appliances. All over the world, utility companies began switching from oil to other fuels. By 1992, the people in my home state (Nebraska) consumed 100 million fewer gallons of gasoline than they did in 1973 (Kotok, 1993: 1). Energy conservation, a consequence of both technological and behavioral changes, proved more powerful than expected, so that by the 1990s the combination of reduced demand for oil and increased supplies made the its real price cheaper in 1993 than in 1973 (Lichtbrau, 1993). Around the world, MDCs tried to establish security measures that would help moderate future crises. These included the creation of the International Energy Agency, an international sharing system, increased communication, the creation of a global oil futures commodity market, and the establishment of prepositioned supply reserves.[1] In fact, the Gulf War of 1992, an event that would have certainly created havoc in the oil markets a decade before, came and went with only a minor blip in the oil futures markets.

The oil shocks of the 1970s were a great historical wake-up call that changed popular and scholarly understanding of energy in many ways. They marked a transition in coming to grips with the environmental and sociopolitical costs of energy. Problems of air and water pollution, many of them associated with energy supply and use, came to be recognized as pervasive threats to human health, economic well-being, and environmental stability (Holdren, 1990: 158). Indeed, energy problems came—perhaps for the first time in history—to be widely recognized as an integral part of environmental concerns. In addition, consciousness of growing dependence on Middle Eastern oil graphically demonstrated the growing economic and geopolitical interdependence among nations and created new foreign policy dilemmas that continue today.

I hope that this historical detour illustrates more concretely some of the ways that energy mediates between human societies and the environment. Before the 1970s were over, the oil shocks had produced significant changes in human societies and world energy markets—by changing (1) the availability of energy, (2) the technical means for converting it into usable forms, and (3) the ways that it was used.

Now that the crisis is gone and energy after the turn of the century is relatively cheap again, you may be wondering just what some problems are with the world's current energy system and why we should be concerned about it. Fair question. For one thing, growth in population and economic growth since the 1970s means that the world's nations are completing for a stake in the last *large* oil reserves on the planet, and problems like those in the 1970s could happen again. Even many oil company executives think that at present rates of consumption, world oil supplies will become very tight (but not exhausted) by the middle of the 2000s. Moreover, there are plenty of reasons for continuing concern about current systems of energy production and use.

ENERGY PROBLEMS: ENVIRONMENTAL AND SOCIAL

There are four different sorts of interacting problems with the world's present energy systems and technologies: (1) "source problems," having to do with energy resource supplies; (2) problems related to population growth and economic growth and development; (3) global economic and geopolitical problems; and (4) sink problems, having to do with the by-products, pollution, health hazards, and environmental impacts of energy systems and technologies.

Source Problems: Energy Resource Supplies

At the turn of the twenty-first century, three nonrenewable fossil fuels supplied about 75 percent of the world's energy needs: Oil (30 percent), natural gas (22 percent), and coal (22 percent). All other sources, such as hydropower, nuclear energy, traditional fuels (wood, dung, plant refuse), solar power, and wind power, together made up the remainder. MDCs used only 10 percent renewable fuels, while the LDCs used 41 percent, mainly traditional fuels (Flavin and Dunn, 1999: 24; Miller, 1998: 396). Those proportions do change, but slowly.

Since the pessimistic estimates of oil reserves in the 1970s, known oil reserves for the world at least doubled (Stanislaw and Yergin, 1993: 88), and energy analysts agree that in the near term the earth's supply of fossil fuels is not a problem. Known reserves of crude oil will last until sometime in the middle of the next century. Natural gas will last at least that long, and even longer if we become willing to pay a higher price to get at the much larger "subeconomic" reserves that are thought to exist. And there is an awful lot of coal in the world, but its use carries extraordinary problems and risks compared to those of oil and natural gas (Miller, 1998: 433, 436, 439–440).

Consider oil. There is a rough consensus among energy analysts that continuing the current rates of oil consumption will deplete the earth's *affordable* reserves somewhere between 2030 and 2072 (British Petroleum Company, cited in Flavin and Lenssen, 1991; World Resources Institute et al., 1996: 276–277). Experts estimate that world oil production will peak sometime between 2010 and 2030 and will decline thereafter (McKenzie, 1997; Podobnik, 1999). If you believe that new oil discoveries will forever push back resource depletion, consider this remarkable fact: Just to keep using oil at the present rate means that we must discover as much oil as there is in Saudi Arabia—25 percent of the world's known reserves—*every 10 years*. Hardly anyone thinks *that* is feasible.[2] Most experts expect little of the world's affordable oil to be left by 2059, the bicentennial of the world's first oil well. Oil company executives have known this for some time, which is why their companies are becoming "diversified" energy companies (Miller, 1998: 433). Several years ago, industry-connected analysts such as Robert Hirsch, vice president and manager of research services for Atlantic Richfield Oil Company, urged beginning an orderly transition to alternate energy technologies in the early to middle twenty-first century (1987: 1471).

All projections about how long it will take to deplete fuel and mineral reserves are expert guesstimates, notoriously dependent on assumptions and contingencies that could change them. Chapters Three and Four discussed the reasons for these resource and scientific uncertainties in some depth, so I won't repeat that here. But I need to mention a few obvious possibilities that could change depletion-time estimates. If trends toward greater MDC energy efficiency continues or resumes with full force, declining demand could stretch out supplies many years beyond current estimates. On the other hand, many things could happen to shorten the estimated years-till-depletion of fuel reserves. These include lack of success in exploring unknown but probable geological sources, increased consumption rates because of greater than expected population growth, greater growth of the world market economy, or strong economic development in the LDCs that elevates world energy consumption. So even though supply constraints are not as constraining as thought in the 1970s, concerns for the mid- and longer-term future accompany even more optimistic supply estimates.

Energy, Population Growth, and Economic Development

Consider the interaction between future energy needs and future world population growth, or the prospect of successful economic development—desperately needed and desired among the masses of poor people in LDCs around the world. In 2000, the world's 6 billion people consumed almost 14 terawatts of energy (a terawatt is equal to the energy in 5 billion barrels of oil). But that aggregated world consumption statistic hid its very unequal distribution among nations. MDCs have about one-fifth of the world's people but consume almost three-fourths of the world's energy. In 1997, the United States alone had 5 percent of the world's people but consumed 25 percent of the world's oil—half of it imported from other nations (Falvin, 1998a: 50). One American consumes as much per capita energy as do three Japanese, six Mexicans, fourteen Chinese, thirty-eight Indians, 168 Bangladeshis, 280 Nepalis, or 531 Ethiopians (Goodland et al., 1993: 5)!

I noted in the last chapter that world population may stabilize in the next century at somewhere between 9 and 11 billion people. (Most demographers think the lower number is unrealistic.) As difficult as stabilizing population may be, it is likely to be much easier than providing energy for the burgeoning numbers of people, to say nothing of their aspirations for economic development and the increases in food, water, and material consumption which that would entail. If the large numbers of Chinese, Indians, Indonesians, and others in the Third World were to become energy consumers living even remotely close to the present living standards of North Americans or Europeans, that would place enormous strains on the supply of global energy resources. Energy is an important constraint on economic development, and if development occurrs in the LDCs in the same way that it did in the MDCs, the planet's energy and mineral supplies would be rapidly depleted. The resulting environmental degradation, toxic wastes, and heat-trapping greenhouse gases would be intolerable.

Policy, Economic, and Geopolitical Problems

By the 1990s, the momentum toward greater energy efficiency stalled, and while some of it lasted, there were disturbing signs of increasing per capita energy consumption (Kingsley, 1992: 115). The rebound in energy consumption was partly a consequence of the marketing of gas guzzling sport utility vehicles and pickup trucks that make up about half of all U.S. new car sales. At a deeper level, the rebound in consumption was a consequence of public policy. In the 1980s, U.S. government policy thinking was dominated by *supply-side policies* that promoted an increased supply of energy resources (oil, coal, etc.), ensuring a low price for energy. These policies undercut much of the potential for conservation to have an effect on energy markets. But concerns with market price ignored some very real externalities not accounted for in the price of oil that are not paid directly by either energy producers or consumers. Here are some important ones:

1. The environmental costs of energy production and use
2. The costs of defending Middle East oil production
3. The costs of purchasing and storing oil to meet emergencies
4. Various costs in U.S. deficit balance of payments between exports and imports (more than one third of which are due to energy imports)
5. The *lost opportunity costs* of tying up capital in costly supply technologies in remote corners of the world when far less costly demand reduction technologies are available
6. The lost opportunity costs of tying up capital in existing infrastructures, impoverishing investment and research in alternative energy sources that are more efficient and/or environmentally benign
7. The costs to U.S. allies and impoverished LDCs for U.S. competition in world oil markets (Gibbons and Gwin, 1989; Kingsley, 1992: 119)

If you really want to get a sense of some of this, imagine factoring into the price of each gallon of gasoline you buy *a share* of other costs. Think about your share of the total and cumulative costs of U.S. military and foreign aid in the Middle East to maintain friendly relations with our suppliers—including the 1990 Gulf War. Indeed, if all of the health, geopolitical, and environmental costs of oil were internalized in its market price and if government subsidies from production were removed, oil would be so expensive that much of it would immediately be replaced by improved efficiency or other fuels (Miller, 1998: 434).

There's a lot of oil and natural gas left in the world, but we have used up much of the cheap and easy-to-get supplies. The depleted oil patch in the United States (in Texas and Oklahoma) is a case in point. It is why the United States continues to rely on Middle Eastern oil—even though the full geopolitical overhead costs are very high. Outside that region, oil and natural gas increasingly come from deep formations in rugged terrain, deep waters offshore, and hostile arctic environments. The search for oil and gas is inherently costly and risky, but it will become even more so in these environments. For instance, a high-stake 1984 exploration for oil in the Alaskan Beaufort Sea cost about $1.5 billion, but no commercial oil or gas was found (Hirsch, 1987: 1467, 1469). Most of the world's oil and natural gas reserves are in the Middle East and Central Asia, though it will take considerable investment in infrastructure to get them (World Energy Council, 1993).

Since the collapse of the Soviet Union, for example, more reliable surveys increased the estimated Russian oil, which in the 1990s ran two to three times the estimates of the 1980s. That is comforting, but much of the Russian oil is in remote and fragile Siberian tundra environments, which will make it very expensive to exploit. In addition, the Russian oil industry is antiquated, inefficient, and has produced devastating environmental abuse (Russian pipelines leak into the soil each year the equivalent of 400 supertanker cargoes!). The Russian oil industry collapsed along with the

Soviet economy, and it will take an estimated investment of $50 billion just to stabilize current (low) production and perhaps another $50–70 billion to increase levels of production. To really modernize and make the industry environmentally cleaner would require even more. Russia does not *have* that kind of money, and the political situation is so volatile and uncertain that Western companies are reluctant to make such investments. In other words, Russian oil will be very expensive to produce for a formidable mixture of political, economic, technical, and geographic reasons (Stanislaw and Yergin, 1993: 86–87).

In the world economy, geopolitical conflicts of interest are likely between the commodity-producing nations and the consuming nations for both fuel and nonfuel minerals. Most disadvantaged will be nations that have neither the money to buy much fuel or the resources to sell. Abstractly, energy is an important part of the patterns of world trade and politics that will determine who is poor and who is affluent, and who is well fed and who is hungry. It is unthinkable to try to understand either current world tensions or environmental problems without considering the importance of the production and distribution of energy around the world.

The important point is not that fossel fuels are becoming absolutely exhausted, but that the era of relatively cheap fuels is coming to an end. It is easily available oil that is scarce, not all oil. Meeting energy needs in the future will require much higher investments than in the recent past. It means extracting fuels from increasingly difficult and marginal sources, accommodating the needs of a growing human population, and paying the geopolitical overhead costs of an orderly energy market in a world system of nations. These costs don't even include the costs of increased environmental damage (Hirsch, 1987; Holdren, 1990: 158; Mazur, 1991: 156).

Sink Problems: Energy and Environment

Though energy supplies are thought to be less constraining now than in the 1970s, environmental problems deriving from the present energy system are thought to be more severe and getting worse (Flavin and Dunn, 1999: 24; Stanislaw and Yergin, 1993: 88). Stated abstractly, the most pressing problems may not be source problems but sink problems.

Burning fossil fuels is a major source of anthropogenic CO_2, a major heat-trapping greenhouse gas. Burning oil products also produces nitrous and sulfur oxides that damage people, crops, trees, fish, and other species. Urban vehicles that run almost exclusively on petroleum products cause much urban pollution and smog. Oil spills, leakage from pipelines and storage, and leakage from drilling sites leaves the world literally splattered with toxic petroleum wastes and by-products. The ecosystem disruption from oil spills may last as long as 20 years, especially in cold climates. Oil slicks coat the feathers and fur of marine animals, causing them to lose their natural insulation and buoyancy, and many die. Heavy oil components sink to the oceans floor or wash into estuaries and can kill bottom-dwelling organisms (e.g., crabs, oysters, and clams) making them unfit for human consumption. Such accidents have serious economic costs for coastal property and industries (such as tourism and fishing).[3] An $8.5 billion accident in Alaska's Prince William Sound by the Exxon *Valdez* supertanker might have been prevented if it had a double hull safety feature. By 1998, virtually all merchant marine ships had double hulls, but only 15 percent of oil supertankers did, even though in theory the Oil Protection Act of 1990 regulated supertankers to reduce the danger of such spills. To get around the law, many oil carriers shifted their oil transport operations to lightly regulated barges pulled by tugboats. This reduction in oil spill safety led to several barge spills. Oil tanker accidents like that involving the Exxon *Valdez* get the most publicity. But experts estimate that between 50 percent and 90 percent of the oil reaching the oceans comes from the land, when waste oil dumped on the land by cities, individuals, and industries ends up in streams that flow into the ocean (Miller, 1998: 527–529).

Coal is hazardous to mine and the dirtiest, most toxic fuel to burn. Mining often devastates the land, and miners habitually suffer and often die from black lung disease. Burning it produces larger amounts of particulate matter and CO_2 than burning other fossil fuels. The combustion of coal accounts for more than 80 percent of the SO_2 and NO_x injected into the atmosphere by human activity. In the United States alone, air pollutants from coal burning kill about 5,000 people, contribute to at least 50,000 cases of respiratory disease, and cause several billion dollars in property damage. Damage to the forests of Appalachia, the northeast United States, eastern Canada, and Eastern Europe can largely be attributed to coal-fired industrial plants. Reclaiming the land damaged by coal mining and installing state-of-the-art pollution control equipment in plants substantially increases the costs of using coal. As with petroleum, if all of coal's health and environmental costs were internalized in its market cost and if government subsidies from mining were removed, coal would be so expensive that it would be replaced by other fuels (Fulkerson et al., 1990: 129; Miller, 1998: 441).

Since the 1970s, coal consumption has grown most quickly in a handful of Asian countries, where energy demands are large and growing (e.g., China, India, and Indonesia). China is the world's leading consumer of coal, followed by the United States as the second largest user. As in Britain in the nineteenth century, coal accounts for about three-fourths of the Chinese energy budget. China's phenomenal economic growth involved parallel increases in the burning of coal because China is rich in coal with few conventional fuel alternatives. By 2000 China accounted for 30 percent of world coal consumption, and energy planners envision a 40 percent increase in the next decade, though extensive health and crop damage are threatening those plans (Dunn, 1998: 52). China is also the third largest producer of atmospheric CO_2 (after the United States and Russia). Because of its dependence on coal, China's economic development has been far more *energy intensive* than that of most nations, meaning that it gets much less economic output from each unit of energy. If both the use of coal and the production of CO_2 continue at historic rates, the Chinese production of greenhouse gas will quadruple in less than 40 years and will surpass that of the United States. Indeed, much of the future of global climate change depends very much on how energy intensive Chinese development will be (Stern et al., 1992: 60–64).

Let me sum up the argument I have been making so far: There is *no* immediate energy crisis. A *crisis* is a rapidly deteriorating situation that, if left unattended, can lead to disaster in the near future. But there is an energy *predicament*, that is, an ongoing chronic problem that, if left unattended, can result in a crisis (Rosa et al., 1988: 168). The energy predicament includes future source constraints and the ways in which the present energy system is intimately connected with environmental degradation, climate change, population growth problems, and the global equity and geopolitical tensions that plague the world. At end of this chapter I will turn to some of the possibilities and options for the transformation of the present system to address our energy predicament. But there are some clues about these possibilities from the relationship of energy to society, and studies of that relationship by scholars, to which I now turn.

THE ENERGETICS OF HUMAN SOCIETIES

The ultimate source of *all* the world's energy is radiant energy from the sun. Fundamental to understanding the energy flows of both ecosystems and human social systems, autotrophic (green) plants transform solar radiant energy into stored complex carbohydrate chemical forms by the process of photosynthesis. These are then consumed and converted into kinetic energy through the respiration processes of other species. Energy filters through the ecosystem as a second species consumes the first, a third the second, and so on (Humphrey and Buttel, 1982: 138). Unlike matter, energy is not recycled but tends to degenerate through the process of *entropy* to disorganized forms such as heat, which cannot be used as fuel for further production of kinetic energy or to sustain respiration. (This entropic property of energy is the second law of thermodynamics.) Such inefficiency means that only a portion of stored potential energy becomes actual kinetic energy.

This inefficiency and wastefulness occurred under conditions that existed long ago in the earth's geological history, when the storage of organic matter in sediment and fossil deposits created the concentrated energy carbon sinks of petrochemical fossil fuels. These fuels became almost the exclusive energetic basis of industrial economies during the last century. Of course, this was a great benefit, because we are now living off the stored energy capital of millions of years ago, but it is also true that the second law of thermodynamics (entropy) means that the relatively plentiful supplies of these fuels are ultimately exhaustible. More precisely, we will never absolutely use them up, but they can become so scarce and low grade that the costs of the energy and investment necessary to extract, refine, and transport them exceed the value of their use. We will have to squeeze the sponge harder and harder to get the same amount of energy, and the damage to the environment will increase as we do so.

Low- and High-Energy Societies

All human societies modify natural ecosystems and their energy flows, but they vary greatly in the extent to which they do so. Human respiration alone requires enough food to produce about 2,000–2,500 calories a day, but people in all human societies use vastly more energy than this minimum biological requirement to provide energy necessary for their shelter, clothing, tools and other needs.[4] Each person in a hunting and gathering society requires about 5,000 kilocalories per day (a kilocalorie is 1,000 calories), but that makes insignificant demands on natural ecosystems compared to the average of 230,000 kilocalories used per person in the United States. (See Table 6.1.)

Table 6.1 illustrates the prodigious growth of world energy consumption since the beginning of the industrial era and the increasing human dependence on petrochemicals. By contrast, the traditional fuels of preindustrial societies (e.g., wood, dung, plant wastes, and charcoal) are still the energy mainstays of many people in poorer LDCs. While the aggregate energy consumption of the world has grown, it is also important to note that most of that growth is accounted for by the MDCs as high-energy societies

TABLE 6.1 Per Capita Energy Consumption in Different Types of Societies

Society	*Kilocalories per day per person*
MDC (U.S.)	230,000
MDC (other nations)	125,000
Early industrial	60,000
Advanced agricultural	20,000
Early agricultural	12,000
Contemporary hunter-gatherer	5,000
Prehistoric	2,000

Source: Adapted from Miller, 1992: 32.

(see Figure 6.1). Indeed, a typical suburban household of an upper middle class American family consumes as much energy as does a whole village in many LDCs!

Industrialization and Energy

Industrialization was possible because new technologies of energy conversion were more efficient than traditional fuels. During the first phase in the early nineteenth century, the dominant technology depended upon coal mining, the smelting and casting of iron, and steam-driven rail and marine transport. The system's components were closely intertwined, and the creation of integrated mining, smelting, manufacturing, and transportation infrastructures made industrialization possible. By the beginning of the twentieth century, the system was being radically transformed again—by electric power, internal-combustion engines, automobiles, airplanes, and the chemical and metallurgical industries. Petroleum emerged as the dominant fuel and "feedstock" for the petrochemical industry.

Withdrawals of so much energy from nature in the United States and other MDCs required substantial modifications of natural energy flows. Industrial era agriculture replaced the diversity of natural species of plants and animals with genetic hybrids that could produce more caloric energy diverted to human use. As noted in Chapter Five, such agricultural monocultures lose stability, lose specialized circuits of energy and mineral recycling, have lower protection against epidemic destruction, and do not

Figure 6.1 Growth in Energy Consumption in the Industrial Era
Source: G. B. Davis, *Energy for Plant Earth,* Copyright © 1990 by Scientific American Inc. All rights reserved. Used with permission.

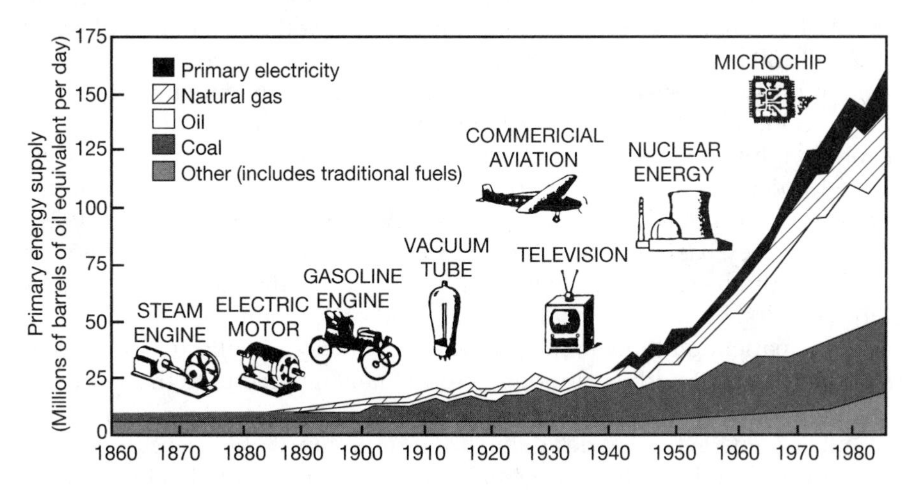

sustain soil fertility. A city alters natural ecosystems even more radically, requiring enormous amounts of energy from remote reserves of fossil fuels to power industry, heating, lighting, cooling, commerce, transportation, waste disposal, and other services. Cities become inert and relatively abiotic. Wastes are no longer naturally absorbed but must be transported to waste treatment plants (Humphrey and Buttel, 1982: 139). In addition, industrial farmers use machinery, fertilizer and fuel manufactured by urban industries, and food is no longer consumed mainly on farms. MDCs thus have integrated agricultural-industrial consumption systems that use enormous amounts of fossil fuels and have vastly modified natural ecosystems and energy flows. In sum, energy has a powerful role in connecting and modifying both ecosystems and social systems, and it is therefore an important topic for the social science understanding of human-environment relationships.

Social Science and Energetics

Remarkably, in spite of the obviousness of the last sentence in the preceding paragraph, until recently the social sciences were not very much concerned with energy and the social and environmental consequences of energy production. That was partly true for reasons I noted in Chapter One: Developing social sciences tended to distance themselves from recognizing the embeddedness of human societies in nature. Even so, scholars made fragmentary attempts to study the society-energy-environment connections.

Early Energetic Perspectives

The first social scientist to study explicitly the energy-society connection was Herbert Spencer, the English theorist of social evolution. Spencer argued that the ability to harness more and more energy was at the foundation of social and cultural evolution (1880). He theorized that the evolution of differences among types of societies could be accounted for by the amount of energy they produced and consumed. But among his voluminous ideas about social evolution, his insights about energy and society passed into oblivion. They were to be rediscovered at intermittent intervals by others, in a not much improved form (Geddes, 1890; Ostwald, 1909; Soddy, 1926; Carver, 1924). Although they basically rediscovered Spencer, these researchers did not do so in exactly the same way. Only Soddy recognized the second law of thermodynamics. Because energy is the lifeblood of economic and social life, he argued, inattention to its limits should be a source of intellectual and political concern. But Soddy's cautions were virtually ignored until the specter of energy shortages a half century later.[5]

Early energetic theories defined the importance of the energy-society connection, but their fragmented and episodic interest provided little cumulative development or elaboration. They shared other flaws. All were developed in the style of large, abstract theories attempting a unified explanation of social life on the basis of a few principles. They provided little guidance for empirical research or policy because they ignored many concrete variations among supposedly comparable societies. Most failed to appreciate the theoretical limits to energy growth imposed by the second law of thermodynamics and viewed social change as a single, linear evolutionary process—often equated with progress. Beyond the notion that energy is the crucial linkage between societies and their biophysical environments, about the only generalization that remains valuable is that increases in energy production and efficiency are related to increases in the structural complexity and the scale of human societies (Rosa et al., 1988: 150–154). That represents precious little in terms of cumulative development of understanding the environment-energy-society relationship!

More Contemporary Perspectives

After World War II, prominent anthropologist Leslie White rekindled interest in energetics by describing the resource and technological bases for social evolution, and sociologist Fred Cottrell developed the notion that available energy limits the range of human activity (Cottrell, 1955; White, 1949). Cottrell tried to demonstrate the pervasive social, economic, political, and even psychological change that accompanied the transition from a low-energy society (preindustrial) to a high-energy society (industrial), and argued that the vast social change to modernity could ultimately be traced to energy conversion (Rosa et al., 1988: 153).

Macrolevel Studies of Low-Energy Societies. Stimulated by the works of White and others, anthropologists in the 1960s conducted a plethora of meticulous empirical studies about environment-energy-society interactions in diverse ecological settings among such cultures as the Tsembaga Maring people of the central New Guinea highlands (Rappaport, 1968), the Eskimos of Baffin Island north of Canada (Kemp, 1971), the !Kung Bushmen of the Kalahari Desert in Southwest Africa (Lee, 1969), and the rural Western Bengali (Parrick, 1969). For a summary, see Kormonday and Brown (Chap. 14, 1998). These studies went far beyond the basic insights of historic energetic theories. Armed with such detailed empirical evidence, scholars for the first time could compare energy flows between societies and look for orderly patterns. Anthropologist Marvin Harris made the most significant attempt to do so and to recast older ethnographic evidence in energetic terms (1971, 1979). He proposed a formula that calculated the productivity of energy in food production in diverse societies.[6] Application of this formula to societies with diverse food production technologies—hunter-gatherers, hoe agriculture, slash and burn agriculture, irrigation agriculture, and modern industrial agriculture—revealed several patterns.

First, while confirming the central insight of historic energetic theories (about the relationship between energy efficiency and societal size and social complexity), these studies cast doubt on the causal sequence suggested by earlier theories. Classic theories had argued that increased technological efficiency led to increased available energy, which in turn led to larger populations and greater social complexity. New anthropological evidence suggested that population pressure was often the driving force of this process, promoting increased technological efficiency of energy conversion to meet rising demands. This evidence was supported by supply-side perspectives on population growth and economic development that were being independently developed by economists (Boserup, 1965). In Chapter Five I noted the ongoing debate and controversy about this perspective on population and food as well as contemporary evidence critical of this view. Second, anthropological studies suggested that high-energy societies would typically replace or assimilate low-energy societies whenever they came into contact. The most obvious example for Americans is the outcome of contact between Europeans and Native Americans, but evidence of this replacement around the world is compelling.[7] Third, these studies of preindustrial societies questioned the long-term outcomes of the process of energy intensification. Because the recurrent response to population pressures was an upgrading of production, preindustrial societies were often positioned to overburden their environments, deplete essential resources at a rate faster than they could be regenerated, and disrupt ecological cycles—and their own long-term sustainability. The anthropological research literature is replete with evidence among preindustrial societies of ecological collapse similar to that of the Mayan and Mesopotamian civilizations described in Chapter Two. Note that this anthropological evidence highlights a central issue of the contemporary energy predicament discussed earlier: the long-term relationship of growing energy and resource consumption with social and environmental sustainability (Rosa et al., 1988: 157).

Macrolevel Studies of High-Energy Societies. Analyses of energy flows in complex MDCs is no easy matter. Economists dominated energetic research after the oil shocks of the 1970s, and they emphasized the importance of energy to the economic performance of societies. Longitudinal research within societies and comparative analyses all suggested a strong relationship between the growth of energy production and the increase in measures of economic growth, such as the gross national product (GNP) (Cook, 1971). (See Figure 6.2.)

These studies interpreted economic indicators such as the GNP as indicative of social well-being, and since economic growth represented improvements in societal well-being, it was but a short step to infer than energy growth was essential to societal well-being. Sociologists first examined directly the relationship between energy growth and measures of social

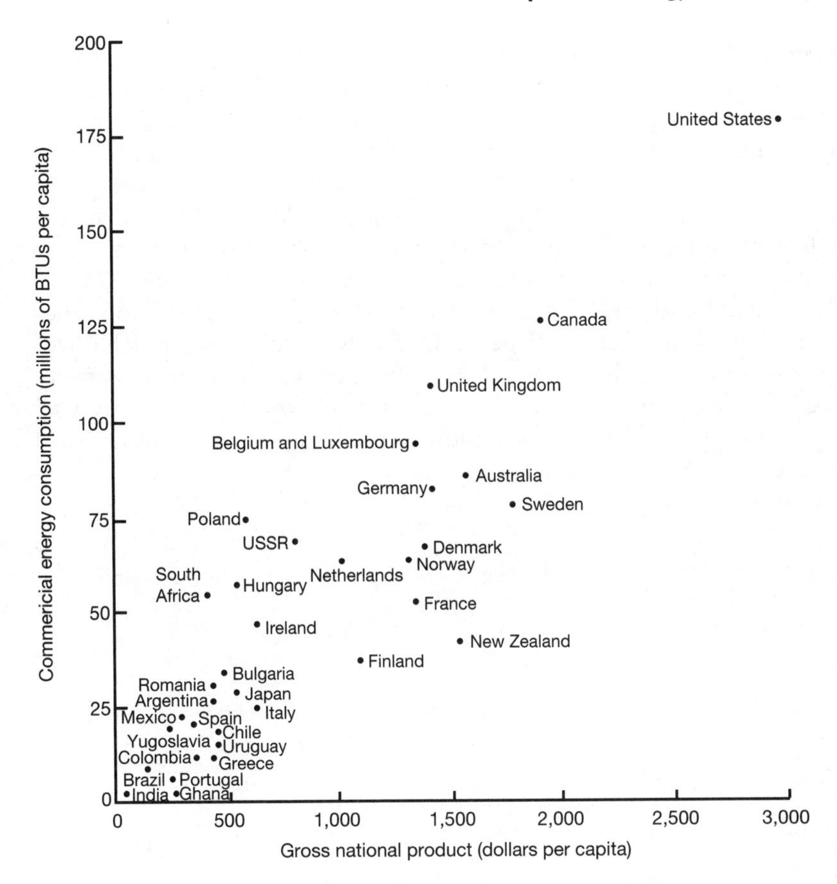

Figure 6.2 The Relationship between Per Capita Energy Consumption and Gross National Product, 1971

well-being (such as improvements in health, education, and nutrition) in a broad array of world societies. Their research generally confirmed a strong positive relationship between energy growth and growth in indicators of social well-being (Mazur and Rosa, 1974). The implication was that constraints placed on energy consumption would lead to a decline in wealth, although much room remained for increased efficiency of energy use (Karmondy and Brown, 1998: 354).

But note: When MDC market economies were separated from the LDCs and nonmarket socialist economies, this relationship virtually disappeared. Many studies supported this finding. These included cross-national longitudinal studies, studies examining the energy use of countries with similar living standards, case study comparisons (such as between the United States and Sweden), and cross-national studies of the relationship between energy intensity, social structure, and social welfare (Rosa et al., 1981, 1988; Schipper and

Lichtenberg, 1976). You can see the "looseness" of this relationship between energy consumption and gross national product measures in Figures 6.2 and 6.3, and particularly in the area marked off with an elliptical field in Figure 6.3.

Responses to the oil shocks of the 1970s, illustrated earlier in anecdotal terms, also illustrate the loose relationship between energy consumption, economic production, and quality of life. *Energy intensity* declined in the MDCs. Energy intensity decline—energy use per unit of economic production—is a means of measuring growing energy efficiency. Between 1973 and 1985, the United States reduced its energy intensity by 25 percent, and other MDCs did so by an average of 21 percent, usually starting from much lower initial energy use levels than those of the United States. But energy intensity continued to slowly increase in the LDCs because population, energy use, and economic production were simultaneously growing (International Energy Agency, 1987; Stern et al., 1992: 120).

Figure 6.3 Per Capital Income and Energy Consumption in 20 Industrial Nations, 1976

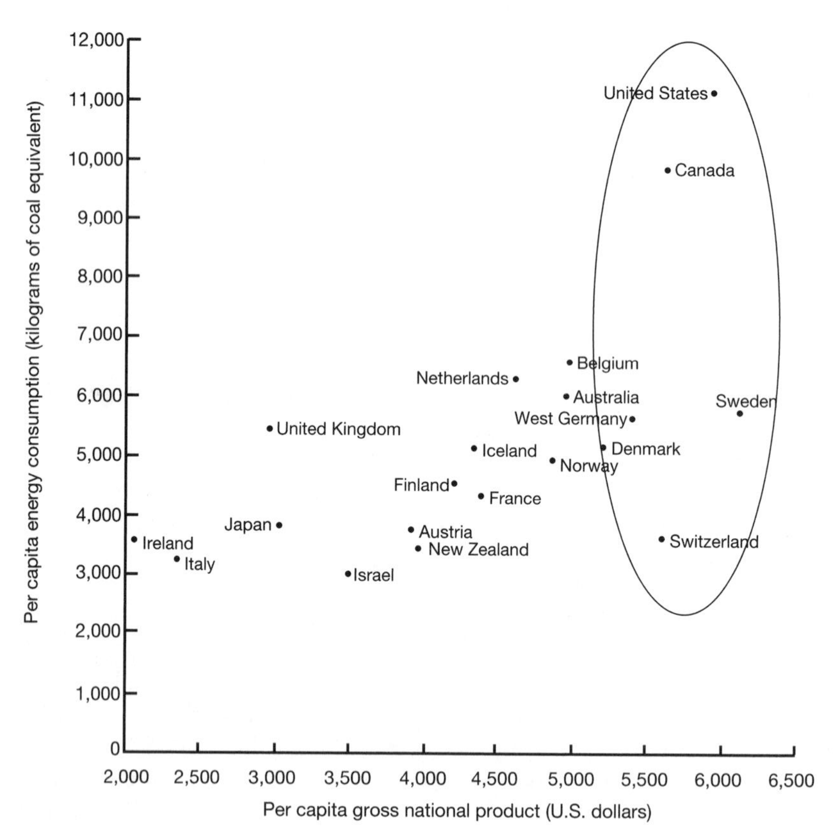

In the United States, reductions in energy efficiency came about in three ways. First, energy users changed the way they used energy-using equipment, curtailing heating, air conditioning, and travel and improving the management and maintenance of industrial equipment and furnaces. By 1986, these behavior changes accounted for about 10–20 percent of the total energy savings. These changes were short lived, however, and easy to reverse when energy prices fell or incomes rose, as they did in the later 1980s. Second, energy users adopted more energy-efficient technologies to provide the same services, either by retrofitting existing equipment (such as by insulating factories and buildings) or by investing in new energy-efficient equipment. These improvements accounted for about 50 percent to 60 percent of America's energy savings by 1986. Unlike the curtailment, management, and behavioral changes noted, these investments were hard to reverse once they were made. Third, the mix of products and services in the economy changed. Demand fell sharply in energy-intensive industries (such as steel and iron production), smaller autos got more of the auto market, and airlines developed a better match between aircraft size and passenger demand on various routes. Together these changes accounted for roughly 20–30 percent of the energy savings up to 1986 (Stern et al., 1992: 122).

Most studies find higher real energy prices to be the most important *single* explanation for these responses, but price increases by themselves are not the whole story. In fact, the actual effects of price changes depend on many other factors as well: technological change, policy choices, change in economic and industrial structure, information processing by energy users, pressure groups and social movements, and—indeed—changes in public attitudes and perception of problems. Because these factors can change independently of energy prices, it seems likely that with appropriate policies in place, energy intensity might have improved much more than it did after the oil shocks of the 1970s—as well as today when energy prices are lower and consumption is higher.

Macrolevel studies and historical data point to the same conclusion, that economic development in the MDCs went through two phases: from (1) rapid industrialization and consumption being highly dependent on increased use of energy from fossil fuels, to (2) economic growth becoming less energy intensive. In the latter phase, economic growth and social well-being could increase with decreasing energy intensity because of shifts in production from industrial to service sectors and because of the adoption of more efficient technologies. In other words, a threshold level of high energy consumption is probably necessary for a society to achieve industrialization and modernity. Once it is achieved, however, there is a wide latitude in the amount of energy needed to sustain a high standard of living. Given that latitude, industrial societies could choose slow-growth energy policies without great fear of negative, long-term consequence to overall welfare (Reddy and Goldemberg, 1990: 113; Rosa et al., 1988: 159; Stern et al., 1992: 64–65).

This evidence has profound implications for understanding and addressing the current world energy predicament. More than these studies, however, physicist and environmental activist Amory Lovins had a greater impact on the popular debates and discourses about energy. From the 1960s until today, his prolific writings popularized the idea that there are connections between energy and social and environmental problems (1977, 1998). Writing after the emergence of the environmental movement in the late 1960s and the oil shocks of the 1970s, his concern was not to discover principles of energetics or social evolution but rather the problems facing contemporary high-energy societies. These, he argued, are on a *hard path* of centralized energy production and high growth that ignores the second law of thermodynamics and inherent limits to growth, and will eventually produce energy crises for which no technological fix is feasible. In contrast, Lovins urged a *soft path* emphasizing energy efficiency rather than growth and a radical restructuring of MDC energy supply systems utilizing renewable energy sources. It would also entail significant social changes to evolve societies with more decentralized and sustainable energy systems. As a scientist-activist, Lovins has been a tireless advocate of these views and plays an important role in contemporary environmental movements. Even though the topic of energy production and consumption remains important, fewer macrolevel studies of energy flows in societies are being conducted today for several reasons. These include

> the assumption of greater stability in a population's energetics than is actually the case, the level of analysis being the population whereas more interest is at the level of the individual, and the notion that energetics is often not a critical area for study in human ecology except where energy is in short supply. (Kormondy and Brown, 1998: 358)

Microlevel Studies: Personal and Household Energy Consumption. The oil shocks of the 1970s also stimulated microlevel studies of energy consumption as well as macrolevel studies. The main goal of these studies was to develop a scientific understanding of whether people could significantly reduce their energy consumption without deterioration in their quality of life. Technical experts concluded that substantial energy savings could be achieved with existing technologies, which some estimated to be as much as 50 percent (Ross and Williams, 1981). Since individuals and households consumed about a third of the nation's energy—roughly evenly divided between transportation and home needs—they were viewed as a vast untapped potential source for energy conservation that would be responsive to social policy.

Engineering perspectives guided early microlevel studies, assuming that energy consumption could be easily explained by physical variables such as climate, housing design, and the efficiency and stock of appliances and vehicles (Rosa et al., 1988:161). As applied to vehicles and transportation, such engineering perspectives caused effective energy savings in the

1980s and early 1990s.[8] The fuel efficiency of American cars and trucks doubled as the cumulative result of many cumulative engineering changes. Given the earlier heavy, overweight, and overpowered gas guzzlers, that may or may not impress you, but it did make a significant contribution to increasing the nation's energy efficiency. Changes such as installing catalytic converters to reduce urban air pollution also addressed other environmental concerns (Bleviss and Walzer, 1990: 103, 106). These were engineering modifications that over time changed the machines driven and the composite fleet of cars and trucks, but not alternations or curtailments in the driving behavior of Americans. The only successful behavior change of the era was the one mandated by law, lowering the federal interstate speed limit from 75 to 55 mph (later, as you know, it was raised back to 65 mph, and 75 mph in some states). Attempts to encourage *voluntary* behavior change and curtailment, such as driving less, car pooling, bicycling, walking, or making greater use of mass transit, were dismal failures—at least on a scale large enough to make much difference.

As with transportation, energy conservation related to housing was dominated by engineering perspectives, emphasizing physical variables like climate, housing design, and the number and efficiency of household appliances. But unlike transportation, the assumption that reengineering homes and appliances would significantly reduce energy use was embarrassed by the first major study that investigated it. The Princeton University Twin Rivers Project, which became a classic social science study of household energy use, was a massive and detailed five-year field research effort. It found that townhouses in similar housing tracts with similar square footage, number of rooms, and appliance packages and occupied by families of similar size varied in energy use by as much as 2 to 1. Furthermore, the energy use of new occupants could not be predicted from that of the previous occupants. The impacts of lifestyle on household energy consumption was so dramatic that the Princeton study helped to justify social science perspectives to a skeptical energy policy establishment still dominated by an engineering orientation (Socolow, 1978; Rosa, et al., 1988: 161).

Research like this was not guided by a particular concept or theory and sought commonsense ways of asking people to reduce household energy consumption, such as turning down their thermostats, closing off unused rooms, or taking shorter showers. As policy-oriented research, it was dismally unsuccessful: Information and education programs, including those providing home energy audits, were consistently unsuccessful. About the only successes of these early post–oil shock studies focused on giving consumers better feedback information on their consumption. These studies recognized a particularly difficult barrier to the self-monitoring of energy use in households: that energy is largely invisible.

Unlike early atheoretic studies, later studies of household energy consumption were guided by two conceptual models: an *economic-rationality model*, favored mainly by economists and engineers, and an *attitude-behavior consistency model*, favored by psychologists and other social scientists. The economic model emphasizes that humans "rationally" respond to changing energy prices, given the presence of more efficient technologies. While escalating energy prices and efficient technologies played an important role in energy conservation, this model has shortcomings. Partly because of the relative inelasticity of energy demand, behavior is slow to respond to price changes, and many energy-use behaviors remain unexplained by price changes. The acquisition of accurate and reliable *information* about energy use, prices, investment costs, expected savings, and other nonprice factors are assumed but ignored by a simple economic-rationality approach (Gardner and Stern, 1996: 100–124; Rosa et al., 1988: 162–163).

The attitude-behavior approach sought to discover the effect of attitudes on energy problems and consumption. Studies understood attitudes broadly as having cognitive, affective, and evaluative dimensions and sought to understand how people understood and attended to energy problems. They also focused on how education and information could change energy-use behavior. But studies often found discrepancies between attitudes and behavior. Attitudes may not overcome barriers to change, price and affordability, lack of knowledge, or energy-use conditions that are embedded in society rather than personal choices (such as the kinds of homes and autos being marketed). One study of household energy-use curtailment analyzed the interaction of price and attitudinal factors. It found that as the kind of energy-saving activity went from easy and inexpensive (such as changing temperature settings) to difficult and expensive (such as insulation and major furnace repairs), attitudes became less powerful as predictors of energy-use (Black et al., 1985, cited in Gardner and Stern, 1996: 77). The conclusion reached by many studies is that while prices and other economic factors play a significant role in household energy behavior and decisions, they can be limited by the importance of social, psychological, and marketing factors. These include the vividness, accuracy, and specificity of information; the trustworthiness of sources of information; institutional barriers to investment; and other noneconomic factors (Stern and Aronson, 1984). I think the power of noneconomic factors relative to economic ones is different for households and big organizations (like governments and corporations) because the latter are likely to possess far more accurate information about energy price changes and investment alternatives than most households.

Unlike earlier studies, studies about values and attitudes that more carefully controlled for differences in information found powerful effects of personal values—moral obligations to change—that often outweighed the power of price incentives (Heberlein and Warriner, 1983). Other studies suggested the importance of involvement in civic and neighborhood organizations as predictive of energy conservation behavior by households, particularly in the contexts of community conservation programs (Olsen and

Cluett, 1979; Dietz and Vine, 1982). Others found that socioeconomic status shapes the modes of energy conservation behavior. More affluent households invest in energy efficiency, while poorer households cope with energy problems by lifestyle modifications and curtailments (Dillman et al., 1983, Lutzenhhiser and Hackett, 1993).

Taking together the macro and micro studies of energetics, one thing is obvious: Energy consumption and intensity are far too complex to be accounted for by either a simple economic-rational or attitude-behavior model. Scholars need an integrated conceptual framework that combines economic, social, and attitudinal factors (Stern and Oskamp, 1987). Such an integrated theory does not exist, but summaries of research literatures provide some policy-relevant clues. Economic incentives for energy conservation are likely to be effective when

- They are directed at specific external barriers, such as costs, access to credit, tax relief, or "inconvenience."
- Significant barriers are not located in the larger social system. These might include urban sprawl with large distances between work, home, and shopping, the "inconvenience" factor, or the unavailability of super insulated houses or efficient autos if they are not on the market.
- They are not counterproductive, such as raising energy prices (without compensatory policies) that force low-income or elderly people to choose between heating homes or buying food in the winter.
- They are combined with other influence techniques, such as information, public campaigns, curbside recycling programs, and moral and ethical arguments. (Gardner and Stern, 1996: 120–122).

Similarly, information and attitude change programs are more effective when they provide

- *Accurate feedback* that ties information directly to people's behavior. One of the successes of early household energy conservation programs was to provide people with information about current energy use.
- *Modeling* that provides illustrations about effective energy-use curtailments (rather than simply discussing the problem). Studies have, for instance, shown people videotapes about effective methods of energy use curtailment rather than resorting simply to moral persuasion.
- *"Framing"* messages to be consistent with people's worldviews and values. North Americans, for instance, are more receptive to arguments about improving "energy efficiency" than to those framed in terms of energy conservation. (Gardner and Stern, 1996: 83–88).

More disturbing than policy complexities is that even with the impressive growth in efficiencies noted from the 1970s to the mid 1980s, the United States still fell far short of reasonably attainable energy conservation goals. Incentive programs in particular were disappointing (Stern and Aronson, 1984). Not only did efforts at increasing the energy conservation intensity of America fall far short of the potential, but the increases in efficiency realized by the mid-1980s stalled and reversed. Most home-insulating incentive programs have been discontinued, speed limits are up again, and gas guzzling vehicles dominate auto markets. Energy efficiency in electricity has essentially been flat since 1980, and energy consumption in the residential and transportation sectors was on the upswing again by 1990, increasing at a rate of 3.3 per year, faster than the population is growing (Fickett et al., 1990: 65; Bevington and Rosenfeld, 1990: 77). Total U.S. energy consumption from all sources reached a new all time-high (79.4 quads) (Energy Information Administration, 1989).

I have discussed research about the relationship between energy and society in some depth because it provides clues about dealing with the present energy predicament. There are more clues when we consider the world's present energy technologies, to which I now turn.

THE PRESENT ENERGY SYSTEM AND ITS ALTERNATIVES

I mentioned that most of the world's present energy needs are supplied by finite or nonrenewable resources, mainly the fossil fuels—petroleum, natural gas, and coal. Another such resource, uranium, fuels nuclear reactors that provide a small portion of the world's energy. Renewable resources also supply about 25 percent of the world's energy needs. This is mainly hydroelectric power. Traditional fuels, such as plant residues, wood, or dung, provide small portions of the world's energy flows. While they may be overused, uneconomic, or environmentally damaging, unlike finite resources these are theoretically renewable. Other renewable energy sources, such as wind power, solar energy, and hydrogen, provide only small portions of current world energy flows but have great potential as alternative sources in the future.

Fossil Fuels

I discussed supply issues and other problems with most of the fossil fuels earlier, so I won't repeat that here. But you should note some of their advantages. Oil is relatively cheap and easily transported, and it has a high yield of *net useful energy*. Net useful energy is the total useful energy left from the resource after subtracting the amount of energy used and wasted in finding, processing, concentrating, and transporting it to users. Oil is a versatile fuel that can be burned to propel vehicles, heat buildings and water, and supply high-temperature heat for industrial and electricity production.

Coal is everybody's least favorite fuel, but there is an awful lot of it around the world. Burning coal produces high useful net energy yield and is the cheapest way to produce intense heat for industry and to generate electricity.

Natural gas, which I did not say much about earlier, is a naturally occurring geological mixture of methane, butane, and propane. In stark contrast to coal, it is clean burning, efficient, and flexible enough for use in industry, transportation, and power generation. It generates fewer pollutants, particulates, and CO_2 than any other fossil fuel: Natural gas releases 14 kg of CO_2 for every billion joules of energy produced, while oil and coal release 20 and 24 kg, respectively. But methane emission from leakage and incomplete combustion is a heat-trapping greenhouse gas 25 times more potent than CO_2. Like oil, natural gas is concentrated in a few parts of the world. The Middle East and Russia contain about 70 percent of the world's known reserves. While natural gas can be shipped by pipeline cheaply on the same continent, it must be converted into liquid natural gas (LNG) and shipped in refrigerated tankers to move it across the oceans—at present a difficult, dangerous, and expensive undertaking (Miller, 1998: 431, 436).

Besides their technical advantages, other advantages of fossil fuels are economic, political, and institutional. Quite simply, whatever their problems, we have an enormous sunk investment in infrastructures to produce, process, and use them. To develop new energy technologies that are economical and practical on a wide basis requires large investments and decades of experimentation. Not surprisingly, the rules of the present energy economies were established to favor the systems now in place, not new possibilities, whatever their advantages. Maintaining the fossil fuel system has short term but very real advantages for both individuals and the powerful corporate interest groups that profit from them. Historically, a powerful set of tax biases and subsidies encourage the use of fossil fuels and favor present operating costs rather than long-term investment in alternatives.

Even so, the fossil fuel age is probably coming to an end sometime in the next century. We cannot see its end, but its decline is already visible. In 1998, world oil use increased by less than 0.8 percent, the lowest annual rate of increase for a decade, and world coal consumption actually fell by 2.5 percent. Only natural gas use continued to expand robustly and is assured a larger future role (Flavin, 1999: 48). What will replace fossil fuels? Fifteen years ago, most experts would have said, with little hesitation, nuclear energy.

Nuclear Energy

Nonmilitary uses of nuclear energy produce electricity. In a nuclear fission reactor, neutrons split Uranium 235 and Plutonium 239 to release a lot of high temperature heat energy, which in turn powers steam turbines that generate electricity. In principle, nuclear reactions are the same kind of fission reactions in the atom bombs of World War II. The complicated systems required to regulate, modulate, contain, and cool such reactions make nuclear plants much more complex to operate than coal plants.

I'm sure you know this is a very controversial way of producing energy. In fact, it looks very much like a technological option that is slowly failing. In the 1950s, researchers predicted that nuclear energy would supply 21 percent of the world's commercial energy. But by 1995, after over 40 years of development and enormous government subsidies around the world, the 430 commercial reactors in 32 countries were producing only 6 percent of the world's commercial energy and 17 percent of its electricity. The United States ordered no new nuclear plants since 1978. After the turn of the century, nuclear energy in America as well as the rest of the world was expected to grow little, if any, as existing reactors wear out and are retired ("decommissioned"). Even in France and Japan, nations that invested heavily in the nuclear option, plans for the expansion of nuclear energy have stalled or been scaled down (Miller, 1998: 443). Why?

The first reason is well known: The risks of nuclear meltdowns and accidents tarnished the public image of the nuclear option. Three Mile Island (TMI), a U.S. nuclear plant in Pennsylvania that allowed radioactive gases to escape, and Chernobyl, a plant in the USSR (now Ukraine) that experienced a complete meltdown, became household words. At TMI, partial cleanup, lawsuits, and damage claims cost $1.2 billion, almost twice the reactor's $700 million construction cost. In 1982, Scandia National Laboratory estimated that a worst-case accident at a reactor near a large city might result in 1,000,000 deaths and $100 to $150 billion in damages. Furthermore, where and how to safely store low-level radioactive wastes accumulating from existing plants created intense political and legal conflict between states, communities, utilities, and agencies like the Nuclear Regulatory Commission (Miller, 1998: 446–447). In your back yard, maybe?

A second, and less widely appreciated, reason is that the planning, construction, and regulation of nuclear plants make them a very uneconomic investment, perhaps inherently so in relation to other options. A state-of-the-art coal-fired plant is a much less costly way of generating electricity. Cold hard economics may be a more potent barrier to the expansion of nuclear energy than negative public opinion or even cadres of antinuclear activists. Nowhere is this clearer than in the vaunted French national system, where growing technical problems in the highly standardized system that had generic flaws led to extensive shutdowns and repair problems costing billions of francs—and a huge cumulative debt (European Energy Report 1993 cited in Lessen, 1993a). Furthermore, dismantling and securing the world's aging stock of spent reactors and the transporting and disposing of nuclear wastes pose safety hazards, political problems, and economic costs that may exceed those of the development and operation of plants (Gibbons et al., 1990: 88). Banks and lending institutions in the United States are leery of financing new nuclear plants, and utility investors have largely abandoned them. A 1996 poll of U.S. utility directors found that only 2 percent would consider ordering a new nuclear power plant (Miller, 1998: 452).

Third, nations that have the technical capacity for nuclear power can also build nuclear weapons. So the diffusion of nuclear energy contributes to the potential proliferation of nuclear weapons and geopolitical tensions. The "nuclear bomb club" has already expanded to include China, Pakistan, India, and Israel. Several international rogue nations, such as Libya, Iraq, and North Korea, are only minimally open to international inspection and treaties and are widely suspected of using the development of nuclear electricity as a cover for developing a covert nuclear weapons capability.

Fourth, nuclear energy's physical potential to contribute to the expansion of world energy needs over the next century is questionable, particularly when compared with the potential of other fuels. Uranium ore is not a plentiful mineral in the earth's crust. German nuclear expert Wolf Häfele estimated that operating only existing plants over the next century, even assuming a 15 percent increase in operating efficiency, would contribute only one-fourth as much to world energy flows as would the potential use of oil and natural gas. They would also exhaust known nuclear fuel reserves and produce mountains of irradiated wastes (1990: 138).

These estimates assume existing nuclear fission technologies, but others are the technological drawing boards. One is the so-called *breeder reactor*, a fission reaction that would greatly stretch fuel reserves, but its practicality, safety, and economic viability are many decades away[9] (Miller, 1998: 454). The ultimate technological fantasy of nuclear scientists is a *fusion reactor*, which would harness the operating principle of hydrogen bombs. A fusion reactor would generate energy by fusing at very high temperature two hydrogen isotopes (deuterium and tritium). Those isotopes are vast and plentiful (and can be produced from ordinary seawater for a few cents a gallon). Fusion reactions leave no toxic wastes. But they require temperatures ten times as hot as the sun (about 100 million degrees fahrenheit!) to stimulate the reaction to even produce energy. No one knows how to safely contain such reactions, and the net energy required far exceeds the energy produced (it has, in other words, a large negative net energy). Research about fusion reactors has been going on for about 44 years now, and a commercial reactor might be built by 2030, but they wouldn't be a significant world energy source until 2100 (Miller, 1998: 454). In the meantime, there are several other safer, cheaper, and environmentally more benign ways of producing electricity.

Renewable Energy Sources

Perpetual and renewable energy sources are both the oldest energy sources used by humans and those with the greatest potential to provide energy and address the many environmental and social problems created by the present system. Taken together, energy from flowing water, biomass (plant and animal remains), wind, and sun *could* meet 50–80 percent of our energy needs by 2030, and perhaps sooner if combined with improvements in energy efficiency (Miller, 1992: 449). But with the exception of hydropower (a mature technology that generates electricity from water-driven turbines), all are potential sources and none makes a significant contribution to the present world energy flows. The principles of generating energy with each source are well established. By 2000 most were not practical or economic on a large scale, but their practicality and affordability are rapidly developing.

Hydropower

Hydropower uses water from dammed reservoirs to turn turbine engines which generate electricity. Hydropower generates about 20 percent of the world's electricity (6 percent of the total energy flow). It produces 50 percent of the electricity in the LDCs, close to 100 percent in Norway, and 9–10 percent in the United States. In 1995, the three largest producers of hydroelectric power were, in order, Canada, the United States, and Brazil (Miller, 1998: 416). Hydropower is highly dependent on topography and annual changes in stream flow, and in much of the world, the potential for hydropower is already developed. It is a mature technology, with a moderate to high net energy yield and fairly low operating and maintenance costs. Hydropower dams produce no emissions of CO_2 or other pollutants. They have an operating life span of two to three times that of coal or nuclear plants. Large dams can be used to regulate irrigation and to provide recreation and flood control. On the other hand, construction costs are high and they are not environmentally benign. They destroy wildlife habitats, uproot people, decrease natural fertilization (resilting) of prime agricultural land and fish harvests below dams, which makes their development inappropriate in many parts of the world, particularly in the LDCs (Reddy and Goldemberg, 1990: 111).

Biomass

Most of the world's people, and about 80 percent of LDCs residents, burn *traditional fuels*, such as wood, charcoal, dung, or plant residues for fuel. These *biomass* fuels account for 4–5 percent of the energy used in the United States and Canada, 35 percent in the LDCs, and about 13 percent of world energy flows (Miller, 1998: 419). Such fuels have a low net energy yield and are dirty to burn, producing a lot of carbon particulate, carbon dioxide, and carbon monoxide as byproducts. Heating a house with a wood or charcoal stove produces as much particulate matter as heating 300 homes with natural gas. But while people in LDC cities may buy wood or charcoal, the great human virtue of traditional fuels is that most people who use them do not purchase them. In rural areas, the women and children usually gather twigs and branches or animal dung for cooking fuel instead of buying wood.

Because most of the huge rural LDC population is poor and depends largely on noncommercial sources of energy, per capita use of commercial energy is much lower than in MDCs. Unlike fossil fuels, biomass is available over much of the earth's surface. In principle, biomass fuels are renewable and environmentally benign. But often the pressure of growing populations stripped the land of trees and vegetation in the search for fuelwood, contributing to deforestation and desertification. The forests of China have been cut down for centuries, and the search for fuel wood today exacerbates desertification, soil erosion, and environmental degradation in much of sub-Saharan Africa, Nepal, and Tibet (Reddy and Goldemberg, 1990: 111).

But traditional fuels do not have to be used unsustainably, and they can be used to produce other fuels. In many LDCs, *biogas digesters* use anaerobic bacteria to convert plant wastes, dung, sewage, and other biomass fuels to methane gas. After the generation of methane, used for lighting and cooking, the solid residue can be recycled for fertilizer for food crops or trees. If allowed to rot naturally, traditional fuels would themselves produce atmospheric methane, which—recall from Chapter Four—is a greenhouse gas much more potent by volume than CO_2. China has about 6 million such biogas digesters, and India has another 750,000, most constructed since 1985. Biomass digesters can be built for about $50, including labor. They can improve the lives of villagers while efficiently recycling plant and animal refuse, reducing deforestation by avoiding the necessity of cutting trees for fuel. They also do not make villagers dependent on expensive energy from big companies, cities, or big power grids. But they have costs and limits. The supply of biomass fuelstock often varies seasonally, and if used in biogas generators it reduces its availability for its usual use as crop fertilizer.

Where low-cost biomass fuel is readily available, there are other possibilities. In Brazil, for example, highly sophisticated industries have been developed to convert plant residues (from sugarcane, sugar beets, sorghum, and corn) into *ethanol*. Large fermentation and distillation facilities now produce the ethanol that powers much of the auto and truck fleets, cutting oil imports and producing thousands of jobs in the ethanol industry. But it required heavy government subsidies, which many poorer LDCs would be unable to capitalize. Some have envisioned cultivating large numbers of rapidly growing plants such as cottonwoods, sycamores, shrubs, or water hyacinths in biomass plantations of "Btu bushes" to produce biomass fuel. But this means the conversion of huge amounts of forest, grassland, or farmland into single-species biomass plantations and further accelerates declining biodiversity (Miller, 1998: 420).

Wind Power

Wind generators basically hook up modern windmills to electric generators to produce power directly. Such power can only be produced in areas with enough wind. When the wind dies down, you need backup electricity from a utility company or some kind of energy storage system. Furthermore, unlike coal or oil, which pack a lot of energy in a small amount of fuel, the amount of wind that blows across each square meter carries only a little bit of power. It would take the combined effort of *many* wind generators installed across large areas of land to produce as much energy as a single fuel burning power plant (Mazur, 1991: 161). Even with these limitations, wind power has a *vast* potential. In some areas the wind blows continuously, such as in the twelve contiguous U.S. Rocky Mountain and Great Plains states from the Canadian border to Texas, a region that contains 90 percent of the wind power potential in the United States. Wind-generated energy in this region could far exceed local demand. Two states, Montana and Texas, have enough wind to satisfy the whole country's electricity needs, and the whole upper Midwest could supply the nation's electricity without siting any wind turbines in either densely populated or environmentally sensitive areas. Similar windswept areas around the world could produce a substantial proportion of world electricity needs. England and Scotland alone have enough wind potential to satisfy half of Europe's electricity needs, and Germany, the Netherlands, and Denmark could easily supply the rest. China's wind power potential could triple the nation's current electricity usage. Wind generators produce no CO_2 or other air pollutants during operation, they need no water for cooling, and their manufacture and use produce little pollution. The land occupied by wind farms can be used for grazing and other agricultural purposes. In sum, wind energy is no longer a research project: It works, and works cheaply and reliably enough to compete with other energy sources.

By 2000, wind power supplied a small proportion of the world's energy flows, but because it is price competitive with fossil fuel energy, it has grown rapidly, by more than 25 percent annually in the late 1990s (compared with rates of growth for fossil fuels of less than 5 percent). In 1980, the world produced 5 megawatts of electricity from wind power, which grew to 2,100 megawatts by 1998, and wind power may produce 10–25 percent of the world's energy budget by 2050. The world's leading producers are Germany, The Netherlands, the United Kingdom, and Denmark and Spain, which produced twice as much wind energy as did the United States. In the Third World, India, China, Mexico, and Egypt have major wind projects. The move of several large companies into wind energy signals a transition underway: By 1997 Enron had purchased wind companies in Germany and the United States and Royal Dutch Shell planned to invest $500 million in renewables, mainly wind power (Flavin, 1998b: 58; Flavin and Dunn, 1999: 28; Miller, 1998: 419).

Solar Energy

The direct use of energy from the sun has the greatest potential as an alternative energy source. An enormous amount of radiant energy falls on the earth's surface, which—if trapped and converted into usable forms—could theoretically supply the energy needs of the world. The total potential of *solar power* is enormous but, like wind power, it is variable, only possible where and when the sun shines, needing storage and backup systems. Solar radiation intensity varies by latitude and with the weather, but still solar energy is available 60 percent to 70 percent of the days in the northern tier of American states, and 80 percent to 100 percent in the southern half of the country (U.S. Department of Energy, 1992). In the sunny regions closer to the equator that include many LDCs, the potential for solar energy is enormous and could supply much of the world.

Solar energy is now practical for space and water heating. The technology of using solar collectors for these purposes is relatively simple. For an investment of a few thousand dollars, using skills possessed by the average carpenter, it is possible to *retrofit* an older home to reduce the use of fossil fuels for heating water or rooms. A *passive solar heating system* captures sunlight directly within a structure through windows or sunspaces that face the sun and converts it into low temperatures heat. The heat can be stored in walls and floors of concrete, adobe brick, stone, or tile, and released slowly during the day and night. *Active solar heating systems* have specially designed collectors, usually mounted on a roof with unobstructed exposure to the sun. They concentrate solar energy, heat a medium, and have fans or pump systems that transmit space heat or hot water to other parts of a building. The potential for reducing America's aggregate heat bill in this manner is very large. On a lifetime-cost basis, solar space and water heating is inexpensive in many parts of the United States. But since subsidies of fossil fuels prices make them artificially low in the United States, such investments were less in the 1990s than after the oil shocks of the 1970s, when energy prices were high and a number of tax incentives existed (briefly). In many warm, sunny nations, such as Jordan, Israel, and Australia, solar energy supplies much of the hot water now, as it does for new housing in Arizona and Florida.

Photovoltaic electricity (PVE) is produced directly when semiconductor cells that create an electric current absorb solar radiation. You are probably familiar with PVE cells that energize small calculators and wristwatches. In many ways PVE is *the* superb energy source to create electricity: It creates no pollution, has no moving parts, and requires minimal maintenance and no water. It can operate on any scale, from small portable modules in remote places to multimegawatt power plants with PVE panels covering millions of square meters. Furthermore, most PVE cells are made of silicon, the second most plentiful mineral on the earth's surface (Weinberg and Williams, 1990: 149). But unlike windmills and solar space heating that are based on mechanical refinements of fairly simple technologies, producing wafer-thin silicon semiconductor solar cells is a high-tech business with considerable costs. Unlike the land around wind generators, land occupied by solar panels cannot be used for grazing or agriculture. But solar panels can sit on rooftops, along highways, and in sun-rich but otherwise empty deserts. Furthermore, the use of land would not be excessive. Hydropower reservoirs use enormous amounts of land, and coal needs more land than solar generators, if you include the area devoted to mining.

The main obstacle to the spread of PVE technology is its price, which is still higher than the cost of fossil fuel-generated electricity. As you might guess, PVE at present accounts for a miniscule portion of world energy flows. Even so, like wind power, PVE is growing rapidly for several reasons. PVE generators found niche markets in the world economy, where they are the cheapest way of delivering electricity to 2 billion rural villagers without having to extend centralized power grids from cities or big regional big plants. By the late 1990s, PVE electricity was growing rapidly in places like Vietnam, Sri Lanka, Colombia, Honduras, Jamaica, and the Dominican Republic, where sunlight is plentiful. Increasingly, PVE cells are used to switch railroad tracks, supply power for rural health clinics, operate water wells and irrigation pumps, charge batteries, operate portable laptop computers, and power ocean buoys, lighthouses, and offshore oil-drilling platforms. Production costs of solar generators continues to drop. Japan, with its troubled nuclear system (mentioned earlier), instituted significant tax subsidies for the installation of PVE generators in both homes and industries. Several European nations are also in the process of removing the traditional subsidies for coal and oil and transferring them to wind power and PVE. As a result, the megawatts produced by PVE generators increased 43 percent in 1997, and sales of PVE generators by American and European firms increased by 40 percent to 50 percent in the late 1990s. If governments have been slow to recognize the huge potential LDC markets for PVE, corporations have not. By 2000, corporations like British Petroleum and Shell Oil were investing heavily in the development of PVE (O'Meara, 1998: 60).

Hydrogen Fuel

If you took high school chemistry and conducted *water electrolysis*, running electricity through water and splitting water molecules into oxygen and hydrogen atoms, you can understand the potential of using *hydrogen gas* as a fuel. You could make hydrogen using solar, wind, or conventionally produced electricity. It is a clean-burning fuel with about 2.5 times more energy by weight than gasoline. When burned, it produces no heat-trapping greenhouse gases, but combines with oxygen in the air to produce ordinary water vapor. Hydrogen can be collected and stored in tanks like propane is today, or it can be transported by pipeline. It is easier to store than electricity. It will combine with reactive metals to form solid compounds called *hydrides,* which could be stored and heated to release hydrogen, as it is needed to fuel a car or furnace. Unlike gasoline, accidents with hydride tanks would not produce dangerous explosions. A versatile fuel, hydrogen could be used for transportation, heating, or industry. *Fuel cells* that combine hydrogen and oxygen gas to produce electricity could power autos, trucks, and buses. Such cells have no moving parts and energy efficiencies several times larger than today's internal combustion engines. Unlike conventional batteries in electric vehicles, fuel cells need no recharging and, as long as hydrogen fuel is available, could be resupplied with fuel in matter of minutes. In 1999, DaimlerChrysler, Ford, General Motors, Honda, and Toyota were spending millions of dollars on experimental hydrogen fuel cell-powered vehicles. At that time, none was perfected or affordable (Associated Press, 1999; Miller, 1998: 424).

Gradually switching to hydrogen and away from fossil fuels as our primary fuel resources would mean a far-reaching *hydrogen revolution* on a profound scale. Technical and social transformations required over the next 50 years could change the world as much as did the agricultural and industrial revolutions. Theoretically, a hydrogen fuel economy could eliminate much air and water pollution, greatly reduce the production of heat-trapping greenhouse gases, reduce the need to use scarce fuel reserves, lower problems associated with fluctuating energy prices, and loosen energy constraints on economic development. The technological vision most attractive to energy experts and environmental thinkers would be to generate electricity by PVE, wind, or some other ecologically benign technology, and use it in electrolysis to create hydrogen fuels for use in industry and transportation. A solar-hydrogen economy would be based on resources that are more abundant and evenly distributed than fossil fuels and could reduce the geopolitical tensions and costs produced by dependence among nations (Weinberg and Williams, 1990: 149; Flavin and Dunn, 1999: 36).

What's the catch? Well, there are some big ones. One barrier is technical: It takes lots of electricity from some source to produce hydrogen by water electrolysis, so the net energy yield is very low. At present, hydrogen costs more to produce than it is worth as a fuel, particularly in highly subsidized fossil fuel markets of the United States. But as with other renewables, the cost of producing hydrogen fuel is rapidly coming down. Furthermore, if all the health and environmental costs of using gasoline were included in its market price (through taxes), it would cost about $4 per gallon, roughly the cost of gasoline in Japan and many European countries. At this price, hydrogen fuel would be competitive. To be of real use in our current energy predicament, hydrogen awaits the development of alternative electrical energy. Other barriers are social and institutional. Switching from a carbon- to a hydrogen-based economy means changing most of the economic infrastructures that are now in place and large investments, even if spread over the next 40 to 50 years. It means convincing investors and energy companies with strong vested interests in fossil fuels to risk lots of capital on hydrogen. It also means convincing governments to put up some of the money for developing hydrogen energy, as they have done for decades with fossil fuels and nuclear energy. In the United States, powerful oil companies, electric utilities, and automobile manufacturers who understandably see it as threats to their (short-term) profits generally oppose such government funding of hydrogen research. Even so, the solar-hydrogen revolution was underway by 2000. German and Japanese governments were spending 7 to 8 times more on hydrogen research as was the U.S. government. German firms planned to market hydrogen systems that would meet home needs and provide fuel for autos, and Germany and Saudi Arabia have each built a large solar-hydrogen plant (Miller, 1998: 424–425).

In sum, hydrogen power has only theoretical potential, but it is an enormously attractive one. It would be particularly valuable when, or if, land or water constraints become serious. Experts estimated, for example, that the PVE hydrogen equivalent of the world's *total* fossil fuel consumption could be produced on 500,000 square kilometers—less than 2 percent of the world's deserts (Weinberg and Williams, 1990: 153–154).

Efficiency as a Resource

Even with these technological alternatives to fossil fuel-based economies, it is important to mention that the cheapest, easiest, and fastest way to change our energy system to address the present energy predicament is to promote energy efficiency (Gibbons et al., 1990: 90). Greater efficiency means reducing demand and using less energy to produce the same services, such as lighted, heated, and cooled rooms, transportation for people and freight, pumped water, and running motors. It means producing the same material quality of life with less conflict over siting energy plants and waste dumps and less foreign debt and geopolitical overhead costs to maintain access to or control over foreign resources. The potentials for greater energy efficiency are enormous in households, transportation, and industrial sectors. Globally, industry accounts for about 45 percent of all energy use, larger than any other single sector. The United States could create greater efficiency, starting with the cheapest measures first. In order of increasing price, such measures include:

- Converting to efficient lighting equipment, which would save the United States electricity equal to the output of 120 large power plants, plus $30 billion a year in maintenance costs
- Using more efficient electric motors, saving half the energy used by such motor systems, which would save the output of 150 large power plants and repay conversion costs in about a year
- Eliminating pure waste electricity, such as lighting empty offices
- Displacing electricity now used with better architecture, weatherization, insulation, and solar energy for water and space heating
- Making appliances, smelters, and the like cost-effectively efficient

Amazingly, these five measures could quadruple U.S. electrical efficiency, making it possible to run the economy with no changes in lifestyles and using no power plants, whether old or new (Lovins, 1998).

Calculations about how much energy could be saved through efficiency depend greatly on the technical and political biases of the people who do the calculating. But on the *conservative* end of the range, it seems certain that the North American economies could do everything they now do with currently available technologies and at current costs, using *half as much energy* (Meadows et al., 1992: 75). This figure is not based merely on speculation. Europeans use about half as much energy per capita as do U.S. and Canadian citizens and have equivalent lifestyles (look again at Figure 6.3). Possibilities for "mining" efficiency are producing a profound shift in thinking about the environment among the business community. Economic thinkers have traditionally viewed protecting the environment and conserving resources as policies of economic restraint and costs. But now some envision a vast future market for efficiency. Reengineering the economies of the world to be more efficient may not only be profitable market for investors, but also the basis for a virtual second industrial revolution. Nations that fail to develop green industrial policies and technologies are likely to lose out economically as well as environmentally (Flavin and Young, 1993; Flavin and Dunn, 1999). I will return to this theme in later chapters.

CHANGING THE WORLD ENERGY SYSTEM: BARRIERS AND POLICY

As noted, we have an energy *predicament*, not a current energy crisis. That predicament has intrinsic causal linkages, even if they are invisible, with other very real social and environmental problems. I mentioned some of these earlier: pollution, loss of biodiversity, environmental degradation, health problems, urban sprawl/congestion, a large national debt and balance of payments problem, geopolitical costs of maintaining access to energy fields, and costly economic dependencies. In LDCs the energy predicament is related to deforestation, desertification, barriers to development, poverty, and hunger. The predicament does periodically boil over into real crises, such as the meltdown at Chernobyl, the Gulf War, and huge oil spills. But most ominously, our present energy system is thought to be the chief culprit in the most serious, if hypothetical, macrothreat to the future of humanity: global warming. The important point for you to recognize is that the web of connections between our energy predicament and the myriad of human social and environmental problems means that *there is some kind of energy transition underway.*

But what kind of transition? And how should society respond? To the point: How can the present energy transition, transforming the historic energy systems that societies created since beginning of the industrial era, be steered consciously toward a more supportive and sustainable relationship among society, energy, and environment?

Barriers

The main barrier to change is certainly not a lack of technical options. I hope you were convinced by the foregoing discussion that we now have a rich menu of technical possibilities for developing efficient, affordable, and environmentally friendly energy sources. Technically workable and economically viable alternatives to our present system are here now, or can be within decades, given sufficient investment. Nor is there much disagreement among energy experts and environmental scientists about what an energy-sustainable society would look like—in broad outline. This is true even when the views of industry-related experts are considered (see, for instance, Davis, 1990: 57).

What, then, are the barriers and circumstances that make a transformational energy policy difficult? There are at least five.

One barrier is that, like many other social problems, the salience of energy problems follows *an issue-attention cycle*, a cycle of rising and falling concern due to energy-related national events and the volume of media coverage they attract (Downs, 1972; Rosa, 1978; Mazur, 1981). Now that the energy crisis of the 1970s is over, supplies have increased and prices have moderated, there is less public concern or media coverage of energy issues. Thus the combination of public concern and media attention that would impel political action is at low ebb. The same was the fate of global warming, a premiere social problem after the long hot summer of 1989, but which "cooled" only to return slowly to public debate and discourse (Ungar, 1992, 1998).

A second obvious barrier is making policy for change when energy is relatively cheap. Investing in efficiency and a transition to new fuels is difficult when short-term market forces run counter to longer-term goals of a more sustainable and environmentally friendly energy system. From the perspective of the individual consumer, it is hardly rational to absorb the costs of conservation and change when no shortage looms. In addition there are problems of bureaucratic interests and organizational mandates. A principal purpose of the utility company is to sell electricity at a profit. Conservation policies that reduce demand and perhaps profitability as well often run counter to managerial objectives. In the short run, then, institutional as well as individual interests can run counter to long-term conservation and transformative policies (Fowler, 1992: 76).

Third, energy policies have been fragmented, contradictory and often paralyzed. In the United States, energy policy was separated by fuel type, with different institutional associations, interests, and regulatory bureaus for each, with few attempts at broader coalition building. Coal interests have dealt with the Bureau of Mines, gas and oil with the Department of Interior, uranium with the Nuclear Regulatory Commission. The net result of government energy policies has been to intervene in market forces unnecessarily with supply-side policies that subsidize costs and increase consumption rather than promote efficiency and alternative fuel development (Switzer, 1994: 138).

A fourth barrier to effective energy policy is that it needs to work on a global basis. Even dramatic improvements in energy efficiency and renewables will not be sufficient to protect the global environment if they are confined to the MDCs. Pleas from MDCs to address global environmental and climate problems through energy restraint will fall on deaf ears in the LDCs, unless the MDCs can find ways to help them achieve increased economic well-being and environmental protection at the same time. Why should LDCs worry at all about saving energy when their prime concern is generating economic growth, which means increasing the availability of energy services? The answer is that energy efficiency reconciles the simultaneous goals of development and environmental protection. I noted earlier, for instance, both the perils of the continued reliance by the Chinese on coal as well as the enormous potential for wind power in China. It is true that the LDCs are currently in a phase of growing energy intensity, but the potential for helping them leapfrog over a fossil fuel–dependent phase are substantial. In the sun-drenched tropics, potentials for increased efficiency as well as the development of wind and solar generated power is enormous. By investing $10 billion a year to tap them, LDCs could halve the rate of growth of their energy demand, lighten the burden of pollution on their environments and health, and staunch the flow of export earnings into fuel purchases. Gross annual savings would average $53 billion for at least thirty-five years, according to experts from the U.S. government's Lawrence Berkeley Laboratory (cited in Lessen, 1993b: 110).

A fifth barrier to a sustainable world energy policy is the one I discussed earlier in relation to both global warming and population problems: the dilemmas of action vs delay in an uncertain world. "If we wait, our knowledge will improve, but the effectiveness of our actions may shrink; damage may become irreversible, dangerous trends more entrenched, our technologies and institutions even harder to steer and reshape" (Holdren, 1990: 160).

Transitions

Even with these barriers to change, a transformation of the energy system is underway in the emerging world system of nations and world market economy. (Chapter Two discussed those concepts.) Some clues about this exist in two previous world energy transformations. Before the industrial revolution, people depended for energy on a combination of traditional biomass fuels (like wood and dung), animal power, and water power. Beginning in the 1800s, a new energy regime evolved around coal, which was the foundation for a steam-powered industrial system.[10] The coal regime diffused around the world in the late nineteenth century; between 1850 and 1913, this single energy resource went from providing 20 percent to more than 60 percent of the world's total commercial energy (Podobnik, 1999). Until 1915, petroleum had a niche market for kerosene to light lamps and was between three and twelve times as expensive as coal in Europe and North America. But under the stimulus of converting naval ships and military vehicles, a petroleum-based energy regime was established more rapidly than could have been by private enterprise alone. The share of world energy provided by oil grew from 5 percent in 1910 to over 50 percent by 1973, and after World War II it was the key resource for transportation, electricity generation, and heating in most of the industrialized world.

As the twenty-first century begins, oil, like coal before it, is entering a state of relative stagnation. Oil production grows at a very slow rate, but experts estimate that production will peak sometime between 2010 and 2030 and thereafter begin to decline (McKenzie, 1997; Podobnik, 1999). In the twenty-first century, a new and more diverse energy regime is emerging. Natural gas production is growing more rapidly than that for oil because it is a cleaner-burning and more plentiful fossil fuel. Natural gas supplies are located mainly in the Middle East and Central Asia, and so, as earlier noted, both the infrastructure investments and the geopolitical overhead costs of exploiting it are very large. In the new energy regime a more diverse mix continues to rely on oil and coal when possible but more natural gas and a rapidly growing decentralized mix of nonrenewables like wind, solar, and solar-hydrogen. Like oil, they are mainly now in niche markets and collectively comprise a small proportion of the world's total commercial energy budget (but more than oil did in 1910). In the medium term, a diffusion of fuel cells would significantly increase the efficiency with which fossil fuels are consumed. In the longer term, switching to solar-generated hydrogen could provide an energy regime that would have few adverse environmental consequences (Podobnik, 1999; U.S. Senate, 1997). A growing number of energy industry officials recognize this regime transition and the inevitable shift away from fossil fuels. In a remarkable speech in 1999, Mike Bowlin, Chairman and CEO of the ARCO oil company, said, "We've embarked on the beginning of the Last Days of the Age of Oil," and went on to say that the world is moving "along the spectrum away from carbon, and heading toward hydrogen and other forms of energy" (cited in Flavin, 1999: 48).

As I noted earlier, many existing big power companies (such as British Petroleum, Exxon, and General Electric) are investing in alternative forms of energy to position themselves for what they see as the emerging regime. At the same time, they protect profits and subsidies from the existing energy regime. But the existing energy regime may be upset by a new breed of "upstart" independent power producers that are increasingly challenging established power monopolies in countries like the United States and the United Kingdom. They are also growing very rapidly in Asian and Latin American countries as established monopolies fail to keep up with demand. A more competitive power industry is likely to diversity its base of power generation to meet changing conditions (Flavin and Dunn, 1999: 32–33). It is an old economic story: Established, lazy monopolies are challenged by small, innovative upstarts, just as IBM mainframe computers were challenged by the diverse manufacturers of personal computers and software.

Making Energy Policy

As industries move from monopoly structures to competitive markets, many U.S. state government regulatory agencies are experimenting with deregulation, and customers will have the opportunity to purchase electricity from less environmentally damaging sources. Indeed, public opinion polls suggest that people would like their future energy mix to be quite different from that dominated by coal, oil, and nuclear energy. According to 700 surveys conducted in the United States over 18 years, strong majorities prefer renewable electricity to current sources. Furthermore, 10 percent to 20 percent were willing to pay a premium for renewable electricity. Consumers in Australia, Canada, the Netherlands, and Switzerland have voiced similar support. Given this potentially large demand, some U.S. utilities are conducting green market research, and some—though still operating in monopoly markets—have begun to offer green pricing programs that allow customers to pay a premium to support planned projects. By the late 1990s, the largest was the Public Service Company of Colorado, which enrolled 7,000 customers who pay premiums (averaging 50 cents per month). Combined with federal grants, those premiums financed the construction of large wind power farms.

But wait. There are both opportunities and risks in deregulation that offers customer choice. As electricity from different sources mixes as it comes through transmission lines, customers need full disclosure such as the mandatory labeling of food so that they know exactly how different proportions are generated and can make reasonable choices beyond what is advertised as "green power" (Flavin and Dunn, 1999: 31–34). Moreover, people may choose the cheapest power rather than the greenest power. Often the cheapest power (under current subsidy conditions) is the most polluting, tempting companies to crank up old polluting coal plants. Businesses use about two-thirds of all electricity in the United States, and their primary goal will be to minimize their electricity prices, not buy green power (Jefferiss, 1998).

Other than deregulating power markets, some utilities have been experimenting with alternatives to the conventional supply-side management that attempts to address energy problems by increasing supplies (finding more oil fields, building more power plants). *Demand-side management* attempts to help customers to use energy more efficiently. They give customers cash rebates for buying efficient lights and appliances, free home-energy audits, low-interest loans for home weatherization or industrial retrofits, and lower rates to households or industries meeting certain energy-efficiency standards. To make such policies feasible, state utility regulators must allow utility investors to make reasonable returns on their money, based on the amount that utilities save. Between 1981 and 1994, Southern California Edison cut electricity demand by an amount equal to that of three large nuclear power plants (Miller, 1998: 402).

What opportunities exist for making effective energy policy? One has to be a broad and effective public education campaign about energy problems. Since concern over energy per se is currently in a low trough, an effective campaign would seek to connect the role of the world's current energy system to environmental, health, economic and geopolitical problems—for which there are substantial levels of public awareness and concern. Second, policy should try to create a large efficiency market. Demand-side management, increasing efficiency, and rebuilding our energy system around renewables should be justified not only as costs or responses to threats, but because they are eventually profitable. This implies an important task for experts and the practitioners of cost-benefit analysis to internalize the full and *long-term* costs of the existing world energy regime. The environment benefits from burning less coal, oil, or gas, with a cascade of other benefits in terms of health care and health insurance costs, less carbon dioxide in the atmosphere, the nation's balance of payments, and so on. A third opportunity would be to gradually shift subsidies from the historic fossil fuels to these emerging technologies with fewer social and environmental costs.

Crises come and go, while predicaments persist. During times such as the present, when energy shortages appear remote and no crisis seems imminent, the challenge for humanity is to create the popular support, political will, technical progress, investments, and global cooperation to steer the emerging energy regime to sustainable world energy system.

PERSONAL CONNECTIONS

Personal Consequences and Questions

Here are some questions to help you think about your personal relation to energy consumption and a variety of social and environmental issues.

1. Following Chapter Four, I asked you to think about your share of the consequences of climate change. Let me turn the tables in the first two questions and ask about your share of energy consumption that produces greehouse gases. On average, for every $3.19 of world economic output, 1 kilogram (2.2 lbs.) of carbon is released into the air. That means the average person sends the equivalent of his or her body weight of carbon into the atmosphere for about every $200 that he or she spends. That figure is based on a world average: You probably contribute much more to the global greenhouse gas production than average as a citizen of a high-consumption MDC.
2. In fact, though the United States has less than 5 percent of the world's population, collectively Americans contribute 22 percent of global carbon emissions. The average U.S. citizen sends about 15,000 pounds of carbon into the air every year. If you spent $2,000 last year (and are of average size) you dispersed 100 times your own weight of carbon into the environment (Kane, 1994: 38). In the MDCs travel accounts for more than 20 percent of energy use, and autos alone for more than 17 percent. And Americans depend on autos far more than the citizens of other MDCs do. So if you are a typical person in the American car culture, you contribute a lot to global warming that way. Each gallon of gasoline used to fuel your car weighs almost 7 pounds, and about 6 of those pounds are carbon. When it's burned, most of it ends up as exhaust. The point of all this is not to try to make you feel guilty, but simply to get you to recognize that people like you and I bear particular responsibility for producing the gases that contribute to climate change. If we paid costs in proportion to our share of the production of such gases, that would be fair. But alas, that is not the case. Those who produce more than their share of climate-warming gases can send the bill to many others at a later time and around the world, including many of those who can least afford to pay.
3. This chapter noted that as nations move from LDCs to MDCs and from early to late industrial (more service-based) economies, they become more energy efficient per unit of economic output. Yet the data displayed in Figure 6.3 demonstrated considerable variation among developed market economies in the relationship between energy input and economic output, with Americans leading the pack in terms of energy inefficiency. Given what you know about conditions and lifestyles in America and other nations, why do you think that is so?
4. The chapter also discussed evidence that residential energy use is determined by the size of homes or apartments, but more powerfully by the energy culture or "lifestyles" of the occupants. Think about that in personal context. For instance, I get some mail from my utility company about how we can be more energy efficient (demand-side management). Those mailers that change attitudes are important, but there are no real incentives attached. So like many households, after insulating the house some years ago, we really haven't done much else. Given your life right now, what are the major barriers, situational, attitudinal, or economic, that prevent you from becoming more energy-frugal?
5. Since transportation is such an important part of our energy budget, here's a pointed question: If you drive a car, how much would gasoline have to cost per gallon to induce you to cut your driving by a meaningful amount? How would that question be answered differently by people with different occupations? What other changes in community life would make it easier for you to do this?

What You Can Do

The small lifestyle changes that relate to possibilities for greater energy frugality are well known:

- Drive less, keep your car tuned up; when possible walk, bicycle, or ride the bus or commuter train; car-pool.
- Insulate your house and turn the thermostat down in winter; adjust to changing temperatures by changes of clothing rather than heating or cooling your house; run appliances frugally, and replace them with more energy efficient appliances when you can.
- Buy "green goods" that have less stored energy used in their production by the time they get to you.
- Spend more of your money on services rather than things.
- Etc., etc. . . . You know the litany of small things you can do. (If many people did them, they would add up.)

The larger and more meaningful lifestyle changes are more difficult and challenging. They require more planning, investment, and integrated lifestyles. What do I mean?

- Plan to live close to where you work, reducing both transportation time and costs. You can find an appropriate job close to where you live, or move closer to where you now work. Either is likely to be a challenge.
- Choose a career that enables you to "walk lightly" regarding energy and other impacts on the environment—in other words, one that rewards frugality. Exactly what kinds of careers would those be? I'm not sure I know, but I think it's meaningful to pose the question. Buddhism emphasizes the notion of "right livelihood" as an ethical imperative. What would right livelihood mean in an ecological sense?
- In general, try to simplify your life in ways that still support your sense of well-being. Doing this is not easy. It raises issues about how you could do this (if you wanted to—and many don't!). It also forces you to examine the exact sources of your sense of well-being.

Real Goods

The bicycle. The most thermodynamic and efficient transportation device ever created, and the most widely used private vehicle in the world. The bicycle lets you travel three times as far on a plateful of calories as you could walking. And they're fifty-three times more energy efficient—comparing food calories with gasoline calories—than the typical automobile. Nor do they pollute the air, lead to oil spills—and oil wars, change the climate, send cities sprawling over the countryside, lock up half of urban space in roads and parking lots, or kill a quarter million people in traffic accidents each year. The world doesn't yet have enough bikes for everybody, but it's getting there quickly: Best estimates put the world's booming fleet of two wheelers at 850 million—double the number of autos, and growing more rapidly than the auto fleet. We Americans have no excuses on this count. We have about as many bikes per person as do the Chinese. We just don't ride them as much.

I admit to being a bike enthusiast (some of my friends have different words for it!). Like many American kids, I grew up riding a bike (a big heavy Schwinn is the one I remember) and didn't discover lightweight bikes with gears until mid-life. I found cycling a life-saving form of exercise and mood enhancer. I enjoy weekend rides through the green fields of the urban hinterlands and discovering the diversity of urban neighborhoods in a more intimate way than I ever could by driving around in my car. I'm fortunate to live close to my work (about 20 minutes away by bike). Usually my car sits at home in the driveway. I don't envy my colleagues, some of whom live thirty miles away in the suburbs (a real drive, by Omaha standards, but a cakewalk in Chicago). They are connected to work by a nerve-racking four-lane auto umbilical cord.

ENDNOTES

1. A futures market for commodities is one that attempts large, unpredictable price swings by allowing investors to commit to buy the commodity at a specified future date for a particular price. They gamble their profits on being right about future prices.
2. Here are two other facts to help you put in context the implications of continuing the growth in oil consumption. Without considering projected higher consumption rates in the future, at present rates of oil consumption, (1) Saudi Arabia, with the world's largest known reserves, could supply all the world's oil needs for only 10 years; and (2) the estimated crude oil reserves under Alaska's North Slope—the largest ever found in North America—would meet world demand for only six months or United States demand for three years (Miller, 1998: 433).
3. The 1990 Exxon *Valdez* accident dumped 11 million gallons of crude oil in Prince William Sound in Alaska. The rapidly spreading oil slick is known to have killed 580,000 birds, up to 5,500 sea otters, 30 seals, 22 whales, and unknown numbers of fish. It oiled more than 3,200 miles of coastline, and the final toll on wildlife will never be known because most of the animals killed sank and decomposed without being counted. Even after the most expensive cleanup in history, the congressional Office of Technology Assessment estimates that only 3–4 percent of the volume of oil spilled by the Exxon *Valdez* was recovered. Beach cleaning crews and their equipment consumed three times the amount of oil spilled by the tanker. The Exxon company shipped 27,000 metric tons of oil-contaminated solid waste to an Oregon landfill (Miller, 1992: 616–617).
4. A *calorie* is the amount of energy needed to raise 1 gram of water 1 degree centigrade.
5. Sir Patrick Geddes was a Scottish biologist, sociologist, city planner, and cofounder of the British Sociological Society in 1909. Unlike Spencer, he sought a

unified calculus of energy flows to study social life (1890/1979). Wilhelm Ostwald and Frederick Soddy were both Nobel Prize winning chemists in the early twentieth century. T. N. Carver was an American economist, who gave energetic theory an ideological coloration. He argued that capitalism was superior because it was the system most capable of maximizing energy surpluses and transforming them into "vital uses" (Rosa et al., 1988: 150–151).

6. Roughly, the ratio of net food energy produced to energy expended, taking into account the number of producers and the hours devoted to production.
7. The most meticulous study of contact between high- and low-energy societies is Pelto's 12-year study of the consequences of the introduction of snowmobiles among the Sami people (Lapps) of northern Finland. The introduction of snowmobiles and repeating rifles were the energy and technological means of the gradual absorption of the Samis into Scandinavian societies. They readily adopted these material culture items, and it transformed their life. It vastly increased the geographic mobility of hunters and the amount of game that could be killed. It shortened the workweek of hunters and trappers, increased their leisure time, increased their earnings, and established a new basis for stratification in their communities (based on who owns and who does not own a snowmobile). It generated a serious ecological imbalance as populations of snowbound game animals were wiped out. It increased their dependence on the Finns, Swedes, and Norwegians for gasoline, consumer goods, and so forth (see Pelto, 1973; Pelto and Muller-Willie, 1972: 95).
8. For example, the typical mid-size car of the 1970s had a boxy design with more aerodynamic drag that reduced efficiency; a large, heavy, iron engine (often a monster power plant with eight cylinders!); rear-wheel drive; and heavy steel body panels and a steel frame (chassis) for structural rigidity. A comparable car by the late 1980s had less aerodynamic drag—which increases fuel efficiency exponentially as speed increases—a small, multivalve aluminum engine, which was significantly lighter and had fewer moving parts; compact front-wheel drive; plastic panels; and a fully stressed (monocoque) body for rigidity (Bleviss and Walzer, 1990: 103, 106).
9. In December 1986, France opened a commercial-size breeder reactor. It was very expensive to build, and the little energy it produced was twice as expensive as that produced by fission reactors. After spending $13 billion, it was shut down for repairs between 1989 and 1995. As a result, Germany, the United Kingdom, and Japan have abandoned similar plans. In 1994, the U.S. Energy Secretary ended government-sponsored research on breeder reactors after spending $9 billion (Miller, 1998: 454).
10. The term *energy regime* means the network of industrial sectors that evolve around a particular energy resource, as well as the consequent political, commercial, and social interactions.

CHAPTER SEVEN

Alternative Futures: Sustainability, Inequality, and Social Change

Americans have had a long love affair with their cars. Indeed, the promise of speed, personal freedom, and convenience that magically conveys you wherever the road will lead is a powerful cultural icon in America and around the world. Cars, in other words, are not just transportation. Particularly in America, they are powerful symbols of personhood and status. Some years ago, when my wife and I had three college-aged youths in our household, I was shocked one night to look out the window and count five cars parked around the house (some were "junkers," but cars nonetheless). How irrational, I thought . . . one for each person in our family!

Many cars are more fuel efficient than they were 20 years ago, but since Americans drive more than ever now, that gain is more than canceled. With about 5 percent of the world's population, Americans have 35 percent of the world's cars and trucks and drive as far as people in the rest of the world combined each year—about 2 billion miles. Furthermore, gas-guzzling "muscle cars" are back. Sport utility vehicles and pickup trucks, which are classified as "light trucks," get outrageously poor gas mileage. They now account for half of all family sales, mostly to people who live in cities. Between 1969 and 1997, the number of vehicles in the United States rose 144 percent. In that same time period, the proportion of households with three or more vehicles rose from 4.6 percent to 18.7 percent. Drivers used to outnumber cars by 30 percent; now the two are equal. Like population growth, the world auto fleet is not growing quite as fast as it was in the 1970s, but the absolute number of autos continues to grow. In the 1990s it did so by about 19 million a year (Lowe, 1991: 57; 1994: 83; Miller, 1998: 321; *New York Times*, 1997).

I'm sure it's not news to you that pervasive problems come with the American car culture. Making autos has a substantial impact on the environment. In 1990, a typical American car contained 1,000 kilograms of iron, steel, and other metals and 100 kilograms of plastic, making the auto industry the leading consumer of metals and plastics. The manufacture of metals and plastics, in turn is a high-impact industry itself: high in energy intensity and in the production of toxic wastes. Fueling passenger cars alone accounts for more than a fourth of world oil consumption, and the refining industry is the world's highest in energy intensity and the fourth highest in total toxic waste emissions. Motor vehicles are the single largest source of air pollution, creating a haze of smog over the world's cities. Such smog aggravates bronchial and lung disorders and is often deadly to asthmatics, children, and the elderly. Automobiles emit a substantial proportion (13 percent) of the CO_2 produced from fossil fuels. Accommodating autos makes substantial impact on land use: Roads, parking lots and other areas devoted to cars occupy half of all urban space in the United States, and more land is now devoted to cars than housing. Cars have also reshaped community life as cities sprawl, public transit atrophies, and suburban shopping centers multiply. Workplaces have begun to scatter, increasing the average commuting time. Despite auto safety improvements, accidents killed more than a quarter of a million people around the world, and several million more were injured or permanently disabled. I'm sure you know that by the year 2000, auto companies were racing to produce experimental efficient, low-emission autos that operate by different technologies. But they are just that: experimental models not in widespread use (Dunn, 1991: 82; Lowe, 1991).

The point is that cars are not only important means of transportation or cultural icon of industrial civilization; they are also powerful symbols of our social and environmental predicaments. How did we get caught up in automania, particularly when forms of transportation exist that are much less socially disruptive and environmentally greener? Furthermore, since autos are so central to our social and environmental predicaments, what are the prospects for an orderly transition to a sustainable society without cooling our passion for cars and, in a sense, regaining control of the auto?

The central concern of this chapter is the relation between the environmental problems discussed and ideas about social and environmental sustainability. *First,* I will discuss the notion of sustainability, its complexities and different meanings. *Second,* the chapter relates sustainability to human impact as framed in Chapter Two, with a special focus on the role of human equity issues. *Third,* it discusses two scenarios that embody the most common but controversial views of human-environment futures and some ethical questions about social choices about such futures. *Fourth,* the chapter describes more concretely some characteristics of human systems that would be sustainable. *Fifth,* I will use sociological ideas about social change to discuss such large-scale social transformations.

SUSTAINABILITY

Ideas about sustainable societies and sustainable development have long and mixed histories. In the last decades, these notions transcended the specialized concerns of scholars to become common goals, or at least irresistible slogans in public discourse and debate about the environmental issues. What is sustainable development? Conceptually and abstractly, the matter is quite simple: *Sustainable* means that the change process or activity can be maintained without exhaustion or collapse; *development* means that change and improvement can occur as a dynamic process (Southwick, 1996: 96). It does not mean profligate use of the natural world without regard to the future, but neither does it imply a static condition. In human terms, it means inventing ways of meeting human needs while preserving the capacity of the biophysical environment to do so. A sustainable society "can persist over generations without undermining either its physical or its social systems of support" (Meadows et al., 1992: 209). In more human terms, a sustainable society is one that "meets the needs of the present without compromising the ability of future generations to meet their own needs" (World Commission on Environment and Development, 1987).

Historically, the notion of sustainable development probably seemed like a nice utopian idea, but not very practical. Nor was it necessary to think about it much. After all, human populations were smaller, economic technologies less powerful, and nature's bounty seemed infinite. But now coming to some approximation of sustainability is not just a nice idea, it is imperative for the future of the world's people—certainly a future that is materially secure, reasonably equitable, and democratic. Who could really oppose sustainability or development? No one wants to dance with the devil. They are important but ambiguous and value-laden ideas, like progress. Their abstract vagueness obscures underlying differences and conflicts about what they would mean in practice.

Sustainability is often spoken of in terms of the *three Es*—economics, ecology, and (social) equity. It invokes a vision of human welfare that takes into consideration both inter- as well as intragenerational equity. It neither borrows from future generations nor lives at the expense of current generations. But lurking just under the surface of these abstractions are substantial conflicts between actors and institutions (Passarini, 1998: 60–63). Consider the conflicts of interest generated by public debate about whether to encourage or discourage material consumption of particular products (like gasoline or inorganic fertilizers). People who sell the products, who immediately benefit from their use, or who see them as dangers to human health or ecological well-being have *very* different outlooks and interests. Similarly, what needs justify the generation of environmental toxins and pollutants, who should pay the costs of abatement, and what resources (physical or biotic) should be kept free of human impact or left for future generations (like virgin forests or wetlands)? In public discourse, *sustainable development* and associated notions like the *carrying capacity* of the earth turn out to be universally acknowledged but inherently politicized concepts. The resulting controversy generates different advocacy organizations and movements with different objectives, resources, and political influence. In the United

States, for example, the Sierra Club, a large environmentalist organization that has existed for decades, and the Sahara Club have similar names. The Sahara Club was formed in the late twentieth century by American interest groups fed up with "pious environmentalists" trying to take away individual freedoms, eliminate jobs, and the nation's economic strength. In its view, humans are masters of the earth and its resources should be exploited for human use, pure and simple (Southwick, 1996: xix).

Scholarly controversy goes to the very heart of the paradigm conflicts that I have been discussing in this book. Think again about potential conflicts integrating the three Es (economy, ecology, equity). For policy, do we start with developing an economy that is less damaging to nature while maintaining rapacious consumption? Do we begin by preserving ecosystems, even if it means sequestering them from human exploitation and restraining consumption? Do we begin with equity, addressing poverty and social inequality to produce the cohesion and social sustainability that make agreements about environmental sustainability even possible (Gould, 1998; Passarini, 1998; Redclift, 1987)? Does this have a familiar sound? It should. Paradigms: Resource allocation? Limiting growth in finite systems? Maldistribution and social stratification?

Similarly, the concept of *carrying capacity,* so useful for population ecologists, is controversial when extended to human systems and the planet earth. Environmental sociologist William Catton is convinced that the earth has a finite carrying capacity and that we have already exceeded it. He argues that there is no such thing as sustainable development, which is a rhetorical and ideological term for those who wish to continue destructive growth and "feel good about it." (1997: 175–178). Lester Brown and Donella Meadows believe that we *may* have already exceeded the earth's carrying capacity, but continue to hedge their bets (Brown and Flavin, 1999; Meadows, 1992). Economist Julian Simon was for decades a tireless advocate of the idea that there is no finite carrying capacity and that development and growth in material consumption should be vigorously promoted to proceed as it has for the last fifty years (1998). In what is probably the most exhaustive review of the history of such ideas, theoretical biologist Joel Cohen argues that notions like sustainable development or the carrying capacity are important, but not concepts with any objective scientific utility (1995). He argues that questions like "How many people can the earth support?" are inherently *normative* and value laden. How many, and at what levels of material well-being? At what material consumption levels, with what technologies? Living in what kinds of biophysical environments? And with what kinds of cultural values, political, and legal institutions? Rather than a benign and participatory sustainability, one could just as easily imagine a sustainability of managed scarcity that is coercively administered by powerful authoritarian elites—resembling a virtual societal slave labor camp (see Schnaiberg and Gould, 1994; Heilbroner, 1974; or go to your video store and rent the video *Soylent Green*, for a powerful malevolent sustainability model based on a science fiction novel by Harlan Ellison).

It may surprise you to learn that I think Cohen is right. Sustainability and carrying capacity are not objectively useful, quantifiable concepts. *But please don't misunderstand*: They are critical as normative social facts and as aids in envisioning worlds we would like or wish to avoid. In the larger picture, they embody the only policy questions that really matter but that require scientists, citizens, and policy makers to address difficult normative and value questions. Natural scientists and neoclassical economists are not accustomed to dealing with normative social facts or policies involving complex normative solutions—but those are sociological specialties! Passerini suggests five sociological bodies of research that contribute to understanding sustainability: time horizons, risk analysis, public and private realms, and social change (1998). I will address some of these issues (time horizons and social change) in this chapter, as I did risk analysis in Chapter Four.[1]

Let me begin with a seemingly objective idea.

CONCEPTUALIZING THE HUMAN IMPACT AGAIN: *I = PAT*

Chapter Two argued that the proximate "driving forces of environmental change" are (1) population size and growth, (2) political economies that stimulate growth, (3) cultural values and belief systems, and (4) technology. Previous chapters examined in depth two causes of environmental modification—human population growth and energy systems that underlie all economic activity in the industrial era. The human causes of environmental change were captured another way in a widely known summary model of the 1970s, when biologist Paul Ehrlich and energy scientist John Holdren created a way of conceptualizing the joint impact of human and environmental forces. They argued that the impact of any population or nation upon its environment and ecosystem is a product of its population (P), its level of affluence (A), and the damage done by particular technologies (T) that support that affluence (Holdren and Ehrlich, 1974). Thus:

$$I = P \times A \times T$$

This is an elegantly simple way of illustrating different but related dimensions of environmental impact: as functions of the number of people, the technologies they employ to produce goods, and the amount of goods they consume. Relative weights of these items are subject to debate, but it is methodologically useful because it is possible to develop quantitative summary measures for each term of the formula (Dunlap, 1992: 464).[2] Since each term multiplies impact independently, it follows that the opportunities for changing them are different in different societies. LDCs have, for example, the most room for improvement in *P*, while MDCs have the greatest potential for improvement in *A* and *T*.

The *I* = *PAT* model is simple, robust, and useful as a framework for research. It is usually applied as an *accounting* or *difference equation*, where values for three of the four terms are used to solve for the fourth (usually *T*), and the relative impact of *P*, *A*, and *T* on *I* is determined by their changes over time. But that assumes the model is linear and the effects of the different terms are proportional. Dietz and Rosa reformulate *I* = *PAT* as a *stochastic* model in which values of the terms of the equation are allowed to vary across observational units (nations). Their reformulation is sensitive to possible nonproportional *threshold effects* that identify diminishing or increasing impacts of the terms of the equation in relation to environmental impact (*I*). Specifically, they used nonlinear regression formulas and other multivariate statistical techniques to study the contributions of population, affluence, and technology on the production of greenhouse gases in various nations. They used existing data from 111 nations about CO_2 emissions (in millions of metric tons of carbon per year), population size, and affluence (gross domestic product per capita). They did not measure technology directly but modeled it as a *residual term*, that is, as a multiplier in the equations to capture all things (physical infrastructure, social and economic organization, culture, and so forth) whose effects were not captured specifically by population and affluence (Dietz and Rosa, 1994, 1997).

Their findings demonstrate the significant utility of the *I* = *PAT* model, so I discuss them here in some depth. Dietz and Rosa found that increasing population among nations increased CO_2 production, and it did so in a linear way without any threshold effects: the more people, the more CO_2. They argue that these finding embarrass the ideas of some economists and Julian Simon in particular that "population growth has little effect or even a beneficial effect on the environment" and lend support to ongoing concern with population growth as a driving force of environmental impacts (Dietz and Rosa, 1994; Simon, 1996). They found, by contrast, that the effects of affluence on CO_2 emissions level off and even decline somewhat at the very highest levels of gross domestic product per capita. They suggest that this decline derives from the shift from manufacturing to service economies and from the ability of more affluent economies to invest in energy efficiency. Unfortunately, this effect occurs at affluence levels above 75 percent of the 111 nations in the sample, so that for the overwhelming majority of nations, continued economic growth can be expected to produce increasing rather than declining CO_2 emissions. In other words, reductions in CO_2 emissions will not occur in the normal course of development and will have to come from targeted efforts to shift towards less carbon intensive technologies (Dietz and Rosa, 1994).

This research used *technology multipliers* to identify some nations for particular analysis because they have values different than one would expect at particular levels of affluence and population size. Bulgaria, Zimbabwe, and Poland, for example, have large multipliers because they emit far more CO_2 than would be expected from their size and level of affluence. In contrast, France, Spain, and Brazil have relatively small multipliers and small CO_2 emissions for their size and levels of affluence. The authors hypothesize about particular "technology-infrastructure" differences that would account for such anomalies. Former Soviet bloc nations consumed a lot of fossil fuels relative to their levels of affluence. Similar factors explain the nations with lower levels of CO_2 production than would be expected at given levels of size and affluence: France's extensive reliance on nuclear power, Spain's use of nuclear and hydroelectric power and its relatively low level of automobile ownership, Brazil's reliance on hydroelectric and liquid natural gas fuels (Dietz and Rosa, 1994, 1997).

Finally, Dietz and Rosa estimated the coefficients of their statistical model to project global CO_2 emissions for the year 2025, using various U.N. projections for population and economic growth. They concluded that to achieve a goal of stable emissions at 1991 levels in the face of economic and population growth, energy efficiency gains per year would need to average 1.8 percent per year from 1990 to 1025. They believe that "such increases are feasible, but will not occur without strenuous efforts" (Dietz and Rosa, 1994). I have discussed this research in some depth because the findings themselves are interesting and significant for both science and policy, but also because it vividly demonstrates the usefulness of the $I = PAT$ model as a research tool. The model is important because it incorporates the major human driving forces of change that almost everyone would agree are important, ones that are at the heart of current debate and discourse regarding environmental problems. It also provides a framework for integrating disparate insights and scholarly traditions. But one of its dimensions requires some elaboration.

AFFLUENCE, SOCIAL INEQUALITY, AND ENVIRONMENT

Not surprisingly, social scientists have been interested in A of the $I = PAT$ equation. They suggest that affluence by itself oversimplifies the social dimension of environmental problems (Dunlap, 1992). It does so because it is a consequence of the operation of more basic human system components: political economy and cultural values. It also oversimplifies because affluence is one end of the whole continuum of social inequality. Social scientists understand not just affluence, but affluence *and* poverty as ends on a continuum of social inequality—both within and between nations—as causes of environmental disruption. If affluence is an environmental problem, poverty may often be just as bad. So the $I = PAT$ model implies social inequality as one of the causes of environmental impact, and I have mentioned it as one of the three basic paradigms related to understanding environmental controversies. So next I examine social inequality in some depth.

Social Inequality and Stratification

Three dimensions of social inequality are relevant here. These include economic and racial or ethnic inequality in the United States, and similar inequalities among nations and people around the world. In the United States, the median family income in 1995 was $41,224 (*median* is the middle of the income distribution.). But a limited number of families and individuals at the top get the largest share of the nation's income and wealth. In that year the richest 5 percent of the nation's families received almost six times as big a share of the nation's income as did the poorest 20 percent of families. It is even more unequally distributed for individuals. So is wealth, the combined economic assets of all kinds, which is different than income. Just 10 percent of the population owns more than one-half of all household wealth (Farely, 1998: 194–195). Clearly, income differences are very large within the United States; they relate to what social scientists call socioeconomic class and form a system of social stratification comprised of layers of differences, from very rich to very poor.

Economic inequality didn't change much for a long time in the United States. It became somewhat smaller between the 1930s and 1940s, the New Deal and World War II decades, and began to become significantly more unequal since the mid-1970s, as the rich became richer and the poor became poorer (Bartlett and Steele, 1992; Fusfeld, 1976: 630). However it is measured, the overall pattern is clear: (1) more, and comparatively richer, rich; (2) fewer in the middle with slowly eroding living standards; and (3) more and comparatively poorer poor (Beatty, 1994; Harper, 1998: 54–55; Starobin, 1993). Growing socioeconomic inequality is visible in most MDCs, not just the United States. In relatively affluent Europe, 3 million are homeless, 10,000 in Paris alone. In Britain, a 1991 census found that between 200,000 and 500,000 people had no permanent address. In relatively affluent America, a 1988 study found that half a million Americans could be found in any given week eating in soup kitchens, living in shelters, or sleeping in the street. Thirty million Americans are chronically malnourished, half of them children and three-quarters of them people of color (Bell, 1998: 27). In fact, by 1990 the American distribution of *wealth* (not income) more closely resembled that of the Philippines, India, and Venezuela than that of Germany, the United Kingdom, or Canada (Sivard, 1993; Durning, 1990: 138)! Such vast economic and social class differences mean that people (obviously) have different levels of material consumption and security, and they impact and experience environmental problems differently. The affluent are to be able to respond to environmental problems with minimal consequences for modifying their lifestyles. They are, for instance, the best able to afford higher prices or energy taxes or to purchase more efficient homes, autos, or appliances. The less affluent classes are less able to do so (Dillman et al., 1983; Lutzenhiser and Hackett, 1993).

Economic inequality is connected with racial and ethnic inequality.[3] In 1995, the median white household was 10 times as wealthy as the median African-American or Latino household. About one of ten white households had *zero or negative net worth,* but one out of four black or Latino households did (U.S. Bureau of the Census, 1995: tables 2, 5). The various forms of racial and ethnic inequality in the United States are so familiar that they require little comment here. But it is important for you to note that not only social class but race and ethnic groups are connected to their biophysical environments in different ways. A 1987 study by the United Church of Christ concluded African Americans and other people of color were two to three times as likely as other Americans to live in communities with commercial hazardous waste landfills. For example, 3 percent of all whites and 11 percent of all minority people in Detroit live within a mile of hazardous waste facilities, a difference by a factor of nearly four. Similarly, over a dozen other studies found that the distribution of air pollution, solid waste dumps, and toxic fish as well as hazardous waste facilities corresponds with either race, social class, or both. An enormous body of research demonstrates that both lower socioeconomic classes and racial minorities bear more than their share of the costs of environmental problems and change. However, some studies that use different units of analysis in research challenge these findings, but they always demonstrate a strong relationship between environmental hazards and social class. Whether related to race or class, such findings, along with the concrete experience of people in communities, stimulated the idea that *environment is a social justice issue,* and helped galvanize environmental justice movements as grass-roots movements for change (I will return to them in Chapter Nine.) Indeed, environmental justice became one of the important civil rights issues in the United States and elsewhere, helping create a political climate for change (Bell, 1998: 22–23; Boerner and Lambert, 1995; Bryant, 1995; Bullard, 1990; 1993).

Growing *global inequality* between people and nations around the world stands out in even sharper relief. In 1960, the richest 20 percent of the word's people had 30 times the income of the world's poorest 20 percent. By 1995, the richest fifth had *82 times* the income of the poorest 20 percent and controlled about 85 percent of the world's income. By contrast, the poorest fifth controlled a measly 1.4 percent (down from 2.3 percent in 1960). You could try to put this breakdown in more positive terms because the percentage of people living in poverty is about the same now as it was in 1960, about 20 percent of the world's population. But the *number* of poor people doubled since 1960, keeping pace with the growing world population. In 1998, about 1.3 billion people were poor, and the economic gap between the top and bottom had widened dramatically (Korten, 1995: 39; Livernash and Robenburg, 1998: 13). Let me underscore the scope of global inequality: In 1997, the world's 225 richest persons had a combined wealth of over $1 trillion, equal to the annual income of the poorest 47 percent of the world's people (2.5 billion) (Postel, 1994: 5; United Nations, 1998: 29–30). Global inequality is graphically illustrated in Figure 7.1 by what some term the "champagne glass of world wealth distribution."

The consequences of this "divided planet," as Tom Athanasiou calls it, are profound (1996). In the 1990s, compared to an average LDC person, an average MDC person consumed three times as much grain, fish, and fresh water; six times as much meat; 10 times as much energy and timber; 13 times as much iron and steel; 14 times as much paper; and 18 times as many chemicals (Durning, 1992: 50).[4] Along with the spread of affluent lifestyles, others live in physical deprivation. In the LDCs nearly 800 million people are malnourished (about 18 percent of the world's population). Pervasive malnutrition means that adults can't work as effectively and that children grow up smaller, have trouble learning, and experience lifelong damage to their mental capacities. One billion of the world's 6 billion do not have adequate shelter to protect them from rain, snow, heat, cold, filth, rats and other pests. One hundred million have no shelter at all. More than a billion lack access to safe drinking water. Considering these facts, it is not surprising that in spite of advances around the world in the *availability* of medical care, people in MDCs live, on average, 15 years longer than people in LDCs (Bell, 1998: 25–26). In sum, both within and between nations, the chasm between the rich and the not-so-rich is growing.

All this illustrates a point made in Chapter Two: In the globalizing world system and market economy, people and nations are becoming more integrated. Total world economic output has increased, but so has severe and slowing growing socioeconomic inequality.

The causes of growing inequality are complex, but in general they have to do with (1) the increasing shift of less skilled labor to low-wage LDCs, and (2) the increasing productivity worldwide through economic and information technologies that reduce the total demand for human labor—skilled and unskilled. In any case, the social consequences of such inequality are profound. It is likely to increase social polarization and political tensions both within and between nations. The world system is a very volatile one so far. As noted in Chapter Five, both legal and illegal migration are increasing, as the world's poor are increasingly desperate to move to where they perceive opportunity for better lives. Furthermore, the "cold war" between superpowers may be over, but regional, civil, and "low-intensity" wars have proliferated. You will remember the names from the news: Ethiopia, Somalia, Haiti, Peru, Liberia, Nigeria, Afghanistan, Algeria, Sri Lanka, Sudan, Rwanda, Bosnia, Chiapas in Mexico, and Kosovo (Renner, 1999). Growing social inequality is not the prox-

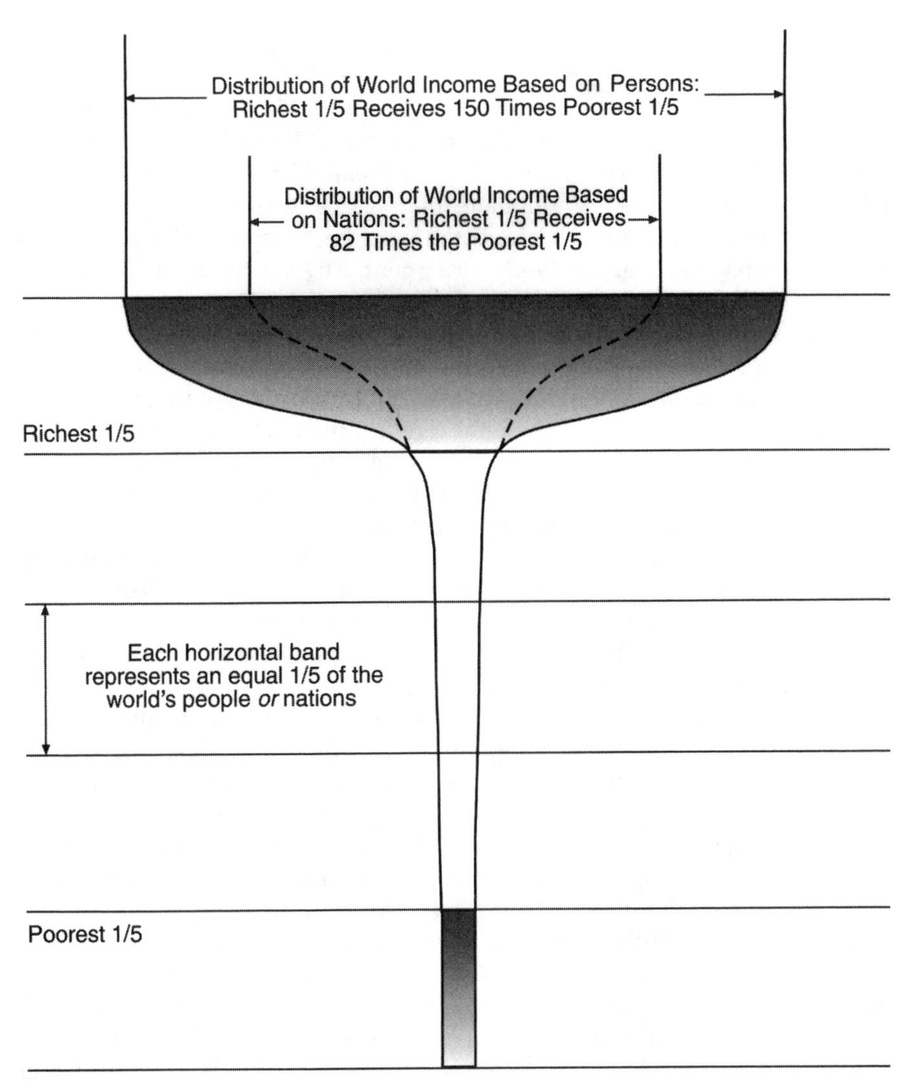

FIGURE 7.1 Champagne Glass of World Wealth Distribution
Sources: United Nations (1998: 219); Korten (1995).

imate cause of any of these, but rather an underlying process that makes social cohesion, stability, and sustainability more fragile and difficult.

Inequality and Environmental Degradation

It might be obvious to you how social sustainability is jeopardized by such gaping domestic and global inequality, but not how, as many contend, growing inequality is *itself* a potent and proximate cause of environmental degradation. Let me explain.

By now ample evidence shows that people at either end of the income spectrum are far more likely than those in the middle to damage the earth's ecological health—the rich because their affluent lifestyles are likely to lead them to consume a disproportionate share of the earth's food, energy, raw materials, manufactured goods and the poor because their poverty drives them to damage and abuse the environment. The poorer classes in MDCs damage the environment not because they consume so much, but because they are able to afford only older, cheaper, less durable, less efficient, and more environmentally damaging products—autos, appliances, homes, and so forth. In other words, the affluent—who can afford the newest and most efficient of everything—damage the environment because of the sheer volume of energy and material they consume. The poor do so because whatever they consume is likely to have a greater per unit environmental impact. It is important to note that it is *not* the poorest among the poor—who have no autos, apartments, or appliances of any kind—who are environmentally most damaging. It is rather the most marginal segment of unskilled workers (lower middle class?) who still have sufficient amenities to have an impact on the environment rather than, for instance, transient or homeless persons, who have virtually nothing.

In LDCs, population pressure and inequitable income distribution push many of the poor onto fragile lands, where they overexploit local resource bases, sacrificing the future to salvage the present. Short-term strategies such as slash-and-burn agriculture, abbreviated fallow periods, harvests exceeding regeneration rates, depletion of topsoil, and deforestation permit survival in the present but place enormous burden upon future generations (Goodland et al., 1993: 7). In fact, with uncanny regularity, the world's most impoverished regions also suffer the worst ecological damage; maps of the two are almost interchangeable. In China, India, Pakistan, and Afghanistan, for instance, the impoverished live in degraded semiarid and arid regions or in the crowded hill country surrounding the Himalayas; Chinese poverty is particularly concentrated on the loess plateau, where soil is eroding on a legendary scale (Durning, 1989: 45).

Often the environmentally destructive behavior of the world's poor is connected with highly skewed land ownership patterns. Rural small landholders whose land tenure is secure rarely overburden their land, even if they are poor. But dispossessed and insecure rural households often have no choice but to do so. Neither hired workers, hired managers, nor tenant farmers care for land as well as owners do (which is also evident in the United States). Being landless is in fact a common condition among rural households in many LDCs: 40 percent of Africans, 53 percent of Indians, 60 percent of Filipinos, 75 percent of Ecuadorians, 70 percent of Brazilians, and 92 percent of rural households in the Dominican Republic are landless or near landless (Durning, 1990: 142). While such poverty impacts the environment, the causality here is not one way. Even before it is degraded, a marginal area by nature does not usually produce enough surplus to lift its inhabitants out of poverty. Poor areas and poor people destroy each other. A reformed land tenure system that gives secure ownership of land, even in small parcels, to the landless peasants of the world would go some distance toward moderating the high birth rates and staunching the destruction of ecosystems by the world's poor.

The affluent classes of the MDCs also threaten the global ecosystem, but not because they are desperate with few alternatives. MDCs have the consumerist culture, the purchasing power, and economic arrangements through the world market economy to consume a disproportionate share of the world's resources, as I noted earlier. They account for a disproportionate share of resource depletion, environmental pollution (including greenhouse gas emissions), and habitat degradation that humans have caused worldwide. A world full of affluent societies that consume at such levels is an ecological impossibility (Durning, 1994: 12).

Returning to the $I = PAT$ model, Tom Dietz reminds us that it is *PA* that matters, not just growth in *P* alone, and I would underscore that *A* models both affluence and poverty. In considering the human impact, it is useful to think of *biospheric equivalent persons* (*BEP*) that account for the per capita impact rather than just growing numbers of people ($BEP = AT/P$). While India or China will contribute much more to future world population growth, every American and Canadian baby will consume and have a vastly greater per capita environmental impact over a lifetime. If we are worried about soil erosion, declining biodiversity, greenhouse emissions, and the like, we should worry more about the per capita impact of North Americans rather than Indians or Chinese (Dietz, 1996/1997).

In sum, affluence *and* poverty threaten the environment, and they increasingly do so as the chasm of social inequality widens around the world. A reduction of social inequality within and between nations would reduce pressure on the environment, both by reducing the resource consumption of the affluent and by reducing the need to overharvest, overgraze, or overfish to meet the short-run subsistence needs of the poor. Furthermore, I think it unlikely that the world's poor and poor nations would willingly agree to environmental agreements and treaties to preserve and restore the natural environment *unless* questions of equity are addressed. To those who live in misery, talk of "saving the environment" by the world's wealthy often sounds like a new form of imperialism, *green imperialism*. Some argue that poverty reduction must come before environmental sustainability while others argue that environmental sustainability is a prerequisite for social sustainability. This is a classic chicken-or-egg question, but how we answer it has important policy implications (Passarini, 1998: 64). Conventional ways of viewing human-environment futures to which I turn shortly embody the ideas of neoclassical economists and natural scientists. They largely (but not entirely) ignore the significance of social inequality.

SOCIAL AND ENVIRONMENTAL FUTURES: TWO VIEWS

Since 1950 the world's human population has doubled, and it recently moved beyond the 6 billion mark. It will undoubtedly rise further. Global economic output has quintupled. The cultural ethos of consumerism that favors high growth and ever-rising consumption is rapidly diffusing around the world. At the same time the chasm of inequality grows and poverty proliferates while the prospects for global equity seem increasingly remote. There are signs that every environmental and ecological system is becoming degraded, and that there is a very real prospect for altering the climate of the planet.

Suppose these trends continue. Can they do so without devastating the planet? Even if humans could use their ingenuity to survive, would it be in terms of such conditions that few would freely choose? Will *Homo sapiens* thereby replicate on a global scale the *outbreak-crash* familiar to ecologists and population biologists (like the bacteria in the petri dish or the reindeer on Matthew Island)? Or are we ingenious enough to invent and "grow" our way into a sustainable high-consumption world for *very* large numbers of people? Can those very large numbers emulate the consumption habits and lifestyles of contemporary Europeans or North Americans, to which they almost universally aspire? The *finite world paradigm* of most ecologists, physical scientists, and many demographers argues not, but the alternative *market resource allocation* paradigm argues that indeed it is possible and probable. This second view represents some (though not all) neoclassical economists and finds powerful political support among business and industry groups, particularly those representing the energy, manufacturing, and extractive industries, and reflects the dominant environmental views of the more conservative majority in the U.S. House of Representatives (since 1994). In this view, if we simply let markets operate, the price mechanism will regulate scarcities and stimulate investment in efficiency and innovation. Given human technological ingenuity and "elasticities of substitutability," things will work out. The *finite world paradigm*, in contrast, views profligate growth as a prelude to disaster and sees technological innovations as allowing rich nations to make only "Faustian bargains."[5] The *resource allocation paradigm* holds that environmentalists and their attempts to dampen material growth and consumption are the real threats to continued human progress. These conflicting views of the trajectory of social and environmental futures have been around in Western intellectual and political circles since the 1940s. Each has sophisticated intellectual articulation, and both have contemporary defenders. Following are two illustrations of the most articulate.

A Future without Limits: Cornucopia?

Since the 1960s, Herman Kahn (the late director of the Hudson Institute) and his colleagues have argued that universal affluence and permanent sustainability are possible and, indeed, the most probable outcomes of present trends. They argue that most people in the world can, in fact, live like contemporary Americans and Europeans without devastating the planet. Kahn and his colleagues argue that "barring bad luck or mismanagement . . . the prospects for achieving eventually a high level of worldwide economic affluence and beneficent technology are bright, and that this is a good and logical goal for mankind" (Kahn and Phelps, 1979: 202). How so?

Taking a very long view, they argued, we are now in part of a *great transition* that began with industrialization in the 1700s. "In much the same way that the agricultural revolution spread round the world, the Industrial Revolution has been spreading and causing a permanent change in the quality of human life. However, instead of lasting 10,000 years, this second diffusion process is likely to be largely completed with a total span of about 400 years or roughly by the late 22nd century" (Kahn et al., 1976: 20). Kahn expects the general pattern of the great transition to follow an S-shaped curve. From the 1800s, there were exponential increases in world population, the gross world product (GWP),[6] and per capita incomes. Beginning in the 1970s, there was, and will continue to be, a leveling of world population growth (rate!) *and* a decline in previously exponential rates of world economic growth, but a continuous spread of affluence so that world per capita incomes will continue to increase. Kahn and his colleagues are at pains to stress that the slowing of economic growth will occur because with the spread of affluence, and so there will be a reduction in the growth of *demand* rather than shortages of *supply*.

Global inequality, I have noted, is what Kahn and his colleagues viewed as a "transitional gap" between the living standards of the poor and the rich nations. They think this phenomenon is inevitable as industrialism spreads and living standards of some parts of the world rise relative to others. But that is analogous to the widespread misery and poverty of early industrialism, which eventually spread better living conditions to many people in industrial societies. See Figure 7.2 for Kahn's depiction of the great transition in terms of his estimates of changes in population growth, the GWP, and per capita incomes.

Kahn envisions the future from now until about 2025—dominated by "superindustrial societies"—as a "somewhat difficult" transition period. It will be a period of slowing down of percentage growth rates for the GWP and a particular period of "malaise" for the rich countries. The present time represents the spread of problem-prone superindustrial economies that experience many difficulties: technological crises, pollution and destruction of the environment, the exploitation of labor in the LDCs, problems of coordination

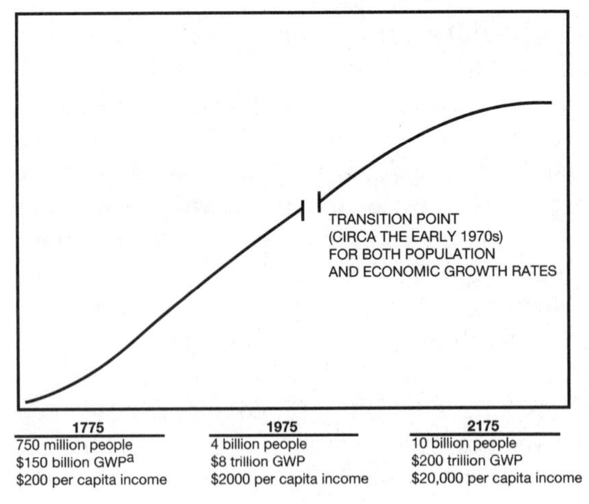

Figure 7.2 The Cornucopian Scenario: The Great Transition
Source: From H. Kahn and J. B. Phelps, 1979, "The Economic Present and Future," *The Futurist,* June: 202–222. Used with permission of the World Future Society.

and control of multinational firms. Because many of the projects of superindustrial societies are so large scale, they are problem prone, and we do not (yet) know how to eliminate, control, or alleviate all these effects. During this period, Kahn predicts that commercial ventures will colonize and initiate economic activity in space, particularly regarding energy and minerals, which will exponentially increase the resource base available to human societies. As superindustrialism spreads among the developed countries, manufacturing industry will spread among the middle-income nations and perhaps eventually to some of the poorer ones. But in Kahn's view, the present environmental problems, the growing social inequality and poverty, and the consequent geopolitical tensions and conflicts are all "transitional" problems. Toward the end of the period (probably sometime in the middle of the next century), the problems of superindustrialism will begin to be successfully managed. There will be the first signs of a world-wide maturing economy (Kahn and Phelps, 1979: 204).

By 2175, Kahn and his colleagues expect superindustrial societies to be everywhere, most likely including a vastly expanded "space economy" and true postindustrial ones rapidly emerging in many places. They predict after this date a slowing down of both population and economic growth rates, not only in percentages but in absolute numbers. But the slowdown of economic growth rates would not mean a decline in standards of living because of (1) the economies of scale that accompany large-scale systems, and (2) intensive technological progress that will provide energy savings and will substitute new resources for scarce ones. Kahn expects that in true postindustrial societies, economic tasks will constitute only a small part of human endeavors. Eventually, "furnishing the material needs and commercial services of a society will be carried out largely by highly automated equipment and complex computers operated by a small professional group" (Kahn and Phelps, 1979: 215). Thus in the context of growing economic efficiency, rapid advances in technology, and a stabilizing world population, unparalleled affluence can be sustained on a global basis. They envision a world of large-scale systems dominated by high technology and technocrats in which the large bulk of humans will be preoccupied with noneconomic pursuits.

While they do not ignore the problems of the present, Kahn and his colleagues have little patience with those who view present problems in apocalyptic terms. Reacting to one such report (Council on Environmental Quality, 1980), they stated that

> Global problems due to physical conditions . . . are always possible, but are likely to be less pressing in the future than in the past. Environmental, resource, and population stresses are diminishing, and with the passage of time will have less influence than now upon the quality of human life on our planet. These stresses have in the past always caused many people to suffer from lack of food, shelter, health, and jobs, but the trend is toward less rather than more of such suffering. Especially important and noteworthy is the dramatic trend toward longer and healthier life throughout all the world. Because of increases in knowledge, the earth's 'carrying capacity' has by now no useful meaning. These trends strongly suggest a progressive improvement and enrichment of the earth's natural resource base, and of mankind's lot on earth. (Kahn and Simon, 1980, cited in Simon, 1983: 13)

In a nutshell, this optimistic view of the future accepts the present trends as basically benign. It is a *cornucopian* view of the future. Kahn and his colleagues have taken a clear human exemptionalist view described in Chapter Two: that humans are essentially exempt from the limits of nature. With faith in human good will and inventiveness, they see no reason to deflect or attempt to change the course of change and growth that has been in effect since the 1600s. As you may imagine, many are attracted to this view, which posits the possibility of continued economic growth, universal affluence, environmental sustainability. (For examples, see Zey, 1994; Simon, 1994, 1998; Naisbett, 1982, 1994. With special reference to America, see Centron, 1994.)

A Future with Limits: Outbreak–Crash?

A counterpoint to the cornucopian scenario argues that present trends are putting us on a collision course with the finite carrying capacity of the planet, which we may overshoot. As noted, some argue that we are already in an overshoot mode. If so, we must dramatically reverse the historic trends of the past 200 years or inevitably suffer a collapse of human civilization because of a collapse of the resource base on which it depends. The most articulate, influential, and controversial statement of this view was by a 1970s futurist thinktank called the Club of Rome, sponsored by a variety of industrialists and multinational corporations. Rather than relying on the mental and intuitive models of Kahn and his colleagues, the Club of Rome used an elaborate computer simulation called a World System Dynamics (WSD) model developed by Massachusetts Institute of Technology (MIT) scientists Jay Forrester, Donella Meadows, and colleagues. They started with what was known about current patterns and trends in population growth, economic growth, resource consumption, food supply, and pollution effects, each of which has been growing exponentially. The WSD model then developed an elaborate set of coefficients for how continued growth in each of these areas would impact the others and attempted to project the sum of these interactions into the future for several hundred years (see Meadows et al., 1972).

The resulting projection by the WSD model was a classic *outbreak-crash model* familiar to population ecologists. The human outbreak-crash pattern predicted by the WSD model argues that current exponential growth in population, resource consumption, and food production will produce such enormous stress on the carrying capacity of the planet by 2100 that the resource and capital inputs to support such consumption levels will not be sustainable. Capital investments can no longer keep up with the growing needs. This prevents increases in fertilizer production, heath care, education, and other vital activities. Without food and necessary services, world population and living standards will undergo a steady decline sometime during the twenty-first century (Humphrey and Buttel, 1982: 97–98). Thus the Club of Rome research group argued that on a global basis, the whole of humanity will replicate the more limited ecological crash experience of the Copan Mayans, the Mesopotamians, the western Roman Empire, and many other preindustrial societies. In their degraded environments, these societies could no longer obtain the investments necessary for social maintenance (see Chapter Two). The views of the Club of Rome research group have been forcefully stated in a variety of technical and popular publications. The group's latest report, using more recent data, is significantly entitled *Beyond the Limits*, and argues that we have already overshot the earth's carrying capacity and are now living with a dwindling resource base (Meadows et al., 1992). This scenario is depicted in Figure 7.3, in what the MIT analysts called their "standard run" reflecting current world conditions.

They produced a large variety of computer runs of the model to reflect more optimistic assumptions (e.g., doubling resource supply estimate, controlling population growth and pollution effects), but the result was always

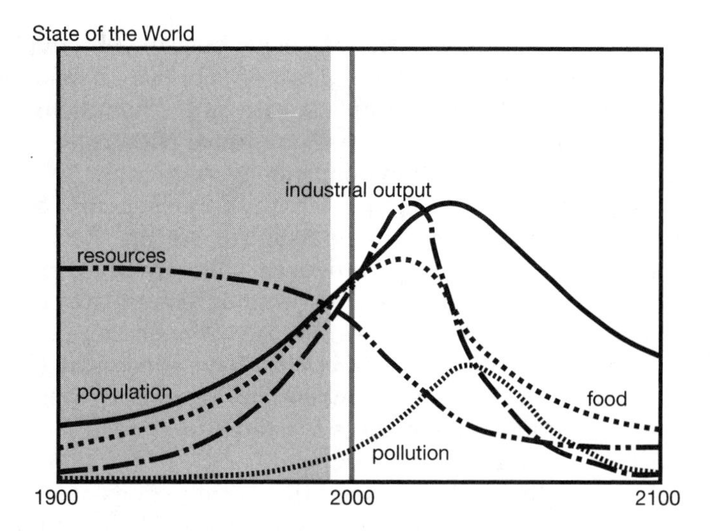

Figure 7.3 The Limits Scenarios "Standard Run"
Source: D. H. Meadows, D. L. Meadows, and J. Randers, *Beyond the Limits*. Copyright 1992. Chelsea Green Publishing Company. Used with permission.

the same: At some time shortly after the turn of the next century (2100), growth will be unsustainable. The problem is not any single dimension but the cumulative effects of the way that they interact. The underlying problem is *continual growth itself*. Hence the MIT researchers emphasized the urgency of global efforts to dampen exponential economic growth itself (not just population growth and pollution side effects) and move toward a *global equilibrium*. The language of the steady state, used in earlier versions of their scenario, has been replaced with the language of sustainability, in which they take pains to point out that certain types of growth are possible so long as there are dramatic reductions in material consumption. In this view, it is not enough to simply wait for markets to adjust to scarcity of food and nonrenewable resources: By that time irreversible declines in ecological equilibrium and resource availability may have already taken place, and a variety of points of no return may have been passed. Nor can technology save us. Technological advances delay the inevitable, since dominant cultural patterns and institutions perpetuate problem solving by growth that is self-defeating in the long term. The spectre raised by this vision is that if present trends continue, after 2100 a smaller human population will be eking out a more marginal existence on an exhausted and polluted planet.

These scholars do not suggest that the collapse of civilization is inevitable. Rather, they believe there is still time to avoid the widespread but gradual collapse of civilization in the next century, even though time is short and we are further along the trajectory to collapse than when their initial report was issued 20 years ago. But achieving sustainability will entail at minimum (1) the establishment of limits on population and economic growth, particularly as the latter implies material consumption; and (2) an emphasis on development tailored to the resource basis of each nation so that economic advances can be environmentally sustainable.

This is indeed a sharp counterpoint to the optimistic cornucopian view of the future. It is a darker and more pessimistic scenario about the future that, and as you might imagine, has provoked a blizzard of commentary and criticism over the years. Criticisms came from different places for different reasons. For instance, earlier versions predicted impending depletion and cost increases of several mineral resources that have not occurred. Many, like copper and petroleum, are more abundant and cheaper than they were in 1970. The subject of mineral resources is treated in a much more nuanced fashion in later publications, emphasizing *sink* rather than *source* problems. On the other hand, the Club of Rome group's projections about the availability of other types of resources (such as water and arable land per capita) seem on target, in the retrospective of 20 years. Other attacks on the perspective of the Club of Rome group have been more political and ideological than empirical. Conservatives and free market economists see the model as an anathema because it provides justification for a planned and rationed world socioeconomic order. The concept of "limits to growth" was also strongly attacked by the political left because stopping economic growth betrays the aspirations of the world's poor for a better life. Indeed, though the Club of Rome recognized the importance of inequality and distributional problems in a transition to sustainability, concrete policy suggestions about them are absent.

Sustainable Consumption: Voluntary Simplicity and Alternative Technology

The "growth in a finite system" view is not a doomsday scenario, pure and simple; it has a bright side. You may be wondering at this point exactly how. It has to do with changing the *treadmill of consumption* by which people consume more material without any real gains in human satisfaction. Some have argued that learning to live sustainably in a world of limits is not just a nasty necessity, but could be a basis for a better, more satisfying social life.

One such suggestion that became a slogan and social movement in the United States urges people in the MDCs to adopt lifestyles of *voluntary simplicity* (Elgin, 1981). By living frugally and more simply, individuals could change patterns of consumption, reduce pollution, and environmental disruption. Personal efforts to live more gently on the earth are not new. Elgin estimated—probably optimistically—that 10 million adult Americans and people from other nations were experimenting with voluntary simplicity in 1981 (Elgin, 1982; Durning, 1992: 137). Simple living *is* less convenient. It values simple rather than fancy habits and appliances (e.g., cooking from scratch rather than eating frozen manufactured convenience or fast foods, clotheslines rather than drying machines, walking, using mass transit, and bicycling more and driving less). It requires more forethought and attention to how life is grounded in the seasons and nature. Lowering consumption need not deprive people of goods and services which *they say* really matter. In fact, it may free them to pursue them—conversation, family and community gatherings, theater, music, and spirituality (Durning, 1992: 140–141). And I need to emphasize that voluntary simplicity is relevant only for the world's affluent consumers: It has little to do with the needs of the poor in the MDCs and around the world, who already live in "involuntary simplicity" of a much more malevolent kind.

Reducing material consumption and adopting more sustainable lifestyles are widely popular among people in the United States, Europe, and other countries. Surveys in 1989 and 1995 found large majorities agreed that Americans consume too much and waste too much; respondents didn't like the competition that comes with overconsumption (Bell, 1998: 62–63). But in fact, relatively few have been willing to *voluntarily* give up the pleasures of life in the fast lane for a simpler lifestyle. Efforts to do so are opposed by powerful and well-financed marketing and media promotions, and people find it hard to integrate simpler lifestyles with the realities of work and family schedules. Many in the MDCs spend more time and energy figuring out how to maintain and extend consumption than practice material frugality and voluntary simplicity! Yet the idea is more than a passing fad, and much evidence exists that advocacy of the idea of voluntary simplicity was growing in the late 1990s, and not only (or even mainly) from ecological concerns (Bell, 1998; Cohen, 1997). I will return to this evidence in Chapter Nine.

The idea of voluntary simplicity is a cultural and lifestyle argument, but there is a related technological one. It urges the increasing adoption of *appropriate technologies* (AT), defined as "technologies that are best able to match the needs of all people in a society in a sustainable relationship with the environment" (Humphrey and Buttell, 1982: 187). ATs are different from the high technologies of the cornucopian scenario. They are often (1) simpler (and can be understood and repaired by the people who use it), (2) less prone to failure (unlike more complex technological systems), and (3) less likely to cause severe ecological side effects. Proponents of AT would advocate, for example, the substitution of organic fertilizers for inorganic ones that have deleterious effects on soil and water, and the substitution of solar space heating for using electricity to heat homes. Heating homes with electricity (perhaps generated by nuclear power plants) when solar space heating is practical, cheaper, and more environmentally benign has been compared to cutting butter with a chain saw (Fickett et al., 1990; Lovins, 1977)! In some senses AT means reintegrating expert knowledge with traditional ways of doing things, as illustrated in Chapter Five in the discussion of traditional, organic, and low input farming. AT is proposed for both the MDCs and LDCs. For the LDCs, AT may provide alternatives to importing inappropriately expensive and complex technologies that have in part been responsible for developmental failures (e.g., Western-style power plants and industrialized agriculture). Rural AT energy and agriculture, not requiring lots of money, machinery or large landholdings, could reduce the migration of displaced peasants to already overcrowded cities. It could thus reduce conflict between rural and urban residents (Schnaiberg and Gould, 1994: 192). The most widely known spokesperson for adopting AT was the late British economist E. F. Schumacher, whose book *Small Is Beautiful* was widely influential (1973).

An Aside: Affluence and Personal Happiness

As you read this, you may be thinking, "Voluntary simplicity may help, if enough people adopted that lifestyle, but can I (or we) really be happy if I were more frugal and refrained from consuming all the stuff and having all the conveniences I want for life in the fast lane?" After all, the dominant social paradigms of MDCs leads us to see affluence as a prerequisite for personal happiness. *But is there, in fact, a relationship between affluence and happiness?* Evidence suggests that even though the desire for more—more money, more stuff—the level of wealth or material consumption has very little to do with happiness, at least beyond a certain minimum. There is not, for instance, as much variation in reported personal happiness among "rich" and "poor" nations as you might expect. Cross-national comparisons of 14 countries in the 1960s did find that people in India and the Dominican Republic—both nations with extensive poverty—expressed the least personal happiness by a large margin. Yet the three "happiest countries" were Cuba, Egypt, and the United States, which differ considerably in per capita consumption. Furthermore, people in Nigeria, Panama, and the Philippines described themselves as only slightly less happy that the substantially wealthier Israelis and Germans. Summarizing this evidence, social psychologist Michael Argyle suggested there to be very small *overall* differences in the levels of reported happiness found in rich and very poor countries. The connection between income and happiness turns out to be *relative* (mainly to social class) rather than *absolute*. That is, the upper classes in any society are more satisfied with their lives than the lower classes are, but they are no more satisfied that the upper classes of much poorer countries or than the upper classes in the much less affluent past (Argyle, 1987, Durning, 1994: 39).

Nor is there any certain link between economic growth and happiness. Since the 1950s, Americans have nearly doubled their consumption, both in terms of GNP and personal consumption expenses per capita. Yet the percentage of Americans who report themselves to be "very happy" peaked in

1957 and has slowly declined since that time (Schor, 1992). Except for the poor, the link between increasing wealth and happiness is not at all clear. Sociologist Michael Bell argues that "you only gain [. . . in an expanding economy . . .] when you are fortunate enough to advance in comparison to others, and by definition that advance must be limited to a few. We can not all be like Bill Gates—there isn't that much money in the world and likely never will be" (1998: 61).

The Two Scenarios: Evidence and Understanding the Debate

I have discussed two strikingly different visions of the future. The cornucopian scenario argues that the good future is to be found in complex technological solutions to resource problems and in the economies of scale that come with the coordinated global management of large-scale, bureaucratic systems. It tends to gloss over the fact that such systems may increasingly become unmanageable, less amenable to democratic control, and more vulnerable to catastrophic blunders, accidents, and disruptions. It takes it as articles of faith that (1) resource limits and environmental decay can and will be overcome by good management, good markets, and technological "fixes," and that (2) without any extraordinary measures, affluence will increasingly diffuse on a global basis. Both of these are plausible but arguable assumptions.

By contrast, the world-with-limits scenario argues that seeking high-tech solutions to sustain growth in a finite world is at best a Faustian bargain that will buy some time but will be ruinous in the end. In other words, what works in the short-term time horizon will add up to a large-scale disaster in the long-term time horizon. Their vision is of a world where life is more comprehensible to ordinary persons, a culture of sufficiency and frugality where quality is deemed more important than quantity of consumption. It is a vision in which mistakes and blunders have less serious systemic ecological and environmental consequences. If nothing else, this vision entails a reversal of the social trends of the last two hundred years. It would involve a deliberate dampening of growth and resource consumption *before* the planet becomes exhausted and polluted.

But there are similarities you should not overlook. Both scenarios envision a global equilibrium emerging after the year 2100 and the possibility of a very good life for much of humanity—albeit very differently defined. Neither view is demeaning to human inventiveness or antitechnological, but they are used for different objectives. Beyond these they disagree dramatically about how we might get there and exactly what a sustainable society would look like.

Neither of these scenarios is likely to be accurate as a prediction. *There are too many unknown and uncontrolled variables, and too many complex causal chains for such large-scale, long-range scenarios to be predictions in any concrete, accurate sense.* As scenarios, they are both normative and objective. They do capture, abstractly and holistically, divergent long-term implications of the contemporary discourse and controversy that I began with in Chapter One, in both its popular and scholarly senses. But if they are not useful as predictions, *what is their value*? It is precisely, I think, that they help us *envision* different possible futures.[7] The future is not entirely within intentional human control, but neither is it something that unfolds, willy-nilly, regardless of how humans envision it or act in the present. The value of such abstract scenarios is that they can help people envision more and less desirable futures, make choices, and accordingly form social policy. Doing so must involve consideration of the most compelling evidence from present trends, however incomplete, murky, and temporally limited.

Evidence??

Since these scenarios embody paradigm assumptions and normative facts related indirectly to power and profits, empirical evidence is not irrelevant to this grand debate about society-environment interaction, but neither is evidence the sole basis upon which the debate becomes part of public discourse. Evidences, of different kinds are typically marshaled by the defenders of each view. The defenders of the cornucopian view note the very great "elasticities of substitution," both historic and potential, in industrial and energy resources. They note that contrary to the Club of Rome predictions, many minerals, including fossil fuels, are more plentiful and cheaper than they were in the 1970s. Very great possible improvements in efficiency will stretch out reserves, and there are many possible alternative technologies for delivering energy for human communities. This mineral elasticity has been painfully learned by nations and communities like mining towns that depended economically on the extraction of particular mineral resources, both LDCs and in America (Freudenburg, 1992a; Freudenburg and Frickel, 1994).

Critics reply that while true, these cost-supply-accounting calculations about minerals and their price do not include the costs of externalities. They note the ongoing pollution of water, declines in biodiversity, and the effect of greenhouse gases on the climate (as evidenced by the recent climate conference in Kyoto). Nor do prices internalize the vastly unequal systems of exchange in the emerging world market with global inequalities. These costs *may* be partly born by taxpayers, or (as of yet) by no one. Most defenders of the limits scenario admit that the most serious problems with industrial minerals have to do not with source but with sink problems (pollutants). Moreover, defenders of the limits scenario point not to the supplies of industrial minerals, but to per capita declines in agricultural resources (arable soil, water) used to produce food. This includes the world's ocean 17 fisheries, now being voraciously overfished by the world's fishing fleets. On a per capita basis around the world, food is becoming less available and more expensive, though that fact may not be noticed among the affluent classes in rich nations for some time—if ever.

With the larger human population that is surely to come, increasing food supply relative to future population growth means appropriating even more of nature for human use, and increasing the current 40 percent of the earth's net primary production (NPP) appropriated by humans and our chosen crop creatures. But what, then, of the way that human survival depends on the *environmental services of natural systems*? These systems are impossible to quantify like the population and food resources data, but they are gigantic and run into the trillions of dollars per year, if we knew how to price them. As more natural systems are appropriated for humans, the environmental services are compromised. At some point, the likely result is a chain reaction of environmental decline—widespread flooding and erosion brought on by deforestation, worsened drought and crop losses from desertification, pervasive aquatic pollution and fisheries losses from wetlands destruction. Stanford University biologist Peter Vitousek and his colleagues argue that those "who believe that limits to growth are so distant as to be of no consequence for today's decision makers appear unaware of these biological realities" (Daily et al., 1997; Postel, 1994: 8; Vitousek et al., 1986).

There you have it: what I think are the most compelling bits of contemporary evidence related to the two scenarios. But you can pick your own evidence and interpret what it means.

Understanding the Controversy

How can different analysts disagree so much about the future? How do they do so, even when they look at the same world and sometimes use the same facts? Which view has the closest approximation to *actual* world futures? The debate is often sterile and unproductive, with each side grasping a portion of the truth, but not the whole truth. Still, it is an important controversy to understand, not "just one for scholars," but for people around the world and their children.

Part of the answer has to do with the selectivity of people's view of the world and its future. No one is completely exempt from selectivity. For example, you can view the United States and see tremendous progress in science, education, economics, and well-being. Or you can see alarming problems of urban ghettos, homeless populations, toxic waste dumps, and social decline. What is the real truth? For whom? Selectivity is even more of a problem in viewing the world. It is perfectly possible to tour the world by jet aircraft and air-conditioned taxi, stay in luxury hotels, and come away with the impression of great progress and prosperity. Likewise, you can tour the world visiting urban slums, refugee camps, exhausted deserts, areas of war and terrorism, and conclude that the visions of doom are here to haunt us today (Southwick, 1996: 88).

For scholars and leaders, it has to do with differences in the paradigm mindsets of the two groups of analysts, who by their training have learned to think about the world in different ways. Scenarios emphasizing more optimistic cornucopian futures were created by neoclassical economists, corporate and business people, technical experts in management, and journalists. They are often supported by elected politicians, who do not wish to be seen as threatening either jobs or profits. Scenarios emphasizing future limits are more plausible to persons from a variety of scientific backgrounds, including population experts, environmental scientists and ecologists, hydrologists, physical scientists and geologists, mathematics and computer modelers, soil scientists, biologists, climatologists, and some social scientists. Clearly, the two groups of analysts look at the same situation from different points of view (Brown, 1991: 5–9; Southwick, 1996: 88).

Considering this, it is no wonder that different analysts can look at the world today, think about the future, and reach totally different conclusions. Between scenarios written by the emerging consensus within scientific communities and those written by business leaders, economists, and journalists, whom do you trust the most to sort through facts and fancies and to come to grips with objectivity and reality?

In fact, there is an emerging consensus among the organized scientific bodies of the world about the importance of attending to the issues of limits and sustainability (Union of Concerned Scientists, 1992; Science Advisory Board, U.S. Environmental Protection Agency, 1990: 17). Consider the following statement by Frank Press, President of the U.S. National Academy of Sciences:

> Human activities are transforming the global environment, and these global changes have many faces: ozone depletion, tropical deforestation, acid deposition, and increased concentrations of gases that trap heat and may warm the global climate. For many of these troubling transformations, data and analyses are fragmentary, scientific understanding is incomplete, and long-term implications are unknown. Yet even against a continuing background of uncertainty, it is abundantly clear that human activities . . . now match or even surpass natural processes of change in the planetary environment. (cited in Silver and DeFries, 1990: iii–v).

Most remarkably, in 1992 the Royal Society of London and the U.S. National Academy of Sciences, two of the world's most prestigious scientific organizations, neither known for taking extreme stands, issued an *unprecedented joint statement* that "advances in science and technology no longer could be counted on to avoid either irreversible environmental degradation or continued poverty for much of humanity" (1992).

I cited official statements by the world's most respected scientific communities. On these issues I, for one, have more faith in them than in the pronouncements of neoliberal economists, industry spokesmen, elected politicians, or environmental journalists. That said, let me enter some caveats.

The data, and its use in simulation models like those of the Club of Rome, often ignore human reflexivity in the late modern world. That is, the high-consequence risks faced today are the result of modernity and technological advance, of our conscious interventions into our own history and nature. The risks associated with global warming, the punctured ozone layer, large-scale pollution, desertification, aquifer degradation, and the specter of global environmental collapse are the result of human activities. They are reflexive in that they are as much humanly produced risks, as are large-scale war, the collapse of the global market economy, overpopulation, and medical *techno-epidemics*—illnesses generated by technological influences (e.g., urban smog) (Giddens, 1994: 78).

Advice from the limits perspective for what LDCs should do sound like little more than attempts to maintain them in permanently subordinate positions within the world system. In short, limits proponents sound like they have little respect for the economic aspirations of most of the world's people. Cornucopians and neoliberal economists, on the other hand, *promise* such improvements but cannot demonstrate how continued growth and free markets will ever bring about universal affluence. They are in fact disingenuous—and many know it—by blaming shortcomings and continued poverty on heavy-handed government interference or environmental restrictions.

The expectation by Kahn and others that a world of six billion people could emulate the living standards of North Americans and Europeans, consuming and polluting at similar levels, is a fantasy that almost all scholars now think is a pure geophysical impossibility. But the cornucopians do provide a key insight: We should focus on human ingenuity to increase limits as well as the limits themselves. Many societies adapt well to scarcities and problems and often wind up better than they were before. The limits the world faces are a product of both its physical context and the ingenuity it can bring to bear on that context. But we may face a widening "ingenuity gap," and some parts of the world are locked into the rising need for ingenuity and the limited capacity to supply it (Homer-Dixon, 1996: 365).

Thinking about High-Risk Choices in Contexts of Uncertainty

Giddens noted that the consequences of large-scale high risks like global warming or those related to sustainability are in a category of their own, in terms of sheer scale. Unlike air pollution in Los Angeles, they are remote and imperceptible to the everyday life of most persons (1994: 219). So "expert" assessments like those of the scientific community can be ignored or challenged in the heat of political debates, particularly when real tax money, jobs, and organizational interests are on the table. Furthermore, even with a growing abstract scientific consensus, great uncertainty remains about the magnitude of particular effects (some of which can't be quantified), and more uncertainty and disagreement about the social consequences of pursing a particular vision of the future. This is particularly true when any policy to promote change creates categories of "winners and losers" relative to those who benefit from the current scheme of things. But it is also true that not to decide to pursue a positive vision of the future is a decision—by default—to let things unfold as they may.

Even if the evidence is murky and inadequate, it is still important to think about the most rational and prudent basis for action and policy in an uncertain future (including doing nothing). What is the most prudent and conservative best bet or gamble for such large-scale and high-risk situations? Which future scenario has the least costs if it fails? If we bet on the cornucopian scenario and it comes to pass, then all is well (universal mass consumption and affluence are indeed, appealing prospects). But if the cornucopian vision is unworkable and there really are limits, then the costs of failure—a degraded planet and social deterioration—are quite high. In the longer term, they are perhaps devastating to *any* dignified human future. On the other hand, what if the finite world scenario is indeed the actual future, and we act on this assumption? What we gain is a much better likelihood of a sustainable relationship between humans and the earth. What if we act on the assumption of a world with limits when in fact it is largely false? What are the costs of this mistake? They are the substantial but manageable investments in social and technical change and global cooperation. For the affluent classes of the MDCs, it would mean a comfortable but certainly more frugal and less affluent life than would be theoretically possible. It would also mean cooperation in addressing growing material destitution in both the MDCs and the LDCs. For myself, I think that assuming that a future with real limits is the best bet, if only because the costs of failure are not nearly so great. If we wait until we find out that the cornucopian future is unworkable, critical points of no return will have been passed.

This is the *no regrets* philosophy that I mentioned as a way of thinking about uncertainty and risks in Chapter Four (about responses to the prospect of global warming) and in Chapters Five and Six (about population and energy issues). It is based on an old strategy for making the most prudent best bets in the face of uncertainty. It was articulated as a rationale for believing in God by French mathematician-philosopher Blaise Pascal (1631–1662). Pascal argued that if one believed in God and was faithful, and God really existed, then all was well. If one was faithful to God and he did not exist, not much was lost; perhaps some forgone pleasures and vices. But if one didn't believe in God and he *did* exist, one was in danger of being in eternal damnation. Pascal's defense of faith was not a proof of what is "true" but has rather to do with the risks and costs of different kinds of errors (for which he was condemned by religious authorities at the time!). Not accidentally, Pascal worked out a number of theorems dealing with probability and became a founder of modern probability theory. Dressed up in the secular language of modern statistics, Pascal's defense of faith has become the "difference between the costs associated with making two different kinds of errors: A 'Type I error,' which falsely rejects a true hypothesis, and a 'Type II error,' which fails to reject a false one" (Siegel and Castellan, 1988: 9).

Both individual choices and social policy are in fact typically guided by such reasoning: We assume major risks where there are low or uncertain probability events with catastrophic consequences. We purchase health, flood, and fire insurance. I pay such premiums with "no regrets" even though my house is unlikely to burn down at any given time. We ban pesticides when their long-term consequences are still uncertain but potentially lethal. We build dams, levees, and flood control systems. We invest in early warning systems for tornadoes and hurricanes. And for over 40 years we have maintained an enormously expensive defense capability as a hedge against the low probability but catastrophic consequence of a nuclear holocaust.

Here's the argument I'm making: Both the most compelling evidence and criteria for dealing with the contingencies of an uncertain future lead me to think that the most prudent and most reasonable choice is to think and act *as if* growing human impacts will indeed at some point push us beyond the limits of ecological/environmental sustainability—to live in ways that few of us would choose to live if given choices. A decline in human well-being could be a slow decline barely noticeable from one year to the next, so that future generations may never know "what they lost," or, as changes accumulate in different dimensions of environmental disruption, they could occur with stunning swiftness. Our most prudent, responsible, and conservative choices should be to promote the transformations to sustainability.

WHAT WOULD A SUSTAINABLE SOCIETY LOOK LIKE?

What kinds of characteristics would more sustainable human sociocultural systems have? What would an approximate template for such a system be, even if it is impossible to describe exactly? Nor can one say by what policies such transformations might come about, but I believe that policies to produce them should be evaluated in terms of (1) the potential for political tyranny, and (2) the potential to reduce inequality within and among nations. Seeking to avert a large-scale overshoot is required, but the choice of methods to do so are probably as important for the future of humans as averting the overshoot itself. Since a truly sustainable society has never really existed (at least since the agricultural revolution), the following description is a hypothetical ideal type. I think there are seven kinds of requirements for sustainability, having to do with (1) population, (2) the biological base, (3) energy, (4) economic efficiency, (5) social forms, (6) culture, and (7) world order.

1. A sustainable society would dampen population growth and stabilize its size. This is particularly important in LDCs. There, dense populations and the momentum of growth, even with a slowly falling growth rate, underlie the poverty and desperation which impel environmental disruption, political conflict, and destabilizing waves of refugees and migrants. Achieving a stable population implies that people have access to contraception and family health care, control the resources necessary to alleviate the worst material insecurity, and significantly reduce gender inequality.
2. A sustainable society would conserve and restore its biological base, including fertile soil, grasslands, fisheries, forests, and freshwater bodies and water tables. Insofar as possible, a sustainable society would design agriculture to mimic nature in its diversity and natural organic recycling rather than overwhelm and degrade agroecosystems with monocultures and industrial chemicals. It would respect and preserve significant wild ecosystems for ethical/esthetic reasons, and the ecosystem services they provide.
3. A sustainable society would gradually minimize or phase out the use of fossil fuels. It would restrict the use of coal and petroleum and shift to natural gas as an interim (but less polluting) carbon fuel, eventually depending more on energy from a wide variety of renewable energy sources as determined by local conditions, including hydrogen fuels, solar, wind, geothermal, biomass, and hydroelectric (Roodman, 1999: 172–173). In the long term, sustainable societies would be powered more by sunlight and hydrogen than carbon. *Cogeneration*, the combined production of heat and power, would be widespread, and many factories would generate their own power, using the waste heat for industrial processes.
4. A sustainable society would work to become economically efficient in all senses. It would invest in the technology and production of efficient vehicles, transportation systems, machinery, offices, and appliances. It would maximize the recycling of material and wastes. More fundamentally, it would reduce waste in processes of production, packaging, and distribution of goods and services. It would reduce waste by decreasing the material component of goods and services. Such *dematerialization* would create a permanent net drop in waste created and resources consumed. Durable goods rather than consumable ones would be emphasized. Thus in a truly sustainable economy, the principle source of materials would be recycled goods. Both producers and consumers would create an economy that functions more like an ecosystem (cyclically) rather than one that only withdraws from sources and throws away junk in environmental sinks (Frosch and Gallopoulos, 1990).
5. A sustainable society would have social forms compatible with these natural, technical, and economic characteristics. A mix of coordinated decentralization and flexible centralization would exist. Thus entrepreneurialism and small-scale networks would flourish along with large organizations and urban life. The latter can produce many economies and efficiencies, and people would come to under-

stand that small is not always beautiful and large is not always ugly (Lewis, 1992: 254). Transportation systems would become an efficient mix of different modes, including autos, ride-sharing programs, mass transit, and bicycles. High-density settlement would be encouraged, and urban sprawl would contract. Multiple units rather than single family dwellings would be encouraged and community networks would evolve to regulate social life and trade, bartering, and cooperative sharing of goods and services that supplement mass markets. Social inequality would persist in a reasonably free sustainable society, but it would establish social policies to inhibit both grinding poverty and redundant material wealth. Economic profits and productivity would be measured more by services related to the quality of life than the volume of "stuff" consumed. Recycling and environmental services themselves would become important industries. New forms of crime would emerge ("exploiting the commons"), and ecological problems would become as politically important as economic ones.

6. A sustainable society would require a culture of beliefs, values, and social paradigms that define and legitimize these natural, economic, and social characteristics. The natural environments of human life would be cognized more as *ecological systems* to be nourished and maintained than as *open environments* to be utilized and exploited at will. Dominant social paradigms that underlie belief and action would change appropriately. The virtues of material sufficiency and frugality would replace the culture of consumerism; materialism simply cannot survive the transition to sustainable societies (Brown et al., 1990: 190). Neither self-worth nor social status would be measured *primarily* in terms of possessions; much of the energy now devoted to accumulating and consuming goods could be directed at forming richer human relationships, stronger human communities, and greater outlets for artistic and cultural expression. Western-style freewheeling individualism would be tempered with a *communitarianism* that balances individual human rights with obligations to community (Etzioni, 1993). There would be social restraints, but truly sustainable societies would not be authoritarian. Tolerance of diversity, social justice, and democratic politics would be valued as necessary to elicit the required responsiveness, cooperation, and coordination of people—both within and between societies (Roodman, 1999: 182–185).
7. In a world where societies are connected with each other and to a shared environment, a sustainable society would be required to cooperate in the negotiation of sustainablity in other societies—in terms of their different circumstances. In doing so, it would participate in regional and international political regimes, treaties, regulatory agencies, and multinational governmental and nongovernmental organizations. It would work to transform the system of global investment and world trade to promote a world of sustainable societies rather than one of growing environmental disruption and inequity. It would promote the development of the LDCs in a sustainable way. In a finite world, it would work to balance the requirements for some sort of global regulatory system with desires of national autonomy (Roodman, 1999: 176).

As you can see, I have described the characteristics of sustainable societies in pure (utopian?) form. But true sustainability may require something close to them. Are today's societies anywhere close to being truly sustainable ones? Of course not! Surely sustainability is relative, and change in that direction may evolve in small, incremental stages. But enough change to be effective eventually requires a dramatic social transformation on a large scale—eventually on a global scale. Given the difficulties of such social transformations, is it reasonable to think that it has even a chance of happening? That depends on how you understand social change, to which I turn next.

TRANSFORMATION AND SUSTAINABILITY: SOCIAL CHANGE

To understand change, return briefly to the three conventional sociological perspectives discussed in Chapter Two. One tradition (*functional theory*) argues that society and change are shaped by the activities and processes required for the viability and survival of the social system itself. These processes are termed *functions*. Another (*conflict theory*) suggests that society and change are shaped by conflict and power relationships among groups, organizations, and classes of people (the "parts" of society) as they compete to control the distribution of limited values and resources. A third tradition (*interactionist or social constructionist theories*) suggests that social action and interaction between persons and groups creates, negotiates, maintains, and revises culture and social definitions that *constitute* both society and social change. These must begin with the particular actions and interaction between real human actors as they create and negotiate definitions of themselves, others, and the world to cope with the circumstances and problems they encounter in life. For purposive human action, *reality* (of all types) *is a social construction*.

These are quite different pictures of society. The first two (functionalist and conflict) are *macro perspectives*; high aerial snapshots of the structures of society and how they operate. The third is a *micro perspective*; a picture from the ground up and more like moving film than a static snapshot. Like photos, none is truer than the others, but they are pictures taken from different altitudes and angles (Wallace and Wolfe, 1991: 74). Both the functionalist and conflict perspectives analyze social life by beginning with *structures and the operation of structures* (although they disagree about what is most fundamental). Interpretive perspectives begin with purposive human thought, action and interaction—in other words, with *human agency*. Besides different levels of analysis (macro-middle range-micro), theories differ in what they assume about the relative importance of *structure* or *human agency* in social life (Giddens, 1984; Sztompka, 1993). What about change?

Functionalism and Change

Early functional thinkers were concerned with the broad evolution of societies (Spencer, 1896; Durkheim, 1893/1964), but in the 1950s functional thinking was dominated by versions that depicted societies as "equilibrium-seeking structures" that avoided change (Parsons, 1951). Similar approaches in anthropology, especially in ecological anthropology, assumed an almost automatic tendency toward self regulation and "adaptation" of human societies to their environments. Both approaches have been discredited (regarding anthropology, see Orlov, 1980).

More contemporary functionalist thought conceptualizes change (Alexander, 1985). An illustration of this thinking goes as follows: Whenever stresses or strains seriously threaten the key features of a system or organization—whatever they might be—the system will very likely initiate "compensatory actions" to counter these disruptions in an attempt to preserve its key features. These may succeed or fail, and in either case they are likely to produce considerable conflict and change throughout the system. *When disruptive stresses are so severe and prolonged that compensatory mechanisms cannot cope with them, the key characteristics of the system being protected will themselves be altered or destroyed.* From this perspective, conflict and change represent a perpetual process of social reorganization (Olsen, 1968: 150–151).

In sum, what's the functionalist take on the prospects for significant and large-scale change? Both hope and warning: Overwhelming problems result in significant attempts to transform the system in adaptive ways, when it becomes impossible merely to prop up or rescue the status quo. Successful adaptation can happen, but it is not a sure thing. Indeed, the historical record is replete with both large-scale successes and failures.

Conflict Perspectives and Change

Conflict processes can reinforce stability and a prolonged stalemate between the dominant and contending parts of a human system. It can promote changes that have the same result, but it can produce a whole "new deal" of power relationships that bring with them new social arrangement that truly benefit a broader spectrum of people. Parliamentary democracy, brought about by conflict between social groups and classes, is a good illustration of this result. So what's the prospect of social transition to real sustainability? Possible, but not very good. The dominant elements of the system will continue to defend continuous economic expansion, from which they benefit. Restraint and regulation will address only the most intolerable environmental conditions, with lots of cosmetic pro-environment activity. When the contradictions between economic expansion and environmental support can no longer be contained or ignored, then the required political support would emerge for radical change and real sustainability. If, that is, we are not by then so far beyond the limits that this kind of change is even physically possible.

Interactionism, Social Constructionism, and Change

These perspectives view *change as not something to "be explained" because social life is fundamentally processes of change*. Social action begins with individual actors but aggregates in the collective action of social movements, publics, organizations, and so on (Touraine, 1978). There are constraints on action at any given time, but society is an ongoing, emergent set of social processes rather than an established "thing" or structure. These processes are so inherently multidimensional and involve such complex chains of cause and effect that predicting their outcomes for stability or change over any long time period is practically impossible. In short, the implications are that a transformation to a sustainable society is very much possible, though not certain or predictable, because that would depend more on the aggregate effects of particular human actions than on how structures or systems operate.

Multiple Perspectives and Change: Agency, Structure, and Time Horizons

As you can see, different sociological traditions depict social change in different ways with different implications. Functionalist and conflict perspectives see change as the operation of structures (differently conceived), and interactionist/social constructionist ones view change as beginning in the actions of concrete humans with structure as the accumulated end product. Put abstractly, much of the disagreement is about the relative importance of structure and human agency.

Many scholars are uncomfortable with this lack of coherence among such different perspectives. Analysts of actual historical change have been especially critical of the macrostructural theories—inherited from the nineteenth century—which assumed that social change is the lawlike, predictable dynamics, development, or evolution of structures. Such theories, from which the actions of real people were strangely absent, could not in fact give a very good account of the particulars of actual historical change. Charles Tilly argued for the liberation of social change theory from entrapment in these assumptions and "pernicious postulates" inherited from the nineteenth century (1984: 12). Tilly thinks that these are misleading and that "we must hold on to the nineteenth century problems . . . [understanding broad historical change] . . . but let go of the nineteenth century intellectual apparatus" (1984: 59).

A consequence of attempts to integrate the relationship between structure and agency is that social theory is beginning to understand social reality and change differently. Here is one cogent summary that you might compare with what Tilly objected to:

1. Society is a process and undergoes constant change.
2. The change is mostly endogenous, taking the form of self-transformation.
3. The ultimate motor of change is the human agency power of human individuals and collectivities.
4. The direction, goals, and speed of change are contestable among multiple agents and become the arena of conflicts and struggles.
5. Action occurs in the context of encountered structures, which it shapes in turn . . . [resulting in] a "dual quality" of both structures and actors.
6. Human interchange of action and structure occurs in time, by means of alternating phases of human agency creativeness and structural determination. (Sztompka, 1993: 200)

Even with this amended view, it is still important to sort out the relationship between levels of social reality and between potential and actual change. Here is a very abstract but insightful set of ideas that do so. *First,* there are two levels of social reality: (1) *individuality,* comprised of individuals or concrete members of groups, communities, and movements; and (2) *totality,* comprised of abstract "wholes," such as societies, cultures, or socioeconomic classes. *Second,* potential for change and actual change differ. The former refers to inherent tendencies, capacities, abilities, or powers; the latter to actual processes, transformations, conduct, or activities (Sztompka, 1993: 213–232). You can see how these distinctions relate to structure and agency in Figure 7.4.

The language used in Figure 7.4 is the language I have been using. Structures have the potential to operate, and agents (individuals) the potential to act. But there is *a middle level of reality* of reality where the two levels come together—the interface between structures and agents and between operations and actions. The combination of agents working within, creating, and being limited by structures is human agency. It is also the combination of the actions of people and the operation of structures in actual, practical outcomes of social interaction and change, or *praxis.* That term, which may be strange to you, comes from the Greek root word from which we get the words *practical* and *practice.* It is

> a dialectical synthesis of what is going on in society and what people are doing. Praxis is the confluence of operating structure and purposely acting agents. It is doubly conditioned, from above by functioning of the wider society and from below by the conduct of individuals and their groups; but it is not reducible to either. This middle level of social reality on which agency and praxis operate is a synthesis of other levels, but it is also the "really real reality of the social world." (Sztompka, 1993: 217)

Figure 7.5 represents these ideas.

Praxis reflects both structures and agents of the social world. Thus this middle level (of agency eventuating in praxis) epitomizes the emergent,

Figure 7.4 Levels of Social Reality by Potential and Actual Change

	Potentiality	*Actuality*
Totality	Structure	Operation
Individuality	Agent	Action

Source: P. Sztompka, 1993, *The Sociology of Social Change*. p. 214. Blackwell Publishers. Copyright Piotr Sztompka, 1993. Used with permission.

dynamic quality of the social world with many possibilities for significant transformation. Looking at the social world through time, as a moving picture rather than a snapshot, we should speak of *social becoming* rather than *society*.

Time Horizons

Such change outcomes related to sustainability are strongly shaped by time horizons from which actors and collectivities operate. How far into the future they are willing to think and plan depends on their willingness to not forgo present benefits for future ones. If actors and collectivities have short time horizons and an orientation toward individuals, they will find it rational to defer the costs of unsustainability to others and to future generations. If, on the other hand, they have long time horizons and a more social orientation, the most rational action may delay immediate gain by contributing to the collective good, in expectation that both they and their communities will benefit in the long run. You can see how important this is: Presently, corporations have time horizons from one business quarter to three years and governments may plan four years in advance, or maybe until the next election. Some individuals and groups may think in terms of 25 to 50 years or their grandchildren's lives (Passarini, 1998: 64–65). "Anticipations of the future become

Figure 7.5 Agency and Practice in Operation

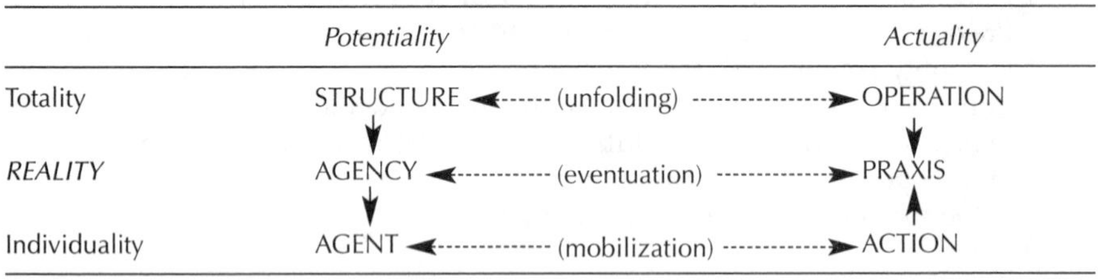

	Potentiality		*Actuality*
Totality	STRUCTURE ←	(unfolding)	→ OPERATION
	↓		↓
REALITY	AGENCY ←	(eventuation)	→ PRAXIS
	↓		↑
Individuality	AGENT ←	(mobilization)	→ ACTION

Source: P. Sztompka, 1993, *The Sociology of Social Change*. p. 219. Blackwell Publishers. Copyright Piotr Sztompka, 1993. Used with permission.

part of the present, thereby rebounding upon how the future actually develops" (Giddens, 1990: 177–178).

CONCLUSION: A TRANSFORMATION TO SUSTAINABILITY?

The whole point of this material about social change was to understand the likelihood of a significant transformation to a truly sustainable social world. So what does it mean? I should be able to answer that question. Let me do so by posing a series of rhetorical questions: Is a major transformation on this scale possible? *Quite simply, yes.* Is it probable? *Who knows? Educated guesses vary widely.* Can the purposive actions of humans shape that process? *Yes.* Can the outcome be mainly as envisioned by any particular human actor or organization? *No.* Is the longer-term outcome of that process really knowable or predictable? *No.* Outcomes of change are no more likely to be positive than negative, *but* we are not really trapped in a particular set of societal structures, institutional arrangements, structures of power and domination, consumption dynamics, and so forth.

There are, in fact, examples of such massive and purposive social transformations. In the nineteenth century, feudalism was abandoned in Japan, as was slavery around the world. This century has seen the retreat of imperialism and the creation of a United Europe. War provides obvious examples. Given the belief that national survival was at stake during World War II, the United States population mobilized and transformed itself in remarkable ways. Equally impressive was the Marshall Plan for reconstructing Europe after the war, and in 1947 America spent nearly 3 percent of its GNP on this huge set of projects (Ruckelshaus, 1990: 131–132). Recently the Soviet system collapsed, largely through the action of agents internal to the system. Most remarkably, by 1993 the Union of South Africa had transformed itself *peacefully and democratically* from an outrageously brutal and authoritarian racial caste system to a multiparty and multiethnic society with a native African as the popularly elected prime minister. None of these changes turned out exactly "as intended" or brought a problem-free social world into being. Such illustrations certainly don't prove that a transformation to sustainability will happen. Before they happened, many people and scholars would have found them highly unlikely or impossible. They only demonstrate that large scale transformations are possible.

The last part of the book focuses on possibilities of transformation to greater sustainablility. The next two chapters examine two elements discussed in this chapter: the operation of structures and the actions of people that combine in praxis and to produce large-scale transformations. Chapter Eight is about structures, and how structures could change to operate in sustainable ways. It is, in other words, about things like markets, politics, policy. Chapter Nine examines human agency, here meaning ideology, action, and the various forms of environmentalism.

PERSONAL CONNECTIONS

Personal Questions and Implications

This chapter discussed affluence as a component of the *I* = *PAT* equation as a way of conceptualizing the human impact. I noted that affluence was an oversimplified way of representing a more complex set of factors and an entire spectrum of inequality into the *I* = *PAT* equation. Chapter Two raised questions about consumerism in the Personal Connections section as an attempt to help you to concretize the very rarefied abstraction of the dominant social paradigm. Here I ask you to return to some personal implications of affluence and the consumerist culture that supports it.

1. *Consumption* in itself is not a problem. Consumption sustains life itself and provides the goods and services that make human life meaningful beyond elementary physical survival. *Consumerism* as a cultural complex is something quite different. It suggests that buying and consuming an ever-increasing supply of things and services will provide security as well as personal happiness and satisfaction. Affluence is an indicator of social power and status. And having the right things (the right makeup, deodorant, or fashionable clothes or autos) is linked to personal and sexual attractiveness. None of the world religions teaches that happiness and fulfillment can be achieved through material acquisition and consumption. In fact, most go to pains to vehemently deny this idea. How, then, did we come to buy the consumerist ethos?
2. There are many forces that impel us to do so: our early socialization, wanting to be liked and accepted, our ability to have burning wants in addition to needs. You might think about how some of these factors have worked in your own life. But an overwhelming reason we become believers in consumerist lifestyles is that there is an industry organized to get us to do so, one that has become increasingly effective, pervasive, and seductive in the last 50 years: advertising. It is certainly not new, and it occurs in many different media, but television has become the premiere advertising medium. The influence of TV cuts across economic classes, literacy levels, and national boundaries. In America, people spend more of their nonwork time watching TV than any other activity (besides sleeping). Sociologist Robert Bellah commented: "that happiness is to be attained through limitless material acquisition is denied by every religion and philosophy known to mankind, but is preached insistently by every American TV set" (1975, cited in Durning, 1992: 147).

   Here's a project for you: Investigate the pervasiveness of advertising in the major forms of mass communication. Figure out what proportion of space in your newspaper is devoted to news that informs or entertains, and how much is taken up by advertisements. Do the same for commercial TV. How much air time is devoted to drama, newscasts, or entertainment, and how much time is devoted to ads? You may be amazed.

3. If advertising were only to inform you about how to find and compare things or services you already wanted to buy, then most ads could resemble the classified ads or the yellow page directory in your phone book. But ads intend to expand your pool of desires. They intend to awaken wants that would lie dormant otherwise, or—critics contend—they manufacture wants that would not otherwise even exist. To illustrate, according to an advertising executive, ads can serve "to make [people] self-conscious about matter of course things such as enlarged nose pores [and] bad breath" (Durning, 1993: 11). But ads don't deal just with small items like mouthwash or facial cosmetics; they also intend to convince you that your automobile or home is in need of fixing, updating, or replacing with a new one. I can think of many things I didn't know I needed until I saw an ad for them or talked with someone who had. What are some examples from your life or from people you know?
4. There are people everywhere whose problems are not a surfeit of material goods, but rather insufficient material goods to maintain a secure, well-nourished, and healthy life. They need more food and material goods that address universal human needs. The problem for affluent consumers is that we have a surfeit of things, but nothing really seems to really satisfy—for very long, anyway. In fact, the multiplication of needs always keeps us unsatisfied. That's the way the consumerist culture is designed to work. Think about it. Think about the times that you had a "burning desire" to have something new. When you got it, did it really make you happy for very long? Probably not. I can think of a lot of examples of this in my life and family experience. I believe, in fact, that the really satisfying best things in life are not things at all, but relationships with people, experiences, and the capacity to find beauty, fascination, and inspiration in the ordinary aspects of everyday life. Do you agree or not?

   Maybe these are just the musings of an aging person, and I have to note that I have not always lived by this outlook, nor did I even think this way when I was younger.
5. *To be fair,* I need to emphasize that advertising is not the only force that promotes consumerism and environmental degradation today. In communist Eastern Europe advertising was illegal, and communist societies devastated their environments by a different dynamic.

   There are other driving forces of the consumerist culture besides advertising. Some claim acquisitiveness to be a part of human nature. (This is debatable.) Social scientists note (1) the erosion of informal neighborhood sharing networks that accompanied urban residential mobility; (2) social pressures to keep up with the Joneses; (3) the proliferation of convenience goods to meet time pressures when work hours expand; (4) national policies that encourage spending rather than saving; and (5) trends in urban design away from compact cities to diffuse ones that encourage malls and sprawl (Durning, 1993: 12).
6. Here's a quote from Gandhi, leader of the Indian independence movement, and inspirer of much of the thinking embodied in the small-is-beautiful movement: "Civilization, in the *real* sense of the term, consists not in the multiplication, but in the deliberate and voluntary reduction of wants." Do you agree with him? Why or why not?

Real Goods

There are two real goods I want to mention. Neither are products or things.

1. *A healthy skepticism about "green goods."* Producers and advertisers got the message about environmentally sophisticated consumers. Not everything that is labeled "green," or environmentally benign, natural, or organic, is. In the United States, about one-fourth of all new products are labeled "ozone friendly, biodegradable, recyclable, compostable, lite, natural, lo cal" or something similar. Sometimes these claims are real and sometimes they are misleading, a fact recognized by consumers as well as environmental scientists. There has been pressure to label the contents of products and display warnings on labels, but reading the labels of products is usually a confusing experience for consumers, who don't have the knowledge to evaluate them as truth claims. Such green deceptions generated counterpressures. In 1994, the U.S. Pure Food and Drug Administration required that food producers use standardized labels to make them easier to understand, detailing the additives, caloric, fat, and mineral content of their products (over howls of protest from some industry groups). Environmentalists in The Netherlands and France have attempted to cut away misinformation by introducing a 12-point environmental advertising code in their national legislatures. In 1994, ten state attorneys general were pushing for similar standards in the United States. Private green-label seals of approval are emerging in several nations (Durning, 1993: 17–18).
2. *Public interest groups:* Such groups monitor marketing campaigns and advocate advertising and media reforms. You can recognize these groups because they are not connected to a particular industry and usually not to a professional community. They are organizations of civic activation. By no means are all concerned with environmental or health matters; some are animated by social justice, religious, family, or political reform issues. Here are some that are concerned with TV ads for children's programming: *Action for Children's Television* is a Boston-based group that won a victory in late 1990 by getting the U.S. Congress to limit commercials aimed at children. The Australian Consumers' Association attacked junk food ads, calling for a ban or restrictions on ads selling unhealthy food to children. In Europe, public interest organizations have been doing the same thing, as has the American Academy of Pediatrics in America. The Center for Media and Values in Los Angeles has been promoting media literacy since 1989 by furnishing parents throughout North America with tips on teaching their children to watch with a critical eye.

ENDNOTES

1. This is, I hope, the only special pleading for a sociological perspective in the book. In fairness, other social scientists (like political scientists and ecological economists) and certainly social ethicists deal with the normative dimensions of social facts. So far, the social discourse about environmental problems has been dominated by neoclassical economists, ecologists, and other physical scientists.

Rather than being a special plea for sociologists, it is meant to open up the debate and discourse to a much broader range of disciplines. (See, for instance, Costanza and Folke, 1997; Gouldner and Kennedy, 1997.)

2. Population, size, and growth rates are obvious indicators; for affluence, measures of the per capita gross domestic product, or per capita consumption of selected goods (copper, meat, steel, timber, cars, plastics, aluminum) are relevant; for technology, the per capita kWh of electricity, or some other energy measure of economic productivity. These are suggestive but obviously simplistic.
3. *Race* is no longer understood by anthropologists or geneticists as a useful scientific term for understanding physical differences *among groups* of people. But certainly race remains a powerful social fact or distinction held by many people around the world, with both positive and negative meanings. *Ethnicity* describes the national or cultural heritage of a group, often related to race but often confused with it as well!
4. These consumption ratios are lower than the income differences already noted because they compare whole countries, not the rich and poor in various nations.
5. The term *Faustian bargain* derives from the monumental drama by German writer Johann Wolfgang von Goethe (1749–1832) about a tragic figure, Faust, who sells his soul to the Devil for pleasure, wealth, and power while he lives but finds himself condemned to hell for eternity. Faust bought short-term gain for long-term damnation.
6. Kahn's term for all the GNPs of the world taken collectively.
7. I have not addressed directly the more primitive ethical question about whether or not one should be concerned about the future at all. Some would argue—and more importantly, behave as if—the important thing to do is to satisfy yourself and don't worry about the future. This kind of total egocentrism may be understandable in the hungry and desperate, who may be too concerned with the prospect of an adequate meal to be concerned about any abstract consideration, but it seems to me ethically inexcusable among those with a surfeit of resources. It demonstrates at minimum a disregard (a callous one?) for the well-being of future generations (including one's children and grandchildren) and the world they will inherit. It also demonstrates a more abstract but very real disregard for the value of other creatures and the human enterprise on the planet earth. For a more articulate argument of different reasons for caring about the future, see Tough (1992).

PART IV

TOWARD A SUSTAINABLE WORLD?

CHAPTER EIGHT

Transforming Structures: Markets, Politics, and Policy

In 1973, when architect Jaime Lerner was appointed mayor of Curitiba, Brazil, it was a sprawling town of 500,000, half full of festering LDC slums (*favelas*). The *favelas* had many problems, not the least of which was garbage that could not be collected because of narrow or nonexisting streets. Because trucks could not get in, and because the garbage was attracting rodents, and disease, Lerner had to come up with a way to get the garbage out. The solution was to pay people for their garbage by placing recycling bags around the *favelas* and giving tokens to the city's transport system for the separated and therefore recyclable trash. The mayor gave tokens that could be exchanged for food for organic waste, which was taken by farmers and made into fertilizer for their fields. It worked spectacularly. Kids scoured the *favelas* for trash and learned to spot the difference between low-density and high-density polyethylene bottles. The tokens gave poorer citizens the means to get out of the *favelas* to where the jobs were while promoting cleanliness, frugality, and the reclaiming and recycling of waste.

The plan was innovative but simple: The money gained from recycling combined with the money saved by not having to take trucks into the narrow streets paid for the tokens. It was a cyclical, waste-equals-food system implemented at the grassroots. Curitiba is now considered a landmark in urban development. But it happened, according to Lerner, because he and others were not afraid to try new things. Not everything worked, but so much did that it has bred an innovative atmosphere throughout the city, which tripled in size by the 1990s. Curtiba is entirely self-sufficient and has decided to no longer accept money from the state because of the red tape involved. It is booming, prosperous, and clean (Hawken, 1993: 213–214).

Shift the scene from Brazil to southern Mexico. When the Zapatista National Liberation Army declared war on the Mexican government on January 1, 1994, the rebels identified as their base the Lacandon, North America's only tropical rainforest, some 3 million acres stretching across the southern state of Chiapas and into Belize and Guatemala. The *rainforest rebellion* sent shock waves through the Mexican government and the ruling party (PRI). The Zapatistas occupied four cities and several villages, and the spirit of revolt spread beyond Chiapas throughout southern Mexico. Populist forces in a half dozen towns and villages locked up local mayors and constables, declaring their behavior to be predatory and oppressive and demanding reform. For the Mexican government, the rainforest rebellion could not have come at a worse time. Mexico had just signed the North American Free Trade Agreement (NAFTA) with the United States and Canada. They made no bones about their intent to press the Mexican government to become more open, democratic, and responsive. PRI, which had been the political party in power since the 1930s was engaged in a national electoral campaign trying to polish its tarnished reputation for political authoritarianism and electoral fraud (Simpson and Rapone, 1996).

The rebels identified their breeding ground as the rainforest, but in truth it was more the dead earth and gutted remains where the rainforest trees once stood. Forty years ago, the forest was largely unoccupied. It is now home to about 200,000 rural people (*campesinos*), many of them indigenous Mayans, as well as another 100,000 refugees from Guatemala who crossed the border to escape the ongoing civil war there. Early in the twentieth century, loggers in search of mahogany and tropical cedar assaulted the Lacandon. Until the 1950s, there were still large stretches of virgin forest, but the pace of logging accelerated in the 1960s. The Mexican government encouraged *campesinos* to settle there, a policy driven by a need to alleviate the overpopulation of central Mexico. The government saw farming the rainforest as more profitable than its preservation. For a while, *campesinos* were able to farm sustainably, raising traditional crops such as corn, beans, squash, chilies and some coffee by slash-and-burn methods, frequently shifting plots to let them lie fallow. But increasing population pressure from continuing streams of migrants and refugees took its toll. Between 1970 and 1990, the population doubled and per capita cropland steadily declined. Chiapas became a kind of dumping ground for marginalized people. In addition, intensified commercial logging took its toll. As around the world, tropical soil quickly loses its fertility when it is stripped of tree cover and farmed intensively. Crop yield declined steadily after 1982, yet the shrinking local harvest had to feed a population growing at 4 percent a year. Chiapas now suffers one of Mexico's highest poverty and illiteracy rates; compared with other regions, people here are more likely to live without running water, electricity, health care, or access to minimal social services. Thus, the rainforest rebellion was an act of desperation—but like all revolts, also an act of hope. The Mexican government, predictably, promised reform but sent soldiers to drive the rebels out of the cities and back further into the forest, or what was left of it (Morris, 1994). The government brutally suppressed the rebellion and expressions of grassroots dissatisfaction, but the combination of external pressure from trading partners and international agencies, widespread dissatisfaction with PRI, and potential revolts spreading across the whole of southern Mexico made a political climate in which real reform was possible. Though PRI's control faced serious opposition in many 1999 elections, real reforms did not happen. In the last several decades, environmental degradation and unequal land distribution stimulated tensions and uprisings of marginalized people in 28 other nations (Renner, 1997).[1]

These stories are very different, and you may wonder what they have in common. They are both about the possibility of structural change. Both illustrate, in very different ways, the connection between environmental conditions and pressures for structural transformation. In terms of the perspective developed in Chapter Seven, this chapter deals with the transformation of structures (markets, politics, and policy) and the next chapter discusses human agency (environmentalism and environmental movements). Both are essential components of the dialectic of social change. The final chapter deals more directly with global dimensions of environmental problems and the possibility of a global transformation.

MARKETS

Humans have obvious needs for an incredible variety of goods and services that are ultimately provided by the resources of the earth. The systems through which such goods and services are distributed that bring investors, producers, sellers, and buyers together are economic *markets*. Think of a city farmers' market or traditional markets in villages around the world in ancient times or in contemporary less developed countries (LDCs). In such markets, people compare quickly and see what the competition is; you can taste a wedge of pear, smell a bunch of roses, or drop an olive on your tongue. You can haggle about prices, compare the quality of goods in different stalls, and, if they are not to your liking, you can walk away. Such pleasures are deeply embedded, richly satisfying, and universally observed (Hawken, 1993: 76). In the longer term, such markets have built-in protections against fraud and misrepresentation. (How many times will you be cheated by the same seller?) In contemporary society, markets are often not concrete places like traditional markets, but rather abstractions to represent the interaction among the costs of production, the asking price, and the price consumers are willing to pay for goods and services. More simply put, real economic values (prices) are determined by the interplay of supply and demand. Markets are important because they can send realistic signals about the actual economic value of goods and services, the work that you do, and the prices that people are willing to pay for a particular product or service in specific circumstances. So there are specific markets for compact discs, Fords, bushels of wheat, books about environmental issues, and the development of more environmentally benign products. All of these products have prices attached that must be paid (by someone) and they have amounts or levels of benefits that you can get for particular prices. To think otherwise is to be either uninformed or naive.

Since neoclassical economic theory views such markets primarily as structures to allocate values, it emphasizes that many human problems (social and environmental) can be understood as market problems and failures. Harvard University economist Theodore Panayotou underlines this point, responding to environmentalists who see unending growth as a problem:

> Resources are limited, but resource use is infinitely squeezeable—well, almost infinitely. Antigrowth advocates [would convince us] that further growth will reduce sustainability, and that we should put a cap on growth and seek greener pastures in qualitative development, self-sufficiency, and other utopian pursuits. . . . The correlation between growth and environmental degradation may simply be a spurious one . . . it is rather the inefficiency and waste that accompanies certain growth paths that is responsible for environmental degradation . . . [caused by] policy and market failures. To put it simply: You get what you pay for, and you lose what you don't pay for. If you subsidize waste, inefficiency, and resource depletion . . . that's exactly what you get. (Brown and Panayotou, 1992: 355–357)

As I noted in earlier chapters, neoclassical economic theory is embedded in an intellectual *resource allocation paradigm* of the human world and its problems because, in its view, free and fully functioning markets would solve problems by allocating resources in an adaptive way—that is to say, to the most (economically) valued ends. Neoclassical theory argues that producers and consumers respond to changing relative incomes, prices, and external constraints, so that—*if* the market signals are allowed to reach individuals and market prices include all the social costs and benefits of individual actions—responses to problems will be rapid and efficient (Stern et al., 1992: 136). *Unfortunately*, these conditions that theory specifies are often *not* met in the real world, and therefore many human social and environmental problems result from *market failures*.

Market Failures

One reason that markets don't always work is because all resources are not owned or used in the same manner. In some types of resource arenas, markets do not work efficiently to send the kinds of real signals just mentioned. Such resource arenas fall into three categories: (1) There are *private-property resources*, which can be owned and used by an individual (or organization). Others can be excluded from using such resources, and since individuals (or organizations) can own them, they are normally more willing to use them frugally and to invest in their upkeep and maintenance. In short, we are more likely to use private property resources with an eye to their long-term sustainability. Private property resources include things like clothing and automobiles, but also such things as privately owned farmland, business equipment, and financial investments. (2) There are *common-property resources*, to which people have virtually free and unrestricted access. They are not really owned by individuals; therefore, few real economic costs exist for individuals (or organizations) for overusing them, and few incentives exist to manage them or pay for their upkeep. Many resources illustrate common-property resources: air, rivers, ground water, international waters, and all the chemical and biological resources that they contain. (3) Somewhere in between private- and common-property resources are *public-property resources*. These are jointly owned by all people of a country, state, or local community, and are managed by a government or public agency. Examples of public-property resources restricted from private ownership are national and state forests, wildlife refuges, beaches, coastal waters, parks, and rangelands. Social institutions can also be understood as resources, and in the United States public-property resources include such things as fire protection, public education, military security, highway systems, and prisons. Obviously people use (or participate in) all these public-property resources, but governments have the exclusive rights to regulate their use. Note that the things provided for you as a member of a family don't come from a private-property resource system, either. So all of your needs are not met through private-property market systems. Precisely how much and what needs are met through private-property and public-property resource pools changes from society to society in terms of different cultural, legal, and political traditions.

This distinction among kinds of resources is an important one for understanding the environmental consequences of economic processes because—as I noted earlier in several chapters—there are particular problems in common-property and public-property resource arenas. Air and rivers have been polluted, water tables drawn down, international fishing grounds depleted. Because they are controlled by the government and subject to pressures from organized political interests, rights of access to timber, grazing land, minerals, and energy resources are often "priced" far below what they would be if they were all treated as private property resources. These difficulties preventing the unsustainable use of common- and public-property resources exist because there are no real economic reasons to preserve them that applies to any concrete human actor. Zoologist Garret Hardin popularized the notion of environmental commons problems as the *tragedy of the commons* (1968). Because social traditions and laws often allow free access, he observed that village pastureland owned in common was often overgrazed, compared to private land. The same principle applies to polluting the atmosphere or overfishing the oceans. In short, common problems produce market failures because of the lack of clearly defined private property rights that leaves no one with the incentives to pay to prevent environmental degradation. Some analysts understand the tragedy of the commons as so pervasive and powerful that it is almost a "law" of the natural world. I will return to this notion later in the chapter.

Other Sources of Market Failure

In the abstract, commons problems represent the generic source of market failure. More concretely, others exist. *First,* I mentioned the problem of *externalities* in earlier chapters. Externalities mean that, for whatever reason, someone pays the "full costs of production and consumption," but they are not calculated into the existing market price. Individuals not involved in buying or selling a good or service may nevertheless be affected. Pollution affects people and other species generally as it flows downstream or drifts in the wind, not just the industries that produce it, or the consumers of products. As I noted in Chapter Six, the full diplomatic, foreign aid, and military costs of keeping crude oil flowing "through the pipelines" are not calculated into the costs of each gallon of gasoline in the United States. In a related illustration of externalities, the costs of decommissioning a nuclear power plant (which has about a forty-year life span or less) could be built into your electric rates but probably aren't. Externalities may be a substantial hidden "tax" on you or others in the future. *Second,* for understandable reasons, governments often impede or supersede the market by providing price regulations or subsidies or creating a sort of *quasi-commons* (public-property resource) from what *could* be privately owned. Examples include the oil depletion allowances and artificially cheap access given to public lands to ranchers and lumber industries. These favors lead to excessive, uneconomic, and environmentally destructive production. In Western parts of the United States, water rights are defined in a way that precludes the emergence of real water markets. Similarly, in China artificially low coal prices and a production quota system that gave no premium for quality led to excessive production of coal compared with other energy options. *Third,* cost accounting problems exist. Markets may not send real signals about complete values/costs because of the difficulty and costs of collecting information about the net value of something that considers its costs to all impacted producers, consumers, and nonconsumers. The accounting problem is particularly intense when we face a dilemma of consuming something now or saving for the future. Neoclassical economists usually "discount" the future and argue that consuming it now is of greater economic value (Rubenstein, 1995; Roodman, 1998; Stern et al., 1992: 85–86, 136).

Environmentally Perverse Subsidies and Market Incentives

You need to understand the powerful and pervasive ways that government interventions distort markets in most nations. They actively encourage public and private decisions that stimulate unsustainable resource use and environmental degradation. Evidence for this is overwhelming. Such interventions flow from the understandable efforts of powerful economic groups and firms to get government leaders to provide protection from the unalloyed discipline of the market and from the desire of politicians to keep people working and prices low. Mechanisms of intervention include tax and fiscal incentives as well as pricing and marketing policies. They also include subsidizing the currency exchange rate with other nations and trade subsidies that make goods cheaper to export. Energy subsidies usually favor large supply projects but undermine funding for innovative and renewable energy development. Such subsidies underwrite the development of coal, oil, and natural gas while ignoring the costs of polluting air, and water. They favor inefficiency and waste. Tax concessions for logging, settlement, and ranching accelerate deforestation, species loss, and soil and water degradation. Brazilian taxpayers subsidize the destruction of the Amazon with millions in tax abatements. Indonesians and Canadians do the same thing, and in the United States taxpayers are subsidizing the destruction of Alaska's Tongass forest. Pesticide subsidies promote excessive use and thereby threaten human health, pollute water, and increase the development of pesticide-resistant species. Subsidies for water resource development lead to agricultural and municipal overuse and discourage conservation. Such subsidies make people pay twice: in taxes and in costs to health and the environment. In the late 1990s, worldwide, subsidies on resource-intensive industries like energy extraction, mining, lumber, and farming added up to at least $650 billion, or about 9 percent of all government revenues, far exceeding subsidies to protect the environment (Roodman, 1998: 35). Agriculture provides the clearest cases of such perverse subsidies.

Virtually the entire food cycle in North America, Western Europe, and Japan attracts huge direct or indirect subsidies. In 1991, these costs to taxpayers and consumers were conservatively estimated to be over $250 billion a year. Such subsidies send farmers far more powerful signals than do the small grants that support soil and water conservation. They encourage farmers to occupy marginal lands, to clear forests, and to encourage profligate use of pesticides, fertilizers, and aquifer water. Moreover, by encouraging vast surpluses at great economic and ecological costs, subsidies create political pressures for more subsidies—for food exports, to "donate" nonemergency food aid to LDCs, and to raise protectionist barriers against food imports. All of these policies are devastating to agricultural productivity in the LDCs, where even the most efficient farmers cannot compete with highly subsidized MDC farmers when their surpluses are "dumped" on world markets (MacNeill et al., 1991: 33–37).

Transforming Market Incentives: Green Taxes and Owning the Commons

Neoclassical economists argue that the prescription for addressing environmental problems is not the kind of environmental protection that takes things out of private markets and makes them common- or public-property resources. This, they argue, is a strategy doomed to have perverse effects by removing any incentive for conservation by actual persons and firms. They would allow the magic of the market to work by creating real incentives that encourage sustainable use rather than profligate and use and subsidized consumption. On the other hand, sociological conflict thinkers like Allen Schnaiberg, who connected markets and capitalism with an environmentally destructive "treadmill of production" would have doubts about allowing the magic of the market address problems of environmental degradation (his ideas were discussed in Chapters Two and Seven). But they also noted the destructive effects of government interventions, as a sponsor of profligate growth (Schnaiberg and Gould, 1994).

How could markets be reformed to produce a more sustainable economy and society? One idea would be to invert the old system of taxes and subsidies to internalize the full costs of doing business and reassign them to the marketplace, where they belong. Doing this would create an economy where business firms prosper by being responsible, both socially and environmentally. In other words, they prosper by competing to be more ecological not because it is the right thing to do, but because it squares with the bottom line of profitability. A common proposal to do this is to shift present taxes on income and payroll to *green taxes*. Governments could gradually and incrementally (not suddenly) decrease taxes on income, savings, and investments ("goods") and increase them on energy and resource use, on polluting emissions to land, air, and water, and on products with a high environmental impact ("bads"). It is important that the purpose of green taxes should *not* be to increase total government revenues (they should be revenue-neutral). Their purpose should be to provide all participants in markets with accurate information about full costs and to undo the perverse distortions produced by the relentless pursuit of low prices. Taxes could then have an environmentally positive impact on consumption patterns and on the cost structure of industry without adding to the overall tax burden on industry and society. Green tax shifts could be graduated so as not to impose an overproportionate burden on low income people, who gain less by lowering income and investment taxes but still consume at market prices. The purpose of imposing green taxes is to give people and companies positive incentives to avoid them, as they now seek to avoid earnings and incomes taxes. Markets would send different kinds of signals. (Hawken, 1993: 167–171). European nations have begun such tax shifting. Germany instituted a toxic waste tax in 1991, and toxic waste production fell more than 15 percent in three years. The Netherlands established a water pollution tax that was the main factor behind a 72–99 percent drop in industrial heavy metals put into water. Norway instituted a CO_2 tax, and emissions appear to be 3–4 percent lower than they would be without the tax (Roodman, 1998: 178).

Germany's Duales System Deutschland (DSD, or green-dot system) illustrates what can be accomplished with green taxes on consumer products. Legislation mandated that manufacturers are legally responsible for packaging material (plastic wrap, cardboard, and bottles). Stores must accept the used material, and producers must accept them back from stores. Producers prepay the costs of recycling of used junked consumer products (such as batteries and old TV frames), and retailers and consumers bear some of these costs (added into the product price). Many companies found ways to reuse or recycle their material, while others opted for simpler packaging. Firms, even many retail firms, now hire "ecology managers." Costs passed on to consumers represented a modest $20 per year. As you can imagine, such a complex system has problems, but it increased the recycling of all materials by 70 percent. In principle, the German law forcibly closes the packaging material loop in the economy, but leaves businesses with flexibility in accommodating the new limit. Austria, France, and Belgium have adopted versions of the DSD system (Roodman, 1999: 175).

Chapter Four noted that an *energy system* most fundamentally and pervasively connects human societies and their biophysical environments. One implication of this is that, of the possible green taxes, taxing energy would be the most fruitful and beneficial, and would provide the greatest short- and long-term benefits. A tax on the carbon content of fuels would give consumers incentives to switch to fuels that produce less pollution or greenhouse gases and would give producers reason to invest in energy efficiency. Besides addressing concerns about global warming, there would be other benefits. A study of the economies of Japan, the United States, the former Soviet Union, and European Common Market nations between 1976 and 1990 found that economic performance was directly correlated with energy prices. Contrary to intuition, the more costly the price of energy resources, the greater the technological innovation and economic growth. But where energy prices were subsidized and below world market value, as they were in the Soviet Union, both innovation and economic growth lagged significantly behind. The United States outperformed the Soviet Union but not the European nations, which taxed energy higher than the United States but not as highly as the Japanese (cited in Hawken, 1991: 180). I hasten to reemphasize that energy green taxes need to be incremental. If they should go up overnight (as they in effect did during the oil boycotts of the 1970s), they would cause inflation and economic and social chaos. But phased in over a longer time (20 years?), producers and consumers would have time to adapt, plan, and reinvent.

Other proposals exist to deal with externalities and commons problems. Because of their fluidity, should a piece of air, river water, or ocean fisheries be "owned" by a person or company? With some ingenuity, quasi-markets could in fact be created where none exists. Examples include proposals to measure industrial pollutants (for instance, sulfur emissions from power plants) and issue "emission permits" to companies based on the volume of pollutants. A company that exceeded its permit would pay a stiff surcharge, but—importantly—a firm that didn't need to use part of it (because it invested in efficiency or pollution control) could sell it to another firm that needed it. In effect, this would create a quasimarket where none presently exists. Permits could be issued in terms of emission levels or practical emission reduction targets for the whole economy or industry. Higher-polluting firms would have to pay surcharges or buy permits; less polluting ones could avoid surcharges taxes or sell permits to others. Chapter Four noted that the Kyoto climate conference in 1998 proposed the creation of a tradable permit system for greenhouse gas emissions among nations. Similar proposals exists for water rights and hazardous recyclable wastes. The efficiency of markets could be harnessed to achieve the implementation of environmental goals through political choices (Roodman, 1998: 194–200).[2]

New Measures of Economic and Social Progress

The economic health of nations is usually measured in terms of changes in the total value of all goods and services bought, a measure called the gross national product (GNP). Economists also use the *real* GNP, which is adjusted for inflation, or the GNP per capita, which is the GNP divided by the number of people in the population. This calculation ignores the fact that in the real world wealth is not evenly divided within the population. Sometimes they use a measure called the gross domestic product (GDP), which factors out the value of imported goods and services. These measures are relatively easy to record and measure, and their *growth* is often taken as a measure of the social as well as economic well-being of a nation. They are established but inadequate measures. They do not deduct from GNP growth withdrawals from or damage to the earth's resources. They treat all goods and services as being alike, whether made producing healthy food, treating sick people made ill by pollution, or cleaning up the damage from massive oil spills or nuclear power disasters. They are not, in fact, good measures of social well-being and do not differentiate between goods produced under safe and remunerative labor conditions from those produced under exploitive and hazardous ones. They tell you nothing about the actual distribution of the value of goods and services among individuals or groups within a nation.

Many economic thinkers have developed alternative measures that are more multidimensional and realistic measures of economic and social well-being to gauge human progress. The United Nations developed a Human Development Index (HDI), which combines economic and social indicators to estimate the average quality of life in a country. Measured on a scale from 0 to 1, the HDI aggregates (1) life expectancy at birth, (2) literacy rates, and (3) real GDP per person, based on data that exist for most of the world's nations. U.N. analysts rank nations and give them standardized scores. The HDI is also a flexible research index. In 1996, for example, when the HDI was adjusted to reflect the status of women, the U.S. rank dropped to fourth place, behind Sweden, Canada, and Norway, while that of China rose significantly (U.N. Development Programme, 1996: 31). World Bank economists Herman Daly and John Cobb developed the Index of Sustainable Economic Welfare (ISEW) as a more comprehensive and ecologically sensitive alternative to the plain GNP. It includes the average per capita GNP adjusted for inequalities in income, depletion of nonrenewable resources, loss of wetlands, loss of farmland from soil erosion and urbanization, the costs of air and water pollution, and estimates of long-term environmental damage from global change such as ozone depletion Their studies concluded that while the U.S. GNP continued to rise between 1950 and 1990, the ISEW rose between 1950 and 1975 but fell by 14 percent between 1977 and 1990 (Daly and Cobb, 1989). A real problem with the ISEW is that it could not be used generally because it depends on information available in only a few countries. Most analysts believe that in the LDCs, where much of the information for these alternative measures is not available, grain consumption per person (for which statistics are usually available) provides a rough estimate of the quality of economic life (Miller, 1998: 708–709).

None of these alternative indicators of economic, social, and environmental well-being is beyond question. They have been criticized as arbitrary in what they include and for the fact that environmental and social costs are notoriously difficult to price (Dietz and Rosa, 1994). Some proposed measures, such as the ISEW, are too complex to be presently useful. Another barrier to their adoption, I think, is that some interest groups would not *want* presently externalized costs incorporated into routine measures of socioeconomic reporting precisely because they would highlight the human and environmental costs of business as usual.

Rational-Choice Theory and Human-Environment Problems

These arguments have a common theoretical thread that is broader than neoclassical economic theory. A wide variety of scholars from diverse disciplines such as behavioral psychology, economics, political science, sociology, and policy studies created a genuinely transdisiplinary perspective on human behavior, now called *rational-choice theory* (Coleman, 1990; Wallace and Wolf,

1991). In this view, humans are rational choice makers. Economic theory argues that they choose economic goods and services in terms of how much they cost and how badly they need them. But rational choice theory argues that—far beyond economic purchases—people make reasoned *social choices*, based on experienced costs and benefits, about all manner of things. These, include, for instance, which politicians to vote for, which member of the opposite—or the same—sex is most attractive, which college major is the right one, whether to obey or violate a law, whether to work hard for some group project or loaf along and get the benefit anyway, whether to stay married or divorce, whether to maintain a social relationship or let it erode, and whether to see a therapist about your problems or deal with them yourself. We choose, rational-choice theory argues, things that have high benefits relative to their costs. When you say, "I don't really have a choice," what that means is that you think the costs are too high to really make a choice. It is not that anyone believes that individuals go around like a cost accountants, meticulously calculating the exact numerical costs and benefits of all manner of choices. The assertion is rather that in some more vague but real sense, humans adapt to life by trying to minimize costs and maximize benefits. Some costs and benefits may be given in nature (e.g., a starving person will do almost anything for food), but others are shaped by culture and perceptions. They may be symbolic as well as material. (People value social honor and spiritual rewards. Think about the religious maxim that it is better to give than to receive.) Rational choices need not operate in the short term. Over time, we develop sense of what are roughly fair exchanges of goods, favors, or obligations to each.

Thus, the human cause of environmental degradation is that we get the benefits of unsustainable consumption, but the costs are often invisible or work in such a delayed time frame that we don't take them into account. Furthermore, rational-choice theory argues that many of the change strategies of environmental movements are precisely the wrong ones to produce significant behavior change. The way to avert ecological disaster is not to persuade people to give up their selfish habits for the common good (often for the benefit of generations yet unborn). Typically, appeals are made in terms of sacrifice, selflessness, and moral shame. A more effective strategy is to tap a durable human propensity for thinking mainly of short-term self-interest. Moral appeals to "be good" do not work very well in the absence of real incentives. We should think about "saving the commons" by privatizing it. Real cooperation, at any level builds up trust from experience with small-scale tit-for-tat exchanges, not from moral exhortation (Ridley and Low, 1993; Low and Heinen, 1993). The most illustrative case in point of ignoring these powerful mainsprings of human motivation is the fate of the Soviet Union, which tried to make a commons of all economic goods (and administer them "morally"). That turned out to be an environmental, social, and political disaster. External costs are somebody else's business, and we can go for "free rides" on commons resources. Or so we think.

I noted in Chapter Seven that most people are aware of environmental problems and agree abstractly with the idea of developing a sustainable society, but the thorny problem is transforming our behavior and the way social systems operate. Rational-choice perspectives suggest that instead of urging us to be good, we transform incentive systems in economic markets and social life to send concrete signals that make self-interest consistent with what is desirable so that people get real rewards for being good. That is the real logic underlying all of those proposals I discussed. It is a powerful and compelling argument based on an undoubtedly protean dynamic of human behavior. I think it is also a slippery and often a misleading one. Let me tell you why.

But Markets Are Not *the* Answer. . .

All of the preceding ideas about internalizing environmental costs, privatizing the commons, creating quasimarkets from common property resources imply that our problems are a variety of market failures and that the prescription is to get markets functioning like they should. Neoclassical economists and more conservative political thinkers are so enamored with market solutions that they believe that the answer to most human and environmental problems is simply to unhook markets from any undue intervention and just let the market work (DiLorenzo, 1993). That is an attractive but a deeply flawed idea. Problems are deeper than market failures, and even fully functioning markets will not, by themselves, solve our problems.

There are at least four recognized limitations of markets. First, markets treat as equal worth (without value judgments) all dollar values, whether generated by cleaning up toxic wastes; producing nuclear missiles; or producing housing, food, or humanly enriching art. Whether a product was made with clean processes or with ones that make a product cheaper by putting carbon, sulfur, chlorine, and other material into the air or water is not counted. Whether a product was made by well-trained workers in a safe environment or by underpaid labor of unhealthy workers or unhealthy children carries no weight and often misrepresents societal preferences by making the less appropriately produced item less expensive. Markets don't care about these things. But people do.

Second, goods that are valued by nonparticipants in formal markets are systematically underpriced. What is the dollar value of a living tree? Usually it is the price at which dead timber can be sold in a market. But what of its value to the person who harvests fruits or nuts from the tree? Or the person who values it for protecting his or her nearby land from being flooded? Or the person who values it because he/she just likes to look at it, or enjoys its shade? The combined net worth of the tree for all these people may be well above its market price as lumber. But barring some cooperative arrangement that incorporates the needs of all those who value the tree, cutting the tree and selling it on the market means that the market will have operated in a way that did not optimally represent its value to all those who valued it (Kane, 1993b: 60, 64). Moreover, none of these human use-values of a tree, alive or cut as lumber, incorporate the myriad functions of a tree for ecosystem maintenance, watershed protection, habitat provision, soil stabilization, and so forth. Market prices don't incorporate the needs of other species in maintaining habitats and sustaining biodiversity.

Third, markets gauge the real value of resources or products only in present actual exchanges. All other attempts to internalize prices or create quasimarkets from common property resources are *shadow prices*, determined in some speculative way by some expert, planner, administrator, or bureaucrat. They are speculative administered prices. Take, for instance, the common practice, mentioned earlier, of discounting future values. Should you consume it today or save it for the future? Established ways of calculating the value of resources in the future assume that inflation and technological change will reduce the future dollar price of a particular resource (forest, mine field, copper ore). In other words, future values are discounted by some percentage for every year that a resource is conserved. This process conflicts with long-term sustainability and reduces the rights of future generations to near zero (Stern et al., 1992: 86).

A fourth limitation of markets has been noted by those on the political left since the days of Karl Marx. Markets may create economic efficiency, narrowly defined, but as they operate over time without some sort of nonmarket restraints, they generate *vast* systems of social inequality that themselves represent significant (but normally externalized) *social costs that effect human welfare and even markets themselves.* The evidence for this effect is overwhelming both within and between nations, as documented in Chapter Seven. Some opposition to the creation of quasimarkets of tradeable emission permits from common-pool resources is on exactly these grounds. Rich firms or nations would have the resources to pay surcharges or buy emission permits from poorer firms or nations (who would be under routine pressure to sell them cheaply). Either way, the rich could still afford to pollute and real reduction would be accomplished on the backs of the poor. In sum, all of these problems with markets mean that for all their virtues, they do not price all things effectively and do not price many things that people care about.

Pure market strategies have an even deeper limitation. People in free-market nations, especially Americans, and neoclassical economists in particular, tend to view markets as somehow natural and real systems that arise spontaneously among all people regardless of differences in philosophy, religion, culture, or political belief. Markets seem almost like a part of nature. We see politics and culture, by contrast, as more obviously socially constructed, arbitrary, whimsical—and often irrational. The GNP is taken as real. The other new measures of social and economic progress are seen as arbitrary (Dietz and Rosa, 1994). Furthermore, when the word *market* is appended to the technical term *economy*, we have the satisfying feeling that we are dealing with forces in the world that function properly without government interference. We think of vast global markets organized by banks and multinational corporations as simply projections of the elemental reality of village markets—even though the scale and connections among market participants is vastly different and the feedback signals about value is much more nebulous and manipulable.

In fact, markets are no more natural than politics and culture, whether traditional face-to-face or the world market economy. There never has been, nor ever will be, a market that operates beyond the specifications and contexts of politics and culture. The traditional village market was consigned to a specific place in the town, and it was conducted on certain days—assigned by cultural tradition. And even traditional markets were protected from marauders, and local constables or soldiers of the local mandarin, caliph, or duke guaranteed orderly commerce. Certainly in contemporary national and international markets, there is really no such thing as a truly free market, unconstrained by political regulation or subsidies. In the global marketplace every nation expects its government to try to create favorable terms of trade for national firms and products. As a relatively free global trade system like that now organized by the World Trade Organization comes into being, it will not be because of the "natural operation of markets" themselves, but by painful and laborious negotiations, politicians, bureaucrats, and corporate representatives—and will be fraught with compromises and opposition. These efforts in the emerging world market system have been going on for more than forty years (I discussed them in Chapter Two, and will return to them in the last chapter). Markets, politics, and culture alike are social constructions of reality.

Why do I spend so much time belaboring this point? Because, if you look again at the market strategies for dealing with environmental problems that I discussed earlier (green taxes, privatizing the commons, and creating quasimarkets with tradeable emission permits), they all require *political action* to reengineer markets that deliver different signals to producers and consumers. It is *not* a case of going from a regulated market to a free one, but of moving from today's environmentally perverse interventions to a new set of less damaging ones. That is *politically* a tough nut to crack. It is all well and good to talk about energy taxes, but what politician in an energy-producing state is going to vote for higher taxes on energy? What senator from Wyoming is going to vote to end subsidies in the form of cheap permits for ranchers to graze their animals on public lands (often destroying them)?[3] In 1995, such subsidies totaled $200 million and the most well-off ranchers owned 90 percent of them (Miller, 1998: 652). The principle of rational-choice theory still holds: Politicians operate in different political resource markets (electoral votes and political action committee [PAC] money). The efforts of the Clinton administration in 1993 to impose a broad but modest carbon tax and to put grazing on publicly owned rangelands closer to private market prices are cases in point. Both initiatives met with utter political failure because they were opposed by powerful coalitions of interest groups that benefit from cheap energy and grazing lands. As Chapter Two noted, the complex division of labor and occupational specialization in industrial societies produce a *quasispeciation*, which means that different economic groups benefit and bear costs very differently, even in the same physical environment. Economic or rational choice perspectives that talk about some sort of overall good ignore this important fact. To take a more positive example, what the Germans have done with the DSD and green-dot system says more about the influence of the German environmental movement (the Greens) and German political culture than anything to do with markets or rationality per se.

So it is one thing to talk about creating a green economy and making doing good consistent with doing well: The premise, I think, is sound. Changing market incentives can change behavior. But changing market incentives means looking squarely in the face of politics.

POLITICS AND POLICY

Like markets, political institutions are also concerned with resource allocation. The classic definition of politics is the process of deciding who gets what, when, and how. But although rational choice theory might understand politics as involving merely a different sort of market (with influence for sale), that is at least partly misleading. Politics involves the mobilization of power to allocate resources for an ostensible collective good and is justified by whether or not it produces public and collective benefits. Markets, on the other hand, are justified in terms of whether or not they produce private gain. Ever since the emergence of nation-states, politics and markets have involved different kinds of cultural legitimation. Historically, in fact, the scope and power of political institutions grew to address precisely those problems either created by or not effectively addressed by economic markets (including those I just noted).

The purpose of all political institutions is to make public policy, or attempts by government agencies to change or control collective patterns of action. But the term *public policy* is a broad umbrella that encompasses an enormous diversity of agents and modalities. Next, I'll outline some different policy options that governments use and their relevance for environmental problems.

Public Policy and Strategies of Social Change

Four broad strategies exist by which public policy attempts to produce change. Environmental sociologist Riley Dunlap described them as different kinds of fixes for environmental problems (1992). They involve using public policy to change technology, behavior, ideas, and laws.

First, and most often identified as a solution for problems, are *technological fixes*. They include, for instance, more efficient auto engines with emission control devices that use less fuel and pollute less, engineering highways and synchronizing traffic lights to reduce auto accidents, better street lights or burglar alarms to discourage crime, insulating your house to cut your fuel bill, genetic breeding of more productive seed hybrids, biotechnology, and so forth. The list of technological proposals to address problems seems endless. Public policy can stimulate the adoption of new technology in a variety of ways, such as public investment, subsidies, tax policies, or regulatory mandates.

Second, and most often contrasted with technology, are the *behavioral fixes* by which public policy provides incentives to get us to behave differently. These are (supposedly) more difficult than technological fixes, which require no behavioral modifications. Examples of behavior fixes include getting people to eat lower on the food chain (for ecological and health reasons); use condoms; stop smoking; wear sweaters; turn down thermostats in the winter; install attic fans and use air conditioners less in the summer; and walk, bike, carpool, or use public transportation as alternatives to driving. Whereas technology requires investment, behavioral changes typically require incentives (or penalties). We are not on new ground: This was the whole point of the rational choice perspective developed earlier as well as of the findings about research on energy conservation in Chapter Six.

Third are the *cognitive fixes*, which attempt to create awareness of problems in people's minds. The assumption is that if you change people's minds, they will change their behavior. Cognitive fixes often rely on public education and media campaigns. Energy conservation ads telling people to "don't be fuelish," or recycling ads reminding us that "if you're not recycling, you are throwing it all away," are cases in point. The appeal of cognitive fixes is that they rely on voluntary change and are compatible with norms of personal freedom. It requires no regulation and little public investment. Unfortunately, there is very little evidence that such strategies work in isolation from others (I mentioned some of this evidence with regard to energy conservation in Chapter Six). Even so, the importance of cognitive change as part of more comprehensive policy change strategies is often underrated.

Fourth are the *legal fixes* that mandate change through laws and regulations rather than incentives, subsidies, or persuasion. Examples include federal speed limits on interstate highways and requirements to remove lead from gasoline, install antipollution devices, or recycle beverage containers or household or industrial wastes. Actually, the first two strategies (technological, behavioral) can be pursued by regulatory or nonregulatory means. Regulatory strategies can be very effective, but they are unpopular in a society that views government regulation negatively. They require great political will, or at least effective mobilization and interest group coalitions, to enact and enforce (Dunlap, 1992). It is a truism among policy scholars that the most effective strategies for change produced by public policy would combine all four approaches. In other words, change could be promoted by simultaneously providing better technical means, changing people's minds, providing material incentives, and regulatory restrictions or targets.

Policy and the Economic Production Cycle

Policy strategies can apply to different domains of social behavior (as the fixes just described), but policy can also be applied at three different stages of the economic production cycle. First, we are most familiar with "end-of-the-pipe" or downstream interventions that work *after consumption has taken place*. Clean air standards, antipollution measures, and recycling are examples. Such end-of-the-economic-cycle strategies obviously work and are, in fact, the way most environmental legislation to date works, either by penalties, pollution standards, or providing incentives. Between 1975 and 1995, the world doubled the volume of paper recycled while the market for air pollution equipment grew to 2 percent of the *world* gross domestic product, easily outclassing the aerospace industry and approaching the significance of the chemical industry (Mattoon, 1998; Renner, 1998). Such strategies, are useful, but they do nothing to reduce unsustainable consumption. In people's minds, they may even constitute a rationale for consuming more! Second, *midstream strategies reduce consumption,* not just encourage frugality with trash and effluents. These include the behavioral changes already noted and industries that use cogeneration processes.[4] Third are policy interventions that work upstream *early in the production process itself,* either to make production more environmentally benign or to reduce waste and materials in the production of products and services. The standout example in the United States is the engineering of more energy-efficient products, ranging from dishwashers to automobiles. Other examples envisioned are products that require less packaging (such as those generated by the German DSD and green-dot system).

History is part of the reason most policy attention has been given to *downstream* interventions that deal with pollution and toxic emissions. Environmental consciousness as it developed in the 1960s focused mainly on pollutants, and awareness of consumption and resource use issues came later. But there are reasons beyond history. Midstream and upstream policies mean intervening in the economy in more fundamental ways than just cleaning up pollution. They mean altering production technologies, consumption patterns, or both. And real upstream policies shift the burden of change from consumers to producers. This is particularly difficult in a political system where producers have more clout than consumers. Even though reducing pollution and waste by end-of-the-pipe controls or recycling is often costly, reducing it through resource efficiency and smart process redesign is usually profitable (Lovins, cited in Miller, 1998: 426). Again, doing well can be combined with doing good. As you might guess, it was probably politically easier to focus on end-of-the-pipe policies. They provided the comforting illusion that we can go on consuming as we like, as long as we clean up the messes.

A catch phrase among those who advocate such midstream and upstream strategies is to *dematerialize the economy* (I mentioned this notion in Chapter Seven). It means using less resources and environmentally damaging production processes per unit of production. Such dematerialization has, in fact, been going on in advanced industrial economies for some time. To illustrate, cars weighed 20 percent less in 1985 than they did in 1975. By 1985, U.S. auto redesign resulted in an annual savings of 250 million tons of steel, rubber, plastic, aluminum, iron, zinc, lead, copper, and glass. In fact, if you look at every durable good you own and use—your car, your TV, your refrigerator, or your house—it weighed more, used more material, and employed greater amounts of embedded energy in its manufacture twenty years ago (Hawken, 1993: 64). Such dematerialization is sometimes the result of market operations (such as the auto industry's response to more efficient imports) but is just as often the result of public policy, such as tax incentives, or mandated fuel consumption or emissions standards. In sum, upstream, midstream, and downstream policies can move us some distance toward a true "industrial ecology" (Frosch and Gallopolous, 1990).

But wait. Without real reductions in consumption, dematerialization, like recycling, will not be sufficient to produce a sustainable economy. In simpler terms, the problem is that while cars, TVs, refrigerators, and houses may use less material and be more energy efficient, there are a lot more of them than there were 40 years ago. Indeed, since 1970, when global materials use was first tracked, material per dollar spent has fallen by 18 percent, but that was entirely canceled out by total materials use per capita (Gardner and Sampat, 1999: 50). We are still on what Alan Schnaiberg called the "treadmill" of production (and consumption!). So the progress so far due to recycling and dematerialization is something of a mirage.

Policy and Social Structure

Political scientist Theodore Lowi worked over decades to develop a conceptual framework depicting how public policy articulates with social structure in different ways (1964, 1972, 1979). Lowi distinguishes between *constituent* and *regulatory policies*. Constituent policies provide benefits to particular constituents, clients, or publics. Providing tax incentives for the lumber or oil industry illustrates constituent policies. The environmental equivalent of traditional constituent policy would be those that provide subsidies for windpower or "gasohol" fuel. Even when they regulate, constituent policies are often—grudgingly—welcomed by particular constituent groups and industries as necessary to police their deviants. Examples include the Securities Exchange Commission that polices the stock market against securities fraud. In the Great Plains states, state legislatures considered enabling legislation to regulate ("meter") water use from the Ogallala aquifer in order to conserve water supplies. While such policies are still embryonic and not very effective, they have met with scattered and surprisingly little opposition from dryland farmers. Constituent policies are politically easy. If they involve subsidies or tax concessions, they are enthusiastically welcomed. If they involve regulation, they are grudgingly welcomed as a necessary collective security measure for an interest group or industry.

In contrast to constituent policies, true regulatory policies are another matter. *Regulatory policies* attempt to control behavior across a broad spectrum of constituent groups, industries, and economic processes. Related to environmental matters, early legislation from the 1960s that established broad air and water pollution standards are examples of such regulatory policies. Other examples are the regulation of utility pricing to encourage a variety of energy conservation measures by organizations and individuals or the 1993 proposals by the Clinton administration to enact broad carbon taxes. Such policies do indeed cast a broad net, and their costs percolate through the economy to affect most groups. Investors, producers, workers, and consumers all eventually share a piece of the costs of true regulatory policies. But precisely because their costs are so pervasive, they are politically unpopular, difficult, and contentious. They raise issues about who really should pay (anyone, it seems, but "us"!). Thus regulatory policies are perceived as inefficient and unjust, as taxes imposed on some by others. Even the environmental policy principle that "the polluter pays" is of little help, because different client groups have very different notions about who the real polluters and beneficiaries of pollution really are (again, anyone but us). Thus regulatory policies instigate political struggles by powerful interest groups to politically capture the agencies responsible for administering them, or at least to capture the fine print of regulations that shape who pays how much. The ill-fated 1993 Clinton proposal for a broad carbon energy tax elegantly illustrates this point. New England politicians made sure that it was a tax on gasoline, not heating fuel oil, which heats many homes in the Northeast. Natural gas producers and transporters, along with many environmental organizations wanted natural gas exempted as the less polluting fossil fuel. The oil industry made sure that the tax was to be paid at the retail pump (by consumers), not at the well head (by producers). The fuel tax that survived this interest-group whipsaw was insignificant. Particularly in the United States, true regulatory policies are increasingly difficult and often turn into de facto constituent policies.

The National Environmental Protection Act of 1969 (NEPA) established the Environmental Protection Agency (EPA) and included language to prevent its capture by regulated industries and the environmental movement alike. Nonetheless, both trade associations and environmental organizations have been active in attempting such capture (Aidala, 1979; Sabatier, 1975). But the pressures against the EPA come as often from environmental organizations as from industry groups. That fact suggests that it has been able to maintain itself as a regulatory agency instead of being captured as a constituent-policy organization. It stands in sharp contrast to the Nuclear Regulatory Commission, which all observers agree became a virtual lobby and propaganda arm for the nuclear power industry. Most national environmental movement organizations have advocated regulatory policies, whether in setting standards for emissions or pollution, screening toxic substances, requiring environmental impact statements, setting aside or protecting ecosystems, or encouraging resource conservation. Such policies have involved the various criteria developed by the technocratic risk establishment (mentioned in Chapter Four) for assessing environmental risks and hazards.[5] In short, American environmental politics have involved a heady and contentious mix of both constituent and regulatory policies (Schnaiberg, 1983).

Second, Lowi distinguishes between distributive and redistributive policies (1979). *Distributive policies* are gifts from the stock of things that governments control. An example is the distribution of the air waves at different frequencies to radio and TV stations by the Federal Communications Commission. Distributive policies allocate a common good, such as logging rights to lumber corporations in national parks or cheap grazing rights to ranchers in public rangelands. Other examples could include incentives for replacement of energy-inefficient equipment for energy-conserving equipment and subsidies to farmers for soil conservation and to promote low-input or sustainable farming practices. As you might guess, distributive policies are very popular. They are perceived as "free gifts" from the governments of things that can be transformed into private income. These illustrations show that distributive policies, while popular, can be connected with moving to more sustainable systems but also with the overuse that attends commons problems.

Redistributive policies involve not just the distribution of goods or resources that government controls, but the redistribution of those that have already been allocated in some way (by governments or markets). Redistributive policies mean using government authority to take traditional benefits, subsidies, or privileges from some and give them to others. You should be familiar with these: the notion that has been around since the 1930s in industrial societies that income taxes should be progressive so that the very wealthy bear a higher tax burden (and pay a higher tax rate) than middle- and low-income groups. This system in effect creates transfer payments of some type, whether direct or in tax concessions, from very wealthy people to create subsidies or social programs for low-income people, who are most disadvantaged in private markets. Redistributive policies have been the hallmark of welfare politics. Earlier I noted that farm subsidies (with whatever purpose or consequence) may be distributive policies. But subsidies and incentives through the Department of Agriculture (DOA) for low-input and sustainable agriculture are better understood as redistributive policies. They involve redistributing limited budgets of the DOA—away from the much larger established subsidies for corporate, agribusiness, and high-input agricultural interests to other priorities and beneficiaries. Another illustration of an environmental redistributive policy proposal is the creation windfall profit taxes to provide energy costs offsets for working- and poverty-class constituents facing higher energy costs. Most proposals for energy and green taxes now include such redistributive clauses. But as you might guess, taking money, traditional benefits, or incentives from some and giving them to others make redistributive policies unpopular, contentious, and politically difficult. Since they constrain the operations of markets and challenge established patterns of wealth and privilege, these policies often stimulate political mobilization among the very groups in the population most able to defend their traditional subsidies and benefits. They are thus politically the most difficult of all types of policy to enact and implement. *So why bother with them?* For reasons of social and economic justice. But if those reasons don't persuade you, there are other reasons. Consider that any market or policy that significantly increases social inequality has very real social costs that will be paid one way or another, perhaps by civil unrest. All nations have found some redistributive policies necessary for social peace.

Politics and the Limits of Policy

The fragility of true regulatory policy, with tendencies to devolve into constituent policy through the capture of legislation and enforcement, and the enormous political difficulties of redistributive policies both mean that policy is indeed rooted in *politics*—the contentious processes of deciding who gets what, when, and how. The legislative politician's dream is to be able to propose only constituent, distributive policy legislation. But the reality in closed and interconnected systems—whether ecological or budgetary—is that politically difficult regulatory and redistribution policies are often required. In democratic nations they require solid bases of electoral support or powerful coalitions of interests groups, lobbies, and movement organizations. Emphasizing that policy is embedded in politics underscores the fallaciousness of the technocratic assumptions that often dominate discussions of public policy: that we can simply devise rational, feasible, and cost-effective market interventions and incentive systems that get us to behave properly and simply enact them. In a pig's eye we can! Not without getting the politics right first.

Political institutions and cultures in different nations are not alike, and the policy process works differently in various nations. In the United States the constitutional separation of powers provides nongovernmental organizations (NGOs) with greater opportunities to shape policy through ligitation and the judicial system. The more centralized political systems of Japan and France limit participation of citizens' action groups in the political process. While citizens of other MDCs are likely to have stronger political party affiliations than Americans, they are less likely to join environmental organizations and other NGOs and are less likely to have direct access to policy debates. Environmental policy is relatively centralized at the national level in Great Britain, Japan, and France and is administered primarily by local governments in Germany. The United States is unusual in providing opportunities for diverse groups of scientists to affect public policy. By contrast, participation by scientists in Europe is more likely to be confined to official channels. The United States is also unusual in having regulatory decisions tied by statute to the outcomes of technical risk analysis studies. Thus it is sometimes easier to have a product or production process banned or restricted in the United States than in many other MDCs (Brickman et al., 1985).

I noted earlier that European nations have taken a substantial lead in improving the human-environment connection. They are the leaders in recycling; in tax and subsidy shifting to promote a greener economy; in promoting alternative energy sources, such as wind and solar; and in supporting international treaties, such as the Kyoto climate treaty. The reasons for the European lead have to do with political structure. For one thing, the American electoral system, with its two-party "winner take all" elections, makes it difficult for reform-oriented groups, factions, and movements to be represented in the executive policy-making process. In Germany, in contrast, parliamentary proportional representation of various electoral parties in the formation of governments provides greater access to the political system for parties and groups committed to social reform (Parkin, 1989). The German and Dutch Green parties, for example, had, in their heyday, political influence out of all proportion to their numbers and resources, which made American environmentalists turn green with envy, so to speak. But important as they are, differences between the United States and Europe run much deeper than the formal differences between a two-party presidential system and a multiparty proportional representational system.

Electoral candidates raising the most money usually win U.S. elections. In 1992, the average U.S. senator spent $3.9 million campaigning for reelection—an amount that requires raising $12,600 per week for 312 weeks (six years!). Unless they are personally wealthy and willing to spend their own money, candidates can only get this money from very wealthy individuals or corporations (Miller, 1998: 740). Thus, the American electoral system has become increasingly driven by the money from corporate political action committees and so-called "soft money" from economic elites. Longworth says that such a system has strong *shareholder control*, that is, control by corporations and economic elites over the U.S. political system. By contrast, other democracies have strong *stakeholder influence*, which give more consideration to the desires and needs of a large array of groups with a stake in the system, such as nongovermental organizations and movements like civic, community, professional, and environmental organizations (1998). Western European nations, for example, give more consideration to stakeholder interests. The German tradition of "codetermination" specifies that labor, local communities, and corporate interests all be represented in the policy process, in about equal proportions (Weinburg et al., 1998). As I write this, there is a presidential election in the offing and a lot of talk about campaign finance reforms to remedy this well-known but universally disliked problem in U.S. politics. But don't hold your breath until they are enacted. After elections, the topic is usually ignored (that is, after the beneficiaries of the present system have been elected).

The U.S. political system seems increasingly less capable of delivering systemwide reforms. One can think of systemwide reforms delivered by public policy in the past: the Progressive Reforms of the 1900s, the New Deal of the 1930s, or the extension of Civil Rights and the War on Poverty of the 1960s. But such political changes in the last few decades as increasing electoral fragmentation, the declining cohesion and power of parties, and emergence of shareholder control all mean that such systemwide reforms are increasingly difficult. Some analysts view these obstacles as politically impossible to overcome in the absence of a clear, immediate, and overwhelming national crisis. The futility of trying to enact such eminently reasonable policies without electoral or political coalitional support is richly illustrated, again, (much to the dismay of environmentalists) by the ill-fated Clinton energy and rangeland policy proposals. The "dirty little secret" about public policy in the United States, known among policy scholars but not often publicly discussed, is that no administration or political party in recent decades has been able to mobilize an effective coalition to support systemwide domestic reforms like those of the American past. You can see this not only in the attempts to create coherent environmental policy, but most graphically in the efforts to create an effective national health system or national health insurance system. Increasingly, American public policy is *retail policy*, that is, constituent policy that addresses the needs of particular organized client groups, rather than wholesale policy in the public interest (Mans, 1994).

THE POTENTIAL FOR STRUCTURAL CHANGE

Considering the limitations of purely market strategies for environmental improvement brought us face to face with politics, and I argued that public policy is often a blunt, limited, and imperfect instrument of change. If markets won't work to do it, and politics often can't, what then? Is all lost? Indeed not. Consider some cases of productive interaction between markets and politics.

Signs of Progress?

Such illustrations are easy. Consider, for instance, the decline in cigarette smoking in the United States. It came at the end of a decades-long consciousness-raising crusade by medical researchers and health organizations. The reduction in smoking involved combined actions by public health officials and state attorney generals willing to use tax and regulatory measures to save tax expenditures, insurers, and businesses who were only too pleased to exclude smokers from their insurance policies, businesses, and stores (for fear of litigation). American tobacco companies now fight a rearguard action: economically diversifying their investments in the United States while promoting their products overseas, particularly in LDCs. Or take changes in the American diet, which exhibited a steady and pervasive decline in per capita red meat consumption. This transformation was similarly produced by a configuration of forces that included activist groups, nutritionists, public education campaigns, willing regulators, and profit-seeking restauranteurs and food producers. Even gigantic food corporations like Con Agra now recognize the trends (and growing markets) for healthy food by producing lines of products with labels like Healthy Choice and Lean Cuisine. This is certainly not the only or even perhaps the dominant trend in the American diet. Nutritionists still find the typical American diet has too much fat and too much meat in proportion to carbohydrates and fruit. Still, in per capita terms, Americans drink less alcohol, eat less red meat, and smoke less than they did two decades ago.

These examples involve aggregate individual behavior, but similar illustrations exist about corporate changes flowing from the interaction of markets, policy, and politics. The most well-known case is tuna. The H.J. Heinz company didn't catch its own tuna, but it was barraged by letters from schoolchildren ("young consumers") pressing them to end fishing techniques using seine nets that encircled and killed large numbers of dolphins. The company announced in 1990 that its Star-Kist brand would buy tuna only from fishing boats using methods that did not kill dolphins. Other companies followed suit. Shortly afterward, Congress passed the U.S. Marine Mammal Protection Act (MMPA), which forbade dolphin-killing techniques not only among U.S. fishers, but also for all tuna imported into the U.S. (File this "tuna-dolphin" case in the back of your mind, because I will return to it.) In December 1992, a network news special charged that Wal-Mart claims to "buy American" were false, and worse, that many Wal-Mart products were made in LDCs under exploitive labor and environmentally damaging circumstances. At about the same time, Phillips-Van Heusen explicitly threatened to terminate orders to apparel suppliers that violated its broad ethical, environmental, and human-rights code. And Dow Chemical, itself certainly no stranger to environmental litigation, asked suppliers to conform not just to local pollution and safety laws but to often tougher U.S. standards. Persistent rumors that McDonald's suppliers grazed their cattle on cleared rainforest land finally led the company to ban the practice in writing (though they claimed that it had never been true). In 1992, Levi Strauss and Co. laid down tough standards of conduct to its 600 suppliers worldwide. After inspecting each one, the company ditched about thirty of them and exacted reforms from an additional 120. The company also pulled out of Myanmar (formerly Burma) for pervasive human-rights violations (McCormick and Levinson, 1993: 48–49). In an effort to shift from a throw-away to a reuse/recycle economy, Atlanta-based Interface, a leading carpet manufacturer with sales in 106 countries, started to shift the firm from the sales of carpets to the sales of carpet services. This involved installing, maintaining, and repairing carpets as desired by clients. When a carpet wears out, Interface simply takes it back to one of its plants and recycles it in its entirety into new carpeting. This process requires no new raw material and leaves nothing for the landfill (Brown, 1999c: 19).

Note that most of these examples concern retail businesses, in close contact with customers and perhaps more amenable to popular pressures to appear "greener." But there are also illustrations from energy and other industries that are removed from direct consumer pressures. Chapter Six noted corporate policy shifts among some of the largest energy firms like British Petroleum, Royal Dutch Shell, and Enron. They broke ranks with other fossil fuel firms to take climate change more seriously and invest in growing markets in renewable energy. In western Canada a giant logging firm, MacMillan Bloedel, startled the world and other logging firms when it announced that it was giving up the standard forest industry practice of clearcutting. That practice would be replaced with selective cutting, leaving trees to check runoff and soil erosion and to provide wildlife habitats to regenerate the forest. In the United States, not always a global leader in recycling, 56 percent of the steel produced now comes from scrap. As a result, steel mills built in recent years are no longer located in western Pennsylvania, where coal and iron ore were close, but are scattered around the country—in North Carolina, Nebraska, or California—feeding on local supplies of scrap. The new mills produce steel with less energy and far less pollution than did the old mills producing from virgin iron ore. A similar shift occurred in paper mills, once almost exclusively near forested areas, now often built near cities feeding on local supplies of scrap paper.

Illustrations of such positive change induced by governments are also easy to see. To take one noted earlier, in the 1990s European countries—including Denmark, Finland, the Netherlands, Sweden, Spain and the United Kingdom—began restructuring their taxes in a process known as *tax shifting*. This process reduces income taxes while offsetting these cuts with higher taxes on environmentally destructive activities (such as fossil fuel burning, generation of garbage, use of pesticides, or production of toxic wastes). Though the reduction in income taxes does not yet exceed 3 percent in any of these nations, the basic idea is well accepted. Public opinion polls on both sides of the Altantic show 70 percent of the public supporting tax shifting. In 1998, a German coalition of the party in power (Social Democrats) and the Greens announced a massive restructuring of the tax system that would simultaneously reduce taxes on wages and raise taxes on CO_2 emissions. This shift was the largest taken by any government and was not bogged down in the complexities of the Kyoto treaty. In that same year, the Chinese government in Beijing acknowledged for the first time that the record flooding in the Yangtze River basin was not merely an act of nature, but was greatly exacerbated by the deforestation of the upper reaches of the watershed. Some state-owned tree-cutting firms were ordered to become tree-planting firms, and the official Chinese view is now that trees are worth three times as much standing as they are cut, simply because of the water storage and flood retention capacity of forests.

Though the U.S. government is often no longer a leader in environmental affairs (for reasons noted earlier), there are signs of such positive changes by some government agencies. The U.S. Forest Service announced in 1998 that after several decades of building logging roads to help logging companies remove timber, it was imposing an 18-month moratorium on building such roads. This ruling restricted a huge public subsidy that had built some 380,000 miles of roads to facilitate clear cutting on public lands. The Director of the Forest Service said the service was refocusing the use of the national forests for recreation and wildlife conservation and to supply clean water and promote tourism as well as supplying timber (Brown, 1999c: 17–20). After mentioning the Clinton administration's ill-fated attempts to introduce environmental reforms without congressional support, it is fair to mention its later efforts. In 1999, the EPA issued mandates to reduce the sulfur content in gasoline, preventing the clogging of cactalytic converters and reducing pollution. The new standard was produced after several years of wrangling among states and opposition from oil companies. At the same time, the EPA announced tighter emission standards for all so-called light trucks, pickups, minivans, and sport utility vehicles, which came to dominate the American new vehicle market partly because they were not held to the same emission standards as passenger autos (*Washington Post*, 1999).

What do these examples mean? Are they evidence that a dramatic transformation to a more sustainable world, as envisioned in Chapter Seven, is underway? *Certainly not.* Illustrations are informative, but multiplying them does not prove a more general point. Furthermore, in the United States and some other nations there is clear evidence of a powerful, pervasive, and well-funded American corporate attack on environmental improvement. As long ago as 1978, corporations were spending close to $900 million per year mobilizing their supporters, an activity that continues today. Trade associations do this by organizing the owners of larger numbers of small businesses to lobby their congressional representatives, while large corporations mobilize shareholders, suppliers, customers, and employees (Beder, 1998: 18). Much of this mobilization was to stop environmental improvement and particularly to attack organized environmentalism. I will return to this point in more depth in Chapter Nine. There are also many companies striving to be known as environmental good citizens whose commitment goes no deeper than glossy ads depicting pristine wilderness surrounding their production facilities, recycling aluminum cans from the company cafeterias, or ceremonial tree plantings. While there is nothing wrong with these things, they are tantamount, in the words of businessman-critic Paul Hawken, to "bailing out the sinking *Titanic* with teaspoons" (1993: 5).

Revisiting the Tragedy of the Commons: Community Resource Management

Earlier this chapter discussed Hardin's famous tragedy of the commons concept, whereby environmental disaster ensues because individuals (or companies) cannot be denied the use of common pool resources. Often the tragedy of the commons is understood as a universal law of nature akin to the law of gravity. Yet there *are* cases of successful community management of common resources (CRM) that mitigate the tragedy of the commons. I note several.

Since the fourteenth century, villagers in Torbel, Switzerland, have practiced rules to successfully manage fragile alpine meadowlands and forests, where cattle were grazed in the summer but not in winter. They decided that alpine lands should belong to the community rather than to private owners. No one was permitted to graze more animals in the summer than they could feed in the winter. Cows were sent to alpine meadows all at once and counted, and trees for harvest were marked once a year by a community forester. To manage viable alpine meadowlands, these rules stood the test of time, population growth, and employment outside the village area (Netting, 1981). Consider an illustration from the United States: the North Atlantic lobster fisheries along the coast of Maine. Unlike many fishing grounds in the North Atlantic, lobster fisheries along the central coast of the state of Maine have been sustainably maintained for decades. Fishers in small boats drop small lobster traps (or "pots") into identifiable shoreline harbors, moving to deeper water in the winter to do lobstering. CRM is possible because the state limits the number, size, and sex of lobsters that can be harvested and requires lobstermen to get a license and display a license number prominently on the line connected to each particular pot. But most of the credit goes to the lobstermen themselves. In order to maintain their livelihood, communities of lobstermen developed strong unwritten rules governing assigned territories that were defended against outsiders. A new fisherman must be accepted by a "harbor gang" and fish only in an assigned territory. Interlopers are sternly warned, and if they persist, their equipment is sabotaged. This CRM preserved lobsters and livelihoods of fishers for decades. Along other parts of the coast it became more difficult to defend well-defined territories with the advent of motorized boats and depth-finding equipment by lobstermen who could afford them. To do so, they had to "invade" many territories and fish far offshore. Both established fishers and interlopers acted with restraint rather than starting an all-out war with disastrous costs for all. Thus CRM worked even when and where control was much looser, and the fisher and lobstermen have survived (Acheson, 1981; Gardner and Stern, 1996: 127–128). Are these cases only unusual exceptions to the tragedy of the commons? No indeed.

Conditions for Successful CRM

Political scientist Elinor Ostrom analyzed many such successful cases of CRM (1990). She focused on the sustainablility of common pool resources that were important for livelihoods and geographically large enough to make it difficult, but not impossible, to exclude individuals from befitting from their use. Ostrom concluded that successful and sustainable CRM systems depend on the characteristics of (1) the resource, (2) the group using the resource, (3) the rules they develop, and (4) the actions of governments at regional and national levels. See a summary of her findings in Table 8.1.

Hardin's tragedy of the commons concept assumes that overriding human motives are always self-centered, and therefore that CRM institutions must always fail. So he believes that government coercion is the only

TABLE 8.1 Conditions Conductive to Successful Community Resource Management

I. Resource is controllable locally
   A. Definable boundaries (land more than water, water more than air)
   B. Resources stay with boundaries (plants more than animals, lake fish more than ocean fish)
   C. Local CRM rules can be enforced (higher-level governments recognize local control and help enforce rules)
   D. Changes in resource can be adequately monitored

II. Local resource dependence
   A. Perceptible threat of resource depletion
   B. Difficulty in finding substitutes for local resources
   C. Difficulty or expense attached to leaving area

III. Presence of community
   A. Stable, usually small population
   B. Thick network of social interaction and relationships
   C. Shared norms ("social capital"), especially norms for upholding agreements
   D. Resource users have enough local knowledge of the resource to devise fair and effective rules

   (*A* facilitates *B*, and both facilitate *C*. All make it easy to share information and resolve conflicts informally)

IV. Appropriate rules and procedures
   A. Participatory selection and modification of rules
   B. The Group controls monitoring, enforcement, and personnel
   C. Rules emphasize exclusion of *outsiders* and the restraint of *insiders*
   D. Congruence of rules and resources
   E. Rules have built-in incentives for compliance
   F. Graduated, easy to administer penalties

Sources: Adapted from Ostrom (1990); Gardner and Stern, (1996: 130).

way to avoid disaster. Hardin is not alone in this way of thinking. Both behaviorist psychology and neoclassical economics view individuals as acting alone and rarely consider how social institutions can shape individual self interest. In recent times, however, economists began to address the questions of institutions. It may make a great difference whether environmental and other human problems are considered in individualistic self-interest terms or are confronted by creating social institutions and using social relationships (Gardner and Stern, 1996: 136–137).

The Social Psychology of Successful CRM

The success of CRM depends on controlling the behavior of individuals. The cases noted earlier were communities with dense social networks and cohesive cultural norms that could shape and restrain individual action. What makes individuals follow rules when they can gain something by breaking them? The key is that most people do what is good for the group (and the resource base) when they *internalize* the group's interest rather than acting from *compliance* with external costs or rewards. Such incentives work only when people expect to be punished, but internalization works all the time. When a CRM system is effective and most people internalize norms, penalties are rarely imposed and the costs to maintain the CRM system are low. Few break the rules, and when they do, mild sanctions are usually enough, making severe ones unnecessary. But without *any* incentives, some could take advantage of the self-restraint of others with impunity. The system's whole basis of trust would begin to unravel, increasing enforcement costs and ending in the tragedy of the commons. People are more likely to internalize group norms when they participate in creating them, when they value these norms for themselves and their community, and when they become a part of the very meaning of community that they share with others. Internalized self control is the ultimate basis for community control. Such informal social control can help manage environmental resources, control crime, and address many other social problems, but they sometimes repress individual desires and may often be in tension with widely held modern values like individual freedom and procedural justice (Kelman, 1958; Gardner and Stern, 1996: 135–136).

Beyond Local Communities: The Role of Central Governments

Ostrom found that the success of CRM also depended on factors beyond local communities, particularly the support of local, regional, and national governments. But government may impede as well as facilitate CRM, particularly if officials accept bribes or political favors to allow some individuals to use more than their share. Such corruption is most likely where government officials have limited ties to local communities and where local resource users do not have enough political power to exercise control over government officials. When they help, governments can provide local rules the status of legally enforceable contracts, and provide support for monitoring the resource in question. The United States and the state of California helped regional water users in both of these ways. Water management institutions were "nested institutions," in which smaller private and municipal pumping and distribution agencies are nested in larger county and regional associations. They negotiated agreements to restrict pumping as an alternative to expensive law suits over water rights. The state helped with the costs of monitoring the agreements, and treated them as legally binding in state courts. Such *comanagement* is a promising new idea in CRM (McCay, 1993).

But in modern developed societies, few people are dependent on local resources like the fishers, woodcutters, and cattle grazers as in the cases just noted. Global markets ensure that people with cash incomes can almost always escape the pain of local shortages and simply buy from elsewhere. That is why the growing costliness of food is not often apparent to people in rich nations. But even in modern societies people are usually dependent on local common-pool resources for water and solid waste disposal. Similarly, on America's dry Great Plains region where I live, water was historically mismanaged by diversion of reservoir and river water as well as by pumping from the large Ogallala aquifer. CRM is not as successful as in southern California, but, as noted in Chapter Three, water problems are a continual source of political squabbles and litigation among states, communities, and use sectors (e.g., agricultural, industrial, municipal).

The waste products of modern societies are almost always disposed of locally. While disposal in running water or the air carries wastes out of communities, thereby weakening support for CRM, solid wastes are almost always disposed locally, in landfills. Almost every U.S. city or large municipality is experimenting with schemes to conserve landfill space, and community-based waste separating and recycling programs—with varying degrees of success. As noted earlier, the United States is behind most Western European nations in this area.

Even where there is not clear local resource dependency, CRM programs have shown promise. Although few U.S. communities depend on local resource for their energy supplies, a number have successfully operated CRM energy programs (e.g., Fitchberg, Massachusetts, and various Minnesota communities instigated by the state Residential Conservation Service Program). Such programs made energy monitoring and efficiency measures easy to practice and, as in these cases, worked mainly by face-to-face communication among people already connected in communities with some degree of trust, and activated personal values about environmental protection (Gardner and Stern, 1996: 139–143).

CRM and Social Trends

As I noted in Chapter Two, the development of the global political economy and world market system stimulated large-scale economic and political systems that rapidly transformed subsistence economies into market economies driven by commodity exchange. Among the many negative environmental consequences of this evolution, a powerful one is that this development process often disrupts local CRM. In this sense, CRM is contrary to powerful social forces of the twentieth century. In fact, these social forces have been weakening two of the major conditions in Table 8.1: local resource dependence and the presence of dense, stable community networks. Family farms and ranches in the United States, owned by individual proprietors interested in sustainable management (and leaving productive land to their children), have been displaced by large corporate farms owned by remote investors. Such investors often have little interest in sustaining productivity beyond the "natural life" of the capital invested (about 10 years). Another illustration comes from the foothills of the Indian Himalayas, where for centuries people locally managed and relied on the forest for cooking fuel, fodder, and food. Forests also helped control the floods that sweep through the region every monsoon season. In the 1950s, commercial lumbering entered the regions, felling the forests. When the area was reforested at the urging of a central government agency, the result was a monoculture of rapidly growing eucalyptus trees that produced lumber but no fruit, little fodder, and few twigs or mushrooms for fuel. By the 1960s, floods in the lower Ganges basin became increasingly serious. The reaction was not long in coming, and it came not from the lumber companies or officials of the central government, but as a popular protest movement (the Chipko movement) initially mobilized by women with experience in Gandhi's nonviolent resistance to British rule. They surrounded and literally hugged trees, staying in place until scheduled tree harvest was postponed or canceled. The movement drew on deeply held spiritual and community values as well as survival needs (see Shiva, 1989: 74–75). Gardner and Stern drew two lessons from the Chipko movement: (1) Commercial development does not "privatize" the commons so much as it shifts community resource control to outside agents, and (2) CRM can sometimes yield social, environmental, and even economic benefits far exceeding commercial commodification that experts and central government officials recognize (1996: 145). Reactions to the intrusions of corporate agriculture into the historic systems of family farms are similarly visible in the United States. Popular movements attempting to protect private farmers lobbied farm state legislatures to make it illegal for absentee-owned corporations to own or operate farms and ranches (my state of Nebraska has had such a law for about 10 years now). The final chapter of the political interactions between this movement in relation to the political influence of large agribusiness firms is still uncertain.

Advantages, Limitations, and the Promise of CRM

Gardner and Stern list six advantages and two limitations about the practice of CRM, which I display summarily. The *advantages* are that CRM

1. Builds on long-standing social traditions.
2. Can internalize externalities.
3. Can be effective over very long time periods.
4. Can encourage people to move beyond egoism or selfishness.
5. Has low enforcement costs.
6. Is often the "forgotten" strategy.

But CRM *is limited* because

1. It works best with a limited range of resource types.
2. Social trends often destroy the basis for its successful practice. (1996: 149–150)

The limitations are serious, indeed. They mean that many world environmental problems are not amenable to CRM and that fewer communities have the necessary skills for practicing CRM. Even with these limitations, CRM, or at least its principles, has great promise for dealing with certain environmental problems as part of a mix of strategies. It could perhaps give new operational meaning to the mantra of the environmental movement: "Think globally, act locally."

CONCLUSION

I began this chapter by analyzing how markets, policy, and politics don't automatically work well to promote sustainability and improve the human-environment relationship. I ended it on a more positive note by providing some illustrations about how they have or are producing that change and a discussion of CRM, which is an antidote to the feeling that we are all somehow forever trapped in the tragedy of the commons. Indeed, it is unthinkable to try to improve the human-environment connections without utilizing the powerful tools of market incentives and public policy. But something was missing from this discussion, namely, the impact of ideas and the power of individuals joining forces to advocate change, apply pressure on governments, or change markets by selective buying or economic boycotts. A close reading of the chapter will find them here, in the form of the schoolchildren who wrote letters to the H. J. Heinz company, the influential Greens of Western Europe, the Chipko movement in India, and many others. This chapter focused only on the structural side of the agency-structure dialectic of change outlined in Chapter Seven. The next chapter takes up the agency side in the various forms of environmentalism.

PERSONAL CONNECTIONS

Implications and Questions

The rational-choice perspective suggests that you do make choices that maximize benefits and minimize costs. Here are some questions to help explore this model in terms of some of the ordinary choices that people make.

1. Earlier I argued that there were some benefits in living close to work. What are some of its costs? What are some costs of living in the suburbs and driving or commuting miles and miles to work? What are some of the benefits? Include in your consideration not only the dollar costs of transportation or the environmental impacts (which is something most people never think about), but things like the social quality of life in various neighborhoods. Are there places close to

where people work that they would not like to live and would bear large costs to avoid? As you can see, deciding what is a net rational choice is not so simple.

2. Many have noted that convenience meals are very expensive per unit price, wrapped in layers of packaging that took an enormous amount of material and energy to produce, and perhaps laced with fat, sugar, salt, chemicals, preservatives, and dyes that make their nutritional and health value questionable. Even knowing this, are there times when the benefits of eating them outweigh the benefits of healthier food? Again, consider costs and benefits broadly: money, costs imposed by job routines, family roles, time constraints, and market availability. Alternately, consider the costs and benefits of cooking the way most nutritionists and environmentalists advocate: buying unprocessed food in larger quantities and cooking as much from scratch as possible.
3. Members of a voluntary organization (a church) once asked me what they could do to increase environmental awareness among their members. They did some things: They insulated the building and didn't heat or cool all of it all the time, they established a paper recycling program and made some utility efficiency improvements, and they featured environmental matters (sometimes) in congregational educational programs. They had a coffee fellowship after services and sometimes served large meals to various groups, often using styrofoam cups, plates, and plastic utensils in voluminous amounts. I suggested that they stop using plastic cups and plates and use ceramic ones and reusable utensils (sometimes they did), and, failing that, that they at least replace styrofoam with recyclable paper cups, even though this gesture was more symbolic that substantive. But maybe that was okay. After checking around town, they reported that an alternative to styrofoam for hot drinks and dishes was not available in local stores except at triple the price. I pointed out that they could order such goods from special environmental goods mail-order catalogues, but again at several times the price. They either had to hire someone for every event to wash an enormous load of cups, dishes, and so forth, stop having coffee fellowships and congregate meals—which had important social functions for the organization—or pay a much higher price for environmentally recyclable goods. *They did the organizationally rational thing. They dropped the whole matter.* To quote Kermit the Frog, "Sometimes it's not easy bein' green"!

   Think of your own examples. There are many things you could do to be more environmentally frugal. Why do they seem difficult? It is easy to talk glibly about changing lifestyles, but this is often difficult for us to do, even when we want to. What are some of the reasons why?
4. You can see the complexities of the rational-choice perspective in action. Some argue that regulatory strategies are indeed necessary for environmental protection, occupational and safety standards, health, social justice, and many other concerns. The National Environmental Policy Act that created the EPA revolutionized the American way of thinking about regulatory policy. Think about this fact concretely. How has your life been impacted, negatively or positively, by environmental or occupational regulation? Talk to some other people for their perspectives: city officials, university administrators, homemakers, your relatives, or small business owners. You will find that hardly anyone likes such regulations. But how do opinions differ about whether they are necessary or not? What do you think shapes divergent opinions?

What You Can Do

If you work for a company or are a student at a college or university, you could help stimulate and conduct a company or campus-wide environmental audit. Such an audit considers all operations and products and includes detailed strategies for making improvements. It can pinpoint opportunities to minimize waste generation, reduce water and air pollution, and conserve energy and water. It can reveal whether a company is investing its financial resources in companies that pollute and suggest alternative investment options. An audit may help company managers or university officials find ways to change their procurement practices so that they buy less plastic and more recycled paper, glass, and aluminum, and more locally or regionally produced materials and services. Such an audit can become the basis for a changed company or campus environmental blueprint (one did at UCLA in 1989). *Caution*: You obviously need people with a great deal of technical expertise to do a meaningful environmental audit, and unless you have this *and* the support or cooperation of people in positions of authority, you will simply be viewed as a bothersome problem they have to deal with! Some U.S. companies shun environmental audits out of fear of lawsuits or prosecution by state and federal environmental agencies. Several states, however, have passed laws protecting the results of a company's voluntary environmental audits as privileged information. Student groups have helped stimulate environmental audits. Administrators and trustees usually sanction well-done audits because while they may suggest initial costs, they save the institution money in the long run (MacEachern, 1990: 138). A step-by-step guidebook to conducting a campus audit is available from the people who organize the Earth Day observances each year. (The last address I had for them was Earth Day, P.O. Box A.A., Stanford, CA 94309. That may have changed. But you can find them.)

Environmental careers? There will be jobs for people with environmental expertise in government, private nonprofit organizations, and companies. There will be opportunities for people with scientific and engineering backgrounds, but also for people with environmental interests combined with backgrounds in other fields, such as business, policy studies, law, the social sciences, ethics, and journalism. An incredible variety of careers that involve environmental and ecological issues exist. Here are just a few:

- *Scientific fields*: environmental health and toxicology, environmental geology, ecology, chemistry, climatology, biology, air and water quality control, solid waste management, energy analysis, energy conservation, renewable energy technologies, agronomy, urban and rural land-use planning, atmospheric science.
- *Resource and land management careers*: sustainable forestry and range management, parks and recreation, fishery and wildlife conservation management, conservation biology.

- *Engineering and architecture*: environmental eningeering, solid and hazardous waste management, environmental design and architecture, product and appliance engineering.
- *Humanities, social sciences, and other fields*: environmental law, law enforcement, policy, consulting, social science, and communications, risk analysis, risk management, demography (population dynamics), environmental economics, psychology, or sociology, environmental communications and journalism, environmental marketing, environmental policy, international diplomacy, public relations, activism, lobbying, and environmental writing and journalism.

For more information, consult Moody and Wizansky (eds.), *Earth Works: Nationwide Guide to Green Jobs* (San Francisco: Harper-Collins-West, 1994), the *Environmental Career Directory* (Detroit, MI: Visible Ink Press, 1993), and publications like *Environmental Career Opportunities* (Bruback Publishing Co., Box 15629, Chevy Chase, MD 20825, tel: 301-986-5545), and *EcoNet* (Institute for Global Communication, 18 DeBoom St., San Francisco, CA 94107, tel: 415-422-0220).

ENDNOTES

1. They were: Brazil, Bolivia, Canada (British Columbia), China, Costa Rica, Ecuador, El Salvador, Ethiopia, Haiti, Honduras, India, Indonesia, Kenya, Mali, Myanmar, Nicaragua, Niger, Nigeria, Papua New Guinea, Peru, the Philippines, Rwanda, Senegal, Somalia, Sudan, Surinam, Thailand, Venezuela, and Zaire (Renner, 1997: 18–19).
2. Such proposals have their critics, who argue that while they might address common problems by creating efficiencies and responsibility of something like private ownership (for the rights to pollute), they also have the same problems with inequality in the distribution of resources that markets do. Relatively wealthy companies could afford to buy the emission permits of more financially strapped ones, but LDCs are deeply suspicious of such proposals, fearing that wealthy MDCs would be tempted to buy the carbon emission rights of poorer LDCs, thus further preventing their economic development.
3. You may be wondering why in the world the federal government would pass laws for subsidies resulting in such destruction of rangelands. The earliest such legislation was enacted in the 1870s to support the settlement and economic prosperity of the dry prairies in western states, which have most of the federally owned land. The same principle gave government subsidies to mining and lumbering companies on government land.
4. Using material or energy produced in one part of production process to serve another, thereby reducing the need to acquire the "virgin materials" of fuels in the total production process.
5. Different criteria include (1) *no unreasonable risk*, as in the regulations in the Food, Drug, and Cosmetic Act; (2) *no risk*, such as the Delaney clause in that act, which prohibits the deliberate use of any food additive shown to cause cancer in test

animals, or the zero discharge goals of the Clean Water Act; (3) *risk-benefit balancing*, such as the regulations that govern the use of pesticides; (4) *standards based on best available technology*, as those embodied in the Clean Air Act, and (5) *cost-benefit balancing*, such as Executive Order 1229, which gives the Office of Management and Budget the power to delay indefinitely, and in some cases veto, any federal regulation not proven to have the least costs to society. All these criteria have been strongly criticized, for reasons similar to those I discussed in Chapter Four.

CHAPTER NINE

Environmentalism: Ideology and Collective Action

Lois Gibbs was a housewife and president of the local neighborhood association of Love Canal, a working-class suburb of Niagara Falls, New York, and she was mad as hell. In the 1970s, she and her neighbors and been complaining to local officials about strange smelly chemicals leaking into their basements, gardens, and storm sewers. Local officials listened but ignored their complaints. Children playing on school grounds and around the old canal got strange chemical burns. The old canal for which the subdivision was named, long deserted by barge traffic, was used by the Hooker Plastic and Chemicals Company as a dumping ground for toxic chemical wastes. Between 1942 and 1953, the company dumped more than 20,000 metric tons of wastes into the canal, mostly in steel drums. In 1953, the company covered the dump site with clay and topsoil and sold it to the Niagara Falls school board for $1 in a sales agreement that specified that the company would have no future liability for injury or property damage caused by the dump's contents. Eventually an elementary school and housing project with 949 homes were built in the 10-square-block Love Canal area.

Informal health surveys conducted by alarmed residents, led by Lois Gibbs, revealed an unusually high incidence of birth defects, miscarriages, assorted cancers, and nerve, respiratory, and kidney disorders among residents. Again, complaints to local officials had little effect. But continued pressure from local residents led New York State officials to conduct more systematic health and environmental surveys, which confirmed the suspicions of the residents (miscarriages were four times higher than normal). They found that the air, water, and soil of the area, as well as the basements of houses, were badly contaminated with toxic and carcinogenic chemicals. In 1978, the state closed the school and relocated more than 200 families living closest to the dump. After outraged protests from the remaining residents and investigations by the Environmental Protection Agency (EPA), President Jimmy Carter declared Love Canal a federal disaster area and relocated all families who wanted to move. About 45 families remained, unwilling or unable to sell their houses to New York State and move. In 1985, former residents received payments from an out-of-court settlement from Occidental Chemical Corporation (which had bought Hooker in 1968), from the city of Niagara Falls, and from the school board. Payments ranged from $2,000 to $400,000 for claims of injuries ranging from persistent rashes and migraine headaches to severe mental retardation. By 1988, a U.S. District Court ruled that Occidental Chemical must pay cleanup costs and relocation costs, which had reached $250 million, but the company appealed that ruling.

Ironically, the dumpsite was covered with a clay cap and surrounded by a drain system that pumps leaking wastes into a treatment plant. By 1990, the Environmental Protection Agency renamed the area Black Creek Village and proposed a sale of the 236 remaining dilapidated houses at 20 percent below market value. However, several environmental organizations filed a federal complaint against the EPA for failing to conduct a health risk survey before moving people back into the Love Canal area. By that time Lois Gibbs had gone on to found the Citizens' Clearinghouse for Hazardous Wastes, an organization that has helped more than 7,000 citizens environmental organizations. About the effort to relocate people in the old Love Canal subdivision, she said, "It would be criminal. . . . It isn't a matter of if the dump will leak again, but when" (cited in Miller, 1992: 561).

About the same time that Lois Gibbs and her neighbors were enraged about the lack of official responsibility for Love Canal, a young Zapotec (one of the Native American tribes in south Mexico) named Eucario Angeles returned from his university education to his home in a rural area south of Oaxaca. It was a tangle of canyons where the Zapotec people eked out a meager existence on parched and eroded soils. National development efforts like roads and electricity had passed them by. Eucario began talking with people in local communities. What were their problems and priorities? Over the weeks of discussion among local residents, a consensus emerged: They should dig ponds at the springs to store their scarce water supply.

Residents assembled work parties, which quickly excavated two rudimentary ponds. Then one thing followed another. A few minnows whimsically thrown in a pool unexpectedly multiplied, which reminded someone that a visitor had once said something about farming fish. Eucario went to town to find out what he could and tracked down the Secretariat of Fisheries. There aquaculture experts supplied him with elaborate specifications for regulation ponds but advised him that uneducated Indians would never succeed. Undaunted, Zapotec work parties set to digging. Despite geological conditions that quickly ruled out the standardized government designs, workers managed to construct an odd assortment of irregular pools. A year later, tired of waiting for a government inspector to bring them the promised fingerlings, Eucario went again to the city, where he convinced the secretariat to bend the rules and give him a plastic bag continuing 175 young tilapia and carp.

By June 1987, when American anthropologist Mac Chapin visited, there were 20 ponds brimming with fish, water supplies were secure the year round, and the risk of crop losses had been reduced with irrigation water conducted through garden hoses. Most impressive, the Zapotecs had organized intricate rotating work schedules for feeding the fish, maintaining the ponds, regulating water flow rates, and harvesting a sustainable yield (Chapin, 1988, cited in Durning, 1989: 17–18).

What do these two illustrations have in common? Two things, really. The *first* is that they concretely illustrate the elements of social change discussed in Chapter Seven: change as concrete and practical outcomes (praxis) of the interaction of the purposive actions of individuals and encountered structures. The *second* is that they both illustrate one type (grassroots) of mobilization for action at its best: In a setting with longstanding problems, a committed organizer and activist arrives on the scene unburdened with blueprints for change and mobilizes the latent talents of community members. From the beginning the community controls the process, but they eventually interact with an expanding web of larger organizations, bureaucrats, experts, and politicians.

This chapter is about such collective action and human agency in change as it relates to transforming the human-environment relationship. It is, in other words, about *environmentalism.* In this chapter I will briefly define environmentalism ideology and collective action (or social movements). Then the chapter will discuss three broad topics: (1) the varieties of American environmentalism, (2) environmentalism and change, and (3) how "successful" environmentalism has been to date. The final chapter will discuss some global dimensions of environmentalism.

Environmentalism is both ideology and action. As *ideology,* it is a broad set of beliefs about the desirability and possibility of changing the human relationship with the environment. I am not using the term *ideology* in a negative or pejorative sense, but merely to denote a set of beliefs about action surrounding an important human activity and perceived set of problems. You could talk equally, as social scientists often do, about the ideologies that surround and justify the operation of economic markets (free-market capitalism) or political systems (democracy, liberalism, conservatism). Environmentalism is rooted in the worldviews, social paradigms, and cognized environments of people discussed earlier, particularly in Chapter Two. But unlike these systems, ideologies are not only abstract beliefs and "models about how the world works." They are beliefs that are used, often deliberately, to justify a desire for change. Environmentalists have produced a social, economic, and philosophical literature of remarkable breadth, depth, and variety that has significantly shaped the political and administrative agendas—if not the actual operation—of many nations of the world.

Although environmentalist ideologies cannot be easily located on a political left-right ideological spectrum, thinking about environmental change must involve thinking about distributional issues that are inherently political, as I hope this book has richly illustrated (Paehlke, 1989: 273).

Environmentalism is also purposive action intended to change the way people relate to the environment. It includes individual purposive action, but more significantly, it means the *collective action* of many individuals as they form groups and organizations intended to transform the way communities, companies, and societies impact their environments. In other words, collective action results in environmental movements. Social movements emerge when problems are defined and framed ideologically to mobilize people in collective action (Snow and Benford, 1988). How does this happen?

A vast literature exists seeking to explain how and when ideology and collective action coalesce as movements for change. There are *social psychological* and *individual-level explanations* that variously emphasize individual psychological problems, relative deprivation, or rational choice in problem solving to explain individual participation in social movements (Harper, 1998: 117–118). But social scientists are more interested in *social-level perspectives* that explain the emergence and significance of movements in society. The earliest of these explanations emphasized the social conditions of modern societies, said to include social disorganization, the rootlessness and mobility of people, and widespread feelings of meaninglessness and powerlessness—*alienation*—of people within society. More recent perspectives emphasize (1) *resource mobilization*, or the presence or absence of community resources which facilitate movement mobilization; (2) *political processes*, such as the importance of shifting political coalitions, opportunities, and encouragement from political elites; and (3) *interpretive* or *constructionist* approaches that focus on the role of ideology. This third perspective emphasizes how problems are defined (or "framed") and popular discourse about problems that constitute the basis for collective action. (For more about perspectives on social movements, see Brulle, 2000; Burton, 1985; Harper, 1998; McAdam and Michaelson, 1994; Oberschall, 1973; and Snow and Benford, 1992.)

As you might guess, these perspectives about conditions in which movements emerge are not mutually exclusive. Researchers found, for example, that relative deprivation,[1] political alienation, and resource availability all had a role in the emergence of community protests about a nuclear waste facilities in America (Kowalewski and Porter, 1992).

AMERICAN ENVIRONMENTALISM

American environmentalism bloomed and changed in the 1960s, but it was the product of over 100 years of collective action and movement organizations that existed in particular historical circumstances. As it developed,

American environmentalism involved not only historically specific organizations, but also different ways of framing environmental problems and different discourses about them, both among environmental movement activists and broader arenas of public discourse (the media and political process). Sociologist Robert Brulle identified eight different environmental discourses that shaped different waves and competing manifestations of environmentalism throughout U.S. history. I list them summarily, in their chronological order, along with an illustrative environmental movement organization (Brulle, 2000).

- *Preservation* (1830s): Nature is important to support both the physical and spiritual life of humans, hence the continued existence of wilderness and wildlife undisturbed by human action is necessary (Wilderness Society, Sierra Club).
- *Conservation* (1860s): Natural resources should be scientifically managed from a utilitarian perspective to provide for the greatest good for people over the longest period of time (Society of American Foresters).
- *Wildlife management* (1890s): The scientific management of ecosystems can ensure stable populations of wildlife, viewed as a crop from which excess populations can be harvested, particularly in recreation and sport (Ducks Unlimited).
- *Reform environmentalism* (1870s, but really flourished in the 1960s): Human health is linked to ecosystem conditions like water quality and air pollution. To maintain a healthy human society, ecologically responsible actions are required, which can be developed and implemented through the natural sciences (Environmental Defense Fund).
- *Environmental justice* (1970s): Ecological problems exist because of the structure of society and its imperatives, and the benefits of environmental exploitation accrue to the wealthy while the poor and marginal bear most of the costs. Hence the resolution of environmental problems requires fundamental social change (Citizen's Clearinghouse for Hazardous Waste).
- *Deep ecology* (1980s): The richness and diversity of life has intrinsic values, so human life is privileged only to the extent of satisfying basic needs. Maintenance of biodiversity requires decreasing the human impact (Earth First!).
- *Ecofeminism* (1980s): Ecosystem abuse is rooted in androcentric ideas and institutions. Relations of complementarity rather than domination are required to resolve conflicts between culture/nature, human/nonhuman, and male/female relationships (World Women in Development and Environment).
- *Ecospiritualism* (1990s): Nature is God's creation, and humans have a moral obligation to keep and tend the creation, including biodiversity and unpolluted ecosystems (National Council of Churches, as well as most denominational bodies).

It is important to mention another human-environment ideological frame, virtually unchallenged in its domination of American environmental discourse from 1620 until the middle of the nineteenth century, that Brulle terms *manifest destiny*. It is a moral and economic rationale for exploiting natural resources, assuming that nature has no intrinsic value, that human welfare depends on the exploitation and development of nature, and that human inventiveness and technology can transcend any resource problem. In effect, the resources of nature are infinitely abundant for human use (Brulle, 2000: 76). We are not exactly on new ground here. He is talking about what I described as the dominant social paradigm for human-environment relations in industrial societies (Chapter Two), as well as what other environmental sociologists termed the evolving *human exceptionalism* paradigm in economics and sociology (Dunlap, 1992). In addition to providing the rationale for the economic development of the North American continent, the concept of manifest destiny continues to serve as the discourse of several waves of countermovements opposed to the goals of environmental movements. Such countermovements share many objects of concern as the movements they oppose and make competing claims on the state and vie for attention from the mass media and the broader public (Meyer and Staggenborg, 1996: 1632).

American Environmental Movements, 1870–1950

Preservation and conservation movements were the first real manifestation of American environmentalism, and they both foreshadowed many contemporary concerns. The swift destruction of America's forests and wilderness in the late nineteenth century by the lumber industry was the greatest public concern. Devastating environmental catastrophes turned public opinion against the cutting of large stands of trees. Cutting left pollution from residual bark, branches, and other waste. Worse, it surrounded small hamlets throughout the country with a virtual tinderbox. Approximately 1,500 persons died and 1,300,000 acres of land were burned in a Wisconsin fire in 1871. Related community disasters, such as the famous Johnstown, Pennsylvania, flood, were attributed to clearcutting, because clear-cut soil does not hold water (Humphrey and Buttel, 1982: 114). Such wanton environmental destruction of America's forests and rangeland produced a broad-based effort to curb the abuses of private ownership and to institute "scientific management" of the nation's environmental resources. There were many individual leaders in this movement (called Progressives or Reformers, as were many leaders for political change in that era). Three were particularly remembered for their influence: President Theodore Roosevelt, John Muir, and Gifford Pinchot. They mobilized public support for conservation and created organizations such as the Sierra Club (founded in 1892 by Muir), the Audubon Society (1905), and many outdoor recreation clubs, such as the Boone and Crockett Club (founded by Theodore Roosevelt).

Environmentalism was given intellectual and ideological shape by the writings of three persons. The first was *George Perkins Marsh* (1801–1882), whose work *Man and Nature: Physical Geography as Modified by Human Action* identified the negative impact of human economic activity on forests and rangeland. It documented the connections between cutting of forests and the erosion of soil, between the draining of marshes and lakes and the decline of animal life, between the forced decline of one species and alterations in the population of others, and even between human activity and climate. Marsh's eerily prescient ecological view is all the more remarkable because it was published in 1864, *before* the automobile, the significant use of oil, and the mechanized clearing of forests or modern mining that were to come (Paehlke, 1989: 15). *John Muir* (1838–1914) reacted angrily toward the anthropocentrism of those who saw humans as above nature. Nature and wilderness were a spiritual experience, and he saw people, at their best, as part of that spiritual whole. Both politically and intellectually, Muir campaigned tirelessly for the preservation of wilderness areas from human intrusion. For him, the notion that the world was made especially for the uses of man was an enormous conceit (Nash, 1967: 131). *Aldo Leopold* (1886–1948) agreed, but his intellectual achievement was a blending of ecology and ethics. He saw the land itself as a living organism. People, he noted, are the only species that can threaten nature as a whole. If we do so, we will, of course, destroy ourselves. Leopold also pointed out that while most humans imagine that they are sustained by economy and industry, these are in turn sustained, as are all living things, by the land. We are therefore but one part of an interactive global ecosystem, and we injure the land at our own peril (Nash, 1967: 182). In short, the intellectual and ideological basis of contemporary environmentalism was well underway in the latter half of the nineteenth century.

The appeal of *conservationism,* as it was termed, was strongest among the upper and upper-middle classes, who were most concerned about outdoor recreation, the shrinkage of the public domain, and the destruction of forests. Conservationists sought to use the legal and political power of the state to protect forest lands from exploitation, resulting in, for example, the Yellowstone Act (1982), the Adirondack Forest Preserve (1885), and legislation to preserve Yosemite (1890) and Mount Rainier Parks (1890, 1899). Such efforts came to be effectively organized by national movement organizations such as Audubon Society and particularly by Muir and the Sierra Club (Humphrey and Buttel, 1982: 113–114).

But government officials constantly struggled to balance two different public interests. Organizations such as the Sierra Club and the Audubon Society urged the *preservation of wilderness,* with a minimum of human use for scientific, aesthetic, and "nonconsumptive" recreational use. Others, such as hunters and fishers as well as large ranching, mining and timber commercial interests, argued for the *utilitarian use of natural resources* subject to "scientific management." The second interest came to be spearheaded by Gifford Pinchot, a private forestry manager on the Vanderbilt Estate in North Carolina. The U.S. Department of Agriculture formed a Forestry Division, helped Congress to pass the Forest Reserve Act in 1891, and hired Pinchot to study the possibilities of the scientific management of forests. He was appointed chief of the Division of Forestry, and his combination of technical and political skills enabled him to form a close relationship with President Theodore Roosevelt, whose domestic policy advocated the "wise use" of natural resources. Pinchot proved politically far more astute than Muir. In short, *the utilitarians won a decisive political victory over the preservationists.* Such policies enabled commercial interests to use public lands, subject to government regulation. They did protect natural resources, but they also reinforced and rationalized the exploitation of public lands by lumber companies and ranchers (Hays, 1959).

After World War I, the United States was confronted with massive environmental calamities such as flooding and soil erosion in the Great Plains "Dust Bowl" as well as by the Great Depression. A second wave of conservationism that developed during the Franklin Roosevelt administration emphasized both protecting and developing natural resources. New Deal programs such as the Civilian Conservation Corps and the Tennessee Valley Authority worked to protect natural resources as well as to stimulate economic recovery. In the 1950s, more emphasis was placed on preservation of natural beauty and wilderness for public enjoyment. This "wilderness movement," spearheaded by older organizations such as the Sierra Club, developed highly publicized campaigns to save the Grand Canyon and Dinosaur National Monument (McCloskey, 1972; Dunlap and Mertig, 1992: 2).

Contemporary Environmentalism

By the 1950s, conservationism was an established social force in American life. The 1970s transformed it into a different and greatly expanded environmental movement. This movement, often called *reform environmentalism,* was a complex system of ideas with origins in two different areas: the utilitarian philosophy of providing for the common good through the application of science and law to public problems, and neo-Malthusian ideas about the crisis of overpopulation. These two perspectives form different components of the current environmental movement (Brulle, 2000: 118). Reform environmentalism was not simply an amplification of conservationism; the transformed environmental discourse viewed problems as (1) being more complex in origin, often stemming from new technologies; (2) having delayed, complex, and difficult-to-detect effects; and (3) having consequences for human health and well-being as well as for natural systems. Because they encompassed both pollution and loss of recreational and aesthetic resources, environmental problems were increasingly viewed as threats to the total quality of life (Dunlap and Mertig, 1992: 2–3; Mitchell, 1980; Hays, 1987).

Like earlier movements, the new American environmentalism had important intellectual and ideological foundations. The first was *Silent Spring* (1962) by marine biologist Rachel Carson—an angry and uncompromising analysis of the toxic effects of modern pesticides on every form of wildlife. Carson focused on the politics of science and the exclusion of the public from knowing what risks they were being exposed to by the development and use of synthetic chemicals. This book reunited the conservationist and preservationist concerns about natural ecosystems with long dormant public health concerns (Brulle, 2000: 125). *Silent Spring* made best-seller lists and sold over a million copies—rare for a serious nonfiction book. Indicative of its impact, the American pesticide industry mounted a $250,000 campaign to prove Carson a "hysterical fool." Carson's work enhanced public awareness of the ecological impact of pesticides, and that awareness helped pass the Pesticide Control Act of 1972 (Sale, 1993: 4).[2] Carson's work really put the issue of pollution on the environmental agenda.

In 1968, zoologist Garret Hardin rediscovered Malthusian ideas in his famous essay *The Tragedy of the Commons* (discussed in Chapter Eight). A more popular and influential work was zoologist Paul Ehrlich's *The Population Bomb: Population Control or Race to Oblivion?* (1968), which forced the issue of overpopulation into public consciousness in an apocalyptic way, claiming that "the battle to feed all humanity is over." Ehrlich's neo-Malthusian work proved to be the most popular environmental book ever, selling over 3 million copies in the first decade. Biologist Barry Commoner, the most durable, most political, and most intellectually sophisticated ideologue of the environmental movement, produced a series of widely read books that publicized the hazards of nuclear wastes and chemical pollution (including *The Closing Circle,* 1971). Commoner argued that the greatest threat to the environment was not population growth per se, but modern technology and the power of corporations that promote it (Humphrey and Buttel, 1982: 122). Commoner's mix of advocacy and science was so influential that *Time* magazine recognized him as the "Paul Revere of Ecology."

Environmental events themselves, when publicized by the media, broadened public awareness of problems. In New York City, eighty people died from smog during an air inversion in the summer of 1966. An offshore oil rig near Santa Barbara poured undetermined millions of gallons of oil along the California coastline in January and February of 1969, killing wildlife and soaking beaches with black, oily goo. The industrially polluted Cuyahoga River near Cleveland burst into flames, and in the summer of 1969 nearby Lake Erie was declared a dying sinkhole as a result of sewage and chemicals pollutants. As the decade of the 1960s wore on, the mainstream media made environmental events high-visibility ones. This attention reached a crescendo by 1970 with a spate of front-page articles and cover stories in *Time, Fortune, Newsweek, Life, Look,* the *New York Times,* and the *Washington Post. Ecology* became a word known—if incompletely understood—by the average citizen. Public outcry about such environmental abuses was loud and widespread. Many were no longer willing to accept pollution and environmental disruption as business as usual and complained about the businesses that produced them and the governments which failed to protect against them (Sale, 1993: 19–25).

The event that symbolized this effervescence of environmental consciousness and activism was *Earth Day 1970*. The idea for this observance began with Senator Gaylord Nelson, who proposed a kind of nationwide environmental teach-in on college campuses, following the model of the 1960s antiwar teach-ins about the Vietnam War. The popular response was overwhelming, and he received a federal grant and support from government agencies (e.g., the Interior Department) to organize the event in spite of opposition from the Nixon administration. Campus activists, then at peak antiwar activism, enthusiastically supported the idea. There are no precise records, but Earth Day 1970 was probably the largest demonstration of the decade. Organizers estimated that 1,500 colleges participated, and in New York, Washington, and San Francisco there were large rallies and street parades. *Time* magazine estimated that some 20 million people took part in what Nelson called "truly an astonishing grass-roots event." There were those who denounced Earth Day 1970 as a subversive communist plot. But the fact that some corporations and established leaders supported it led others to see it—using the rhetoric of the 1960s—as subtle manipulation by the establishment to promote a more efficient rape of resources. Nonetheless, Earth Day 1970 was a surprising demonstration of the depth of feeling about environmentalism at that time (Sale, 1993: 34–35).

The immense political momentum of the environmental movement continued to build. In 1972, the Apollo 17 Crew took a series of photographs of the earth from 22,000 miles away. One of these is on page 305 of this book and has become an icon of the environmental movement. Using the metaphor "spaceship earth," this picture came to represent a fragile earth with a finite limit and delicate natural balance—to which the fate of humanity is collectively linked (Brulle, 2000: 127).

Older national conservation organizations were invigorated by this upsurge of environmental consciousness as they attempted to incorporate it. They did so by adapting their historic concerns to new ones. The National Audubon Society, for instance, enthusiastically supported antipesticide campaigns, while the National Wildlife Federation began a legal challenge to polluters (Sale, 1993: 19–20). But older organizations could not adequately respond to or contain newer concerns. The founding of the Environmental Defense Fund (EDF) in 1967 marked the beginning of a new wave of environmental movement organizations based on the environmental discourse whose appearance was marked by Rachel Carson's book. EDF action methods, using scientific research and legal action to protect the environment and human health, served as an exemplar for many newer organizations, including the Environmental Defense League (1968), the Natural Resources Defense Council (1960), Zero Population Growth (1969), and Friends of the Earth (1969) (Brulle, 2000: 126). To make the mixture even headier, new local and regional organizations were established, such as the Save-the-Redwoods League (California), Citizens for a Better Environment, and the Adirondack Council (New York) (Schnaiberg and Gould, 1994: 149).

Reform Environmentalism and the Environmental Lobby

What emerged was a national network of transformed older and newer environmental movement organizations that dominated the movement's presence in Washington. Their interests and strategies differed: Some engaged in pro-environment lobbying; some developed the expertise and scientific capability for educational programs and advocacy research; some specialized in litigation to shape the development and enforcement of environmental policy; some purchased land to set aside for wilderness preserves. Among these organizations, an informal coalition of lobbying groups called the "Group of 10" met periodically to discuss common strategies and problems. Other organizations worked with this core coalition. See Table 9.1, which suggests the growth of this network.

As you can see from Table 9.1, since 1960 most national organizations experienced tremendous growth. It came in four spurts during which there was heightened public concern about environmental problems. The *first* was in the years just prior to Earth Day 1970, and the *second* resulted from the pub-

Table 9.1 National Environmental Lobbying Organizations

Organization	*Year Founded*	*Membership (thousands)*		*Budget ($ million)*
		1960	*1990*	
Sierra Club	1892	15	560	35.2
National Audubon Society	1905	32	600	35.0
National Parks and Conservation Association	1919	15	100	3.4
Izaak Walton League	1922	51	100	1.4
The Wilderness Society	1935	10	370	17.3
National Wildlife Federation	1936	NA	975	87.2
Defenders of Wildlife	1947	NA	80	4.6
Environmental Defense Fund	1967	—	150	12.9
Friends of the Earth	1969	—	30	3.1
Natural Resources Defense Council	1970	—	168	16.0
Environmental Action	1972	—	20	1.2
	Total	123	3,153	217.3

Source: Adapted from Mitchell et al., 1992: 13.

licity surrounding Earth Day. Between 1960 and 1972, national environmental organizations grew in membership by 38 percent. Though the rate of growth slowed during the 1970s, significant growth continued, confounding those who saw the movement as a fad. Ironically, most observers believe that attacks of the conservative Reagan administration on environmentalism stimulated a *third* wave of growth during the 1980s. The Sierra Club grew by 90 percent, the Defenders of Wildlife and Friends of the Earth each grew by 40 percent, and the Wilderness Society grew by a whopping 144 percent. A *fourth* surge in membership occurred at the turn of the 1990s, stimulated by the visibility of ecological problems, such as toxic waste problems, beach contamination, the Exxon *Valdez* oil spill, and global warming. It was also stimulated by the publicity and mobilization efforts surrounding continuing Earth Day celebrations. All told, the national movement organizations claimed more than 3 million members (or at least "checkbook supporters") by the 1990s, and 9 million in 2000 (including preservation, wildlife management, and conservation groups) (Mitchell et al., 1992: 2–3; Brulle, 2000: table 10.3).

Reform Environmentalism, Public Opinion, and Legislation

The mobilization of reform environmentalism channeled and amplified environmental awareness and concerns among broad segments of the population. For instance, national public opinion poll trend data between 1965 and 1970 demonstrate a growing willingness to define air and water quality as significant problems. Increasingly, they were seen as (1) deserving government attention (from 17 percent to 53 percent), (2) serious in respondents' vicinity (from 28 percent to 69 percent), and (3) government spending areas they would *least* like to see cut (from 38 percent to 55 percent). People were even more willing, by smaller increases, to pay modest taxes to address pollution problems (44 percent to 54 percent). The growth of pro-environment attitudes can be illustrated even by the most abstract and contentious environmental issue, global warming. In a 1982 survey, only 12 percent of a national sample saw the greenhouse effect as "very serious," but by 1989 41 percent did so, while another 34 percent said it was "somewhat serious" (Dunlap and Scarce, 1991: 661). As noted in Chapter Four, by 1998 80 percent of Americans (including 79 percent of self-identified independents, 84 percent of Democrats, and 73 percent of Republicans) thought that the MDCs should go forward with the Kyoto treaty, even if not joined by LDCs (Ecology USA, 1998b). In the language of public opinion analysts, by 1970 environmental protection had become a *consensual issue*, meaning that large majorities expressed pro-environmental opinions while only small minorities expressed anti-environmental ones. Even so, public opinion polls did not report *how salient* environmental concerns were. The most reasonable interpretation suggests that a majority of the public had accepted environmentalists' definition of environmental issues as problems and had become sympathetic to environmental protection, but only a minority saw the environment as one of the nation's most important problems (Dunlap, 1992: 92–96; Smith, 1985).

In the 1970s, when Republicans as well as Democrats attempted to govern from the political center, newly elected Republican President Nixon, reflecting the growing national mood, announced that he was an environmentalist and supported the development of legislation to protect the environment. Early results were impressive, including the National Environmental Policy Act (NEPA, 1970), the Federal Water Pollution Control Act (1972), the Coastal Zone Management Act, the Federal Pesticide Control Act, and the Marine Mammal Protection Act (all in 1974). NEPA was the legislation for the creation of the Environmental Protection Agency (EPA), with a broad mandate to consolidate federal programs. The legislation that created the EPA created a far-reaching transformation of the regulatory powers of the government to represent and institutionalize environmentalism within the federal government and was written in a way to prevent the capture of the agency and its programs by businesses and other regulated interests. It required an *Environmental Impact Statement* (EIS) of every federal agency project and had the power to approve or veto projects. Environmental legislation usually included so-called hammer clauses intended to produce strict compliance through mandatory deadlines, explicit and detailed procedural prescriptions, provisions for citizen participation, and citizen legal standing to sue agencies. In fact, almost every major piece of environmental legislation has been challenged in court by industry, environmentalists, or community groups—sometimes simultaneously. But NEPA and the EPA helped to create a transformed new era of administrative law, characterized by the expanding participation of environmental and nontraditional groups in administrative decision making (Rosenbaum, 1989: 214–219; Miller, 1992: 680–681). Environmental legislation continued to be passed in the 1980s, even after the arrival of the conservative Reagan administration in Washington and the growth of more conservative congressional majorities (Nuclear Waste Policy Act, 1982; Clean Water Act, 1987). By 1992, 36 states also had laws or executive orders requiring environmental impacts for state projects. Between 1976 and 1990, Congress passed more than 55 pieces of environmental legislation related to a broad spectrum of environmental, energy, and conservation problems. Importantly, this influence spread to other agencies concerned with environmental issues, such as the Department of Energy, the Bureau of Land Management, and the Nuclear Regulatory Commission.

The Limits of Reform

Reform environmentalism constituted the dominant ideological frame for American environmentalism with the possibility of examining every conceivable environmental or ecological issue. Yet with the exception of the

Montreal Protocol related to ozone depletion, the most dramatic successes of reform environmentalism were in the early years (the 1970s) and its limitations became more apparent by the 1990s. New ecological problems, including the proliferation of endocrine disrupters, biodiversity loss, and global warming, have been identified without any meaningful efforts to address them. Why?

Reform environmentalism was based primarily on the writings of natural and physical scientists. With the exception of Barry Commoner, the vast majority of ecological scientists did not examine the social and political causes of ecological degradation. Therefore, while they may have great competence in their specific areas of expertise, the discourse obscures the social driving forces of environmental degradation (see Chapter Two). The problem is not that the analysis is wrong, but that it is partial. Hence, reform environmentalism has been unable to develop a meaningful political vision of how to create a more sustainable society. Without such a vision, reform environmentalism is politically naive and perhaps irrelevant (Taylor, 1992: 136). Environmental reform takes the form of piecemeal reform efforts, continually mired in technical and legal debates and carried out within a limited community of lawyers and scientists (Brulle, 2000: 130).

Additionally, reform environmentalism fostered practices that, however effectively shaping public attitudes, limited its capacity for political mobilization. By practicing piecemeal science-based reform, reform environmentalist organizations exhibit an oligarchic style because scientists play such a prominent role. Politicians and the public are only bit players—heeding the advice of scientists. There is no real need to involve the public, except for financial support, creating what Brulle calls "astroturf" rather than "grassroots" organizations. Over two-thirds of such organizations have an oligarchic top-down management style (the Sierra Club is a notable exception, a national movement organization that maintains effective grassroots local and regional chapters). Without effective grassroots support, such movement organizations can become distant from the very constituencies they claim to represent and often find their capacity for independent action compromised. In such compromises, they are often manipulatable by corporations, politicians, and foundations—whose support they find essential (Brulle, 2000: 130).

In sum, reform environmentalism has been a potent force in national policy making as it evolved from a loosely coordinated social movement to a cohesive public interest lobby. As it grew from amateur enterprises to organizations run by scientists and lawyers, its political clout was accompanied by conservatizing pressures to play by the rules of the game in the world of Washington, D.C. politics. By the 1980s, simmering tension existed between the nationals and other forms of environmental movement organizations (Mitchell et al., 1992: 24).

Environmental Justice Movements

Reform environmental organizations in their Washington offices took the soft political road of negotiation, compromising with others about the amount of pollution or environmental disruption that was acceptable. But people living directly in polluted communities took a hard political road of confrontation, demanding not that the dumping of hazardous waste be slowed, but that it be stopped (Sale, 1993: 58). Grassroots mobilization expanded dramatically in the 1970s and 1980s. As you might think, precise numbers of people involved in grassroots environmental movement organizations are themselves informed guesses, but by 1989 national networks that work with local groups reported 8,300 existing groups. Local environmental activism was stimulated by clear and present community health hazards rather than by abstract concerns such as protecting wilderness areas or declining biodiveristy. Grassroots organizing has been triggered by toxic waste dumps, radioactive wastes, nuclear plants, and proposals to build garbage incinerators and hazardous waste disposal facilities and a variety of other hazards. Local groups typically document a hazard and link it to a current or potential health problem, such as a cluster of cancer cases or a series of adverse reproductive outcomes. Since corporations usually cause these problems, such environmentalism means redefining environmental hazards as corporate crime. In demanding changes in corporate practices, local activists are inevitably drawn into interaction with public health officials, lawyers, and scientists (Cable and Benson, 1993; Freudenberg and Steinsapir, 1992: 29).

The key organizing frame or discourse for such movements is *environmental justice* in that it integrates both social and ecological concerns more than reform environmentalism. That frame pays particular attention to questions of distributive justice, community empowerment, and democratic accountability. It does not treat the problem of social exploitation as separable from environmental degradation, but instead argues that human societies and the natural environment are inextricably linked, and that the health of one depends on the health of the other (Taylor, 1993: 57). Hence, addressing environmental problems effectively means fundamental social change based on the empowerment of local communities.

Reform movement organizations often found their supporters primarily among white middle-class persons. Environmental justice movement organizations have a much broader and well-developed social base. Since they are particularly impacted by environmental hazards, minorities of all kinds—African Americans, Native Americans, and Latinos—as well as working-class homeowners are drawn into grassroots environmental movement organizations. In Warren County, North Carolina, a primarily African-American group struggled to block dumping of PCB-contaminated soil in a landfill. On the Pine Ridge Native American reservation in South Dakota, the Women of All Red Nations sought to force cleanup of contaminated water and land. In California, Mothers of East Los Angeles organized a Mexican-American community to block construction of an oil pipeline through its neighborhood (Freudenberg, 1984). In Boyd County, Nebraska, one of the poorest rural counties in the state, some local residents struggled in a highly polarized community against siting a low-level nuclear waste facility there. As noted in Chapter Seven, the unequal impact of environmental hazards is pervasive. A large and growing body of empirical studies document this fact. Studies do not agree about the exact sequence of causes by which this happens, but no study failed to document the pervasive presence of unequal hazard impacts (for instance, see Bryant, 1995; Bryant and Mohai, 1992; Bullard, 1990; Daniels and Friedman, 1999; Goldman and Fitton, 1994; Schnaiberg and Gould, 1994: 153; Mitchell et al., 1999).

It should occur to you that environmental justice movements illustrate a basic environmental paradigm (centering on stratification and inequality) discussed several times earlier, particularly in Chapters Five and Seven. German theorist Ulrich Beck, for instance, observes that

> the history of risk distribution shows that, like wealth, risks adhere to the class pattern, only inversely. Wealth accumulates at the top, risks at the bottom. . . . It is especially in the cheaper residential areas for low-income groups near centers of industrial production that are permanently exposed to various pollutants in the air, the water, and the soil. . . . It is not just this social filtering or amplification which produces class specific afflictions. The possibilities and abilities to deal with risks, avoid them or compensate for them are probably unequally divided among the various occupational and educational strata. (1996: 35)

Lois Gibbs, the organizer of the Love Canal Protest, put it in less sociological language:

> [They] all knew that those poisons were in my backyard. . . . And they made a decision . . . because my husband made $10,000 a year, because we were working class people, that it was OK to kill us. (cited in Brulle, 2000: 140)

Thus, minorities and working-class persons are drawn into grassroots environmentalism because environmental hazards are more likely to be located in communities with substantial minority or blue-collar populations. African Americans, for instance, have higher blood levels of carbon monoxide and pesticides, and African-American children have a rate of lead poisoning six times that of white children (National Center for Statistics, 1984; Radford and Drizd, 1982). Not only are they more exposed to environmental hazards, but there is a growing recognition that they were the targets of hazardous and dangerous projects that more affluent communities were able to resist. Their circumstances are, in other words, related to the pattern of economic and political power in society, and for grassroots groups environmental issues become issues about social and racial justice. *Environmental justice* thus involves questions about political power as well as about public health hazards (Capek, 1993; Freudenberg, 1984).

Unlike the national organizations, environmental justice movements depend almost entirely on volunteers. Women tend to be overrepresented in both the membership and leadership of such grassroots organizations. Experienced community activists may become involved, but it is a distinguishing characteristic of grassroots environmental movements that new leaders arise, often housewives with no previous organizing experience. Over time, the effect of this experience was the development of considerable scientific and organizational skills and a transformed political consciousness. The reason for this is the typical process through which grassroots efforts develop. After defining environmental hazards as corporate law violation, local activists usually turn to agencies of the state to enforce or create environmental regulation. In doing so, they find that the state is not a neutral player in the process, often being more responsive to corporate than community interests. Therefore, they challenge the democratic responsiveness, credibility, and effectiveness of officials—from city hall to the EPA (Cable and Benson, 1993). As they enter arenas where more is required than personal experience and anecdotal reports, they seek scientists as allies, but their connections with experts and scientists is often ambivalent. Activists in environmental justice movement organizations learn that both government and science can be used against them, often deflecting and trivializing their claims. The outcome is typically significant skepticism and mistrust of both science and officialdom. They tend to reject the image of science as the "objective pursuit of truth" (Freudenberg and Steinsapir, 1992: 29; Pellow, 1994). Grassroots activists learn quickly, perhaps more quickly than most, that environmental problems are not purely, or even primarily, technical problems.

Environmental justice movement organizations are mobilized by people who are frustrated, morally outraged, and passionate about protecting their homes and families, particularly when they feel that the officials have been unresponsive or that they have been manipulated and outmaneuvered by established powers (as illustrated earlier by the quotation from Lois Gibbs). Given their experience and the scarcity of resources to organize, grassroots movement organizations often move beyond the respectable rules of the game available to reform environmentalism (research, lobbying, electoral efforts). They often use the only methods available to them, including *direct action tactics,* such as picketing the homes of key opposition figures or holding sit-ins that block the construction of new facilities. At Love Canal, Lois Gibbs and other residents held two EPA officials hostage for several hours; two days later, President Jimmy Carter declared Love Canal a disaster area (Gibbs, 1982).

Local environmental justice movements did not stay local. Such grassroots movement organizations need allies, both for scientific and technical expertise and political support. Larger movement structures emerged from networks and coalitions between local movement organizations. Most of these are regional, such as the Grassroots Environmental Organization in New Jersey, Citizens Environmental Coalition in California, Texans United, and the New York Coalition for Alternatives to Pesticides. These educate members about scientific and political issues, provide forums for exchanging experiences, and develop broader policy and advocacy strategies. A small number of national organizations have emerged from grassroots struggles. These include the Citizen's Clearinghouse for Hazardous Wastes, founded in 1981 by Lois Gibbs, and the National Toxic Campaign. These groups organize national conferences, offer leadership training, publish newsletters and manuals, and provide technical assistance to local groups (Freudenberg and Steinsapir, 1992: 30–31).

Grassroots movement mobilization is inherently difficult and fragile to maintain over time, partly because they always wind up fighting city hall in one way or another. Some scholars argue that the most likely outcome of grassroots environmentalism is to be defeated, coopted, or at best achieve victories that are more symbolic than real (Pellow, 1994). Yet without making any judgment about their net effects, Nicholas Freudenberg, who specializes in the study of such grassroots groups, finds many examples of positive achievements.[3] Critics of grassroots movements often charge that their concerns are narrow and self-interested, ignoring broader obligations to society (hence the derogatory acronym NIMBY, "Not in my back yard!"). What the critics of NIMBYism fail to recognize is the substantial contribution to improved public health by such groups. In addition, they fail to recognize that participation in NIMBY groups is often a consciousness-raising experience. Local activists often graduate to broader concerns. What started out as an attempt to clean up the Love Canal wound up being a national toxic waste campaign, and many groups that started by blocking the construction of garbage incinerators become advocates for recycling and waste reduction measures. NIMBY often undergoes a transcendence to NIABY ("Not in *anybody's* back yard!").

Like reform environmentalism, grassroots environmental justice movements have limitations. In spite of what I just said, environmental justice movements may be successful only in particular instances, and governments may be able to resist pressures for more general environmental reforms. Significant societal change in environmental standards requires a coordinated, nationwide coalition of local environmental organizations. Another limitation is that environmental justice is an exclusively anthropocentric discourse. Its concern with nature is limited to examining how ecological degradation affects the human community. Hence environmental justice movements cannot inform a cultural practice that could protect nature or biodiversity outside of human focused utilitarian considerations (Brulle, 2000: 150).

Other Voices: Deep Ecology, Ecofeminism, and Ecotheology

As the limitations of reform environmentalism to resolve environmental problems was becoming apparent by the 1980s, other environmental discourses were taking shape. Even though none has a significant mass base (even an astroturf one), they have shaped the intellectual and ideological texture of contemporary environmentalism. This is true because they were often formulated by intellectuals and scholars or by influential refugees from reform environmentalism. I think that some have a very limited potential for a significant mass base, while others, particularly the last variety, has a considerable potential because of its connections with religious institutions.

Deep Ecology. *Deep ecology*, originally formulated by Norwegian philosopher Arne Naess in the 1970s, was brought to the United States primarily by philosopher George Sessions and sociologist Bill Devall, who coauthored its first popular versions (1985). In contrast to what they view as the "shallow environmentalism" of most of the environmental movement, deep ecology thinkers are *biocentric* or *ecocentric* rather than anthropocentric. Deep ecology emphasizes that (1) the richness and diversity of all life on earth has an intrinsic value, which is threatened by human activities; (2) human life is privileged only to the extent of satisfying vital needs; (3) maintaining biodiversity requires a decrease in human impacts on the natural environment and substantial increases in the wilderness areas of the globe; and therefore (4) economic, technological, and cultural changes are necessary (and perhaps eventual reduction in the human population size) (Devall and Sessions, 1985). Deep ecology also emphases the *"self realization"* of humans as belonging to nature, referring to the process whereby one strives for "organic wholeness" in nature. Those influenced by deep ecology were also determined to reclaim their spiritual identity with nature, some in terms of Buddhist traditions, and some in terms of reviving native "tribal rituals." (Devall, 1992: 56; Sale, 1993: 63).

Although diverse lifestyles and social policies are potentially compatible with deep ecology thinking, its literature emphasizes decentralized and small-scale human communities, self-sufficiency, participatory democracy, and lifestyles that minimize material consumption and maximize the richness of nature. Again, we are not on new ground. Deep ecological thinking was very much influenced by the limits scenario of the future that I discussed in the last chapter, and particularly by the positive possibilities of voluntary simplicity lifestyles. Deep ecologists support the protection of ancient forests and other wild ecosystems, the restoration of biodiversity, and some advocate vegetarianism. Most advocate nonviolent direct action strategies for change of the sort emphasized by Gandhi. They support green consumerism that would minimize the environmental impacts of consumer goods and green politics, meaning the formation of political movements and parties to advocate ecological principles. But most fundamentally for deep ecologists, the path to ecological freedom requires cultivation of an ecological consciousness that permits humans to see through the erroneous and dangerous illusions of Western cultures that justify human dominance over the nonhuman environment (Devall and Sessions, 1985). As complicated and abstract as these ideas are, they gained a diverse following, primarily among intellectuals and activists in the United States as well as in Canada, Australia, and northern Europe. Indeed, many people know about or accept some of these ideas without recognizing the discourse that connects them.

Deep Ecology Organizations and Action. Earth First! (EF!), the most widely known deep ecology organization, was founded by Dave Foreman and a handful of other people disillusioned by their experience in national environmental organizations. EF! advocated civil disobedience combined with *absolute nonviolence* against humans and other living things, and *strategic violence* against "things" such as bulldozers, powerlines, and whaling ships (Miller, 1992: 689). Although not committed to any specific political tactics, speaking in many voices, and disavowing bureaucracy, centralized decision making, sexism, and hierarchy, EF! was a vortex of radical environmental action during the 1980s (Devall, 1992: 57). The founders of EF! were inspired by Edward Abbey's novel *The Monkey Wrench Gang* (1975) and advocated militant tactics in defense of nature, eventually including guerrilla theater, media stunts, and civil disobedience. Unofficially, they included *ecotage* (also called *monkey wrenching*): sabotaging bulldozers and road-building equipment on public lands, pulling up survey stakes, cutting down billboards, and, famously, "spiking" trees at random to prevent their being cut and milled. The advocacy of such tactics alarmed many, but investigations of actual actions largely found EF! not guilty of the most damaging accusations (Sale, 1993: 66).[4]

Never large in comparison with national reform environmentalist organizations, by 1980 EF! had 75 chapters in 24 states, mostly in the Southwest and on the West Coast, and some in Canada and Mexico. As you might guess, advocating such militant tactics made EF! very controversial and elicited considerable opposition. They were described variously as anarchists, ecowarriers, a tribe, a collection of social deviants, ecoterrorists, and visionaries. The group's founder, Dave Forman, said that "from one side have come efforts to mellow us out and sanitize our vices; from another efforts to make us radical in a traditional leftist sense, and there are on-going efforts by the powers that be to wipe us out entirely." Indeed there were. The FBI spent three years and $2 million infiltrating EF!, and in 1989 a trumped-up federal suit charged Foreman with conspiracy for helping to finance the destruction of an electric power tower near Phoenix, Arizona. In 1990, two EF! activists were car bombed in California. Such events led to Foreman's dropping out of the organization, which splintered into several rival groups in the early 1990s (Devall, 1992: 57; Sale, 1993: 57; Miller, 1992: 689–690).

EF! was not the only manifestation of deep ecology. First, it inspired *bioregionalism*, a movement that advocates changing political boundaries of human communities to boundaries defined by ecology. Existing political boundaries were defined by many historical accidents of human settlement and control. A *bioregion* is a geographical area defined by ecological commonalites, including soil characteristics, watersheds, climate, and native plants and animals. Thus this perspective would reframe human existence as a part of a natural ecosystem, not apart from it. Many problems could be understood in an ecological focus, such as dealing with water shortages among communities that are located within the same watershed. Second, deep ecology inspired the formation of the academic discipline of *conservation biology*, understood as the unification of evolutionary biology and ecology with a normative commitment to preserve biodiversity. That field emerged in 1986 at a national conference in Washington, D.C. sponsored by the National Academy of Science and the Smithsonian Institute, attended by 14,000 people. At this conference a group of eminent biologists redefined and publicized the problem of endangered species, coining the term *biodiversity* that I have been using all along (see especially Chapter Three). They did this to spur political action, based on the belief that "humans and other species with which we share the earth are imperiled by an unparalleled ecological crisis"(Takacs, 1996: 9). The leading organization of this movement among scholars is the Society for Conservation Biology. Conservation biologists formed alliances with financiers to create the Foundation for Deep Ecology in 1989 that funds activities in support of rainforest preservation, grassroots activism, and indigenous Third World people's efforts to protect their natural environment from destruction. In 1997, it had assets of over $35 million. By the 1990s, groups of forest activists were trying to "reinvigorate" the Sierra Club by getting it to lobby for a ban on clearcut logging on public lands (known as the Zero Cut Campaign). The proposal was passed by a vote of Sierra Club members nationwide. Whether such efforts will revitalize reform environmental movements is not clear, but the Sierra Club and a number of forest protection groups introduced the National Forest Protection and Restoration Act in the House of Representatives in 1997 (HR 2789). Its purpose is to reverse the policy and legislation that since 1897 has allowed commercial logging in national forests (Brulle, 2000: 134–137).

Limits of Deep Ecology. Deep ecology's core themes (biocentric equality and "self-realization" that widens the human understanding of the self to include the natural world) promote overcoming a narrow egoistic (and anthropomorphic) understanding of self-interest and simultaneously developing a sympathy with other living things. But the discourse is limited because movement organizations shaped by deep ecology do not generally focus on issues outside of the protection of natural areas, which limits their range of issues and concerns. Political actions shaped by deep ecology focus almost exclusively on the defense of wilderness, with virtually no efforts being extended to reform society. Indeed, deep ecology often vacillates between overt hostility toward the human community and vague appeals to extend that community to the broader natural world. These characteristics limit deep ecology's ability to shape alternative political and social practices (Taylor, cited in Brulle, 2000: 138–139). They also, I think, limit its appeal to a large popular base.

Ecofeminism. The ecofeminist discourse is a blend of feminist and ecological thought that that emphasizes conceptual connections between the domination of women by men and the domination of nature. It sets the problems of women and the ecological crisis in a common framework by linking the domination and exploitation of women by men and the exploitive domination of nature (Merchant, 1981). Both are seen as products of a patriarchal society in which *domination* has emerged as a pervasive cultural theme and social paradigm. The Western worldview, with its abstract science and the impulse to control "nature" that deep ecologists hold responsible for the ecological crisis is, in fact, a historical product of the thinking of men in patriarchal societies. But the domination of both nature and women is not caused by generic human nature but rather by specific institutional arrangements developed and controlled by men.

But was this not always so? Not according to ecofeminist scholars. They sought evidence for this view by retrieving the "full" historical record from our commonly understood history (in patriarchal societies, not surprisingly, a history largely written and interpreted by men). Ecofeminist scholars found evidence that European neolithic societies were largely peaceful, minimally stratified, harmonious, and goddess worshipping until the invasion of patriarchal, militaristic, Indo-European pastoralists in the fourth millennium B.C.E. From this turning point in the prehistory of Western civilization, the direction of our cultural evolution was quite literally turned around. The cultural evolution of societies that worshiped the life-generating powers of the universe was interrupted. There appeared invaders on the prehistoric horizon from peripheral areas of the globe who ushered in a very different form of society. Unlike the neolithic communities they conquered, who considered the earth as a mother and left statues representing fertility all over Europe, the invaders worshiped the "lethal power of the blade." In other words, they emphasized the power to take rather than give life as the ultimate power to establish and enforce domination (Gimbutas, 1977; Eisler, 1988). "Domination" is thus a pervasive mindset that applied to other men, women, and to nature. It is not human nature but a worldview with a history.

By providing a plausible—but still arguable—history, ecofeminist prehistorical scholarship provides an evolutionary depth lacking in deep ecology. But even if you don't accept this historical argument, a more indisputable set of facts supports ecofeminist views: the connection among women, development, and environmental disruption in the LDCs. There, economic development projects replaced ecologically sustainable subsistence agriculture by cash crop monocultures that often appropriated and destroyed the natural resource base for subsistence. Although LDC men as well as women suffer from the devastation of their environment, women are the greater losers because as the primary producers of food, water, and fuel women were more likely to lose their livelihood. Moreover, in powerfully patriarchal LDCs, they have less access that do men to land ownership, technology, employment for wages, and small business loans should they desire them. In short, men have been the prime beneficiaries of development, from the colonial era to the contemporary world market system (Shiva, 1988: 1–3; Mies, 1986). In Chapter Five I noted the consensus among demographers that a key to slowing world population growth was to improve the status of women. Ecofeminists argue more broadly that the ecological crisis is inextricably bound with patriarchal domination and the subordinate status of women—both in its long history and around the world today. Furthermore, survey research over the last decades consistently demonstrates that U.S. women are more concerned than men about health threats posed by environmental and technological problems. This has been found not only in the United States, but also around the world. Indeed, when compared with age, class, education and other social variables, gender is the most powerful predictor of environmental concern (Dunlap, 1998: 486).

Ecofeminism is one of the newest and most unique environmental discourses, and its ability to service as a model for a significant social movement has not been developed. There are a few ecofeminist movement organizations, but that is not really the point, I think. Ecofeminism has a considerable following, particularly in university circles and women's studies programs. Philosophers and historians have issued a torrent of books about the subject. Moreover, the voice of women is increasingly being heard at international gatherings with environmental implications (such as the 1994 Cairo Conference about population or the subsequent Beijing Conference about the global status of women). Like deep ecology, ecofeminists are not all of one voice, reflecting schisms in the larger feminist movement. They disagree, for example, about whether to emphasize the female "nature of nature" or whether to attempt to transcend gender roles altogether. Given the dominance of patriarchy around the world, the acceptance of the ecofeminist discourse is problematic. But ecofeminism is important as a pervasive set of ideas and critiques deep ecology and reform environmentalism, and is applicable to a wide variety of environmental concerns.

Ecotheology. The beginnings of *ecotheology* are found in reactions to Lynn White's landmark essay (1967). White argued that the Western biblical tradition, on which both Judaism and Christianity are based, was the root of the modern environmental crisis because man was viewed as the master of and apart from the rest of God's creation. According to White, "more science and more technology are not going to get us out of the present ecologic crisis until we find a new religion, or re-think our old one" (White, 1967: 1206). Although an arguable view (as I noted in Chapter Two), it did create a problem for Western theologians and religious thinkers. Rejecting radical anthropomorphism, they sought to develop a spiritual vision of the environment combined with the imperative for humans to preserve God's creation.

Several versions of this perspective developed. One stems from the African-American churches of the United States, which linked a spiritual view of the environment with the environmental justice and environmental racism movements. They have been major forces in these movements. For example, the first protest in 1982 against a toxic landfill in North Carolina was led and organized by a local African-American church, and the early influential empirical study of environmental racism was sponsored by the United Church of Christ (Bullard, 1990). Caring for the environmental preservation was thus linked to the creation of just and caring human communities (Brulle, 2000: 157). Another perspective, known as *Christian stewardship*, focuses on an evangelical interpretation based on a biblical mandate to care for God's creation. Founded on conservative Christian theology, it creates a moral imperative to preserve God's creation and makes minor adjustments in Christian theology to accommodate environmental concerns. It still sees God as a transcendent being and human nature as fallen, sinful, and in need of redemption. According to an early statement, "Christians who should understand the creation principle, have a reason for respecting nature, and when they do, it results in benefits to man. Let us be clear: it is not just a pragmatic attitude; there is a basis for it. We treat it with respect because God made it" (Schaeffer, 1970: 76, cited in Brulle, 2000: 158). A third view, usually called *creation spirituality*, sees the need to go beyond the Christian tradition (hopelessly beyond redemption in light of ecological problems) to develop alternative notions of the creation. Creation spirituality seeks a new synthesis of religion and science. One of its founders, Matthew Fox, advocated "the overcoming of dualisms of the western worldview so that we can see the creation as a whole" (Oelschlaeger, 1994: 169). Because he accepted all religions as revelations of the sacred in different contexts, Fox was excluded from the Catholic Church. Popular writer Thomas Berry advocated a "new story" for humanity, uniting Genesis with scientific knowledge. That "new story" would remove humanity from a position of privilege in the universe. The resonances with deep ecology are obvious.

In the mid 1980s, a movement emerged within religious communities that had traditionally not been involved in environmental issues. In 1989, Pope John Paul II wrote an encyclical entitled *The Ecological Crisis: A Common Responsibility.* Following were statements about religion and the environment by other American denominations, culminating in a unified statement by leaders of 24 major religious bodies entitled *Statement by Religious Leaders at the Summit on the Environment.* By 1993, virtually all major religious bodies had issued a proclamation on environmental degradation. Also in that year a *National Religious Partnership* was formed, which united the major Protestant, Catholic, Jewish, and evangelical communities into one organization focused on developing and implementing religious approaches to combating environmental degradation (Brulle, 2000: 156–160).

Ecotheology is a new and yet emerging discourse. Since the majority of Americans have religious affiliations (at least nominal ones), it would seem to have a huge potential to mobilize a popular base. It is also connected with established institutions that have considerable infrastructure resources facilitating mobilization. I think it has two limitations. First, ecotheology will be limited to the extent that its carriers are agents for legitimizing the status quo rather than agents of social and political transformation. Second, since it calls for humans to value nature not for its own sake, but rather because of its divine origins, it may have limited acceptance outside of communities of believers.

To summarize, I have sketched the varieties of contemporary environmentalism in rather broad strokes. As "environmentalisms" based on different ideological frames (or discourses), with different organizational structures and action strategies and diverse clienteles, they can all claim successes, but, as you can see, all have limitations. One thing that is needed is a *metanarrative,* or master frame, that could enable the varieties of environmentalism to work in complimentary fashions, rather than in contradictory ways that blunt the effectiveness of the larger movement. Such an environmental metanarrative would provide common discourse, enabling people to unite around actions creating a just, democratic, and sustainable societies—and world order. It would not destroy existing discourses, but rather would incorporate them, creating a larger capacity for collective action. Lest you think that is impossible, consider a more limited version of a metanarrative that did just that. Recall the concept of biodiversity, discussed earlier as a fairly recent invention of conservation biologists. It incorporated separate preexisting environmental discourses about deforestation, overfishing, habitat destruction, the introduction of exotic species, and endangered species. It did not destroy other discourses. It rather expanded the concerns of various groups to see their common purpose, and created the potential for greater collective action (Brulle, 2000: 188). Before saying more about the successes and failures of environmentalism, I turn now to the one remaining human-environment discourse that has been a durable framework for opposition to American environmentalism.

Manifest Destiny: Anti-Environmentalism and Countermovements

The oldest and most pervasive human-environment discourse derives from the dominant environmental social paradigm of America and industrial societies discussed in Chapter Two. Called *manifest destiny*, it provides a moral and economic rationale for exploiting the natural environment. Manifest destiny assumes that (1) nature has no intrinsic value, (2) nature is unproductive and valueless without human labor that transforms it into commodities upon which human welfare depends, and (3) there are abundant natural resources for humans, who have rights to use it to meet their needs. In addition to providing a rationale for the development of the North American continent by European settlers, manifest destiny provided the ideological discourse or frame for several waves of countermovements opposed to the goals of environmentalism (Brulle, 2000: 76). As I noted at the beginning of this chapter, countermovements share many of the objects of concern as the movements they oppose and make competing claims that vie for attention of the media, politicians, and broader publics.

The most significant early manifestation of manifest destiny countermovements erupted when President Grover Cleveland created 23 new national forests (bringing the total to 39 million acres) in 1897. Protests happened in the West, including a mass rally that attracted 30,000 people in Deadwood, South Dakota. A Montana senator accused the president of "contemptuous disregard" for people's interest, and a Washington state senator asked: "Why should we be everlastingly and eternally harassed and annoyed and bedeviled by these scientific gentlemen from Harvard College?" (Robbins, cited in Brulle, 2000: 79). Another series of western protests over federal land policy happened in 1925–1934. Known as the Stansfield Rebellion, it was led by Oregon Senator Stansfield and focused on opposition to the imposition of grazing fees in the national forests. The Taylor Grazing Act as amended in 1939 resolved the issue in a series of compromises between the interests of ranchers and conservationists. I noted earlier the opposition to Rachel Carson's book *Silent Spring* and industry attempts to defame her character in the media. That publicity campaign backfired. The sales of the book soared, and legislation subsequently banned the most objectionable pesticides.

By the 1970s, laws governing land use had shifted in favor of environmental movements, with wilderness protection and endangered species acts that permanently locked up parts of national forests, and placed limits on land use, even by private property owners. Groups representing western economic interests, primarily ranchers and miners, made major efforts to return control of federal lands to local economic interests. The Sagebrush Rebellion, as it was known, was a reincarnation of earlier land use issues. Several movement organizations emerged by the late 1970s, such as the Center for the Defense of Free Enterprise, which protested environmental restrictions being imposed on the free enterprise system, and the Mountain States Legal Defense Fund, a business-supported anti-environmental law firm. Its president and chief legal officer was James Watt, later to become director of the EPA under President Reagan (Brulle, 2000: 76–83).

Such countermovements are a measure of *both* the power and success of environmental movements *and* their difficulties in that they elicit powerful and sustained opposition. Protecting the environment may well be a consensual value, as public opinion polls indicate, but the costs and benefits of doing so are very unevenly distributed. You need to understand the magnitude and distribution of those costs to appreciate the rationality of groups and organizations threatened by environmental protection.

Ranchers, miners, corporations and the business community bore the most obvious and most direct costs of environmental protection. The previous chapter said quite a lot about *internalizing* environmental costs that are presently unaccounted for. What does internalizing full environmental costs mean for the business community? Companies were required to meet standards regulating waste discharges and power plants to comply with air quality standards, and corporations were responsible for the cleanup of leaking landfills. Some were able to combine environmental regulation with increased efficiencies in the use of fuels and material, but many, particularly in the 1970s, simply added control devices to existing plant and equipment. This mostly threatened the capital investments of companies, less so their revenue flows, but nonetheless reduced their net profitability unless they could pass those costs through to consumers. That was often not permitted by power regulatory agencies. In the late 1970s, for instance, the costs of coal-fired electric plants increased 68 percent, and 90 percent of that cost was attributable to antipollution measures required by the EPA (Farber and O'Connor, 1989). As you can guess, they did not enthusiastically and voluntarily embrace such changes.

By the late 1970s, when it was obvious that environmentalism was not a passing fad and faced with threats to profitability, many corporations took an offensive position. Companies spent massively on public campaigns to depict themselves as good environmental corporate citizens, while environmentalists were depicted as unrepresentative nuts who needlessly threatened economic prosperity, jobs, and human well-being. Corporate-sponsored attacks on environmentalists took a variety of forms: political lobbying and public relations campaigns with hired public relations firms and advertising. By 1995 American firms were spending $1 billion a year on anti-environmental activities (Beder, 1998: 108). Such campaigns also included "sponsored research" designed to refute the claims of environmentalists and regulatory agencies.[5] Except for the direct action tactics of grassroots movements and the radicals, corporate strategies were often a mirror image of those of the environmentalist groups. Increasingly, corporations are likely to use litigation tactics against anyone who publicly opposes their use of the environment (Clyke, 1993: 87–88; Sale, 1993: 102). Such suits, called *strategic lawsuits against public participation* (SLAPPS), have the intention of silencing activists or diverting their attention away from the issues themselves. SLAPPS charge environmental groups and grassroots activists with defamation of character, interference with contracts or business, or conspiracy. About 25 percent of SLAPPS are about development and zoning issues, and another 20 percent surround pollution and animal rights issues. In the 1990s, the average SLAPP was for $9 million, and some have been for as much as $100 million. Most of these suits are unsuccessful and eventually dropped, but in the meantime they create many problems for activists (financial troubles, and fears of retribution and exercising constitutionally guaranteed free speech) and corporate America has found SLAPPS an effective strategy for neutralizing opponents (Dold, 1992).

Where legal methods did not suffice, environmentalists were harassed by extralegal methods—offices trashed, cars smashed, homes entered, death threats. The home of a Greenpeace worker in Arkansas was burned, and two EF! workers were firebombed in California. Karen Silkwood, a worker in an ARCO nuclear fuel processing plant in Oklahoma, was harassed and possibly killed by agents of the company before she tried to "blow the whistle" on fraudulent safety reporting procedures. In the most egregious case and well-documented case, a congressional committee discovered in 1991 that the corporate managers of the Trans-Alaska Pipeline paid hundreds of thousands of dollars for a nationwide hunt to find and silence critics of the Alaska oil industry—complete with eavesdropping, theft, surveillance, and sting operations—and harassed its own employees to cover up leaks about its environmental and safety errors. The point is, I think, that not all of the illegal "radical" action is in defense of the environment. Corporations' actions are sometimes supported by the government agencies. Government agents have infiltrated environmental groups (I mentioned the case of EF! earlier) and have used tactics of surveillance, intimidation, anonymous letters, phony leaflets, telephone threats, police overreaction, and dubious arrests (Sale, 1993: 102–103).

By the 1980s and 1990s, a renewed and coordinated countermovement was underway, termed the *wise use movement*. Movement organizations emerged with curious names:

- National Wetlands Coalition (oil drillers and real estate developers)
- The U.S. Council on Energy Awareness (the nuclear power industry)

- Friends of Eagle Mountain (a mining company that wants to create landfills in open pit mines)
- Wilderness Impact Research Foundation (logging and ranching interest groups in Nevada)
- American Environmental Foundations (a Florida property owner's group)
- Global Climate Coalition (corporations opposed to regulations to control global warming)

From their names, you would think that *these* were environmental organizations! It is a weakness of anti-environmentalism that its movement organizations *must* often be given deceptive names, if they are to succeed publicly (a phenomenon that environmentalists termed *greenscamming*). Imagine the public relations problems of a group calling itself something like the "Coalition to Trash the Environment for Profit." In 1989, the wise use movement adopted a formal agenda and created the Center for the Defense of Free Enterprise. The agenda statement included 20 items, such as that "all public lands including wilderness and national parks shall be open to mineral and energy production," and that the endangered species act be amended to "exclude relic species in decline before the appearance of man" (Gottleib, cited in Brulle, 2000: 84).

In November 1991, 125 business groups and other coalitions formed the Alliance for America, heavily funded by timber cutters, oil drillers, ranchers, and other anti-environmental corporations who aim, according to its leader, to "destroy environmentalists by taking away their money and their members." Mary Bernhard of the U.S. Chamber of Commerce said that there was a renewed commitment by the business community to "turn things against the environmental movement" (Sale, 1993: 104).

Tough opposition indeed! Business opposition has not derailed, but it has certainly deflected and blunted the impact of environmental movements. Part of the problem is that it has always been difficult for anti-environmental movements to coopt the moral high ground. Who wants to "dance with the devil" and openly adovcate economic practices that palpably destroy the planet's natural resource base, or kill humans? Which is why anti-environmental movements must often operate under a cover of deception, as noted, that makes them vulnerable to public disclosure. Greenscamming is a strategy that exposes a vulnerability. Moreover, given the consensual nature of pro-environmental attitudes, such countermovements are likely to backfire. Such was the case in the first part of the conservative Reagan administration. At the urging of environmental countermovements, legislation was proposed to weaken environmental laws, and reduce budgets of programs protecting the environment. James Watt, director of the EPA with a mandate to destroy his own organization, was quoted as saying: "America's resources were put here for the enjoyment and use of people . . . and should not be denied to the people by elitist groups." But this generated a successful surge of mobilization by environmental movements and a blocking of many initiatives proposed by the Reagan administration. Reagan's first two EPA directors, James Watt and Ann Gorsuch, were forced to resign under allegations of corruption. Furthermore, it is important to note that business reaction has not been uniform. For understandable reasons, opposition was always stronger in capital-intensive and extractive industries and weaker in retail industries that are more directly connected to public opinion.

ENVIRONMENTALISM AND CHANGE

I will conclude this chapter with a "reading" of the net impact and effectiveness of American environmentalism. But first, given extensive mobilization over several decades around the cause of environmental protection, what have the actual impacts been? For what kinds of actual change has environmentalism been least partly responsible? Fair questions.

The Question of Durability

The first imperative of a social movement is to survive. Many don't. Common wisdom predicted that after the 1970 Earth Day hoopla was over and media attention diminished, public environmental concern would significantly decline. Renowned American sociologist Amitai Etzioni pronounced:

> This new commitment (to fighting pollution) has many features of a fad: a rapid swell of enthusiasm (most ecology groups are less than 6 months old), fanned by the mass media. And the commitment is rather shallow. (1970)

In this, he was dead wrong (as well as ill informed about history). Environmentalism grew rapidly in the 1960s, perhaps declined in the early 1970s, but underwent revitalization in the 1980s, and continues vigorously today. It was not, as some had feared (or perhaps hoped), a passing fad. Environmental movement organizations have become a stable feature of American institutional arrangements (Dunlap, 1992: 96–102; Mitchell, 1980: 423).

Changing Attitudes and Beliefs

Remarkably, even after two presidential administrations (Reagan and Bush) and a continuing congressional majority under the Clinton administration with political mandates from their supporters to attack environmentalism and weaken existing legislation, the movement continued to grow and public pro-environment attitudes continued to deepen. A 1989 Harris national poll found that 94 percent said that the country should be doing more than it now does to protect the environment and curb pollution. Overwhelming evidence suggests that environmental protection again became a consensual issue commanding support from the vast majority of the population (Dunlap, 1992: 107; Harris, 1989: 3). Similar evidence suggests that this is a global phenonmenon, but I will discuss that in the final chapter. While such supportive attitudes are wide and deep, they are not evenly held among the American population. What kinds of people are likely to be more aware and concerned about environmental problems?

In general, research suggests that educated people are more concerned than less educated people, younger adults more concerned than older adults, and persons whose lives are more directly impacted by environmental hazards are more concerned than those whose lives are less directly impacted. *These differences are not statistically very powerful ones* (Van Liere and Dunlap, 1980; Gould et al., 1988; Samdahl and Robertson, 1989). As I noted, gender is the most powerful predictor of environmental attitudes. Women are more concerned than men about health and safety issues of a given technological risk, but evidence of similar gender differences about other concerns (such as biodiversity) is weak. Furthermore, women are not more likely than men to participate in environmental movement activities (Stern et al., 1993; Freudenburg and Davidson, 1996). Most research shows urban people to be more environmentally concerned than rural people, possibly because urbanites live in more degraded and polluted environments. But recent studies suggest a more complex picture. Rural farmers are more concerned than nonfarm rural residents are, and owner-operator farmers are more concerned than absentee owners or farmers who lease land. Furthermore, depending on the *kind* of environmental issue, some urban people are less concerned than rural people are. In short, urban rural differences exist, but in complex rather than simple ways (Freudenburg, 1991; Williams and Moore, 1991; Constance et al., 1994). Compared to nonactivists, *environmental activists* have higher levels of concern, often greater exposure to environmental risks, greater political interest, more contact with other activists, and—most importantly—greater personal resources, such as knowledge about particular issues or how to participate more effectively (Stern, 1992: 287). The widespread growth of these beliefs provided a depth to environmental concerns that was lacking in the 1960s (Dunlap, 1992: 112). Environmentalism has become, in other words, a stable component of American culture. This raises an interesting question. How much has the dominant social paradigm (DSP) described in Chapter Two changed? Since I have said quite a lot about the importance of paradigms and the question has been addressed with great sophistication by a team of American sociologists over the last decades, I will discuss this research in some depth.

Paradigm Change?

Recall that Chapter Two depicted the DSP of industrial societies as (1) assuming that the environment is an open-ended resource system for human use, (2) nature for its own sake is not to be valued, (3) that growth, consumption, and the accumulation of wealth are important values, (4) that there are no "real" limits to growth, and (5) that science and technology are effective means to address all manner of human, economic, and environmental problems. What evidence is there that pro-environment attitudes signal a fundamental and pervasive cultural transformation rather than simply an accumulation of unconnected attitudes and beliefs? What evidence is there, in other words, for the emergence of a new ecological paradigm (NEP) within industrial societies?

Dunlap and Van Liere began to concretize the question for empirical research (1978, 1984). They constructed scales to measure the DSP, and with survey data from about 1,500 people in a Washington state sample they used statistical factor analysis techniques to empirically isolate stable dimensions of the DSP. The four that emerged were (1) faith in the efficacy of science and technology, (2) support for economic growth, (3) faith in material abundance, and (4) faith in future prosperity. All of those were found negatively related to support for pollution control, resource conservation, and environmental regulation.

Similarly defining the NEP concept for research, Dunlap and Van Liere constructed a 12-item NEP scale of statements for people to agree or disagree with. This scale contained such item as (1) the importance of maintaining the balance of nature, (2) the reality of limits to growth, (3) the need for population control, (4) the seriousness of anthropogenic environmental degradation, and (5) the need to control industrial growth. Not surprisingly, they found that members of environmental movement organizations scored higher than others on these items but that the general public tended to accept the context of the emerging environmental paradigm much more than they had expected (1978). Subsequently, other researchers found support for the NEP and similar concepts in a variety of public samples in the United States and Canada (e.g., Edgell and Nowell, 1989; Caron, 1989).

Olsen, Lodewick, and Dunlap (1992) completed the most sophisticated empirical analysis of the growth of ecological thinking among Americans, based on survey data from samples in Washington in 1982 and Michigan in 1988. They constructed survey items of both beliefs and values related to the DSP and NEP (or, in their words, "technological" vs. "ecological" beliefs and values). Table 8.2 depicts some of the items for values (they had similar items for beliefs) and the distribution of responses in the Washington state sample.

Olsen and colleagues constructed an index of strong-to-weak support for statements related to technological vs. ecological values. They found that the combination of technological beliefs *and* values was not very widely held:

Table 9.2 Ecological and Technological Values, with Distribution of Responses, Washington State Sample

Questionnaire Items	*% Responses*						
Ecological Values	*5*[a]	*4*	*3*	*2*	*1*	*Technological Values*	*N*
People should adapt to the environment whenever possible.	43	39	11	5	1	The environment should be changed to meet people's needs.	691
Natural resources should be saved for the benefit of future generations.	39	36	19	5	1	Natural resources should be used primarily for the benefit of the present generation.	699
Nature should be preserved for its own sake.	29	26	14	20	11	Nature should be used to produce goods and services for people.	692
Environmental protection should be given priority over economic growth.	21	30	23	18	8	Economic growth should be given priority over environmental protection	688

[a]5 = strongly agree, 4 = mildly agree, 3 = undecided, 2 = mildly disagree, 1 = strongly disagree.
Source: Olsen et al., 1992: 43, 66.

Only a minority of people (27 percent) accept them either completely or moderately, while more than half (53 percent) held a few technological beliefs but none of the values and 19 percent rejected them entirely. By contrast, they found ecological beliefs and values to be very widely held. Over half of their sample (56 percent) was committed to both beliefs and values, either completely or moderately; another quarter (27 percent) accepted them weakly or partially; and a minority (17 percent) rejected them completely (1992: 53, 75). They concluded, consistent with earlier research, that core elements of the DSP are not widely held by the public, while the NEP seems to be very widely held. Perhaps the "dominant" part of the DSP has become a misnomer!

Most discussions of the NEP assume that people who hold it also accept many other elements of the postindustrial worldview—about such things as population control, economic stability, meaningful work, government and public responsiveness, political participation, international cooperation, decentralized ("human-scale") organizations, nonmaterial lifestyles (voluntary simplicity), individuality, and renewable energy. Olsen, Lodewick, and Dunlap, examining the relationship between high NEP scores and many of these, found extensive positive but *very weak* correlations between them. The scores were strongly connected only to two clusters of items, and the authors concluded that while the NEP is indeed a part of a broader postindustrial worldview, the linkage between "ecological thinking" and that broader worldview varies and cannot be assumed just because they seem like logical connections (1992: 171–172).

This research may represent an important contribution to understanding how paradigm changes occur. Most scholars, following the ideas of philosopher of science Thomas Kuhn who coined the term, argued that a *paradigm shift*, whether in scientific or public thinking, was a seemingly sudden and discontinuous two-stage shift in a whole pattern of thinking. If this is the case, the researchers assumed few people who accept one paradigm would also hold elements of the other. But in fact, their data showed a considerable number of people who held elements of both—particularly beliefs, but less so values. They believe that instead of sudden paradigm shifts, there is a more gradual dialectical change, in which two paradigms contend and exist as contradictions (even in the minds of the same people). But the eventual outcome is a synthesis of the two previous ones. Olsen and colleagues find, for instance, that a considerable number of people continue to believe in the efficacy of science and technology, but also that industrial society has caused great damage to the environment and that population growth and material consumption should be limited. Most people rejected the notion of a "steady-state" economy popular in the ecological thinking of the 1970s. People see the need for the enhancement of social life, but they do not see it as dependent on constantly rising industrial production. They speculated about the emergence of a synthesis that they termed (using one of the buzzwords of the times) a *sustainable development paradigm*, because of the overwhelming acceptance in their sample among those who accept elements of both technological and ecological thinking. This is intriguing—speculative and arguable—and the authors rightly caution against concluding too much from limited cross sectional data. Real confirmation of these ideas depends on genuinely longitudinal data. But the main conclusion that I think is now beyond doubt from this and other researches is that core elements of the old DSP are being rejected fairly rapidly in the United States and probably other industrial countries in favor of ecological thinking. With the accumulation of pro-environment attitudes and evidence of some transformation of the DSP about environmental questions, you would expect some evidence about visible social trends or behavior changes. I think there are some.

Voluntary Simplicity Again?

Chapter Seven mentioned the notion of *voluntary simplicity* that was popular in the late 1970s. Philosopher Duane Elgin coined the term to describe and promote a movement away from materialism, excessive consumerism, and unsustainable consumption that degrades the environment. That early voluntary simplicity movement was linked to New Age spirituality, the themes of the 1960s counterculture, and "moving" back to nature. Its rhetoric was primarily moralistic, presenting idyllic images of romantic asceticism and living in nature or rebuking consumers for overindulgence. As you might guess, it had a limited appeal and was overwhelmed by the more powerful forces promoting living on credit, shopping at malls, and buying SUVs. But the idea of voluntarily simplifying lifestyles was more than just a passing fad. It returned in the 1990s along with a movement to promote it. In 1995, a national survey found that 85 percent of respondents thought that Americans "consume more than they need to," 93 percent agreed that "the way we live produces too much waste," and 91 percent thought that "we focus too much on getting what we want now and not enough on future generations." A wide variety of groups sprang up dedicated to plain living, frugality, "downshifting," and living lightly on the earth. Amy Dacyczyn's newsletter *Tightwad Gazette,* devoted to spending less and living more, had 100,000 subscribers before she got tired of producing it. Similar publications exist, such as *Frugal Gazette, Miser's Gazette,* and the *Something for Nothing Journal.* People interested in voluntary simplicity make contact through a wide variety of local discussion groups called "simplicity circles," publications, and web pages. Simplicity circles, which started in Seattle, now exist in many states.[6] In addition, the Center for a New American Dream (CNAD) seeks to change North American attitudes about high consumption in a "throw-away" culture (Bell, 1998: 62–63).[7] So there is a second visible movement, but its power and pervasiveness still pales compared to the opposite urgings of marketers, advertisers, and the mass media.

If the new voluntary simplicity movement is to have greater appeal and impact, it needs to be promoted differently than was the earlier movement. High mass consumption is more complex than promoters of simple lifestyles often think. Consumerism is related to the creation of meaning and transmission of meaning in modern societies. Consumption defines identity, status, and privilege more easily than production in increasingly large and uncontrollable global systems of production. It is not just a psychological blight or "sick culture" (Cohen, 1997). Most importantly, voluntary simplicity would have broader appeal if seen as a means of improving the quality of life in a society where many experience life as overly complex and hectic and in which "life in the fast lane" provides little true leisure. Voluntary simplicity could be seen as a means of reclaiming something important rather than depriving one self and one's family.

Changed Economic Structures

As part of the dialectic of social change, movements and collective action can create and transform structures as well as the way structures operate. Chapter Eight provided several illustrations of corporations transforming themselves in the direction of greater sustainablility. I only want to underline that these positive transformations are not happening without pressure, from the public and, in turn, often from environmental movements and activists. MacMillan Bloedel, one of western Canada's largest logging firms, stunned the world when it announced that it was giving up the standard practice of clearcutting forests. It did this only after a long period of negative publicity, protests, and litigation by environmental organizations. San Francisco's Rain Forest Action Network has probably had the most impact over time. It helped persuade Burger King to cancel $35 million in beef contracts with forest-killing cattlemen in Central America, persuaded Mitsubishi to use tree-free paper, and took two years to convince Home Depot to switch from old-growth-cut lumber to lumber from sustainably produced sources (Motavalli, 2000: 26).

McDonald's was probably no more wasteful than other fast food chains, but to many Americans McDonald's became a symbol of a consumer society run amok. The company was criticized and pressured for slapping food in elaborate wrappers that were thrown away minutes later, for producing litter and congestion, and for dark rumors (never really substantiated) that the tropical rainforest was being cleared for its cattle to graze. Some city councils banned its styrofoam burger boxes, calling them "McToxics." Responding to such pressures, McDonalds went from villain to role model in one of the great turnarounds in modern times. Turning to environmentalists for advice about how to reduce wastes, McDonald's reduced its waste flow by 30 percent since 1990, became one of the country's leading buyers of recycled material, and cut its energy use. Fred Krupp, director of the Environmental Defense Fund, said, "They've made a long series of commitments and they've hit every target (Allen, 2000: 1e). Not all fast food chains have responded so positively, and menus that rely so heavily on beef are still questionable for both health and ecological reasons, because of the food needs of cattle and the waste they produce.

CONCLUSION: HOW SUCCESSFUL IS ENVIRONMENTALISM?

Given these examples and evidence of cultural transformation and structural changes, amid the continuous mobilization of environmentalism in the last half century in American (and the world) you would think that the earth's vital signs would surely have improved. But it is really hard to make a convincing case that this has happened. To be sure, there are visible signs of progress: Standards have been raised, rivers and air are getting cleaner, and there are signs of economic transformation. And yet the environment inexorably deteriorates. The 6 billion people bring with them anthropogenic megaphenomena like global warming, desertification, global soil erosion, water overdrafts, ocean pollution, rapidly encroaching urban development, biodiversity declines, and resource depletion. Even normally optimistic

Dennis Hayes, the organizer of Earth Day 1990, admitted that "by any number of criteria you can apply to the sustainability of the planet, we are in vastly worse shape than we were in 1969, despite twenty years of effort" (1990: 56). The Worldwatch Institute noted in 1998: "We have overwhelmed the natural systems from which we emerged and created the dangerous illusion that we no longer depend on a healthy environment . . . We need to restore the balance with nature while expanding economic opportunities for the billions of people whose basic needs . . . are still not being met" (cited in Motavalli, 2000: 26).

Has environmentalism then failed? Some argue that it has, faced with overwhelmingly powerful counterforces. Others argue that environmentalism has been a modest success. Whether or not it is so depends on what is meant by "success."

In one sense, environmentalism has been a resounding success. Within the space of a single generation, environmentalism has become embedded in law and custom, text and image, classroom and workplace, practice and consciousness in such a way that suggests that, far from eventually fading away, it will continue to have durable impact some time to come. Environmental movement organizations have proved enormously durable and successful compared with many other movement organizations. Citizens in America and around the world are vitally concerned about the quality of their physical world in a qualitatively new way. Environmental issues do not dominate politics, but they are on every political agenda and are now firmly connected to global development and national security issues. So if the objective of the environmental movement were only to increase public consciousness and create self-sustaining organizations and lobbies, it would be rated a smashing success. But, of course, the ultimate goals of environmentalism are to improve the conditions that gave rise to and continue to drive the movement. By that criterion its success is clouded with ambiguity, to put it mildly. But if the forces driving environmental deterioration are without historical precedent, so are the forces driving a transformation to a more sustainable society and world order. Both sets of forces operate on a global basis, and it is to these that I turn in the final chapter.

PERSONAL CONNECTIONS

Questions and Implications

1. Have you or your acquaintances ever demonstrated support for environmental concern by taking part in an environmental event, meeting, rally, celebration, or the like? What kinds of things? Did they change or reinforce your thinking? Your behavior? Did they change the composition of your social network of friends and acquaintances? Your knowledge about "people resources" on your campus, or in

your community? If you were a part of an organized activity, did you feel a need to explain to others in your surroundings what you were doing and why? How did they react?

2. How do your acquaintances view environmental activists or organizations? Ask a variety of people in different walks of life (ask your friends, several professors, clergy, businesspeople, your relatives). What kinds of different responses and perceptions do people have? Do people differentiate between "responsible" and "radical" environmentalism? Do people see environmentalism as a threat? To what?
3. What images of environmental organizations and activists are portrayed in the media?

What You Can Do

While this book has treated the abstract big issues, I hope you recognize that there has been another theme: that individuals matter. Often significant social change comes from the bottom up, not from the top down. I began this section in the introductory chapter by noting (with some misgivings) the established slogan of environmentalism: "Think globally, act locally." Individuals can matter in several ways. You can change your own lifestyle to be more environmentally benign (which is a good in itself but may also demonstrate to others the possibility of positive change). Many of the "what you can do" suggestions that I have offered so far are about these kinds of possibilities. If the idea of voluntary simplicity intrigues you and you want to find out more about how people are doing it or make contact with others, try these websites:

www.simplicitycircles.com; www.newdream.org; www.adbusters.org.

Changing yourself is important, but I also think it's important that you join with others to act or raise consciousness about human and environmental issues. How so?

1. The contemporary environmental movement was greatly amplified on hundreds of college campuses thirty years ago on Earth Day 1970. Campuses are still crucibles for important environmental activity. What people learn and do in college they are likely to carry into other settings after graduation (one hopes!). You can help organize or join an ecology or environmental group on campus. That organization could be involved in a number of projects. For instance—
   - Hold an ecological film festival or some activity to appreciate nature locally. (A group at my university organized student groups to go to a nearby wildlife reserve to see migrating Canadian geese, which are there in the thousands.)
   - Create a coalition with other campus groups to promote environmental issues. (Groups have worked with fraternities and student government organizations to organize campaigns among students to recycle and carry with them reusable pop/coffee containers rather than use throwaway ones.)
   - Organize lectures, discussions, and teach-ins about ecological matters. Use resource persons from academic departments and the community (e.g., a local or regional Sierra Club or Audubon Society group may provide resources).

- Affiliate with national student environmental/ecology organizations. The largest student environmental organization in the United States is the Student Environmental Action Coalition (SEAC), established in 1988 at the University of North Carolina–Chapel Hill (address: SEAC, 217A Carolina Union, University of North Carolina, Chapel Hill, NC 27599; tel: 919-962-0888).

2. Join an existing environmental/ecology organization in your community and support its activities.
3. Join one of the many conservation/environmental/ecological national organizations. You will pay (normally not more than a magazine subscription, but whatever you are willing to above that). In return, you get information and support research, public education, litigation or lobbying efforts. Some are membership organizations and others are not. (There are *hundreds,* and would take a book itself just to publish the names and addresses.)
4. All attempts to create change (about anything important) involve politics and political action. Here are a few *guidelines for effective grassroots organizing for action* from John Gardner, former cabinet officer and founder of the citizen's organization Common Cause.
   A. Have a full-time continuing organization.
   B. Limit the number of targets and hit them hard. Most groups dilute their efforts by taking on too many issues.
   C. Form alliances with other organizations on particular issues.
   D. Get professional advisers to provide you with accurate, effective information and arguments.
   E. Have effective communication that will state your position in accurate, concise, and moving ways.
   F. Persuade and use positive reinforcement—don't attack. Confine your remarks to the issue; don't make personal attacks on individuals. Try to find allies within the institution, and compliment individuals and organizations when they do something you like.
   G. Do your homework, and then privately approach public officials whose support you need. It's best not to bring up something at a public meeting unless you have the votes lined up ahead of time. Most political influence is carried on behind the scenes through one-on-one conversations.
   H. Instead of fighting with your opponents, respect their beliefs and work with them to achieve your goals when possible.
   I. Organize for action, not just for study, discussion, or education. Minimize meetings. Have a group coordinator, a series of task forces, a press and communications contact, legal and professional advisers, and a small group of dedicated workers. A small, well-organized group can accomplish more than a large, unwieldy one.
   J. Work in groups, but keep in mind that people in groups can act collectively in ways that individuals know to be stupid. (adapted from Miller, 1992: 688)

ENDNOTES

1. Social psychologists use the term *relative deprivation* to define the deprivations people have as they compare their problems and experiences "relatively" with others in their social environments. It is different than *absolute deprivation,* which

means being objectively deprived in terms of malnutrition, illness, safety, or some other threat to physical security or well-being. Most scholars suggest that relative deprivation is a more powerful cause of movement participation than absolute deprivation and that people who are absolutely deprived may not have the energy or resources to organize movements for change. They are likely to be more absorbed with more elemental needs.

2. Among similar important works, M. Bookchin's *Our Synthetic Environment (1962)* was concerned with environmental deterioration due to industrial pollution, and Secretary of the Interior Stewart Udall's *The Quiet Crisis (1963)* advocated a program for the protection and preservation of the natural environment (Brulle, 2000: 125).
3. Freudenberg and Steinsapir list seven kinds of outcomes: (1) blocking the construction of waste disposal facilities, forcing the cleanup of toxic dumps, banning aerial spraying of pesticides, etc.; (2) altering corporate practices, particularly through product liability suits; (3) applying increased popular economic and political pressure for the prevention of environmental hazards; (4) winning legislative victories about rights for citizen participation in environmental decision making; (5) increasing community mobilization; (6) linking environmental problems with problems of social justice; and (7) promoting broader public environmental consciousness (1992: 33–35).
4. The only known injury from tree spiking was to a millworker at a Louisiana Pacific mill in California in 1987, when a bandsaw struck an embedded spike. The company blamed it on EF! and the media broadcast the allegation with great intensity. EF! was never charged or even investigated, however, and no evidence ever connected it to the spiking. Furthermore, the tree was not in an old-growth-area activists had been defending, nor was it even standing when it was spiked, not a monkey wrench tactic (Sales, 1993: 66).
5. Consider a recently distributed article entitled "Environmental Effects of Increased Atmospheric Carbon Dioxide," in March 1998, in a format closely resembling a reprint of the *Proceedings of the National Academy of Sciences* (which it was not). The article concluded that predictions of global warming are in error and that increased CO_2 levels greatly benefit plants and animals. In fact, this article was in a nonrefereed publication funded by the George C. Marshall Institute, an anti-environmental think tank (Brulle, 2000: 85)
6. In a few minutes on my computer, I located twenty-two in California, seven in Minnesota, four in New York, and four in Missouri. The website address for the voluntary simplicity network is www.simplicitycircles.com.
7. CNAD's website is www.newdream.org.

CHAPTER TEN

Globalization: Trade, Environment, and the Third Revolution

Until about 1990, two powerful human change efforts, *environmentalism* and the *amplification of world trade*, unfolded in largely unconnected ways. International trade is, of course, much older that self-conscious environmentalism. Trade between distant people has played a major role in human social and cultural integration since ancient times. It brought pasta to Italy, silk to France, Columbus (and all that came with him) to the Americas, polio vaccine to the world, and Coca-Cola, Marlboros, and Ben and Jerry's ice cream to Moscow and Beijing (Zalke et al., 1993: xiv). Even though they are not historically connected, you need to connect the emergence of the world political economy (discussed in Chapter Two) with global environmentalism to understand our current environmental situation and problems. For a homey illustration, consider the cup of coffee sitting beside my computer right now.

Since coffee doesn't grow in the United States, it began its journey with the picking of coffee beans from trees grown in high tropical mountain regions in Colombia (or in similar regions in other nations, such as Costa Rica or Kenya). There, natural forests were cleared to grow coffee trees, and as a result much of the world's tropical "cloud forests" became among the most endangered ecosystems. Growing the coffee trees required several doses of insecticides, probably manufactured by industries in either in the Rhine River valley in Europe or in the Mississippi River basin in Louisiana, both of which have become the world's premiere toxic waste zones. As the coffee trees were sprayed, some pesticides got in the lungs of farmers and residues washed down mountainsides and collected in remote streams and rivers. The beans were shipped to New Orleans in a freighter constructed in Japan from Korean steel. The steel was made from ores mined on tribal lands in Papua New Guinea, and the tribal people received little or no compensation for their ruined forests and contaminated water. In New Orleans the beans were roasted and packaged in four-layer bags constructed of polyethylene, nylon, aluminum foil, and polyester. The three layers of plastic were made from oil shipped by tanker from Saudi Arabia. The plastics were fabricated, again by factories in Louisiana's famous toxic waste corridor, where cancer rates are disproportionately high among the people—mainly African Americans—who live and work there. The aluminum layer of the package was made in the Pacific Northwest from bauxite strip-mined in Australia and shipped across the Pacific on a barge fueled by oil from Indonesia. The mining of bauxite displaced Australian aboriginal peoples from their ancestral lands, and the bauxite was refined with energy from a hydroelectric dam on the Columbia River, which has altered an entire ecosystem. Among other things, the dam has diminished the ability of Native Americans to pursue their traditional livelihood—salmon fishing.

Bags of roasted coffee were shipped by trucks from New Orleans to Omaha, where I live. Those trucks used gasoline from oil extracted from the Gulf of Mexico and were refined at a plant near Philadelphia, where heavy air and water pollution has been linked to cancer clusters, contaminated fish, and a decline of marine wildlife in the Delaware River basin. I made coffee using filter paper manufactured from paper pulp from trees cut in some forest. Altogether, my cup of coffee required several direct uses of fossil fuels and several indirect uses that I have not mentioned—such as the energy costs of the high-rise offices where company executives produce those lusciously attractive ads for coffee and other products. My coffee contributed to the degradation of ecosystems in several regions—Colombia, Papua New Guinea, Australia, Louisiana, and Washington state—and has disrupted the livelihood of indigenous people in several areas of the world. While you cannot trace the exact inputs to any particular cup of coffee, the facts on which this account is based are true (see Durning and Ayers, 1994: 20–22).

I must admit that I hardly ever think about these things when I enjoy a cup of coffee, which is one of the simple pleasures of life. But as you can see, getting it here is not so simple and involves the participation of a vast array of material, processes, people, and industries as well as a variety of forms of envionmental degradation in worldwide locations. Nor is the point to depict coffee as an especially villainous product. If you wanted to do the research, you would find the same to be true for many other products. Indeed, I find it increasingly hard to find anything in my house, my office, or on my back that wasn't made—at least partly—with labor, resources, or financing in another nation.

This final chapter continues the concern of the previous ones about understanding *social change* as it relates to environmental issues and problems. It extends that focus to global dimensions of change by discussing (1) the growth of trade and the world market system, (2) the emergence of global environmentalism as a parallel phenomenon, and (3) problems and strategies for negotiating international treaties (and "regimes") that seek to regulate globalization in the interests of human well-being and environmental protection. The chapter ends with a summary assessment and a discussion about long-term human transformations and the prospect of a "third" human revolution. It is a brief but important summary of the book and I urge you not to skip it.

THE GROWTH OF WORLD TRADE AND THE WORLD MARKET ECONOMY

While historically ancient, the volume of world trade accelerated exponentially since the 1950s. Not only did it grow much faster than population, but also the value of traded goods and services routinely outpaced the total world production of goods and services. In 1997, goods traded around the world were worth $5.9 trillion. The MDCs (particularly the United States, Japan, and the European Union) now dominate most world trade. However, the World Bank projects that between 2000 and 2020 about half of further growth in exports from the MDCs will be traded with several large populous nations among the LDCs (particularly China, India, Indonesia, and Brazil) as well as Russia (Strauss, 1998b). As Chapter Two noted, the growth of the world market system was made possible by technical innovations in manufacturing, transportation, and information technologies but was also a result of deliberate policy. Conferences of diplomats and state bankers formulated policies after World War II (the first at Bretton Woods, New Hampshire). They were determined not to replicate the world depression and financial instability of the 1930s, when high trade barriers and tariffs were reciprocally erected by nations, each trying to protect home industries from foreign competition and get themselves out of the great world depression. It may shock to you to realize that the average world tariff on imported goods was then about 60 percent of the cost of the product. Such stiff tariffs constricted world trade, deepened the economic depression, and prolonged economic hardships that most believe was a major cause of World War II. After the war, increasing trade between nations was viewed as a way of producing continual market expansion, sales, profits, and mutually beneficial economic growth as well as improved living standards of the world's people. That, at least, is the main understanding of the *neoliberal* economic thinking that dominated economic thought during the expansion of the world market economy. But as I hope previous chapters demonstrate, deep environmental and social problems exist within the emerging world market and political economy. The explosion of production and consumption among many of the world's growing population has degraded the most elementary life support systems (like water and soil) and is decreasing the diversity of living things on the earth. It is probably changing the climate of the earth. As noted in Chapter Seven, growing affluence exists alongside mushrooming inequality and poverty, and over 20 percent of the world's people are mired in poverty so deep that they cannot meet the most basic subsistence needs (Giddens, 1995: 98). For better or worse, international commerce is increasingly shaping our everyday lives, our livelihoods, and environmental conditions and problems with which we live. One response to the growth of these problems was environmentalism. It was not only an American phenomenon but is a global phenomenon.

GLOBAL ENVIRONMENTALISM

As in the United States, environmentalism grew in virtually every MDC nation during the 1960s and 1970s, but particularly in Western Europe. I mentioned the influence of the German "Greens" on German environmental politics and policy, but in fact, environmentalism in Germany has a long history reaching back to the turn of the last century in "conservation" movements similar to those in the United States (Dominick, 1992). Environmentalism visibly emerged in Eastern Europe and the Soviet Union as those systems began to unravel in the 1980s. A major Russian environmental movement organization was appropriately named Zelenni Svet (Green World). As in the United States, heightened public awareness everywhere led to collective action and pressure on public officials. A multitude of environmental and other private nongovernmental organizations became internationally linked or networked as nongovernmental organizations (NGOs in the parlance of diplomats and international agencies).

One important stimulus for *global environmentalism* was increasing cooperation among the world's scientific communities to study things at biospheric and planetary levels. Starting with the International Geophysical Year (1957–1958) and the International Biological Programme (1963–1974), the International Council of Scientific Unions established the Scientific Committee on Problems of the Environment (SCOPE). Other agencies, such as the World Meteorological Organization (WMO) and the Food and Agriculture Organization (FAO), were sponsored by the United Nations along with some international NGOs, such as the International Union for Conservation of Nature and Natural Resources (IUCN). At both national and international levels there was scientific progress in atmospherics, soil science, oceanography, environmental toxicology, and ecology. Although organized by MDCs, scientists from LDCs emerged as leaders in the diffusion of environmental awareness in their nations, and in 1971 SCOPE convened a meeting of LDC scientists in Australia to consider environmental issues from their perspective (Caldwell, 1992: 65).

Another stimulus for the globalization of environmental consciousness was the emergence of problems that transcend national boundaries. These included concerns about nuclear radiation, pollution of the air and water, transborder shipment of hazardous materials, the global reduction of biodiversity (especially tropical deforestation), the spread of contagious disease, and dilemmas about use of outer space, the seas, and the Antarctic region. Popular environmental awareness was also amplified by high drama media-enhanced disasters. These include, for instance, the disastrous escape of toxic gases at a Union Carbide plant in Bhopal India in 1984 in which a First World corporation killed people in Third World communities. Other examples include the vast diffusion of radioactivity from the atomic nuclear reactor disaster at Chernobyl in the USSR (1986), visible pollution of the Rhine River running through much of Western Europe, a massive chemical spill into the Rhine from a Sandoz plant in Switzerland (1986), and vividly reported ocean oil spills in *many* places.[1] All of these disasters were widely reported in the international press, dramatizing that environmental vulnerabilities are ultimately global. Equally important to the emergence of global environmentalism were the more general threats of the 1990s: the international shipment of hazardous material, chiefly wastes; the disintegration of the stratospheric ozone layer; and the threat of global climate change. In each case scientific findings, publicized in the media, preceded public awareness and eventually led to popular calls for governmental and international action (Caldwell, 1992: 67).

Environmentalism was not limited to the affluent MDCs or stimulated only by global problems. As in the United States with grassroots and environmental justice movements, environmental movement organizations proliferated in the LDCs, most remarkably among the world's poorest and most marginal to the modern world economy. The most acclaimed LDC grassroots movement grew in the hills of Uttar Pradesch, India. In 1993, a local timber company headed for the woods above an impoverished village, and local men, women, and children rushed ahead of them to *chipko* (literally "hug" or "cling to") the trees, daring the loggers to let the axes fall on their backs. Maybe you've heard the word "tree hugger" as American slang for environmentalists protecting forests. That's where it came from. The Chipko movement has gone beyond resource protection to ecological management. The women who first guarded trees from loggers now plant trees, build soil-retention walls, and prepare village forestry plans. Similarly, communities of traditional fishers in Brazil, the Philippines, and the Indian states of Goa and Kerala organized to battle commercial trollers and industrial polluters who depleted their fisheries.

The people of the world's disappearing tropical forests around the world have begun to defend their homes as well, despite a pace of destruction that made their task a daunting one. In the late 1970s, a union of 30,000 Brazilian rubber tappers[2] in the rain forest decided to draw the line. Their tactics were simple and direct; where the chainsaws were working, men, women, and children would peacefully occupy the forest, putting their bodies in the path of destruction. This action was met with violent reprisals that continue today. In 1988, opponents gunned down Chico Mendez, the national leader of the rubber tappers—later to become a powerful symbolic martyr of rainforest protection for environmentalists around the world. At a high price, the tappers made modest gains: They helped reshape World Bank development policy, and international environmental organizations have called on the Brazilian government to set off large "extractive reserves" where tappers could continue their traditional livelihood. Across the Pacific, Borneo's Dayak tribe was less fortunate. Living in Malaysia's dense forest, they opposed cutting the tropical hardwood, the heart of Malaysia's export strategy. The Dayaks wanted it cut only sustainably and have battled timber contractors by constructing roadblocks and appealing to European consumers to boycott Malaysian lumber. Government intransigence stymied their effort.

An African federation known as Naam is among the most successful of the world's grassroots movements at mobilizing people to protect and restore overused agricultural resources. Building on precolonial self-help traditions, Naam taps vast stores of peasant knowledge and creativity to halt the deterioration of the drought-prone Sahel (in western sub-Saharan Africa). Naam originated in Burkina Faso and now extends under different names into Mauritania, Senegal, Mali, Niger, and Togo. Each year during the dry season, thousands of Naam villages undertake projects that they initiate with minimal external assistance. They build large dams and a series of check dams to trap drinking and irrigation water and to slow soil erosion. Hundreds of Naam farmers adopted a simple technique of soil and water conservation, developed by Oxfam UK (a multinational NGO) in which stones are piled in low rows along the contour to hold back the runoff from torrential rains. While halting soil loss, these structures dramatically increase crop yields (Harrison, 1987; Durning1989: 34–39).

In Kenya, East Africa, the idea of a grassroots environmental movement was conceived during political campaigns in the early 1970s. In 1974, an articulate and enthusiastic Kenyan woman, Wangari Maathai, attended an international U.N. meeting where she met and developed connections with women active in environmental and social justice NGOs. Returning to Kenya, she started a reforestation "Green Belt" movement, primarily among Kenyan women, for small farmers to plant trees. The goals of the movement were not only to staunch soil erosion, but to educate people about the interrelations of the environment to other issues, such as food production and health. The Green Belt movement intended to improve the income and sense of efficacy of women in particular. Remarkably, the Green Belt movement slowly thrived in a hostile and authoritarian political climate by using consensus and nonconfrontational strategies (vs. conflict strategies) and by developing support among international organizations. By 1992, 10 million trees had been planted that survived (a 70–80 percent survival rate) and as many as 80,000 women were involved in work at nursery sites. Gradually people learned that trees prevented soil erosion and loss of fertility. They came to see the link between loss of soil fertility, poor crop yields, and famine. At the local level, the Green Belt movement increased farm income and also helped to transform Kenyan communities by empowering them to help themselves. At the societal level, environmental degradation gained widespread recognition as an important issue. In 1989, environmental leaders from other African nations attended workshops conducted by the Kenyan Green Belt movement, and in 1992, after discussions at the U.N. Earth Summit in Rio, a Pan-African Green Belt Movement was launched (Michaelson, 1994). As you can see from these illustrations, protection of the environment in the LDCs is clearly and directly linked to the need for improved economic and material security. For obvious reasons, improved material security is often a more pressing issue among the poor in the LDCs than environmental protection per se. But people everywhere have come to understand a relationship between protecting and improving material security and protecting the environment.

The important point here is that increasing scientific understanding of global problems, the obvious impacts of environmental resource degradation, the roles of people in the emerging world market economy, and the emergence of environmental activism through much of the nonindustrialized world resulted in a truly global concern about environmental hazards. As evidence, consider the Gallup Organization's survey about the Health of the Planet (HOP) conducted with their international affiliates. HOP compared the public's views of environmental issues and environmental concerns among representative samples in 24 economically and geographically diverse nations. The study found significant levels of environmental concern among people in all nations, including rich ones (like the United States and Denmark) and poor ones (like Nigeria and India). But concerns were not identical around the world. Surprisingly, people in poor nations were more likely to rate the environment as a serious problem than their counterparts in rich nations. Understandably, however, they rated it less serious relative to other national problems (Gallup et al., 1993). Some have argued that poor people are not as concerned about the environment as the affluent because they have yet to experience the material security and the "postmaterialist values" that spawn environmentalism. But as the last chapter demonstrates, that is not true in the United States. Nor is it true around the world. The preponderance of evidence from HOP and other research contradicts the common idea that environmental concerns are a luxury reserved for affluent people and nations (Dunlap and Mertig, 1995). Reflecting both activism and public opinion, researchers estimated that 100,000 NGOs existed around the world that worked on some issue related to environmental protection. Most of these organizations were started as recently as the 1980s (Runyan, 1999: 13). I will return to the impacts of international environmental

NGOs later as part of an emerging world civil society, but before doing so, I return to the other important thread of this chapter, the growth of world trade—particularly treaties and their environmental implications.

ORGANIZATIONS, TRADE, TREATIES, AND THE ENVIRONMENT

To state the obvious, a phenomenon as large and important as the emergence of the world market and world system of nations was connected with many multilateral organizations, conferences, and treaties. They attempt to promote or regulate it to address the interests of people, organizations, and nations around the world. Undoubtedly you have heard of many of these, but here are some of the more important ones.

1. *The World Bank* (WB) and the *International Monetary Fund* (IMF) were created by negotiation of the gathered bankers and diplomats at Bretton Woods, New Hampshire, in 1944 to create a new international order after World War II. MDCs funded the WB to provide loans that increased international economic cooperation, economic development, standards of living around the world, and world peace; the IMF was to stabilize the international monetary system. The WB became the largest and most powerful international financial agency, providing massive development loans around the world for its clients among the LDCs (required to be states).[3] The WB's involvement with a client was a seal of approval that often freed additional monies from private banks and governments. The WB shaped the *kinds* of development projects for which clients sought support. It emphasized "getting money out the door" to new clients rather than monitoring ongoing consequences of projects. More importantly, its institutional bias was for large-scale, capital-intensive, centralized projects and for quantifiable economic returns rather than less quantifiable social or ecological benefits.
2. *General Agreement on Trade and Tariff* (GATT) was the major multinational agreement on trade rules, policy and dispute resolution developed in 1948 between major MDC war powers after World War II. Since 1948 there have been many rounds of GATT talks, each resulting in freer flows of international trade, and the average world tariff rates fell from about 40 percent after the war to about 5 percent today. By 1994, there were over 100 signatories to GATT, and another 30 countries abide by its rules while not being signatories. The last meeting, known as the Uruguay Round, was negotiated in 1994 and extended GATT regulations to most goods. The last and hardest to negotiate are (still) to get nations to remove their cherished agricultural subsidies and create international copyright protection for intellectual property. In 1993, GATT became the World Trade Organization (WTO) that truly represents the culmination of 50 years of painfully negotiated world trade policy. Still, as a grand political compromise, the WTO is still full of loopholes, special exemptions, exclusions, projections, and preferential clauses that protect the interest of various nations. It regulates trade but gives much less attention to issues about human welfare

and protecting the environment. As such, the WTO has become a symbol for much controversy and conflict about globalization.

3. *United Nations Conference on Trade and Development* (UNCTAD) is an official body of the U.N. formed in 1964 to represent the interests of the LDCs seeking an alternative to GATT. Though the WTO now attempts to address LDC interests, UNCTAD convenes every four years, primarily as a forum for the exchange of views, but does not negotiate treaties. In 2000 it met in Thailand. True to its origins, UNCTAD promotes trade preferences for LDCs and focuses primarily on commodity trade and agreements.
4. Other *United Nations programs* deal with environment and development issues, including the United Nations Environmental Program (UNEP), the United Nations Population Fund, (UNPF), the United Nations Development Program (UNDP), as well as older but more specialized agencies such as the World Health Organization (WHO), the Food and Agriculture Organization (FAO), and United Nations Children's Fund (UNICEF). Of these, UNEP has been the agenda-setting agency for global environmental politics because of its unique mandate. UNEP attempts to be the catalyst and coordinator of the environmental activities of diverse international agencies and has also been the dominant international organization to initiate and manage global negotiations. Indeed, since the 1970s most of the international environmental agreements resulted from negotiations sponsored by UNEP.
5. *European Union* (EU), formerly the European Common Market, is the largest and most developed regional trade zone in the world. Encompassing most Western European nations in a vast system of tariff reductions, cross-border labor arrangements, and passport and license agreements, it is now experimenting with a common currency. Since its beginning shortly after World War II, it gradually evolved into a political as well as an economic union; some national government functions are assumed by a European Parliament elected from all member nations. The EU became the world's largest (relatively) free trade zone, encompassing the mostly duty-free trade of Portuguese oranges, British oil, Danish herring, Finnish lumber, Italian Fiats, German industrial machinery, and most everything in between. Negotiating the EU was not easy: The stubborn French, for instance, refused to give up their cherished grain subsidies, while the environmentally finicky Danes haggled about almost everything. Interestingly, the more developed northern countries gave special concessions for 15 years to the less developed southern nations (Spain and Portugal) to enable them to catch up to the economic standards of the north. But the EU has become a true political economy that attempts to speak for Europeans on social, political, and environmental policy as well as tariffs and trade.
6. *North American Free Trade Agreement* (NAFTA) was both a response of the North American nations to the EU and an attempt to create a regional agreement. Based on a U.S.-Canada trade agreement in 1989, it metamorphosed into three-way negotiations among the United States, Canada, and Mexico, and was ratified in 1993. NAFTA does not aim for the level of political and economic integration of the EU. Rather, it is limited to a free exchange of products, commodities, services, and investments—along the line of the proposals the United States has made to expand world trade through a more powerful GATT

(Walthen, 1993: 8). NAFTA has become an established part of the economic institutions of the three nations involved, but like the WTO it was (and is) politically difficult and controversial. NAFTA controversies about such complex changes relate to perceptions of who benefits and loses among the nations involved and whether or not corporations and investors benefit at the expense of workers, communities, and environmental protection.

7. *Other regional agreements* and customs unions existed by 1994 (at least 60 of them). These include a Latin American one (MERCOSUR) that includes Argentina, Brazil, Paraguay, and Uruguay. There is one in West Africa. Another significant one for Americans is the large Association of Pacific Rim nations (APEC, of which the United States and Canada are members). Following the idea of U.S. President Bush, President Clinton convened in 1994 leaders from the western hemisphere's 34 democracies (excluding Cuba) to discuss the creation of a Free Trade Area of the Americas by 2005, creating a huge tariff free zone from the north shore of Alaska to Tierra del Fuego at the southernmost tip of Latin America and tighter economic integration among 850 million people. As I write, this zone has not become a reality, but it is still an idea afloat.

Approaches to International Trade

Such treaties are struck in the interest of increasing the total volume of international trade. But there are three very different existing approaches to international trade: free trade, protectionism, and managed trade.

1. *Free trade* means the unlimited exchange of commerce between buyers and sellers across national borders. It is often wrongly associated with total deregulation of trade, but in fact it does not require the elimination of product standards, worker protection rules, or environmental regulation. It simply requires that existing trade regulations be administered in a way that does not discriminate against foreign companies. The heart of free trade is the venerable neoclassical economic principle of *comparative advantage*, which holds that a country should specialize in those goods it produces most efficiently and trade with other countries for the goods they produce most efficiently, even if both countries can produce comparable goods at home. This theoretically enhances the overall level of economic activity in all trading nations. In addition to such economic advantages, free trade is understood eventually to promote political cooperation and intercultural understanding and to reduce the likelihood of political conflict. The EU, for example, was forged in part to solidify Europe and mute the historic animosities between European nations (which after all, did instigate two world wars).
2. *Protectionism*, on the other hand, seeks to protect domestic industries from foreign competition. Protectionism is an old and common way of understanding the interests of nations in relation to world trade. Otherwise known as economic nationalism, the idea was called *mercantilism*, in preindustrial Europe during the 1700s, when kings administered it. Protectionism was the dominant trading policy of nations until after World War II, and the idea still

governs much of the world trading system in spite of the growth of other ideas. It can take the form of tariffs on imported goods, quotas for imported goods (such as most countries have in the trade of textiles), or outright bans on certain imports, such as the Japanese ban on imported rice, or it can impose *export subsidies* that make national goods cheaper than unsubsidized foreign goods. Both American and European agricultural exports have been surrounded by layers of subsidies. Protectionism can also take softer forms of requests for *voluntary import restraints*, which the United States historically made on the import of Japanese autos—at the behest of powerful political pressures from American automakers. Protectionism may protect domestic industries, but it also stimulates other nations to retaliate. Protectionist policies may shelter domestic producers and jobs, but over time they result in higher domestic consumer prices, globally inferior goods, and uncompetitive industries. While a protectionist nation may be able to export goods that are unique or scarce, it may overall become a self-constrained domestic economy in a world where most things are available from a variety of sources. For example, the fiercely protectionist Soviet economy was able to export diamonds even as the rest of its economy remained isolated and certainly not competitive in the larger world market economy.

3. *Managed trade* is somewhere in between free trade and protectionism. Managed trade means that governments allow extensive international trade but seek to intervene through tariffs, subsidies, and other policies to make domestic products more attractive at home, to nurture new industries, and to stimulate domestic research and development. Managed trade describes the U.S. efforts to set numerical targets for Japanese imports of designated U.S. goods like supercomputers and auto parts. It is behind currently fashionable "domestic content" policies that exist in many industries (Walhten, 1993: 5–6).

While three "approaches" govern international discourse about trade, I think *in practice* they devolve into various forms of managed trade. Protectionism has lost much of its intellectual luster and legitimation, but it is certainly not a spent force in world trade policies. Enormous pressures exist for hard-pressed industries to seek such shelter from their governments in harshly competitive world markets (as, for example, Japanese rice farmers, French wheat farmers, and American auto and textile manufacturers have done). But it is still true that the evolution of the world economy toward greater global integration penalizes protectionism. Countries that persist in protectionist policies fear, and reasonably so, that if they resist the powerful centripetal forces of global integration, they will become economic backwaters. But while the idea of free trade dominated neoclassical economic thinking, it is very hard to practice. In fact, evidence to date suggests it is nearly impossible to practice in a pure form—although trade is certainly "freer" than it was in the 1930s. The reasons are not hard to comprehend. I argued in Chapter Eight that there are in fact no free markets unconstrained by political and cultural forces. The same goes for international trade. No trade agreement to date—WTO, NAFTA, the EU, or any other—is in fact completely "free." Importantly, transnational corporations have been among the most important backers of free trade agreements. They are also, ironically, a major reason why it is impossible to practice. How so?

Transnational Corporations (TNCs) and World Trade

If you read the fine print in WTO agreements, you would find it not "free" but still full of loopholes, concessions to special-interest groups, variable tariffs, and outright giveaways to industries that happened to be sufficiently wealthy and strongly represented in the negotiations. It was, for instance, executives and officials from TNCs—including Nestlé, Pepsico, Phillip Morris, Monsanto, and DuPont—that served in an advisory capacity to the U.S. WTO and GATT negotiators. Other nations were similarly "represented" and advised. There were *no* representatives from small businesses, farms, churches, unions—or environmental organizations (Hawken, 1992: 97).

TNCs are the structural skein of the world market system and the entities that governments most represent in that system. A TNC is a corporation chartered in one nation, but with branch offices, operating divisions, and/or subsidiary companies operating in many nations. The various divisions of a TNC can buy and sell and loan money back and forth at special prices far below real-world market prices and interest rates. It can declare economic "losses" in transactions between its divisions as tax writeoffs. A TNC can establish a headquarters in a nation with low or zero taxation while doing much of its business in nations with higher corporate tax rates. It can sell products in one country that are banned in others, for health or environmental reasons. TNCs scour the world for profits, cheap resources and labor, tax havens, and lax environmental regulation, thus manipulating the comparative advantages of many nations. Financial capital, not production, is at the heart of the TNCs, and their geographical amorphousness means that their control and regulation is increasingly less exercised by nations than by TNCs themselves. They facilitate the cross-border flows of capital, labor, and technology and tend to make geography or national identity irrelevant. This phenomenon produces economic concentration of assets, conglomeration, mergers, and megacorporations. As Chapter Two noted, many TNCs now have greater economic assets that do the nations in which they do business. Ironically, in the world market system, powerful TNCs partly and intentionally *nullify* comparative advantage by integrating and transcending the comparative advantage of many nations—both natural and legal. Indeed, the growth of systems of TNCs means that to a large extent national differences are no longer competitive factors.

The TNC-driven world market system has extended the global pursuits of profits by making the entire world an economy where all goods are exchangeable on a moment's notice. The arithmetic of this growth is simple: By the 1990s, the ten largest businesses in the world had collective revenues of $801 billion, greater than the smallest 100 nations of the world together. The 500 largest companies in the world controlled 25 percent of the world's gross output but only employed .05 percent of the world's population (Hawken, 1993: 91–92). As vast abstract connections, they lose much of the humanizing social regulation of face-to-face relations in traditional markets.

The largest TNCs grew at such rapid comparative rates that they resemble, at least in their political and economic power, separate nations without boundaries. Indeed, TNC and world trade advocates sometimes argue that they *are* the nations of the future. I hope not, because there are still important differences between the purposes of nations and corporations. Private corporations exist to make profits, and their primary concerns begin and end with that goal. That is not, in itself, evil, but you need to recognize that it is governments, civic organizations, and movement organizations (NGOs) that raise issues about social well-being, the quality of civic life, and environmental sustainability. Corporations usually do not—if such concerns conflict with profits.

The Environmental Impacts of TNCs

As you might guess, TNCs have many critics among religious, labor, social justice, and environmental groups. To my knowledge, no TNC has ever initially asked whether the introduction of its product into another nation is a social or ethical good without being bludgeoned to do so by pressure groups. At their outrageous worst, TNCs have promoted pharmaceuticals, pesticides, infant formulas, and contraceptives in LDCs after they were banned as unsafe in their home country. They have imported vegetables for American tables grown in LDCs and sprayed with banned pesticides—thus completing the circle of toxins. They have brokered the international sale of solid and toxic wastes to cash-strapped nations when rich nations regulated against them. Shipments of toxic industrial and pharmaceutical residues arrived in African nations from Europe and in Central America, the Caribbean, and Latin America from the United States. Where organized local resistance grew, the governments and corporations involved simply moved such dumping elsewhere. TNCs brokered the cutting of rainforests in Indonesia and Malaysia for export to Japan. Similarly, Texaco made a real mess in the Ecuadorian rainforests, where it dominated the nation's burgeoning oil industry for 20 years. Using environmental management techniques far below international standards, the company severely polluted Oriente—the Amazonian region in the eastern third of the country—and destroyed the lives of the indigenous people and small farmers who live there. Scholars documented 30 spills from breaks in the Texaco-built trans-Ecuadorian pipeline, and a gigantic spill of 16.8 million gallons resulted from these breaks—50 percent more than the much publicized Exxon *Valdez* Alaskan spill (Fierro, 1994).

TNCs face strong incentives to behave badly. Thus those in the natural resource and mining businesses often cosy up to whichever regime is in power, however nasty, in order to protect their investment. Those making consumer goods often flit to whichever country offers the best deal on labor costs at the moment. You could create a large catalogue of TNC human and environmental horror stories, for which case material is abundant. But you should also understand the more complex big picture of their impacts. According to OECD research,[4] TNCs around the world often pay better than do domestic firms. That is true even in LDCs, for example in Turkey, where wages paid by foreign firms were 124 percent above average wages. Their substantial power in the world is also more complex than it seems. Huge corporations come and go rapidly. Among the big American firms (in the Fortune 500 list) in 1980, 40 percent were gone by 1995. Globalization not only threatens small firms. The mergers among the lumbering giants that get so much attention are often defensive moves. Furthermore, don't overestimate the extent to which corporations will move *for lower environmental regulations alone*. Labor costs are generally more important as a component of total production costs. In the 1990s, environmental compliance costs of OECD TNCs were about 2 percent of sales income. A 1991 U.S. government interagency task force study found that in Mexico, "U.S. firms, particularly the larger multinational firms most likely to undertake large investments [in Mexico] often hold subsidiaries to a worldwide standard, usually . . . as high as they must comply with in the U.S." An example is the Ford Motor Company, which applied U.S. environmental practices in its Mexican subsidiary (Lasch, 1994: 55). More generally, research suggests that environmental standards often converge upwards, not downwards. TNCs do leave at least some capital in cash-strapped nations, and they often practice higher occupational and environmental regulatory standards than host nations themselves demand.

Again, I caution that citing positive and negative illustrations proves no general case. Whether you consider TNCs in general as irredeemably perverse structures or as potentially progressive, humane, and responsible is a function of ideology as well as evidence. My own view is that TNCs are neither the chief causes of evil in the world nor the harbingers of a brave new utopia of some sort. The whole truth is more complex.

Environmentalists, Neoliberal Free Traders, and LDCs

Revisit with me the famous "tuna-dolphin" case discussed earlier. Recall that the Marine Mammal Protection Act (MMPA) forbade the import of tuna that were caught with "purse-seine" nets that often (unintentionally) ensnared dolphins along with tunas. In August 1991, a three-member panel of negotiators for GATT ruled the MMPA embargo provisions to be an unfair trade barrier because they banned the importation of tuna from countries such as Mexico, whose tuna fishermen kill more dolphins than U.S. standards allow. GATT (and later WTO) rules do contain a specific rule that allows countries to have environmental protection policies that may contradict trade policies, but the panelists ruled that this exception pertained only to efforts of counties to protect the environment *within their own borders*. The ruling set off shock waves in the organized environmental community because it highlighted the long-suspected conflict between free trade and environmental protection. Environmentalists around the world viewed this decision as a smoking gun proving the environmental insensitivity of trade officials. On the other hand, trade officials—particularly those outside the United States—saw the decision as an important rebuff to U.S. efforts to unilaterally impose environmental policies that actually were a protection for national industries. While the tuna-dolphin case achieved great notoriety, there are many other cases of unilateral national actions and standards that became international irritants (Esty, 1993: 46; Zalke et al., 1993: xiii).

In 1996, the U.S. Court of International Trade ordered the United States to begin enforcing provisions of the Endangered Species Act designed to protect sea turtles. Fishing nets, particularly shrimp nets, killed 150,000 sea turtles per year. The law in question closed the lucrative U.S. shrimp markets to countries whose shrimpers were not required to use "turtle excluder devices" (TEDs), required by U.S. shrimpers since 1988. Spurred by the U.S. embargo, 16 nations moved to require TEDs, but India, Malaysia, and Pakistan chose a different tack—to launch a WTO challenge. In 1998, the WTO ruled against the U.S. law banning imports from foreign shrimpers without the use of TEDs that would reduce unintended killing of sea turtles resulting from shrimp trawling. Consider another illustration. After the Europeans banned the import of U.S. and Canadian meat (particularly beef) treated with growth hormones for both health and environmental reasons, the WTO ruled in favor of the North Americans, who suspected that the ban was really to protect the European cattle industry. When the Europeans defied the WTO and refused to revoke the ban, the United States imposed 100 percent tariffs on about $117 million worth of EU food imports, including fruit juices, pork, mustard, and some cheeses. Europeans, particularly the French, reacted angrily and symbolically, picketing McDonalds restaurants and doubling the price of Coca-Cola in places. If you add this to the widespread resistance of Europeans to consuming and importing North American genetically altered foods for similar reasons (noted in Chapter Five), a serious trade war (a real "food fight") is brewing among the MDCs (French, 1999: 23, 24).

Why Free Trade Advocates Dislike Environmental Regulation

Free trade advocates fear that stronger measures to protect the environment will stifle international trade between nations and corporations. For example, companies from LDCs may not have the technology or expertise to meet advanced environmental standards in MDCs and may lose access to those markets. Furthermore, free trade advocates suspect that environmental regulations are often designed and administered to protect domestic industries, as you can see from the previous illustrations. For these reasons, free trade negotiators advocate the "harmonization" of local, state, and national environmental regulations throughout the world so that such rules will not put any company at a competitive disadvantage when competing with the companies of another nation. Free trade theory argues that the level of environmental regulation is not important so long as there is a level playing field. In practice, however, there will be an intense political struggle to determine the stringency of harmonized international standards. The same legions of corporate lobbyists who advocate minimal environmental regulation in the United States can be expected to descend on the WTO and lobby aggressively for *low* global environmental standards. If standards are set too high, LDCs may be unable to afford appropriate technology, even in their own countries. But lower standards may unacceptably reduce environmental protection (French, 1993: 173–174).

Why Environmentalist, Labor, and Community Groups Are Suspicious of Free Trade

Environmentalists have long had some abstract misgivings about the consequences of deregulated free trade. *First,* free trade magnifies the environmental impacts of production by expanding markets for minerals and commodities far beyond national boundaries. *Second,* international trade allows nations that have depleted their own resources bases, or that have passed strict laws protecting them, to reach past their borders for desired products, effectively shifting the environmental impacts of consumption to someone else's backyard. *Third,* the free trade principle of comparative advantage, to the extent that it is practiced, pushes every national economy toward specialization. Environmentalists fear that increased specialization will cause LDCs to base their economies narrowly on export commodities like timber or certain agricultural crops (e.g., coffee or bananas) that can be environmentally damaging. Intensive specialization undermines efforts to promote diversified, community-based economies that are less resource intensive and more in harmony with local ecological constraints (French, 1993: 159; Walthen, 1993: 11).

The concerns of many environmental, labor, and citizens groups are far more concrete that these concerns. They worry that deregulated free trade will undercut existing environmental protection laws, health and worker safety provisions, and laws guaranteeing rights of workers to organize for better wages and benefits. Such laws can be declared "nontariff" restrictions on trade, and overturned. As companies seek to reduce production costs, many industries may shift production to nations with weak environmental, labor, and safety laws, or weak enforcement of all of them, as in the relocation of U.S. industries just on the other side of the Mexican border (the so-called *maquiladora* industries). They did so seeking cheap labor, but also in part to avoid the costs of U.S. regulations on the use, transportation, and disposal of toxic wastes (Walthen, 1993: 11). This is a reasonable fear, particularly since U.S. companies spend a lot of time complaining about the costs of environmental regulation. Still, it is important to say again that corporations do not inevitably exploit either people or the environment when they become international, as illustrated earlier in this chapter and at various places in the book. Furthermore, some respond to pressures from public opinion, activists, and official sources to become greener and fairer, and some find ways to turn "being good" into profits.

Why LDCs Are Ambivalent about Both Free Trade and Environmentalism

After the demise of the communist system, leaders of LDCs welcomed free trade as a mechanism of economic and social development. But think of LDCs in three layers. The most inspirational model for LDC free market development was the newly industrializing nations of Asia (e.g., Korea, Taiwan, Thailand, and Singapore), which provided models of development by investment and trade. Yet those nations are in very different circumstances than other LDCs. Below them are a level of LDCs where many are poor but there is a substantial nonpoor population and significant economic and technical capacity (e.g., India, China, Brazil, Costa Rica and Colombia, Malaysia). At the bottom is a tier of LDC nations of vast poverty, ruled by small elites, where development has not succeeded on almost any terms you want to imagine. These nations are less integrated in the world market economy and—not accidentally—more densely populated, and more environmentally degraded (Bangladesh, Nepal, Rwanda, Haiti, Somalia) (World Bank, 1991). International trade often allows wealthier and more educated classes of even the poorest nations to prosper. They become the local brokers for TNCs. But the poor and small rural landholders—of which there are vast numbers—have very little reason to be enthusiastic about the growth of world trade. It shifts investment to production for export, consolidates land holdings in large, more modern ("efficient") farms, increases the landlessness and poverty of peasants, and exacerbates migration to already overcrowded cities. Cruelest of all, producing food for export may increase the cost of food in local markets.

Even though LDC leaders want to make money by increasing exports to the MDCs, they know that historically the world market system has been rigged against them, because it is a system in which control remains in the MDCs and TNCs based in MDCs. LDCs were encouraged to utilize their comparative advantage and remain largely the exporters of minerals and commodities rather than striving for diverse industrialization, as have the MDCs. Yet as you can see from the foregoing, TNCs have often been busily reducing that comparative advantage to dirt cheap (exploitive?) labor and an environment to be plundered. The economic fates of LDCs are often tied to small fluctuations in the world price of products like coffee, hardwood, or mineral ores. LDCs have long advocated reform of the world market system to give them better prices for their mineral and commodity products (UNCTAD was formed with that goal). Even though LDCs participate in the WTO, they view it with suspicion. They allege that the WTO represents the interests of the MDCs and protects those interests while reducing the value of LDC commodity products—thereby perpetuating the LDC disadvantage in the world market system. It creates a kind of lottery system, where low-wage countries compete to make products for high-wage countries and hope that by allowing their workers to be exploited by multinational corporations, they too can hit the jackpot and eventually become high-wage countries (Hawken, 1992: 97).

LDC perception of environmentalism is similarly ambivalent. I noted earlier that environmentalism and grassroots environmental movements are indeed global, but there remain deep suspicions about environmentalism orchestrated and led by MDCs. To the educated classes in LDCs, the interconnected web of problems with dense population, massive poverty, environmental degradation, and related refugee and political instability are obvious. Countries at the Cairo conference in September 1994 almost universally recognized population problems, unlike earlier meetings. Yet while viewing population problems as real, they argue that mainly the MDCs have the luxury of the resources and technology to deal with them. They are afraid that MDCs will define environmental problems narrowly as population growth problems and are fond of pointing out that the average Canadian, Japanese, or German consumes much more and contributes much more to global environment degradation than the average person in Nepal, Bangladesh, or China. They believe that the MDCs will promote strategies for the mitigation of ecological problems that will defeat their aspirations for increasing economic growth and human well-being. They fear that MDCs will try to pass the buck to them; given their historic experience as colonies or dependent economies, they are probably right about this. A 1989 conference of LDCs (the "nonaligned" nations) underlined the issue of what has been termed *green imperialism*. The poor nations of the world are unwilling to enter into bargains with the MDCs in which they trade human well-being and development for global sustainability. They will drive hard bargains with the advocates of both free trade and environmentalism. My guess is that during the next century they will be the ones that MDCs can refuse only with increasingly difficulty.

Trouble in the World System: Seattle and the WTO

As you can see, the emerging world system and economy is pervasive and powerful but also permeated with contradictions, suspicions, and intense conflicts of interest. The WTO is something of a symbol and lightening rod for anger, fears, and critics of globalization. Three disconnected worldviews about world trade and globalization compete for dominance in the world today. *First* is the view of TNC and trade officials and the WTO that goods flowing across borders and oceans without tariffs promise economic redemption for all. *Second* is the view of environmentalists, who fear that what happens to sea turtles or the rain forest in one place can become legal precedent and accumulate to damage the viability of the planet; nothing can be protected or set aside if it undermines trade and TNC profits. A *third* worldview is more complex. It encompasses all those who are primarily concerned about economic and social justice and human rights—both a newer understanding of universal human rights (endorsed by the UN) as well as their particular rights for productive work, health, safety, dignity, and human well-being. This third worldview includes, I think, those for whom environmental problems are primarily problems of social justice. These views were all spinning around the same world, but not on the same axis. When the promises of abundance and the easy consumer life for all meet the realities of environmental constraints and deterioration, deepening income inequality, and the destruction of human well-being, intensifying conflict emerges, not sustainable world peace.

It all came to a head in December 1999 in Seattle, Washington, where the WTO, trade ministers, and corporate representatives were to meet (behind closed doors) and negotiate another round of agreements. From America and around the world, a motley collection of people representing diverse interests were there—labor, environmentalist, human rights, advocates of openness and democracy, and LCD nations. They had been mobilizing for more than a year and came to Seattle to demonstrate in the streets. Korean farmers protested trade rules that would flood their nation with cheap American and Canadian wheat and beef. Environmentalists from around the world protested the destruction of nature in the name of trade and profits. American labor unions protested the flight of their jobs in the global search for cheap labor. Advocates of democracy protested the closed nature of discussions about such important issues in which most of the world's communities and people were not represented. Activists opposing exploitive and child labor were there, as were a diverse collection of LDC groups protesting the idea that there should *be* some kinds of minimal labor and occupational safety standards (which would disadvantage whatever economic leverage they had). There were also those who came simply to redeem the antiwar activism of the 1960s. Most of the demonstrators were orderly and disciplined, but some were not, smashing windows and looting stores held to be symbols of the global hegemony of TNCs (especially McDonalds and Starbucks). That led to a nasty confrontation between demonstrators and Seattle police (dressed in riot gear like "robocops"). It was all reported as high media drama. The confrontation on the streets reflected disagreements among the trade ministers inside the hall (over the same issues and cleavages), and the meeting was terminated early and inconclusively without reaching *any* new agreements. The world was stunned by the diversity and effectiveness of the opposition, which derailed and shut down the WTO. How had such effective opposition developed, particularly when arrayed against the most powerful national and corporate actors in the world system?

A GLOBAL CIVIL SOCIETY

Earlier I spoke about the proliferation of environmental movement organizations around the world. They should be seen as part of an emergent global civil society along with feminist, human rights, and activist organizations of all kinds. And increasingly, they are not fragmented and isolated but are networked in such a way as to make the kind of mobilization that happened in Seattle possible. It is tempting to view the world as being shaped only by international political and business elites, and diplomats, but that is a lopsided view of things. Powerful forces for change and transformation also emerge from the ground up, so to speak.

What do I mean by civil society? Since the French Enlightenment, social theorists have divided the social world into three distinct realms:

1. *Political society,* which includes the state, public bureaucracies, and political parties
2. *Economic society,* which includes corporations, private business enterprises, professional and labor unions
3. *Civil society,* which includes a diverse ensemble of organizations that grow from community, religious, cultural, and ideological interests and affiliations (Gramsci, 1971)

To be more concrete, I am talking about the whole melange of community and civic organizations, social movement organizations, voluntary associations, and even recreational groups—what policy scholars call NGOs. Political society is about power and governance. Economic society is about livelihood and making money. The civil society sector is viewed as promoting norms of *civility* (such as fairness, equity, social obligation, and civil rights, and environmental protection). Such civil norms are also the ultimate source of the social consensus that undergirds law.

As long ago as de Tocqueville in the 1830s, observers noted the importance of organizations as the structural basis of democracy. Some later theorists, such as Max Weber, feared the erosion of civil society by a powerful hegemonic coalition of state and economic bureaucracies. The civil society mediates between individuals and the social order by giving form to the aspirations of individuals and protecting them from the intrusiveness and power of large bureaucratic institutions. After decades of neglect, there is a revival of scholarly interest in the notion of the civil society, as an important social formation for inhibiting self-centered individualism, restraining the predations of powerful states and corporations, and a force which, at its best, can "civilize" both behavior and institutions (Shils, 1991; Alexander, 1998). What is important here is the *embryonic global-level civil society* (of networked INGOs) that has been emerging in tandem with the global market system.

You can see this new phenomenon by considering the network of human rights groups in various nations that monitor abuses of human rights or the impressive and diverse groups that assemble to monitor the honesty of elections where the global community has reason to doubt the honesty of local officials. Human rights monitoring groups exist around the world and have monitored elections in South Africa, Haiti, Mexico, and countless other nations, often led by squeaky clean international personalities like American ex-president Jimmy Carter or South African Bishop Desmond Tutu.

I noted earlier that there are at least 100,000 environmental NGOs around the world. But the number of recognized INGOs grew from 176 in 1909 to more than 20,000 by 1996 (Runyan, 1999: 14). More significant than their numbers, are their international linkages and mobilization capacities. An impressive example is the International Union for the Conservation of Nature (IUCN). It includes NGOs in organizations from 60 countries, 120 government agencies, 350 NGOs and can draw on the world of six international commissions composed of over 3,000 volunteer scientists. The IUCN has a major influence on global agreements regarding wildlife conservation and species loss. Other international umbrella organizations exist working on rainforest preservation, pesticide control, poverty and hunger, climate change, and many other issues. The Pesticides Action Network unites 300 environmental, consumer, farmworker, union, and religious organizations in more than 50 nations and claims millions of indirect members (Porter and Brown, 1991: 59). INGOs can influence treaties governments in diverse ways, by lobbying or pressuring governments on issues, by publicizing information of strategic importance, or by lobbying at international conferences. The Natural Resource Defense Council (NRDF) successfully lobbied the U.S. ban on CFCs in spray cans as early as 1987, years before treaties banned it. An international coalition of conservationists and animal welfare groups organized a boycott of Icelandic fish in 1988 because Iceland was blatantly violating an international moratorium on whaling. It succeeded in getting 100 school districts and many fast food and supermarket chains to refuse to buy Icelandic fish. The $50 million loss of U.S. sales brought an end, temporarily, to Iceland's violation of the global whaling moratorium. Environmental INGOs convinced the World Bank and the U.S. government not to fund China's proposed Three Gorges Dam, which would have caused widespread ecological destruction and displaced 1.3 million people.

MDC civil society organizations increasingly pressure corporations themselves, rather than government agencies, by organizing shareholder resolutions and investor boycotts. Such have caused Home Depot, Shell, Enron, Citigroup, McDonalds, Quaker Oats, and Monsanto to modify or back away from environmentally and socially destructive behavior. Monsanto, a promoter of bioengineered crops, lost significant stock value (25 percent in 1999) by "not listening to what the market was saying," according to Monsanto president Robert Shapiro. He went on to say, "We thought it was our job to persuade, and we forgot to listen. Our enthusiasm for (biotechnology) has been widely seen—and understandably so—as condescension and arrogance" (*Bloomberg News*, 2000: 5m). In 2000, Kellogg, Coca-Cola, Safeway, and McDonalds were fighting off nineteen shareholder resolutions filed by thirty-three institutional investors urging them to get out of the genetically modified food business. Shareholder activism, particularly in combination with the growing popularity of social investment funds made up of corporations that pass human rights and environmental criteria, is an increasingly powerful force in shaping corporate behavior (*Bloomberg News*, 2000).

The international civil society grew in response to the global dominance of TNCs in the world. They reflect a growing trend toward democratization and particularly the collapse of authoritarian regimes around the world during the 1980s and 1990s (Diamond and Plattner, 1993; Harper, 1998: 253). They also reflect the growing limitations of national governments in a world system that often transcends national boundaries and the sovereignty of national institutions. INGOs were also facilitated by the new information technology just as the world market was. INGOs are often more adept and effective than both governments and businesses at responding to social and environmental problems that threaten human security—including access to food, shelter, employment, education, and health services. In Bangladesh, for example, a child is more likely to learn to read with the assistance of one of the 5,000 NGOs working on literacy programs than through a state school or other organization (Runyan, 1999: 16). Agencies like Oxfam have been more effective at addressing hunger and poverty than governments or TNCs by providing access to credit and establishing market cooperatives among low-income farmers around the world, from Mozambique to the rural southern United States (Georgia), thus addressing issues of poverty and hunger. The Gameen Bank in Bangladesh was legendary for pioneering a "micro-lending" approach to alleviate poverty that simply bypassed governments and large lenders (like the World Bank). INGO networks can emerge suddenly. One of the more stunning events of the late 1990s was the emergence of a coalition of 350 groups to promote a treaty banning the manufacture, use, transfer, and stockpiling of antipersonnel land mines. The treaty went into effect in 1999, over the strong opposition of Russia and the United States. Even so, the growth and power of the global civil society is quite remarkable.

But there are limitations. *At their worst*, INGOs can be unaccountable and opaque, deliberately misleading in their pursuit of narrow goals, and funded by hidden interests. Furthermore, the resources of even the largest NGOs and INGOs that can number their supporters in the millions, pale in comparison to the resources, capital, and technical resources of other world actors, particularly states and TNCs. INGOs work against powerful forces, and they often only dimly perceive how their effectiveness is limited by the institutional contexts within which they operate (e.g., fiscal policies, laws, tax structures, and so forth) (Fisher, 1993: ix). The key to their effectiveness is certainly not overwhelming resources, but their suppleness, flexibility, and a sharply growing learning curve for effective advocacy (e.g., knowing which buttons to push). *At their best*, INGOs represent concerned publics and are democratic representatives of communities that fill the voids and interstices left between corporations and governments. Cumulatively, they have helped to counter what NGO researcher Julie Fisher called the "narrow political monopolies" of governments and entrepreneurs. Compared to NGOs, governments are rigid, limited by political monoculture, and painfully slow to react (except militarily), change, and grow. When effective, they can prod businesses, states, and multinational organizations into action in four ways: setting agendas, negotiating outcomes, conferring legitimacy, and implementing solutions (Runyan, 1999).

For illustrations, consider the presence of civil society groups at recent world conferences sponsored by the United Nations: at Rio in 1992 about the environment, Cairo in 1994 about population and development, Beijing in 1996 about the global status of women, and Kyoto in 1997 about global warming. The United Nations encouraged civil society and advocacy groups to attend the proceedings, even though there were official delegates representing governments. However, they were surrounded and influenced by a vast collection of NGOs and INGOs representing environmental, labor, human rights, religious, and women's groups as well as some representing indigenous people. They had opportunities to represent their perspectives and interests. The presence of such groups made the conferences noisier, more chaotic, and more colorful than previous United Nations conferences. But they were also vastly more open and democratic than the WTO negotiations. The presence of such groups undoubtedly lent an aura of legitimacy to such proceedings.

TOWARD A "GREENER" GLOBAL SYSTEM?

As you can see, globalization, with all its problems, is a powerful and pervasive force, but so are global environmentalism and emerging international civil society networks. They represent a clash between different spheres of international relationships and law. They present the world with major legal challenges because it is not clear which kind of treaty trumps another when two are in conflict. How do such international agreements evolve? What are some of the substantial obstacles and difficulties in creating a more just and environmentally sustainable ("greener") world? What are some processes and strategies by which such progress can happen?

Evolving Treaties and International Regimes

An international *treaty* is a document of intentions signed by nations, which may or not be ratified by their legislative bodies. *Regimes* are the resulting systems of action and organizations that monitor compliance. They have timelines and quantitative targets related to goals, and legal sanctions for noncompliance, which vary in specificity and effective enforcement. The WTO is a long-developing regime about world trade. Many global environmental treaties have been agreed to in recent decades. Some give only lip service to good intentions, but others have effective developed regimes. To enumerate and describe such treaties in any detail would take another book.[5] To continue progress about environmental and human problems in a world context, a variety of kinds and levels of international regimes must continue to evolve. Minimally, an *international regime* is a set of norms, rules, or decision-making procedures that produce some convergence in actors' expectations in a particular issue area. An international regime may be a *convention* containing all the legally binding obligations on parties or a *framework convention* that imposes few specific, binding obligations but rather states goals, norms, and general principles, with legal obligations yet to be negotiated. Complete regimes include *protocols* that spell out specific, detailed, and binding obligations on all parties (Porter and Brown, 1990: 17).

The 1973 Convention on Trade in Endangered Species of Wild Fauna and Flora illustrates a complete convention with protocols that was negotiated, so to speak, at one sitting. The Montreal Ozone Protocol (1987), discussed in Chapter Four, was initially a framework convention that was followed in quick succession by a series of legally binding protocols (e.g., London, 1990). Earlier I mentioned the United Nations Conference on Environment and Development (UNCED), held in Rio de Janeiro, Brazil in 1992. UNCED was the most comprehensive international environmental conference in recent times. Over 100 heads of state attended, making it the largest environment and development summit meeting ever. It resulted in a framework convention, the nonbinding Rio Declaration on Environment and Development.

The Rio Declaration placed human beings at the center of sustainable development concerns, and emphasized that environmental protection is an indispensable part of the development process. It stressed that the eradication of poverty must be a concern of the whole world and that the LDCs deserved special consideration. It argued for global consensus rather than unilateral action by individual nations. It was nonbinding because a small number of nations—including the United States—refused to sign it as a legally binding convention, to the great disappointment of most government delegations, environmentalists, and NGOs in attendance. Two legally binding conventions resulted from UNCED, the *Biodiversity Convention* and the *Framework Convention on Climate Change*. They provided for the transfer of technology and scientific cooperation between the MDCs and LDCs. While these conventions recognized the sovereignty of each nation, they nevertheless moved environmental concerns toward a system of global governance. The climate change convention led to subsequent conferences about climate change in Kyoto and Buenos Aires, discussed in Chapter Four. Ironically, an important achievement of the UNCED was *Agenda 21*, drafted in preparation for the meeting. Though nonbinding, it is a long-term strategic agenda setting an action plan to reverse environmental deterioration and promote sustainable development (Moss, 1993: 118–119). As you can see from the current problems of the WTO, the UNCED Rio conference outcomes, and the more recent Kyoto climate conference, the creation of effective international regimes is fraught with difficulties. What are some of them?

Hard Rocks: Problems in Global Regime Formation

To be effective, international regimes must be negotiated by and binding on diverse multitudes of nations and agencies, including the largest and most powerful and the poorest. At best, this involves long and contentious political negotiations. I mention only three kinds of difficulties. *First* there is the problem of *veto states* and *veto coalitions*. While effective conventions require unanimity, states do have different environmental circumstances and sociopolitical dynamics that result in very different environmental, social, and economic interests. The actual costs and risks of environmental protection are never equally distributed among nations, nor do nations have the same perceptions of equitable solutions to environmental problems. For every global environment issue, there is at least one and sometimes more groups of states whose cooperation is essential (as was that of the United States in the Rio UNCED negotiations). On a global whaling moratorium, for instance, Japan and three other states account for three-fourths of the world whale catch and therefore had veto power over any agreement to save the whales. Brazil, India, and China can block an international agreement on climate change by refusing to curb fossil fuel consumption in their development programs. If they continue in becoming "dirty coal economies," their combined carbon emissions could eventually overwhelm reductions carried out by the rest of the world. When such states oppose or block agreements, they become veto states or coalitions.

Economic and geographic factors impel some nations to be veto states. Because Norway's coastal population is dependent on fishing and whaling, Norwegian fishermen prevailed on Prime Minister Gro Bruntland to defend Norwegian whaling despite her role as chair of the first U.N. environment conference (at Stockholm, 1972). Norway, Japan, and Greece have tended to be swing or blocking states on questions of marine pollution from oil tankers because of the importance of their shipping industries. Brazil's agroindustrial elite has long been at odds with any global regulation of deforestation. The United States and Canada, both large producers and exporters of petrochemicals, blocked the initial agreement called for by European states in 1990 to limit emissions of greenhouse gases. Similarly, at the 1997 Kyoto conference, the United States was a major impediment to forming a more demanding and more enforceable convention about greenhouse emissions. (As of 2000, the U.S. Senate has not ratified the anemic treaty resulting from Kyoto.) Such factors can work to make nations leading states as well. For example, nations with large populations in low-lying coastal regions (such as the Netherlands, Australia, Bangladesh, and all island republics such as the Maldives) have been leading advocates of climate conventions because of their particular vulnerability to rising sea levels (Porter and Brown, 1990: 38–43). Pollution "downwind" nations, such as Canada, have been leading nations in promoting conventions about acid rain, while "upwind" exporters of airborne pollutants (such as the United States) have opposed them. As late as 1989, the United States continued to call for more research about the causal connection between acid rain and forest dieback, which at the present state of science is akin to tobacco companies continuing to deny a causal connection between cigarettes and lung cancer.

A second kind of difficulty has to do with problems of competition, conflict, and overlapping jurisdictions among public agencies dealing with environmental issues. Among U.N. agencies, for example, large, well-entrenched, bureaucracies of the FAO and WHO believed that they were dealing adequately with the environmental aspects of their mandates, and resisted UNEP's oversight attempts. The FAO is one of the most powerful international organizations within the U.N. system. It has had a major impact on the global environment—critics allege much of it has been negative—through its promotion of export crops and heavy use of chemical inputs. The FAO was one of three international organizations designated to establish national tropical forest action plans (TFAP). UNEP was excluded from any role in these plans. The consequence of the FAO's role has been to weaken the conservation thrust of such plans, because the FAO's main constituency has been government forestry departments who often promote the commercial exploitation of forests (Winterbottom, 1990). Lacking the large staff and budgets of the older more specialized agencies, UNEP is not always treated as a bureaucratic equal. And the U.N. Development Program (UNDP) seems to have been competing with UNEP for the authority and financial support to deal with environmental problems in LDCs. The UNDP, for example, entered into a relationship with the Inter-American Bank and produced a document about the relationship between poverty and environmental degradation at the same time that UNEP Latin American ministers were organizing Agenda 21 (Porter and Brown, 1990: 52).

A *third* kind of difficulty in forming effective global environmental treaties/regimes stems from the need to change the priorities and operating styles of other important multinational agencies, particularly the "Bretton Woods" institutions (the WTO, the IMF, and the World Bank). Having examined the case of the WTO earlier, consider the World Bank (WB). In the 1970s and 1980s, it supported rainforest colonization schemes in Brazil and Indonesia, cattle ranching projects in Central and South America, and tobacco projects in Africa. The WB prefers massive physical projects, such as large hydroelectric dams, highways, power plants, and other large scale public works projects that—as often as not—enrich LDC elites at the expense of ecosystems and the politically powerless. In fact, in its first 50 years, the planet's biggest benefactor has loaned hundreds of billions of dollars to LDCs, yet much of that world remains desperately poor with badly plundered natural resources (Porter and Brown, 1990: 54; French, 1994). Worldwide criticism of the WB began to mount in the 1980s from environmentalists, social activists, NGOs, and—more significantly—from donor MDC parliamentary bodies threatening to cut its funds.

In response, WB reforms created a central environmental department in 1987, announced that all loans were to be screened for their environmental impact, and promised that NGOs would be more actively consulted in both donor and borrowing nations. In actual practice, the WB often continued to ignore or violate these goals. But by the 1990s there was some progress: The WB made "freestanding" environmental loans and some loans earmarked for the alleviation of poverty. But the bulk of its lending still went to the kinds of large-scale infrastructure projects as those just mentioned. At its fiftieth year in 1994, popular pressure grew for fundamental reforms. Twenty-three U.S. development, environmental, and other civic organizations formed a coalition provocatively entitled "50 Years Is Enough" and forged ties with similar NGO networks in Europe, Asia, Africa, and Latin America. A network of parliamentarians from Europe, the United States, and Japan pledged to work through their respective legislatures to encourage the WB to amend its charter to incorporate environmental, human rights, and social concerns (French, 1994: 18). Results have been mixed but promising.

Effective Global Treaties and Regimes: Bargains, Strategies, and the World System

Having discussed problems, let me turn the coin to discuss conditions and strategies for effective global treaties/regimes. Agreeing to treaties requires bargains among parties to such negotiations. The word *bargain* conjures up many images, some positive and some negative. A bargain involves at least two actors and two issues, and it implies a tradeoff between parties on issues. Global bargains are more complex when there are many parties and issues. A bargain is an exercise in negotiated policy reform. It forces everyone to ask: "How much are we willing to give up in the short term to realize longer term goals? What do we want in return?" MDCs will approach this question in the context of increasing domestic political pressure to "do something" about the environment. LDCs are under equal intense pressure to reduce poverty and foreign debt and open markets for their products without further degrading human or ecological capital. *Effective bargains require two conditions:* (1) Each party at the table must have, directly or indirectly, those with the power, knowhow, and finances actually to make changes, and (2) negotiations must establish how changes can be linked to promote the net interest of all concerned. It is in the differences between parties that there are opportunities for tradeoffs (MacNeill et al., 1991: 81–85).

There are examples of such bargains being struck. The 1987 Montreal Protocol (on substances and depletion of the ozone layer) is a case in point. During negotiations the major MDCs were unable or unwilling to deal with certain issues of crucial importance to LDCs, such as those of sharing of environmental resources, financial burden sharing, and access to technology. As a result, a number of LDCs, notably China and India—with a huge potential to increase CFC emissions—refused to adhere to the protocol. At a second meeting of the parties in London in June 1990, the MDCs agreed on amendments to the protocol that established a fund of $160 million to help LDCs finance the cost of phasing in substitutes for CFC ($240 million if India and China participated). In response, China and India reconsidered their positions and signed the protocol after being granted a grace period during which they could continue to produce CFCs while substitutes were being phased in (MacNeill et al., 1991: 64).

Most thinking about effective treaties and their ratification emphasizes that states (or "parties" to the negotiations) are interested actors that choose rationally among various proposals that distribute costs and benefits in accordance with their interests. Rational-choice theory, as discussed in Chapter Eight, described this pervasive way of understanding the choices of individuals and organizations. But other factors are at work. Frank analyzed the ratification of international environmental treaties among a cross section of nations from 1900 to 1997. He found the strongest predictor of whether or not nations would ratify international environmental treaties was their linkages to the world society. Linkages were understood and measured by how common, relative to size, were memberships and affiliations of a nation's NGOs to INGOs.[6] The more frequent the international affiliation of a nation's NGOs, the more likely a nation was to ratify international environmental treaties, and this linkage was a more powerful predictor variable than economic development, population size, and political democratization variables. This finding supports the idea that nation-states exist within a wider world system in which discourse about environmental protection is central and legitimate. It also supports the notion of the importance of a global civil society, as I argued elsewhere (Harper, 1998).

Turning from empirical conditions for effective treaties to strategies, three broad strategic themes exist to promote more effective global environmental regimes:

1. A continuation of the political process that has brought *incremental change* in global diplomacy during the last several decades
2. An effort to create a new level of *"North-South" partnership* on economic, developmental, and environmental issues
3. An attempt to create new institutions of *global environmental governance* that would reduce the power of individual states to block or weaken environmental agreements and ensure enforcement. (Porter and Brown, 1990: 144)

Incremental Change

Incremental change assumes that reasonable progress is possible without radical change in policy or institutional frameworks at the global level. Incrementalism denies the need to address the interrelatedness of global issues and forces, dealing as it does with issues on a case by case basis. And the incremental approach would do little to bind LDCs to global action agreements because it would leave unresolved the reluctance of MDCs to divert major resources to LDCs in return for their participation. Incremental strategies would not demand any fundamental changes in lifestyles or domestic institutional arrangements. It would settle for modest progress toward effective regimes, on the assumption that further increments of progress will follow later. But reasonable projections of greenhouse-gas emissions, tropical deforestation, diversity loss, or toxic chemical pollution over the next two decades suggest to many that incrementalism is unlikely to build the momentum necessary to reverse the serious trends (Porter and Brown, 1990: 143-147).

The North-South Partnership

The North-South partnership seeks to avoid dealing separately with issues of debt, trade, financial flows, and technology transfer and would make cooperation on such issues a central feature of environmental diplomacy. The so-called Brandt Commission that convened in Cancún, Mexico, has made such proposals as early as 1969. It need not be a single, all-encompassing agreement. It would start from the assumption that the environment and natural resources can only be conserved under conditions of sustainable global development, and that the present world market system makes this impossible. A North-South partnership would recognize that, particularly after the cold war, the fundamental global polarity is no longer East-West, but North and South, as defined by relations between MDCs and LDCs. Achieving such a partnership would require a series of new arrangements covering a range of economic, geopolitical, and environmental issues, which could take many years. It would require that MDCs display a new willingness to address the primary economic concerns of LDCs as well as the objective obstacles to environmental and resource management. It would require LDCs—particularly the largest and most important resource-holding states such as Brazil, Mexico, China, India, and Indonesia—to conduct development plans in more environmentally benign ways with technical and financial support from the MDCs. Concerns about climate change and deforestation could be linked with population issues. The MDCs would be required to open markets fully to LDC products and to remove presently high MDC agricultural subsidies. They would be required to provide the capital and technology transfer to enable LDCs to leapfrog over the dirty energy–intensive phase of economic development.

All of this requires the recognition of links between mutual dependence and self-interest among nations, which often does not exist in the current zero-sum approach to international negotiation. Indeed, strong resistance exists in the United States, Japan, and Germany for the kinds of resource transfers implied in this approach, as well as for the removal of still existing protectionist barriers (particularly on agricultural products). North-South partnership would require a complex and delicate set of arrangements that could collapse for many reasons. Its critics claim it is too costly. Proponents counter that the costs of not working toward some such arrangements could prove fatal to global transformation toward sustainability (Porter and Brown, 1990: 150–152).

Global Environmental Governance

Global environmental governance has been increasingly advocated by both unofficial and official observers. It is founded on the widespread perception that existing national and international institutions as well as international law are inadequate to the challenges facing the world in the coming decades.

In the words of New Zealand's Prime Minister Geoffrey Palmer, the current system of "small incremental steps, each of which must subsequently be ratified before it comes into effect is mismatched with the earth's fast moving environmental problems" (Palmer, 1989). Furthermore, advocates of the governance approach maintain that the absence of effective enforcement mechanisms is a cardinal weakness of the present system (MacNeill, 1990: 23).

The global environmental governance approach suggests that only far-reaching institutional restructuring at the global level can stem the tide of environmental degradation. The most ambitious proposal, which surfaced at an international conference at the Hague in 1989 sponsored by Dutch, French, and Norwegian prime ministers, was to create a global environmental legislative body with the power to impose environmental regulation on nation-states. As you might guess, there was significant resistance to this idea, particularly from the United States, Britain, China, and Japan, who were reluctant to yield their national sovereignty. As a result, no explicit plan for such a body was passed. The final declaration, a framework convention, now signed by 30 nations (but not the United States, China, or Japan), called for a U.N. authority that could take effective action "even if . . . unanimous agreement has not been achieved" (Porter and Brown, 1990: 153).

Another component of the global governance approach is the development of a body of legal concepts that would further reduce the zone of absolute sovereignty of individual nations in issues affecting the global environment. Three legal norms were suggested for global environmental governance:

- The *precautionary principle*, which would site the burden of proof from the opponents of a given activity to those engaged in the activity in question, e.g., potential polluters
- The concept of the *common heritage of mankind*, which would overturn the old legal assumption that the sovereign states could do whatever they pleased outside the jurisdiction of other states
- The concept of *intergenerational equity*, that would give future generations legal standing

To some extent, these norms are already operative. The common heritage principle is embodied in the existing law of the seas, and large proportions of the MDCs accept the precautionary principle. Yet none of them has acquired the force of law to qualify as international customary law.

Such a global environmental governance regime would be very difficult to create: It would come into being only if the strongest international actors (*hegemons*) assert the necessary power to create it. So far, proposals have come from less powerful states. Furthermore, weaker states and LDCs would continue to have veto power over it. In other words, its creation would require a true global consensus. The arguments against it are obvious: higher costs of large-scale monitoring and policing mechanisms, threats to national sovereignty, and, of course, objections of those who argue that the world faces no real global ecological crisis that would require such a global governance mechanism (Porter and Brown, 1990: 156). Indeed, the global governance approach seemed hopelessly idealistic only a few years ago, but in the 1990s it was given new legitimacy by the support from many quarters: nations, multilateral agencies, and civil society groups. But there is powerful opposition rooted in the reluctance to give up national sovereignty over environmental policy (or, for that matter, any policy). More fundamental in determining the fate of global environmental politics in the coming decades is the spread of popular consciousness and activism about environmental threats. If there is a single force that could sweep away the formidable obstacles to strong new global environmental regimes, it is the support of voters, grassroots activists, and an increasingly dense, vocal, and effective civil society sector throughout the world (Porter and Brown, 1990: 157). On this score, there is reason for optimism.

It should not be lost on you that such an environmental global governance regime would immediately run into a headlong collision with the aims and criteria for dispute resolution of the WTO—as that regime is presently constituted. This prospect raises a much larger question. In 1999, some of the street protestors in Seattle and some media commentators argued that the shutdown of the WTO conference sounded the death knell of globalization and signaled a return to economic nationalism. But I think the world market, global cultural diffusion, and transboundry character of environmental problems make it far too late in the day for that. The problems at Seattle suggest, rather, the futility of creating a global trade regime without the other broadband dimensions of global governance that people care about: environmental protection, human rights, and the dignity of human labor. What we need, I think, is not just a global environmental regime, but an extension of the broad spectrum of global governance, dealing with the full range of issues—economic, social, cultural, human rights, and environmental—with which nations now contend. Such global governance can grow without implying the creation of a centralized world government, which is probably not in the cards. The Rio accords on global warming and biodiversity, without enforcement mechanisms, are notable for the blithe disregard with which signatories have treated them. The history of international labor accords tells the same story. The International Labor Organization (of the U.N.) existed for 81 years and without the force of sanction has been unable to do much of anything. Political and environmental globalization needs to catch up with economic globalization. For an American sitting president (Clinton) to say as he did in Seattle that "global laws on the treatment of workers should be enforced with real sanctions authorized by a worldwide body" was a milestone in the evolution of global governance (Wright, 2000: 23).

PROGRESS AND PROBLEMS

Is the main story in human-environment relations one of progress or one of progressive deterioration? It is tempting to try to invent a sort of box score, but that would be an artifice. So let me merely mention some signs of progress and some signs of deterioration, and let you draw your own conclusions as to which is the main story. Among the signs of progress, anthropogenic carbon emissions dipped slightly in 1998, even though concentrations of atmospheric CO_2 continued to increase. How much of this effect is attributable to increasing energy efficiency and how much to the ongoing free fall of the Russian economy and Asian recession is not clear. Energy efficiency is palpable, however; the global growth in compact fluorescent light bulbs since 1990 saved enough energy to close eight large coal-fired power plants. Bicycle production continues to triple auto production on a global basis, and bicycle use dwarfs auto use in India and China and is an important component of transportation in Europe. Bicycle commuting even tripled in the United States in the last decade (but from a very small base!). More significantly, world energy use is undergoing a transformation toward a declining use of coal, a leveling off of petroleum use, and a growth of natural gas—cleaner burning and lower in carbon content. Nonfossil renewable energy (wind, solar) has made unexpected progress. Major energy companies are preparing and positioning themselves for a transition away from a petroleum based energy system (Brown, 1994a: 17–18). All of this bodes well for stabilizing climate change, but changes in efficiency will not be sufficient if the aggregate energy demand of a larger population continues to increase.

But there is still plenty of bad news to go around. Scientists still expect the earth's climate to warm significantly over the next century. The main questions are how much and how fast. Human activity continues to produce prodigious declines in biodiversity, extinguishing species that evolved over long geologic time, and few areas of the earth are now really wild: We live in a "socialized" ecosystem. Migratory species are threatened because of the fragmentation of habitats by human uses. The accumulation of waste and toxic pollutants is still a global problem. Grain production slowed over the last decade, caused by limited availability of new cropland and fresh water for irrigation and by the declining response of crops to additional fertilizer. Limits are beginning to constrain the production of beef and sheep on the world's rangelands, and ocean fisheries are being depleted. Water scarcity is becoming commonplace, stirring competition among industrial, residential, and agricultural sectors. The growth rate of the human population is slowing everywhere, but population momentum is a powerful threat to environmental sustainability, as nearly 84 million people a year are added to the world's population (mainly in LDCs). But profligate consumerism among MDCs is probably a more palpable environmental threat. Socioeconomic inequality continues to grow, both within and between nations in the world system, and a significant proportion of the world's people (perhaps 25 percent) lives in absolute material destitution. Environmental activist Mark Dowie captures the mixed picture this way:

> There's clearly measurable progress. People on a broad level are more aware of environmental issues. But awareness may not be enough. It's hard escaping the conclusion that it's all too little, too late. We protect one forest, and lose five, one species is brought back from the brink, but 100 quietly disappear. We control the damage to the ozone layer, but lose ground on global warming. (cited in Motavalli, 2000: 29)

Progress and Problems: Political Security

At some point these interlocking problems will force their way onto the global political agendas in an even more forceful way than they have already done. But using the words "at some point" might encourage you to think that these interlocking environment-human problems are abstract, distant, and must be dealt with by somebody else, but not us here and now. *That is an illusion.* There is a dimension of these problems that I have not mentioned—the way they affect our national security in the world. I'm not talking about some abstraction but about the kind of national security for which we mobilize troops and send soldiers to distant corners of the world. The protracted Cold War between the Soviets and the West is over. But in recent years Americans and Europeans have mobilized troops to send to fight or keep the peace in the post–Cold War world (to Iraq, Bosnia, Serbia, Somalia, Rwanda, and Haiti). With the exception of the conflict in Bosnia and Serbia, human-environment conditions are related to all of these conflicts at a deep level, below the obvious political conflicts of nation, faction and tribe. The Gulf War obviously had to do with our dependence on oil for transportation and the world distribution of that resource. And it took place in a region with rapidly growing populations desperately short on water and agricultural resources. Water shortages play a critical role in Middle East tensions and conflicts. If you look below the level of ethnic and factional hatred in Somalia and Rwanda and the self-devouring political culture in Haiti, there are similar characteristics. They are all countries where the majorities are illiterate and live in absolute poverty. They are nations that have high birthrates and dense populations, overused agricultural bases, and a devastating drop in per capita food production. When the political conflagration finally came in Somalia, Rwanda, and Haiti, it instigated vast refugee flows that caused serious problems to their neighbors (in the case of Haiti, it was the United States).

Here's the point: Dense populations of desperately poor people on an overused biophysical environment can jeopardize the political stability of regions and the world. This means not just hypothetical problems, but real ones that require real intervention (by the U.N. or alliances of nations). And real tax money (yours). And sometimes the lives of real soldiers on the line (your sons or daughters?). Our response to such problems has typically been too little too late, and our ability to deal effectively with them in long-term ways is ambiguous. But in an interdependent world the MDCs can't just walk away from such problems, either—even ones of only regional scope. Rwanda, Somalia, and Haiti are small nations that pose no serious national security threat to the MDCs. But what if such conditions spread, even in lesser degrees, to many nations with large populations (say, China, India, Egypt, or Mexico)? In the post-Cold War world, if human-environment conditions progressively deteriorate, the nations of the world may have to invent new international regimes to control such situations and pay for the costs of doing so (Brown; 1994b; Gleick, 1991; Kelly and Homer-Dixon, 1996; Mathews, 1991).

Goals for Change: What Needs to Be Done

This whole book has attempted to convince you that human-environment problems are indeed very real problems. They shape our quality of life today and will determine the kind of world our children and their children inherit. I have also tried to convince you that we should act—individually, in local communities, nationally, and internationally—to address these complex, interlocking problems. We should do so with all the resources, intelligence, moral courage, and willingness to make reasonable bargains with others that we can muster. We should not fool ourselves—the costs of addressing such problems are very great, yet they are not simply costs, but rather rational choices that invest our energies and resources in a sustainable world, rather than in short-term gain with disastrous long-term consequences. The sooner we act, the more options and maneuvering room we have, and the more chances to fail and try it again.

We know *what* needs to be done. It's a remarkably simple list of things to describe, things I discussed earlier in more depth. Most basically, human survival in a sustainable world requires that we broadly promote three sets of ideas among people around the world:

1. That cohabitation with nature is necessary.
2. That there are limits to human activity.
3. That the benefits of human activity need to be more widely shared. (Kates, 1994: 118)

More specifically, we need to

1. Stabilize human populations.
2. Reduce excess material consumption.
3. Change damaging technologies into environmentally more benign ones.

We must focus on those things that indirectly affect these three goals. We must abolish absolute poverty and work towards greater equity within and between nations: Extreme inequality is one of the powerful causes of overconsumption and environmental degradation. Insofar as possible, we must develop agricultural and industrial systems that mimic nature. Furthermore, we must focus as much of our energies and resources on education, health, and family planning as we have on industry and agriculture—which have often failed us. We must cushion or remove the weight of debt, which inhibits progress among the poor nations of the world. LDCs must be encouraged not to seek the kind of economic development that repeats the environmental history of the MDCs. Above all, we must create reasonable equality between men and women in education, politics, law, property rights, and economic opportunities.

We must strengthen institutions and build alliances to shorten the delays in perceiving and acting on environmental problems. We must improve the capacity to monitor changes in the environment. We have the technology to do so: We need to commit the funds and the manpower to do so. We need to educate farmers, foresters, manufacturers, and citizens to spot early warning signals of environmental damage.

We must promote democracy where it does not exist. Where it does, we must strengthen it by improving education and creating the right of access to government and company information on environmental impacts. We must introduce free markets where they do not exist. And where they do, their blind spots and perverse subsidies should be removed: Environmental and social costs should be fully reflected in prices. We must increase security of tenure or ownership over productive land. And on forests, rangelands, rivers, oceans, and the atmosphere, we must strengthen community, national, and international controls (Harrison, 1993: 298–299).

We must improve our understanding of environmental processes. We are only beginning to understand the incredible complexity of ecosystems and the ways they interact with human societies. We need a comprehensive understanding of how human populations, institutions, and technologies interact with the biophysical changes in the environment itself.

We know what needs to be done. We will have to invent how do these things. Doing so will require that we transform our consciousness, our behavior, and the structures and institutions that organize human social life. These are indeed tough challenges. If we are unable to meet them, we may demonstrate that *Homo sapiens* are indeed inventive and clever but fools as well. I believe that we will not prove ourselves to be fools.

CONCLUSION: LONG TRANSFORMATIONS AND THE THIRD REVOLUTION

In a felicitous phrase, physical scientist and futurist visionary Gerald Platt said that we stand at the "hinge of history." By that he meant that we now find ourselves on the cusp of one of the great transformations in the history of humans on the earth, with a door closing on one epoch as it opens on another. There were only two such great transformations in human history that fundamentally transformed human livelihoods, social life, and connections with nature. They were truly revolutionary in historic retrospect but probably did not seem so from the vantagepoint of people living them out one day at a time.

You know these; I discussed them earlier in some detail (in Chapter Two), so for this conclusion I review them only briefly. The first was the *agricultural revolution* of Neolithic times, when most humans gradually evolved from hunter-gatherers to horticulturalists and the cultivation of grain. It was based on the diffusion of certain discoveries and innovations but also on the tendencies of hunter-gatherers to consume beyond the limits of their wild food supply. Agriculture magnified people's impact on the environment and changed their attitudes as wilderness changed from a source of sustenance to areas of weeds, pests, and predators to be cleared for gardens and fields. Thus the agricultural revolution was a fundamental transformation in human-environment relations. Because it enabled populations to grow large and remain settled, it produced a fundamental social transformation of increased differentiation, a revolution that transformed simple system organized around kinship to complex systems of stratified social classes and specialized institutions. The second great transformation, the *industrial revolution,* is only 200 years old. Like the agricultural revolution, the industrial revolution was stimulated by critical inventions and environmental shortages (e.g., the virtual clear cutting of English timber by the late 1600s). Like the agricultural revolution, the industrial revolution enabled human populations to grow much larger. Like the first revolution, industrialism revolutionized social forms. These included the *urban revolution,* as increasing proportions lived in cities and engaged in nonagricultural occupations; a *state revolution,* as the loosely organized states of agricultural societies evolved into powerful national governments; and a *bureaucratic revolution,* as huge complex organizations came to organize and dominate many spheres of life. Like the agricultural revolution, the industrial revolution magnified the human impact on the environment, separated them from the biophysical world, and led humans to believe that they were not dependent upon it. But in reaction the industrial degradation of nature also led people to romanticize Nature and produced environmental movements. Most LDCs are somewhere in transition between agricultural and industrial societies, which is why their potential for environmental destruction is truly massive if they repeat historic development trajectories (Olsen, 1991: 564–566; Harrison, 1993: 300–301).

The first two human revolutions are still with us and have not yet run their course, but many observers of our times believe that already underway is what in retrospect will be called the *third revolution*, which will be a reworking of industrialism even as it transformed agricultural societies. Intellectuals and scholars have been observing this phenomenon for several decades in terms of very different social and theoretical perspectives.[7] Some describe it as the evolution of postindustrial societies. Some describe it as an information society. Some have depicted the emergence of societies that will transmute the assumptions and social forms of modern societies ("modernity") into postmodern ones. Some see an emerging world of New Age spirituality. And around the world, fundamentalists of all sorts, religious as well as cultural, recognize that their traditional world is under attack by currents of change. As you can see, there are so many versions of the emerging third revolution that it is hard to know now how to fundamentally describe it. My guess is that within the next century there will be an intellectual consensus on what it is. But in all of these perspectives there is an agreement that massive pressures for social change are developing in industrial societies and that the result will not be minor alterations but major transformations (Olsen, 1991: 568).

In spite of various descriptions, some dimensions of the third human revolution are becoming clear. First, it will be, as is obvious, an information revolution, meaning that the creation, storage, transmission, and utilization of all kinds of information—especially scientific and technical knowledge—will become a primary force in human societies as growing food and making things were in the past. We will still, of course, need food and things, but increasingly specialized information will drive the creation of wealth. The old productive forces will be increasingly controlled by information-driven systems (as in computers), and a majority of workers will handle information, not plants, animals, or machines (Olsen, 1991: 565). This is not a new argument, but I need to note that this is not necessarily the gateway to utopia, as some observers have implied. In terms of human-environment relations, it may be neutral. The information revolution enhances our capacity to plunder the environment on a large scale and won't necessarily make us more environmentally frugal, but it also suggests that the future wave of wealth making will involve less material consumption than it did in the industrial age. The correlation between economic growth and ecological destruction is weakening.

The second dimension of the third revolution that is increasingly clear is the *interdependence revolution*. There are increasingly elaborate, dense, powerful, and volatile relationships among people in the same society and among societies. We are being pulled together into integrated webs of interdependencies that have many dimensions: economic, political, cultural, demographic, and so on. And recently we have again become aware of our common dependence on a shared physical environment and ecosystems. It used to be possible to study communities and cities, theoretically at least, as distinct entities with their own internal dynamics without paying much attention to the political, economic, demographic, and environmental forces that impinge upon them from their environments. No more. It has also been common in the social sciences to focus only on one society or culture (except for anthropologists, usually one's own). But the very notion of *society* becomes problematic as societies throughout the world are becoming interrelated and interdependent within the world system (the same could be said about the notions of *nation* and *economy*). Growing global interdependence does not mean that regional or local differences and concerns will disappear into some bland globalism. Far from it: The global system is full of local people ("new tribalisms"?) struggling for autonomy. Differences will not disappear, but we can no longer behave as if other people around the world don't matter (Harper, 1998: 304-315). As with the information revolution, growing human interdependence has mixed implications for reshaping human-environment relations. Since that was what this chapter is all about, I don't really think I need to say more about it. The important point is that whatever else the third revolution is, these two dimensions will be so interwoven in it that it will be an informational-interdependence revolution (Olsen, 1991: 566).

A *third dimension* of the third revolution may be a *sustainability revolution*. I say "may" be, because this is not as certain to be a dimension of the forthcoming human transition as the informational-interdependence revolution. It may evolve in response to our increasingly severe environmental predicament. It now seems unavoidable that the world that you and your children live in will be warmer, more crowded, more interconnected, more socially diverse, with less wilderness, fewer species, and a more constrained food-producing systems.

In an extraordinarily short period—a matter of decades—we will need to feed, house, nurture, educate and employ at least as many more people as already live on the earth. If in such a warmer, more crowded world environmental catastrophe is to be avoided, it can be done only by (1) maintaining *severe inequities* in human welfare, or (2) by adopting very different trajectories (Kates, 1994: 118). Those different trajectories must entail the transition to sustainable societies. The basic requirement of a sustainable society is to survive on a permanent basis by avoiding the kinds of ecological and organizational crises that could destroy society itself. A sustainable society must evolve (or be deliberately instituted, as the case may be) to sustain human life indefinitely. I have described the dimensions of sustainability at some length in Chapter Seven and some prescriptions for what we must do to move toward sustainability in previous sections of this chapter, so I won't repeat any of that here.

That a sustainability revolution is a part of the coming third revolution is likely, but is also a hope. The deepening environmental crisis, our awareness, and rational choice-making capability all make it likely. Our challenges are historically unprecedented, but so is the alignment of forces pushing us across the threshold toward global sustainability. Listen to Lester Brown, sometimes described as the premiere Cassandra of our times:

> More and more people in both the corporate and political worlds are now beginning to share a common vision of what an environmentally sustainable economy will look like. If the evidence of a global awakening were limited to one particular indicator, such as growing memberships in environmental groups, it might be dubious. But with the evidence of growing momentum now coming from a range of key indicators simultaneously, the prospect that we are approaching the threshold of a major transformation becomes more convincing. (1999c: 22)

The likelihood of such a threshold transformation, like all social change, is not "determined," but depends on contingencies and human agency. The happy ending will not "naturally evolve," but neither are we trapped by "powerful forces" that will destroy us. I told you in the first chapter that I was not a very true-blue Cassandra. Perhaps now you can see why. I grant you that my outlook is as much about faith and hope as about evidence. If I am pessimistic, I remain a hopeful pessimist. I've been thinking about how to explain that when I discovered this statement by senior and eminent environmental geographer Robert Kates, and I want to end by sharing it with you:

> Fifteen years ago Lionel Tiger of Rutgers University suggested that there was a "biology of hope," an evolutionary human tilt toward optimism that compensates in part for our ability to ask difficult questions such as "Can human life on the earth be sustained?" Although unpersuaded by his somewhat tenuous chains of argument, I share his inclination. Not because I have excessive confidence in the invisible hand of the marketplace, or of technological change, or even of James Lovelock's Gaia principle, in which life itself seems to create the conditions for its own survival. Nor is it just the wisdom and energy of my grandchildren and their enormous cohort of wise and energetic children around the world. Rather it is because hope is simply a necessity if we as a species, now conscious of the improbable and extraordinary journey taken by life in the universe, are to survive. (1994: 122)

PERSONAL CONNECTIONS

1. You can easily get a personal sense of the emerging world market economy and international trade system. Look at the labels on your clothes and shoes. Where were they made, or at least sewn together? Look at labels on frozen and canned goods at your grocery store, particularly seafood. From what nations are they imported? Do the same for your computer, CD player, or small tape player and

headphone set. Look at the manufacturer's labels on the parts of your car engine, even if it is an American auto.

2. As I did with my cup of coffee at the beginning of this chapter, try to trace out how some product that you use is produced in a variety of locations and how various resources and ecosystems are impacted by its manufacture, transportation, sale, consumption, and disposal. This trail is not an obvious one, and you will have to give this some careful thought or study, but it is an intriguing thing to do. A good resource for such a project is Ryan and Durning (1997), who discuss the origins and environmental impacts of coffee, newspapers, T-shirts, shoes, bikes (and cars), computers, hamburgers, french fries, and colas.
3. *A summary reflection:* This book discussed ecosystems and human systems, the state of earth's resources, some human causes of environmental problems (population and energy), sustainability and social change, markets, politics, movements, and global problems. Thinking back over your reading and discussions, what stands out as some of the most important particular things that you learned? More generally, how have your perspectives and opinions about environmental issues and controversies changed since the beginning? In general, are you optimistic or pessimistic about our ability to cope with all of the problems that we face in the near future? As I was finishing writing it, I realized, surprisingly, that I was more optimistic than I thought, and I discovered that I really wasn't a true-blue Cassandra. Are you more or less optimistic than you were before reading/discussing this material? How or why so or not so?

ENDNOTES

1. Off the coasts of Cadiz, Spain; the New Hebrides islands north of Scotland; Texas and Mexico; and Alaska.
2. Rubber tappers make their living in the Amazon basin by harvesting latex from rubber trees spread throughout the region. They also gather Brazil nuts, fruits, and fibers in the forest and cultivate small plots near their homes. A 1988 study showed that sustainable harvesting of such nonwood products *over 50 years* would generate twice as much revenue per hectare as timber production and three times as much as cattle ranching (Miller, 1992: 261).
3. World Bank lending increased from $1.3 billion in 1944, when it was founded, to more than $22 billion annually in 1992, in constant 1992 dollars (French, 1994: 12). Officially named the International Bank for Reconstruction and Development, the WB is the largest such financial institutions, but there are now others, such as the Inter-American Bank, the Asian Development Bank, and the North American Development Bank.
4. OECD means Organization for Economic Cooperation and Defense, including most European nations as well as the United States and Canada.
5. International treaties and regimes exist about fishing and polluting the oceans, logging, climate modification, international transportation and disposal of hazardous wastes, international trade in endangered species, Antarctic protection, depletion of the atmospheric ozone layer, biodiversity, and so forth. They reflect an emerging international legal climate of regulatory regimes to address global issues.

6. In order not to obtain artifactual results, environmental NGOs per se were excluded from the analysis.
7. For a sample of such works, see Daniel Bell's *The Coming of Post-Industrial Society* (1973), John Naisbett's *Megatrends* (1982), Kirkpatrick Sale's *Human Scale* (1980), Alvin Toffler's *The Third Wave* (1980), and Daniel Yankelovich's *New Rules* (1981).

References

ABRAMOVITZ, J. (1998). Forest decline continues. In L. Brown, M. Renner, and C. Flavin (Eds.), *Vital signs: Environmental trends that are shaping our future* (pp.124–125). New York: W. W. Norton.

ABRAMS, E., and RUE, J., (1988). The causes and consequences of deforestation among the prehistoric Maya, *Human Ecology*, 16 (4), 377–396.

ACHESON, J. (1981). The lobster fiefs, revisited: Economic and ecological effects of territoriality in the Maine lobster industry. In B. McCay and J. Acheson (Eds.), *The question of the commons.* (pp. 37–65). Tucson, AZ: University of Arizona Press.

ALEXANDER, J. (1988). *Real civil societies: Dilemmas of institutionalization.* Thousand Oaks, CA: Sage.

ADAMS, R. M. (1981). *Heartland of cities: Surveys of ancient settlement and land use on the central flood plains of the Euphrates.* Chicago: University of Chicago Press.

AIDALA, J. V. (1979). *Regulating carcinogens: The case of pesticides.* Paper presented at the annual meeting of the Society for the Study of Social Problems, Boston.

ALEXANDER, J. (1985). *Neofunctionalism.* Beverly Hills, CA: Sage.

ALEXANDER, S., SCHNEIDER, S. and LAGERQUIST, K. (1997). The interaction of climate and life. In G. Daily (Ed.), *Nature's services: Societal dependence on natural ecosystems* (pp. 71–92). Washington, DC: Island Press.

ALLEN, S. (2000, January 30). McDonald's cleans up its environmental act. *Boston Globe,* Cited in the *Omaha World-Herald,* pp. 1E, 5E.

ALTIERI, M. (1995). *Agroecology: The science of sustainable agriculture.* Boulder, CO: Westview Press.

ALTIERI, M. (1998). Ecological impacts of industrial agriculture and the possibility for a truly sustainable farming. *Monthly Review, 50* (3) 60–71.

AMANO, A. (1990). Energy prices and CO_2 emissions in the 1990s. *Journal of Policy Modeling,* 12: 495–510.

ANDERSON, C. (1993). The biotechnology revolution: Who wins and who loses? In H. Didsbury (Ed.), *The years ahead: Perils, problems, and promises.* Bethesda, MD: World Future Society.

ANDERSON, E. (1984). Who's who in the Pleistocene: A mammalian beastiary. In H. Martin and H. Klein (Eds.), *Quaternary extinctions: A prehistoric revolution* (pp. 40–89). Tucson, AZ: University of Arizona Press.

ARGYLE, M. (1987). *The psychology of happiness.* New York: Methuen.

ARMILLAS, P. (1971). Gardens on swamps. *Science, 174*; 653–661.

ASSOCIATED PRESS. (1993, March 1). Third World birthrates dropping; more using contraception. *Omaha World-Herald,* p. 4.

ASSOCIATED PRESS. (1994a, October 20). EPA reports improvements in air quality. *Omaha World-Herald,* p. 19.

ASSOCIATED PRESS. (1994b, December 11). Leaders vow trade pact by year 2005. *Omaha World-Herald,* pp. 1, 11.

ASSOCIATED PRESS. (1999). Middle East urged to conserve water. *Omaha World-Herald,* 3:3.

ATHANASIOU, T. (1996). *Divided planet: The ecology of rich and poor.* Boston: Little Brown.

BALAAM, D., and VESETH, M. (1996). *Introduction to international political economy.* Upper Saddle River, NJ: Prentice Hall.

BARASH, D. (1979). *Sociobiology: The whisperings within.* New York: HarperCollins.

BARNET, R. J., and MULLER, R. E. (1974). *Global reach: The power of multinational corporations.* New York: Simon and Schuster.

BARTLETT, D., and STEELE, J. (1992). *America: What went wrong?* Kansas City, MO: Andrews and McMeel.

BAYLIS, R. (1997). INFOTERRA: Water for food. Cited in *New Scientist, January 2, 1997.* Bayless@CARDIFF.AC.UK

BEATTY, J. (1994). Who speaks for the middle class? *Atlantic, 273* (5) 65–78.

BECK, U. (1986). *Riskogesellshaft: Auf dem Weg in eine andere Moderne.* Frankfurt, Germany: Suhrkamp.

BECK, U. (1996) World risk society as cosmopolitan society? Ecological questions in a framework of manufactured uncertainties, *Theory, Culture, and Society, 13,* 4.

BEDER, S. (1998). *Global spin: The corporate assault on environmentalism.* White River Junction, VT: Chelsea Green.

BELL, D. (1973). *The coming of post-industrial society.* New York: Basic Books.

BELL, M. (1998). *An invitation to environmental sociology.* Thousand Oaks, CA: Pine Forge Press.

BENDER, W., and SMITH, M. (1997). Population, food, and nutrition. *Population Bulletin, 51,* 4.

BERGER, P. L., and LUCKMANN, T. (1976). *The social construction of reality.* New York: Doubleday.

BEVINGTON, R., and ROSENFELD, A. (1990). Energy for buildings and homes. *Scientific American, 263*(3), 76–87.

BIRDSALL, N. (1980). Population and poverty in the developing nations. *Population Bulletin, 35*:1–48.

BLEVISS, D. L., and WALZER, P. (1990). Energy for motor vehicles. *Scientific American, 263* (3) 103–109.

BLOOMBERG NEWS (2000, February 13). Companies are lending an ear to activists. *Omaha World-Herald,* 5–M.

BOERNER, C., and LAMBERT, T. (1995). Environmental injustice, *Public Interest 95,* 118: 61–82.

BOOKCHIN, M. (1982). *The ecology of freedom: The emergence and dissolution of hierarchy.* Palo Alto, CA: Cheshire Books.

BOOKCHIN, M. (1986). *The modern crisis.* Philadelphia: New Society.

BOSERUP, E. (1981). *Population and technological change: A study of long-term trends.* Chicago: University of Chicago Press.

BOULDING, E. (1992). *The underside of history: A view of women through time* (2nd ed., vol. 2) New York: Sage.

BOYER, B. (1991). *No safe place to hide: Great Lakes pollution and your health.* Buffalo, NY: State University of New York at Buffalo Press.

BRICKMAN, R., JASANOFF, S., and ILGEN, T. (1985). *Controlling chemicals: The politics of regulation in Europe and the United States.* Ithaca, NY: Cornell University Press.

BROOKFIELD, H. (1992, January). The numbers crunch: Is it possible to measure the population "carrying capacity" of the planet? *UNESCO Courier*, pp. 25–29.

BROWN, L. (1988). *The changing world food prospect: The nineties and beyond.* Worldwatch Institute paper no. 85. Washington, DC: Worldwatch Institute.

BROWN, L. (1991). The new world order. In L. Starke (Ed.). *State of the world 1991.* New York: W. W. Norton.

BROWN, L. (1994a). Overview: Charting a sustainable future. In L. Brown, E. Kane, and D. M. Roodman (Eds.), *Vital signs, 1994: The trends that are shaping the future* (pp. 15–21). New York: W. W. Norton.

BROWN, L. (1994b). Who will feed China? *World Watch, 7*(5), 10–22.

BROWN, L. (1997). Can we raise grain yields fast enough? *World Watch.* July/August, 10, 4: 8–17.

BROWN, L. (1998). Overview: New records, new stresses. In L. Brown, M. Renner, and C. Flavin (Eds.), *Vital signs 1998: Environmental trends that are shaping our future* (pp. 1–24). New York: W. W. Norton.

BROWN, L. (1999a). Overview: An off-the-chart year. In L. Brown, M. Renner and B. Halweil, *Vital signs 1999: The environmental trends that are shaping our future.* New York: W. W. Norton.

BROWN, L. (1999b). Feeding nine billion. In L. Starke (Ed.), *State of the world, 1999.* New York: W. W. Norton.

BROWN, L. (1999c, March/April). Crossing the threshold: Early signs of an environmental awakening. *World Watch, 12* (2) 12–22.

BROWN, L., and FLAVIN, C. (1999). A new economy for a new century. In L. Starke (Ed.) *State of the world 1999* (pp. 3–21). New York: W. W. Norton.

BROWN, L., FLAVIN C., and KANE, H. (1992). *Vital signs: Trends that are shaping our future.* New York: W. W. Norton.

BROWN, L., FLAVIN, C., and POSTEL, S. (1990). Picturing a sustainable society. In L. Starke (Ed.) *State of the world, 1990.* New York: W. W. Norton.

BROWN, L., KANE, H., and AYRES, E. (1993). *Vital signs 1993: The trends that are shaping our future.* New York: W. W. Norton.

BROWN, L. and PANAYOTOU, T. (1992). Roundtable discussion: Is economic growth sustainable? *Proceedings of the World Bank annual conference on development economics* (pp. 353–362). Washington, DC: World Bank for Reconstruction and Development.

BROWN, L., and WOLF, E. C. (1984). *Soil erosion: Quiet crisis in the world economy.* Worldwatch Institute paper no. 60. Washington, DC: Worldwatch Institute.

BRULLE, R. J. (2000). *Agency, democracy and nature: U.S. environmental movements from the perspective of critical theory.* Forthcoming. Cambridge, MA: M.I.T Press.

BRYANT, B. (1995). *Environmental justice: Issues, policies, and solutions.* Washington, DC: Island Press.

BRYANT, B., and MOHAI, P. (1992). *Race and the incidence of environmental hazards: A time for discourse.* Boulder, CO: Westview Press.

BUCHHOLZ, R. A. (1993). *Principles of environmental management: The greening of business.* Englewood Cliffs, NJ: Prentice Hall.

BULLARD, R. D. (1990). *Dumping in Dixie: Race, class, and environmental quality.* Boulder, CO: Westview.

BULLARD, R. D. (Ed.). (1993). *Confronting environmental racism: Voices from the crossroads.* Boston: South End Press.

BURINGH, P. (1989). Availability of agricultural land for crop and livestock production. In D. Pimentel and C. W. Hall (Eds.), *Food and natural resources.* San Diego, CA: Academic Press.

BURNS, T. R., and DIETZ, T. (1992). Cultural evolution: Social rule systems, selection and human agency. *International Sociology, 7* (3), 259–283.

BURTON, M. G. (1985). Elites and collective protest. *Sociological Quarterly, 25,* 45–66.

BUTTEL, F. H. (1978). Social structure and energy efficiency: A preliminary cross-national analysis. *Human Ecology, 6,* 145–164.

BUTTEL, F. H. (1986). Sociology and the environment: The winding road toward human ecology. *International Social Science Journal, 109,* 337–356.

BUTTEL, F. H. (1989). Resources, institutions, and political-economic processes: Beyond allegory and allegation. *Social Science Quarterly, 70,* 2, 468–470.

BUTTEL, F., and TAYLOR, P. (1992). Environmental sociology and global change: A critical assessment. *Society and Natural Resources,* 5, 211–230.

CABLE, S., and BENSON, M. (1993). Acting locally: Environmental injustice and the emergence of grass-roots environmental organizations. *Social Problems, 40* (4), 464–477.

CAIRNCROSS, F. (1991). *Costing the earth: The challenge to government, the opportunities for business.* Boston: Harvard Business School Press.

CALDWELL, L. K. (1992). Globalizing environmentalism: Threshold of a new phase in international relations. In R. E. Dunlap and A. G. Mertig (Eds.), *American environmentalism: The U.S. environmental movement, 1970–1990* (pp. 63–76). Philadelphia: Francis Taylor.

CAMP, S. L. (1993, Spring). Population: The critical decade. *Foreign Policy, 90,* 126–144.

CAMPBELL, B. (1983). *Human ecology: The story of our place in nature from prehistory to present.* New York: Aldine.

CAPEK, S. (1993). The "environmental justice" frame: A conceptual discussion and application. *Social Problems, 40* (1), 5–24.

CARBON DIOXIDE ASSESSMENT COMMITTEE (CDAC). (1983). *Changing climate.* Washington, DC: Natural Resources Council, National Academy Press.

CARON, J. A. (1989, Spring). Environmental perspective of blacks: Acceptance of the new environmental paradigm. *Journal of Environmental Education, 20,* 21–26.

CARPENTER, W. (1992). Cited in T. Miller, *Living in the Environment,* 7th edition. p. 636, Belmont, CA: Wadsworth.

CARSON, R. (1962). *Silent spring.* Boston: Houghton Mifflin.

CARVER, T. N. (1924). *The economy of human energy.* New York: MacMillan.

CATTON, W. R. (1993/1994). Let's not replace one set of unwisdoms with another. *Human Ecology Review, 1* (1), 33–38.

CATTON, W. R. (1997). Redundancy anxiety. *Human Ecology Review, 3* (2), 175–178.

CATTON, W. R. (1998). Darwin, Durkheim, and mutualism. In L. Freese (Ed.), *Advances in human ecology,* vol. 7, (pp. 89–138). Stamford, CT: JAI Press.

CATTON, W. R., and DUNLAP, R. (1978). Environmental sociology: A new paradigm? *The American Sociologist, 13,* 41–49.

CATTON, W. R., and DUNLAP, R. (1986). Competing functions of the environment: Living space, supply depot, and waste repository. Paper presented at the 1986 meeting of the Rural Sociological Society, Salt Lake City, Utah.

CENTRON, M. (1994). *American Renaissance* (2nd ed.). New York: St. Martin's Press.

CHARLES, D. (1999, OCTOBER 15). Hunger in America, *The Morning Edition.* Washington, DC: National Public Radio.

CHASE-DUNN, C. (1989). *Global formation: Structures of the world-economy.* Oxford, England: Blackwell.

CHOMSKY, N. (1989). *Necessary illusions: Thought control in a democratic society.* Boston: South End Press.

CICCANTELL, P. (1999). It's all about power: The political economy and ecology of redefining the Brazilian Amazon basin. *Sociological Quarterly, 40* (2), 293–315.

CITIZEN ACTION. (1992). *Water at risk: Pesticides* [pamphlet]. Lincoln, NE: Author.

CITIZENS' ENVIRONMENTAL COALITION (CEC). (1986, November 3–5). Climate change and associated impact. *Proceedings of the Symposium on CO_2 and Other Greenhouse Gases: Climatic and Associated Impacts.* Brussels, Belgium

CLARK, M. E. (1991). Rethinking ecological and economic education: A gestalt shift. In R. Costanza (Ed.), *Ecological economics: The science and management of sustainability* (pp. 400–414). New York: Columbia University Press.

CLARK, W. C. (1990). Managing planet earth. In *Managing planet earth: Readings from Scientific American* (pp.1–12). New York: W. H. Freeman.

CLARKE, L. (1988). Explaining choices among technological risks. *Social Problems, 35* (1), 22–35.

CLARKE, L. (1991). The political economy of local protest groups. In S. R. Couch and J. R. Kroll-Smith (Eds.), *Communities at risk: Collective responses to technological hazards* (pp. 88–111). New York: Peter Lang.

CLARKE, L. (1993). The disqualification heuristic: When do organizations misperceive risk? *Research in Social Problems and Public Policy,* 5.

CLYKE, F. C. (1993). *The environment.* New York: HarperCollins.

COHEN, J. E. (1995). *How many people can the earth support?* New York: W. W. Norton.

COHEN, M. (1997, Fall). Sustainable consumption and society. *Environment, Technology, and Society, 87,* 3.

COLEMAN, J. S. (1990). *Foundations of social theory.* Cambridge, MA: Harvard University Press.

COLLINS, R. (1975). *Conflict sociology: Toward an explanatory science.* New York: Academic Press.

COMMONER, B. (1971). *The closing circle.* New York: Knopf.

COMMONER, B. (1992). *Making peace with the planet.* New York: The New Press.

CONDORCET, M. DE (1979). *Sketch for a historical picture of the progress of the human mind.* (J. Barraclough, Trans.) London: Wiedenfield and Nicholson. (Original work published in 1795.)

CONSTANCE, D. H., RIKOON, J. S. and HEFFERNAN, W. (1994). *Groundwater issues and pesticide regulation: A comparison of Missouri urbanites' and farm operators' opin-*

ions. Presented at the annual meeting of the Midwest Sociological Society, St. Louis.

COOK, E. (1971). The flow of energy in an industrial society, *Scientific American, 224* (3), 134–147.

COSTANZA, R. (ED.). (1991). *Ecological economics: The science and management of sustainability*. New York: Columbia University Press.

COSTANZA, R., CUMBERLAND, J., DALY, H., GOODLAND, R., and NORGAARD, R. (1995). *An introduction to ecological economics*. Boca Raton, FL: St. Lucie Press.

COSTANZA, R., and FOLKE, C. (1997). Valuing ecosystems services with efficiency, fairness and sustainability as goals. In G. Daily (Ed.), *Nature's services: Societal dependence on natural ecosystems*. Washington, DC: Island Press.

COTGROVE, S. (1982). *Catastrophe or cornucopia: The environment, politics and the future*. New York: John Wiley and Sons.

COTTRELL, F. (1955). *Energy and society*. New York: McGraw-Hill.

COUNCIL ON ENVIRONMENTAL QUALITY. (1980). *The global report to the president of the U.S.: Entering the 21st century*. New York: Pergamon Press.

CRAIG, J. R., VAUGHAN, D. J., and SKINNER, B. J. (1988). *Resources of the earth*. Englewood Cliffs, NJ: Prentice-Hall.

CROSSON, P. R., and ROSENBERG, N. J. (1990). Strategies for agriculture. In *Managing the planet: Readings from* Scientific American. New York: W.H. Freeman.

DAILY, G., MATSON, P. and VITOUSEK, P. (1997). Ecosystem services supplied by soil. In G. Daily (Ed.), *Nature's services: Societal dependence on natural ecosystems* (pp. 113–150). Washington, DC: Island Press.

DALY, H. E., and COBB, J.P., JR. (1989). *For the common good: Redirecting the economy towards community, the environment, and a sustainable future*. Boston: Beacon Press.

DALY, H. E., and. TOWNSEND, K. N. (Eds.). (1993). *Valuing the earth: Economics, ecology, and ethics*. Cambridge, MA: MIT Press.

DANIELS, G., and FRIEDMAN, S. (1999). Spatial inequality and the distribution of industrial toxic releases: Evidence from the 1990 TRI. *Social Science Quarterly, 80* (2), 244–262.

DAVIS, G. R. (1990) Energy for planet earth. *Scientific American, 263* (3), 55–62.

DE BLIJ, H. J. (1993). *Human geography: Culture, society, and space* (4th ed.). New York: John Wiley and Sons.

DE JONG, G., and FAWCETT, J. (1981). Motivations for migrations: An assessment and a value-expectancy research model. In G. De Jong and R. Gardner (Eds.), *Migration decision-making*. New York: Pergamon Press.

DER SPIEGEL. (1992). May 25, p. 77, reprinted in *Utne Reader*, May/June 1993, 57.

DEVALL, B. (1992). Deep ecology and radical environmentalism. In R. E. Dunlap and A. G. Mertig (Eds.), *American environmentalism: The U.S. environmental movement, 1970–1990* (pp. 51–62). Philadelphia: Francis Taylor.

DEVALL, B., and SESSIONS, G. (1985). *Deep ecology*. Salt Lake City, UT: Peregrine Smith.

DIAMOND, J. (1987). The world mistake of the human race. *Discover* reprinted in A. Podolefsky and P. Brown (Eds.) *Applying Anthropology: An Introductory Reader*, 5th Edition (pp. 80–84). Mountain View, CA: Mayfield.

DIAMOND, L. and PLATTNER, M. (1993). *The global resurgence of democracy*. Baltimore, MD: Johns Hopkins.

DICKINSON, R. E. (1986). In B. Bolin et al. (Eds.), *The greenhouse effect: Climate change and ecosystems* (pp. 207–270). New York: Wiley and Sons.

DIETZ, T. (1987). Theory and method in social impact assessment. *Sociological Inquiry*, 57: 54–69.

DIETZ, T. (1994). Evolutionary theory and environmental politics. *Human Ecology Review*, *1* (1), 46–53.

DIETZ, T. (1996/1997). *The human ecology of population and environment: From utopia to topia*, *3* (3), 168–171.

DIETZ, T., BURNS, T., and BUTTEL, F. (1990). Evolutionary thinking in sociology: An examination of current thinking. *Sociology Forum*, *5*, 155–185.

DIETZ, T., FREY, S., and ROSA, E. (1992). Risk, technology, and society. In R. E. Dunlap and W. Michelson (Eds.), *Handbook of environmental sociology*. Westport, CT: Greenwood Press.

DIETZ, T., and ROSA, E. (1994). Rethinking the environmental impacts of population, affluence, and technology. *Human Ecology Review*, *1*, 277–300.

DEITZ, T., and ROSA, E. (1997). Effects of population and affluence on CO_2 emissions, *Proceedings of the National Academy of Science, United States*, *94*, 175–179. Also online at http://www.pnas.org

DIETZ, T., STERN, P., and RYCROFT, R. (1989). Definitions of conflict and the legitimization of resources: The case of environmental risk. *Sociological Forum*, *4*, 47–70.

DIETZ, T., and VINE, E. L. (1982). Energy impacts of a municipal conservation program. *Energy*, *7*, 755–758.

DILLMAN, D., ROSA, E., and DILLMAN, J. (1983). Lifestyle and home energy conservation in the United States: The poor accept lifestyle cutbacks while the wealthy invest in conservation. *Journal of Economic Psychology*, *3*, 299–315.

DILORENZO, T. J. (1993, October/September). The mirage of sustainable development. *The Futurist*, 14–19.

DOLD, C. (1992, July/August). "Slapp back!" *Buzzworm*, 34–41.

DOMINICK, R. H. (1992). *The environmental movement in Germany*. Bloomington, IN: Indiana University Press.

DORNER, P. (1972). *Land reform and economic development*. Baltimore, MD: Penguin.

DOWNS, A. (1972). Up and down with ecology–the "issue-attention cycle." *Public Interest*, *28*, 38–50.

DUNCAN, O. D., and SCHNORE, L. F. (1959). Cultural, behavioral, and ecological perspectives for the study of social organization. *American Journal of Sociology*, 65, 132–146.

DUNLAP, R. (1980, SEPTEMBER/OCTOBER). Ecology and the social sciences: An emerging paradigm. *American Behavioral Scientist*, *24*, 1–149.

DUNLAP, R. (1983). Commitment to the dominant social paradigm and concern for environmental quality: An empirical examination. *Social Science Quarterly*, *65*, 1013–1028.

DUNLAP, R. (1992). From environmental to ecological problems. In C. Calhoun and G. Ritzer (Eds.), *PRIMIS: Social problems*. New York: McGraw-Hill.

DUNLAP, R. E., and CATTON, W. R. (1983). What environmental sociologists have in common. *Sociological Inquiry*, *33*, 113–135.

DUNLAP, R. E., and MERTIG, A. G. (1992). The evolution of the U.S. environmental movement from 1970 to 1990: An overview. In R. E. Dunlap and A. G. Mertig (Eds.), *American environmentalism: The U.S. environmental movement, 1970–1990* (pp. 1–10). Philadelphia: Francis Taylor.

DUNLAP, R. and MERTIG, A. G. (1995). Global concern for the environment: Is affluence a prerequisite? *Journal of Social Issues*, 51, 121–137.

DUNLAP, R. E., and MICHELSON, W. (Eds.). (in press). *Handbook of environmental sociology*. Westport, CT: Greenwood Press.

DUNLAP, R., and SCARCE, R. (1991). The polls–poll trends: Environmental problems and protection, *Public Opinion Quarterly, 55*, 651–672.

DUNLAP, R. E., and VAN LIERE, K. D. (1978, Summer). The new environmental paradigm: A proposed measuring instrument and preliminary results. *Journal of Environmental Education*, 9, 10–19.

DUNLAP, R. E., and VAN LIERE, K. D. (1984). Commitment to the dominant social paradigm and concerns for environmental Q: An empirical examination. *Social Science Quarterly, 65*, 1013–1028.

DUNN, S. (1998a). Carbon emissions rise. In L. Brown, M. Renner, and C. Flavin (Eds.), *Vital signs, 1998: Environmental trends that are shaping our future* (pp. 66–67). New York: W. W. Norton.

DUNN, S. (1998b). After Kyoto: A climate treaty with no teeth. *World Watch, 11* (2,4), 33–35.

DUNN, S. (1999). Automobile production drops. *Vital signs 1999: The environmental signs that are shaping our future*. New York: W. W. Norton.

DURKHEIM, E. (1964). *The division of labor in society*. (G. Simpson, Trans.). New York: Macmillan. (Original work published in 1893.)

DURNING, A. B. (1988). *Action at the grassroots: Fighting poverty and environmental decline*. Worldwatch paper no. 88. Washington, DC: Worldwatch Institute.

DURNING, A. B. (1989). *Poverty and the environment: Reversing the downward spiral*. Worldwatch paper no. 92. Washington, DC: Worldwatch Institute.

DURNING, A. B. (1990). Ending poverty. In L. Starke (Ed.) *State of the World 1990*. New York: W. W. Norton.

DURNING, A. B. (1991). Asking how much is enough. In L. Starke (Ed.), *The state of the world, 1991*. New York: W. W. Norton.

DURNING, A. B. (1992). *How much is enough? The consumer society and the future of the* earth. New York: W. W. Norton.

DURNING, A. B. (1993). Can't live without it: Advertising and the creation of needs. *World Watch*, *6* (3), 10–18.

DURNING, A. B. (1994, March/April). The seven sustainable wonders of the world. *Utne Reader*, *62*, 96–99.

DURNING, A. B. and AYERS. E. (1994). The history of a cup of coffee. *World Watch*, *7* (8), 20–22.

ECKHOLM, E. P. (1976). *Losing ground: Environmental stress and world food prospects*. New York: W. W. Norton.

Ecology USA. (1998a, October 5). *Slants and Trends, 27*, 19.

Ecology USA. (1998b, November 2). *Slants and Trends, 27*, 21.

Ecology USA. (2000, January 24). *Panel of experts calls climate change evidence indisputable, Slants and Trends*, 16–17.

EDGELL, M. C. R., and NOWELL, D.E. (1989). The new environmental paradigm scale: Wildlife and environmental beliefs in British Columbia. *Society and Natural Resources*, *2*, 285–296.

EDWARDS, B. (2000, November 19). Genetically engineered rice, *The Morning Edition*. Washington, DC: National Public Radio.

EHRLICH, P. R. (1968). *The population bomb*. New York: Ballantine Books.

EHRLICH, P. R., and EHRLICH, A. H. (1992). *The population explosion.* New York: Doubleday.

EHRLICH, P. R., and HOLDREN, J. P. (1974). Impact of population growth. *Science, 171,* 1212–1217.

EHRLICH, P. R., and HOLDREN, J. P. (1988). *The Cassandra conference: Resources and the human predicament.* College Station, TX: Texas A & M Press.

EISLER, R. (1988). *The chalice and the blade.* New York: Harper and Row.

EITZEN, S., and BACA ZINN, M. (1992). *Social problems* (5th ed.). Boston: Allyn and Bacon.

ELGIN, D. (1982). *Voluntary simplicity: Toward a way of life that is outwardly simple, inwardly rich.* New York: Morrow.

ENERGY INFORMATION ADMINISTRATION. (1989). *Annual energy outlook 1989.* Publication no. 0383, 89. Washington, DC: Department of Energy/Energy Information Administration.

ESTY, D. C. (1993). Integrating trade and environment policy making: First steps in the North American free trade agreement. In D. Zalke, P. Orbuch,and R. F. Housman (Eds.), *Trade and environment: Law, economics, and policy* (pp. 45–55). Washington, DC: Island Press.

ETZIONI, A. (1970, April). Editorial. *Science.*

ETZIONI, A. (1993). *The spirit of community: Rights, responsibilities, and the communitarian agenda.* New York: Crown.

FALKENMARK, M., and WIDSTRAND, C. (1992). Population and water resources: A delicate balance. *Population Bulletin, 47,* 3.

FARBER, D., and O'CONNOR, J. (1989, Summer). The struggle for nature: Environmental crises and the crises of environmentalism in the United States. *Capitalism, Nature, Socialism, 2,* 12–37.

FARELY, R. (1998). *Sociology* , (4th ed.). Upper Saddle River, NJ: Prentice Hall.

FICKETT, A. P., GELLINGS, C., and LOVINS, A. B. (1990). Efficient use of electricity. *Scientific American, 263* (3), 64–75.

FIERRO, L., (1994, September/October). Ecuador: The people vs. Texaco. *NCLA Report on the Americas.*

FINSTERBUSCH, K., and FREUDENBURG, W. (1993). Social impact assessment and technology. In R. E. Dunlap and W. Michelson (Eds.), *Handbook of environmental sociology.* Westport CT: Greenwood Press.

FINSTERBUSCH, K., LLEWELLYN, L., and WOLF, C. (Eds.). (1983). *Social impact assessment methods.* Beverly Hills, CA: Sage.

FISCHER, C. (1976). *The urban experience.* New York: Harcourt Brace Jovanovich.

FISCHOFF, B. (1990). Psychology and public policy: Tool or toolmaker? *American Psychologist,* 45, 647–653.

FISHER, J. (1993). *The road from Rio: Sustainable development and the nongovernmental movement in the Third World.* Westport, CN: Prager.

FLANNERY, J. A. (1994, October 27). Trade group report says Omaha water contains herbicides. *Omaha World-Herald,* 10.

FLAVIN, C. (1986). Moving beyond oil. In L. Starke (Ed.) *State of the world, 1986.* New York: W. W. Norton.

FLAVIN, C. (1988). How many Chernobyls? *World Watch, 1* (1), 14–18.

FLAVIN, C. (1997). Storm damages set record. In L. Brown, M. Renner, and C. Flavins (Eds.), *Vital signs 1997: The environmental trends that are shaping our future* (pp. 70–71). New York: W. W. Norton.

FLAVIN, C. (1998a). Last tango in Buenos Aires, *World Watch, 11*(6), 10–18.

FLAVIN, C. (1998b). Wind power sets record. In L. Brown, M. Renner, and C. Flavin (Eds.), *Vital signs 1998: The environmental trends that are shaping our future*, pp. 58–59. New York: W.W. Norton.

FLAVIN, C. (1999). Wind power blows to new records. In L. Brown, M. Renner, and B. Halweil (Eds.), *Vital signs 1999: the environmental trends that are shaping our future* (pp. 52–53). New York: W. W. Norton.

FLAVIN, C., and DUNN, S. (1999). Reinventing the energy system. In L. Stark (Ed.), *State of the World, 1999*. New York: W. W. Norton.

FLAVIN, C., and LENSSEN, N. (1991). Designing and sustainable energy system. In L. Starke (Ed.), *State of the world, 1991*. New York: W. W. Norton.

FLAVIN, C., and YOUNG, J. E. (1993). Shaping the next industrial revolution. In L. Starke (Ed.), *State of the world, 1993*. New York: W. W. Norton.

FOWLER, H. (1992). Marketing energy conservation in an environment of abundance. *Policy Studies Journal, 20* (1), 76–86.

FRANK A. G. (1997). *Capitalism and development in Latin America*. New York: Monthly Review Press.

FREESE, L. (1997). Evolutionary connections. Greenwich, CT: JAI Press. Cited in A.R. Maryanski, Evolutionary sociology. In L. Freese (Ed.), *Advances in human ecology*, vol. 7 (pp. 29–30). 1998. Stamford, CT: JAI Press.

FRENCH, H. F. (1993). Reconciling trade and the environment. In L. Starke (Ed.), *State of the world, 1993: A Worldwatch report on progress toward a sustainable society*. New York: W. W. Norton.

FRENCH, H. F. (1994). The World Bank: Now fifty, but how fit? *World Watch 7* (4), 10–18.

FRENCH, H. F. (1999, November/December), Challenging the WTO. *World Watch, 12*, 22–27.

FREUDENBERG, N. (1984). *Not in our backyards! Community action for health and the environment*. New York: Monthly Review Press.

FREUDENBERG, N., and STEINSAPIR, C. (1992). Not in our backyards: The grassroots environmental movement. In R. E. Dunlap and A. G. Mertig (Eds.), *American environmentalism: The U.S. environmental movement, 1970–1990* (pp. 27–37). Philadelphia: Francis Taylor.

FREUDENBURG, W. R. (1984). Boomtown's youth: The differential impacts of rapid community growth on adolescents and adults. *American Sociological Review, 40* 697–705.

FREUDENBURG, W. R. (1988, October). Perceived risk, real risk: Social science and the art of probabilistic risk assessment. *Science, 242*, 44–49.

FREUDENBURG, W. R. (1990). A "good business climate" as bad economic news? *Science and Natural Resources, 3*: 313–331.

FREUDENBURG, W. R. (1991). Rural-urban differences in environmental concern: A close look. *Sociological Inquiry, 61* (2), 167–198.

FREUDENBURG, W. R. (1992a). Addictive economies: Extractive industries and vulnerable localities in a changing world economy. *Rural Sociology, 57* (3), 305–332.

FREUDENBURG, W. R. (1992b). Nothing recedes like success? Risk analysis and organizational amplification of risks. *Risk: Issues in Health and Safety, 3* (1), 1–34.

FREUDENBURG, W. R. (1993). Risk and recreancy: Weber, the division of labor, and the rationality of risk perceptions. *Social Forces, 71*, 909–932.

FREUDENBURG, W.R., and DAVIDSON, D. (1996). Gender and environmental risk concerns: A review and analysis of available research. *Environment and Behavior, 28*, (3), 332–339.

FREUDENBURG, W.R., and FRICKEL, S. (1994). *Digging deeper: Mining-dependent regions in historical perspective*. Paper presented at the 1994 meeting of the Midwest Sociological Society, St. Louis, MO.

FREUDENBURG, W.R., and FRICKEL, S. (1995). Beyond the nature/society divide: Learning to think about a mountain. *Sociological Forum, 10*, 361–392.

FREUDENBURG, W. R., and PASTOR, S. (1992). Public responses to technological risks: Toward sociological perspective. *The Sociological Quarterly, 33* (3), 389–412.

FROSCH, R. A., and GALLOPOULOS, N. E. (1990). Strategies for manufacturing. *Managing planet earth: Readings from Scientific American*, (pp. 97–108). New York: W. H. Freeman.

FULKERSON, W., JUDKINS, R. R., and SANGHVI, M. K. (1990). Energy from fossil fuels. *Scientific American, 263* (3), 128–135.

FUSFELD, D. (1976). *Economics*, (2nd Ed.) Lexington, MA: Heath.

GALLUP, G., DUNLAP, R., and GALLUP, A. (1993). *Health of the planet*. Princeton, NJ: Gallup International Institute.

GARDNER, G. (1998). Organic waste reuse surging. In L. Brown et al. (Eds.), *Vital signs, 1998: Environmental trends that are shaping our future* (pp. 130–131). New York: W. W. Norton.

GARDNER, G. and SAMPAT, P. (1999). Forging a sustainable materials economy. In L. Starke (Ed.), *State of the world, 1999* (pp. 41–59). New York: W. W. Norton.

GARDNER, G. T. and STERN, P.C. (1996). *Environmental problems and human behavior* . Needham Heights, MA: Allyn and Bacon.

GEDDES, P. (1979). *Civics as applied to sociology*. Leicester, England: Leicester University Press. (Original work published in 1890.)

GELBARD, A., HAUB, C, and KENT, M. (1999). World population beyond six billion. *Population Bulletin, 3* (54), 1.

GIBBONS, J. H., BLAIR, P. D., and GWIN, H. L. (1990). Strategies for energy use. *Managing the planet earth: Readings from* Scientific American, (pp. 85–96). New York: W.H. Freeman.

GIBBONS, J. H., and GWIN, H. L. (1989). Lessons learned in twenty years of energy policy. *Energy Systems and Policy, 13*, 9–19.

GIBBS, L. (1982). *Love Canal: My story*. Albany, NY: State University of New York Press.

GIDDENS, A. (1991). *Modernity and self-identity: Social and society in the late modern age*. Stanford University Press.

GIDDENS, A. (1995). *Beyond left and right: The future of radical politics*. Stanford, CA: Stanford University Press.

GIMBUTAS, M. (1977, Winter). The first wave of Eurasian steppe pastoralists into Copper Age Europe. *Journal of Indo-European Studies, 5*, 281.

GLACKEN, C.J. (1967). *Traces on the Rhodian shore: Nature and culture in western thought from ancient times to the end of the eighteenth century*. Berkeley, CA: University of California Press.

GLEICK, P. (1991, April). Environment and security: The clear connection. *The Bulletin of Atomic Scientists*, 17–21.

GOLDEMBERG, J. (1990). How to stop global warming. *Technology Review, 93*, 25–31.

GOLDMAN, B., and FITTON, L. J. (1994). *Toxic wastes and race revisited*. Washington, DC: United Church of Christ Commission for Rural Justice.

GOODMAN, H., and ARMELAGOS, G. (1985). Disease and death at Dr. Dickson's mounds. *Natural History, 94*, 9.

GOODLAND, R., DALY, H., and KELLENBERG, J. (1993, June 27-July 1). *Burdensharing in transition to environmental sustainability*. Paper presented at the seventh general assembly of the World Future Society, Washington, DC.

GOULD, K. (1998, Spring). Nature tourism, environment, and place in a global economy. *Environment, Technology, and Society, 89*, 3–5.

GOULD, L.C., GARDNER, G. T., DELUCA, D. R., TIEMANN, A. R., and DOOB, L. W. (1988). *Perceptions of technological risks and benefits*. New York: Russell Sage Foundation.

GOULDNER, L., and KENNEDY, D. (1997). Valuing ecosystem services: Philosophical bases and empirical methods. In G. Daily (Ed.), *Nature's services: Societal dependence on natural ecosystems*. Washington, DC: Island Press.

GRAMSCI, A. (1971). State and civil society. In Q. Hoare and G. N. Smith (Eds.), *Selections from the prison notebooks of Antonio Gramsci* (pp. 210–276). New York: International Publishers.

HÄFELE, W. (1990). Energy from nuclear power. *Scientific American, 263* (3), 137–144.

HALWEIL, B. (1998). Bio-serfdom and the new feudalism. *World Watch, 11* (6), 9.

HALWEIL, B. (1999). The emperor's new clothes. *World Watch, 12* (6), 21–29.

HAMMOND, N. (1982). *Ancient Mayan civilization*. Rutgers, NJ: Rutgers University Press.

HANNIGAN, J. (1995). *Environmental sociology: A social constructionist perspective*. New York: Routledge.

HARDIN, G. (1968). The tragedy of the commons. *Science, 162*, 1243–1248.

HARDIN, G. (1993). Second thoughts on the tragedy of the commons. In H. E. Daly and K. N. Townsend (Eds.), *Valuing the earth: Economics, ecology, and ethics*. Cambridge, MA: M.I.T. Press.

HARMAN, W. W. (1979). *An incomplete guide to the future*. New York: W. W. Norton.

HARPER, C. L. (1998). *Exploring social change* (3rd ed.). Englewood Cliffs, NJ: Prentice Hall.

HARRIS, L. (1989, May 14). Public worried about state of environment today and in future. *The Harris Poll*, 21, 1–4.

HARRIS, M. (1971). *Culture, man, and nature*. New York: Thomas Crowell.

HARRIS, M. (1979). *Cultural materialism*. New York: Vintage.

HARRISON, P. (1987). *The greening of Africa*. New York: Viking/Penguin.

HARRISON, P. (1993). *The third revolution: Population, environment, and a sustainable world*. London: Penguin.

HAUB, C. (1993). Tokyo now recognized as world's largest city. *Population Today, 21* (3), 1–2.

HAWKEN, P. (1993). *The ecology of commerce: A declaration of sustainability*. New York: HarperCollins.

HAWLEY, A. (1950). *Human ecology: A theory of community structure*. New York: Ronald Press.

HAYES, D. (1990, April). Earth Day 1990: The threshold of the green decade. *Natural History*, 55–70.

HAYS, S. P. (1959). *Conservation and the gospel of efficiency*. New York: Atheneum.

HAYS, S. P. (1972). *Conservation and the gospel of efficiency: The progressive conservation movement, 1890-1920*. New York: Atheneum.

HAYS, S. P. (1987). *Beauty, wealth, and permanence: Environmental problems and human behavior*. Needham Heights, MA: Allyn and Bacon.

HEBERLEIN, T. A. (1975). Conservation information: The energy crisis and electricity consumption in an apartment complex. *Energy Systems Policy, 1*, 105–118.

HEBERLEIN, T. A., and WARRINER, G. K. (1983). The influence of price and attitudes on shifting residential electricity consumption from on-to-off peak periods. *Journal of Economic Psychology, 4*, 107–131.

HECHT, S. B., and COCKBURN, A. (1989). *The fate of the forest*. London: Verso.

HECHT, S. B. (1989). The sacred cow in the green hell: Livestock and forest conversion in the Brazilian Amazon. *The Ecologist, 19*, 229–234.

HEILBRONER, R. L. (1974). *An inquiry into the human prospect*. New York: W. W. Norton.

HEILBRONER, R. L. (1985). *The making of economic society* (7th ed). Englewood Cliffs, NJ: Prentice-Hall.

HENDRY, P. (1988). Food and population: Beyond five billion. *Population Bulletin, 43*, 2.

HIRSCH, R. L. (1987). Impending United States energy crisis. *Science, 235*, 1471.

HOLDREN, J. P. (1990). Energy in transition. *Scientific American, 263* (3), 156–164.

HOMER-DIXON, T. (1996). Environmental scarcity, mass violence, and the limits of ingenuity. *Current history, 95* (604), 359–365.

HRABOVSZKY, J. P. (1985). Agriculture: The land base. In R. Repetto (Ed.), *The global possible: Resources, development, and the new century* (pp. 211–254). New Haven, CT: Yale University Press.

HUMPHREY, C., and BUTTEL, F. (1982). *Environment, energy, and society*. Belmont, CA: Wadsworth.

HUTCHINSON, E. P. (1967). *The population debate*. Boston: Houghton Mifflin.

HUTCHINSON, G. E. (1965). *The ecological theater and the evolutionary play*. New Haven, CT: Yale University Press.

INTERGOVERNMENTAL PANEL ON CLIMATE CHANGE (IPCC), (1999). Climate Change 1996: IPCC Second Assessment Report. http://www.ipcc.ch/cc95/cont-95.htm

INTERNATIONAL ENERGY AGENCY. (1987). *Energy conservation in IEA countries*. Paris: OECD.

JEFFERISS, P. (1998). Power switch, *Nucleus, 20* (2), 1.

JOHNSON, D. G. (1973). *World agriculture in disarray*. London: Fontanta.

JOLLY, C. L. (1993, May). Population change, land use, and the environment. *Reproductive Health Matters, 1*, 13–24.

JORDAN, S. (1993, September 3). Analyst sees global opportunities for U.S. agriculture. *Omaha World-Herald*, 16.

KAHN, H., BROWN, W., and MARTEL, L. (1976). *The next 200 years*. New York: Morrow.

KAHN, H. W., and PHELPS, J. (1979, June). The economic present and future. *Futurist*, 202–222.

KAHNEMAN, D., SLOVIC, P., and TVERSKY, A. (1982). *Judgment under uncertainty: Heuristics and biases*. Cambridge, CT: Cambridge University Press.

KAHNEMAN, D. and TVERSKY, A. (1972). Subjective probability: A judgment of representativeness. *Cognitive Psychology, 3* (3), 430–454.

KANE, H. (1993a). Photovoltaic sales growth slows. In L. Brown, H. Kane, and E. Ayres (Eds.) *Vital signs 1993: Trends that are shaping our future* (pp. 52–53). New York: W. W. Norton.

KANE, H. (1993b). Managing through prices, managing despite prices. In D. Zalke et al. (Eds.), *Trade and the environment: Law, economics, and policy* (pp. 57–70). Washington, DC: Earth Island Press.

KANE, H. (1994). Put it on my carbon tab. *World Watch, 6* (3), 38–39.

KATES, R. W. (1994). Sustaining life on the earth. *Scientific American, 271*, (4), 114–122.

KATES, R. W., AUSUBEL, J. H. and BARBERIAN, M. (Eds.). (1985). *Climate impact assessment: Studies of the interactions of climate and society.* New York: Wiley.

KELLER, E. A. (1992). *Environmental geology* (6th ed.). New York: Macmillan.

KELLY, K. and HOMER-DIXON, T. (1996). Environmental security and violent conflict: The case of gaza. Toronto: The Environment, Population, and Security Project and The American Association for the Advancement of Science.

KELMAN, H. C. (1958). Compliance, identification, and internalization: Three processes of attitude change. *Journal of Conflict Resolution, 2,* 51–60.

KEMP, W. B. (1971). The flow of energy in a hunting society. *Scientific America, 224,* 104–105.

KENT, M. (1984). *World population: Fundamentals of growth.* Washington DC: Population Reference Bureau.

KEYFITZ, N. (1990). The growing human population. *Managing the planet: Readings from Scientific American* (pp. 61–72). New York: W. W. Freeman.

KINEALY, C. (1996). How politics fed the famine. *Natural History, 105* (1), 33–35.

KING, T., and KELLY, A. (1985). *The new population debate: Two views on population growth and economic development.* Population trends and public policy Paper no. 7. Washington, DC: Population Reference Bureau.

KINGSLEY, G. A. (1992). U.S. energy conservation policy: Themes and trends. *Policy Studies Journal, 20* (1), 114–123.

KLEIN, D. R. (1968). The introduction, increase, and crash of reindeer on St. Matthew Island. *Journal of Wildlife Management, 32,* 350–367.

KLEINSCHMIT, L., RALSTON, D., and THOMPSON, N. (1994). *Evaluation of relative impacts of conventional and sustainable systems on rural communities.* Walthill, NE: Center for Rural Affairs.

KNORR-CETINA, K. D. (1981). *The manufacture of knowledge: An essay on the constructivist and contextual nature of science.* Oxford, England: Pergamon.

KORMONDY, E., and BROWN, D. (1998). *Fundamentals of human ecology.* Upper Saddle River, NJ: Prentice Hall.

KORTEN, D. (1995). *When corporations rule the world.* West Hartford, CT: Kumarian.

KOTOK, D. C. (1993, December 6). Arab oil embargo changed habits. *Omaha World-Herald,* 1–2.

KOWALEWSKI, D., and PORTER, K. (1992). Ecoprotest: Alienation, deprivation, or resources? *Social Science Quarterly, 73* (3), 523–534.

KRAUSE, F., BACH, W., and KOOMEY, J. (1992). *Energy policy in the greenhouse.* New York: John Wiley & Sons.

KUHN, T. (1970). *The structure of scientific revolutions* Chicago: University of Chicago Press.

LAPPÈ, F., COLLINS, J., and ROSSET, P. (1998). *World hunger: Twelve myths.* New York: Grove Press.

LASCH, W. H., III (1994, May/June). Environment and global trade, *Society,* 52–58.

LASHOF, D. A. (1989, January). The dynamic greenhouse: Feedback processes that may influence future concentrations of atmospheric trace gases and climate change. *Climate Change.*

LEE, D. R. (1992). The perpetual assault on progress. *Society, 29* (3), 50–54.

LEE, R. B. (1969). !Kung bushmen subsistence: An input-output analysis. In A. Vayda (Ed.), *Environment and cultural behavior* (pp. 47–78). Garden City, NJ: Natural History Press.

LENSKI, G., and NOLAN, P. (1999). *Human societies: An introduction to macrosociology.* New York: McGraw-Hill.

LESSEN, N. (1993a). Nuclear power at virtual standstill. In L. Brown et al. (Eds.), *Vital signs 1993: Trends that are shaping our future* (pp. 50–51). New York: W. W. Norton.

LESSEN, N. (1993b). Providing energy in developing countries. In L. Brown et al. (Eds.), *State of the world, 1993* (pp. 101–119). New York: W. W. Norton.

LEWIS, M. W. (1992). *Green delusions: An environmentalist critique of radical environmentalism.* Durham, NC: Duke University Press.

LEWIS, M. W. (1994). Environmental history challenges the myth of a primordial Eden. *The Chronicle of Higher Education, 40* (35), A56.

LIPTON, M. (1974). *Why poor people stay poor: Urban bias in rural development.* Cambridge, MA: Harvard University Press.

LIVERNASH, R., and RODENBURG, E. (1998). Population change, resources, and the environment. *Population Bulletin, 53* (1), 20–21.

LONGWORTH, R. (1998). Global squeeze: The coming crisis for first-world nations. Skokie, IL: NTC/Contemporary Books.

LOTKA, A. J. (1922). Contribution to the energetics of evolution. *Proceedings of the National Academy of Sciences, 8,* 147–151.

LOTKA, A. J. (1924). *Elements of physical biology.* New York: Williams and Wilkins. [Republished in 1956 as *Elements of mathematical biology.*] New York: Dover.

LOTKA, A.J. (1945). The law of evolution as a maximal principle. *Human Biology, 14,* 167–194.

LOVINS, A. B. (1977). *Soft energy paths.* Cambridge, MA: Ballinger.

LOVINS, A. B. (1993). Letter to the editor. *Atlantic, 272*: 6.

LOVINS, A. B. (1998). Energy efficiency to the rescue. In T. Miller, *Living in the Environment* (10th Ed., p. 378). Belmont, CA: Wadsworth.

LOW, R. S., and HEINEN, J. T. (1993). Population, resources, and environments: Implications of human behavioral ecology for conservation. *Population and Environment, 15,* 7–40.

LOWE, M. (1991). Rethinking urban transport. In L. Starke (Ed.), *State of the world, 1991.* New York: W. W. Norton.

LOWI, T., JR. (1964). American business, public policy, case-studies, and political theory. *World Politics, 16,* 677–715.

LOWI, T., JR. (1972). Four systems of policy, politics, and choice. *Public Administration Review, 32,* 298–310.

LOWI, T., JR. (1979). *The end of liberalism* (2nd ed.). New York: W. W. Norton.

LUHMANN, N. (1993). *Risk: A sociological theory.* New York: Aldine De Gruyter.

LUTZENHISER, L. and HACKETT, B. (1993). Social stratification and environmental degradation: Understanding household CO_2 production. *Social Problems, 40* (1), 50–73.

MACEACHERN, D. (1990). *Save our planet: 750 everyday ways you can help clean up the earth.* New York: Dell.

MACK, R., and BRADFORD, C. P. (1979). *Transforming America* (2nd ed.). New York: Random House.

MACNEILL, J. (1989–1990, Winter). The greening of international relations. *International Journal, 45.*

MACNEILL, J., WINSEMIUS, P., and YAKUSHIJI, T. (1991). *Beyond interdependence: The meshing of the world's economy and the earth's ecology.* New York: Oxford University Press.

MAMDANI, M. (1972). *The myth of population control.* London: Reeves and Turner.

MANN, C. (1993). How many is too many? *Atlantic 271,* (2), 47–67.

MANS, T. (1994). Personal communication.

MARSH, G. P. (1874). *The earth as modified by human action.* New York: Scribner and Armstrong.

MARTIN, P. (1984). Prehistoric overkill: The global model. In P. Martin and R. Klein (Eds.), *Quaternary extinctions: A prehistoric revolution* (pp. 354–403). Tucson, AZ: University of Arizona Press.

MARTIN, P. and MIDGLEY, E. (1999, June). Immigration to the United States. *Population Bulletin, 54,* 2.

MARX, K. (1920). *The poverty of philosophy.* Chicago: Charles H. Kerr.

MARYANSKI, A. R. (1998). *Evolutionary sociology.* In L. Freese (Ed.), *Advances in human ecology, vol. 7* (pp. 1–56). Greenwich, CT6: JAI Press.

MATHEWS, J. T. (1991). Implications for U.S. policy. In J. T. Mathews (Ed.), *Preserving the global environment: The challenge of shared leadership* (pp. 309–324). New York: W.W. Norton.

MATOON, A. (1998). Paper recycling climbs higher. In L. Brown et al. (Eds.), *Vital signs, 1998: Environmental trends that are shaping our future* (pp. 144–145). New York: W. W. Norton.

MAURITS LA RIVIERE, J. W. (1990). Threats to the world's water. In *Managing planet earth: Readings from Scientific American* (pp. 36–48). New York: W. H. Freeman.

MAZUR, A. (1981). *The dynamics of technical controversy.* Washington, DC: Communications Press.

MAZUR, A. (1991). *Global social problems.* Englewood Cliffs, NJ: Prentice Hall.

MAZUR, A., and ROSA, E. A. (1974). Energy and lifestyle: Cross-national comparison of energy consumption and quality of life indicators. *Science, 186,* 607–610.

MCCAY, B. J. (1993). Management regimes. Presented at the conference on Property Rights and Performance of Natural Resource Systems. The Biejer Institute. Stockholm, Sweden.

MCCLOSKEY, M. (1972). Wilderness movement at the crossroads. *Pacific Historical Review, 41,* 346–364.

MCCORMICK, J., and LEVINSON, M. (1993, February 15). The supply police: The demand for social responsibility forces business to look far beyond its own front door. *Newsweek,* 48–49.

MCGINN, A. (1999). Charting a new course for the oceans. In L. Starke (Ed.), *State of the world, 1999* (pp. 78–95). New York: W. W. Norton.

MCKENZIE, R. B. (1992). Sense and nonsense in energy conservation. *Society, 29* (3), 18–22.

MCNAMARA, R. S. (1992, November/December). The population explosion, *The Futurist,* 9–13.

MEAD, G. H. (1934). *Mind, self, and society: From the standpoint of a social behaviorist.* Chicago: University of Chicago Press.

MEADOWS, D. H., MEADOWS, D. L., RANDERS, J., and BEHRENS, W. W., III, (1972). *The limits of growth: A report for the Club of Rome's project on the predicament of mankind.* New York: New American Library.

MEADOWS, D. H., MEADOWS, D. L., and RANDERS, J. (1992). *Beyond the limits: Confronting global collapse and envisioning a sustainable future.* Post Mills, VT: Chelsea Green .

MERCHANT, C. (1981). *The death of nature: Women, ecology, and the scientific revolution.* San Francisco, CA: Harper and Row.

MERRICK, T. W. (1986). World population in transition. *Population Bulletin, 41,* 2.

MEYER, D., and STAGGENBORG, S. (1996). Movements, countermovements, and the structure of political opportunity, *American Journal of Sociology, 101,* 6.

MICHAELSON, M. (1994). Wangari Maathai and Kenya's green belt movement: Exploring the evolution and potentialities of consensus movement mobilization. *Social Problems, 41* (4), 540–561.

MIES, M. (1986). *Patriarchy and accumulation on a world scale.* London: Zed Books.

MILBRATH, L. W. (1989). *Envisioning a sustainable society: Learning our way out.* Albany, NY: State University of New York Press.

MILLER, T. G., JR. (1992). *Living in the environment* (7th ed.). Belmont, CA: Wadsworth.

MILLER, T. G., JR. (1998). *Living in the environment* (10th ed.). Belmont, CA: Wadsworth.

MITCHELL, J. (1998). Before the next doubling. *World Watch, 11* (1), 20–27.

MITCHELL, J., THOMAS, D. and CARTER S. (1999). Dumping in Dixie revisited: The evolution of environmental injustices in South Carolina, *Social Science Quarterly, 80* (2), 229–243.

MITCHELL, R. C. (1980). Public opinion on environmental issues. In *Environmental quality: The eleventh annual report of the Council on Environmental Quality.* Washington, DC: U.S. Government Printing Office.

MITCHELL, R. C., MERTIG, A. G., and DUNLAP, R. E. (1992). Twenty years of environmental mobilization: Trends among national environmental organizations. In R. E. Dunlap and A. G. Mertig (Eds.), *American environmentalism: The U.S. environmental movement, 1970–1990* (pp. 11–26). Philadelphia: Francis Taylor.

MORRIS, V. (1994, May/June). Rainforest revolt. *Audubon,* 16–17.

MOSS, A. H., JR. (1993). Free trade and environmental enhancement: Are they compatible in the Americas? In D. Zalke, P. Orbuch, and R. F. Housman (Eds.), *Trade and environment: Law, economics, and policy* (pp.109–132). Washington, DC: Island Press.

MOTAVALLI, J. (2000, January/February). Flying high, swooping low, *E Magazine, 11,* 1.

MYERS, N. (1989). *Deforestation rates in tropical forests and their climatic implications.* London: Friends of the Earth.

MYERS, N. (1997). The world's forests and their ecosystem services. In G. Daily (Ed.), *Nature's services: Societal dependence on natural ecosystems* (pp. 215–236). Washington, DC: Island Press.

NABHAN, G. and BUCHMANN, S. (1997). Services provided by pollinators. In G. Daily (Ed.), *Nature's services: Societal dependence on natural ecosystems* (pp. 133–150). Washington, DC: Island Press.

NAISBETT, J. (1982). *Megatrends.* New York: Warner Books.

NAISBETT, J. (1994). *Global paradox: The bigger the world economy, the more powerful its smallest players.* New York: Morrow.

NASAR, S. (1993, January 17). It's never fair to just blame the weather. *New York Times.* Reprinted in *Themes of the times: Sociology,* 13.

NASH, R. (1967). *Wilderness and the American mine.* New Haven, CT: Yale University Press.

NATIONAL ACADEMY OF SCIENCES. (1991). *Policy implications of greenhouse warming.* Synthesis Panel. Washington, DC: National Academy Press.

NATIONAL CENTER FOR STATISTICS. (1984). Blood levels for persons 6 months to 74 years of age: United States, 1976–1980. *Vital Statistics, No. 79.* Hyattsville, MD: National Center for Health Statistics.

NATIONAL RESEARCH COUNCIL. (1986). *Population growth and economic development: Policy questions.* Committee on Population, Working Group on Population and Development. Washington, DC: National Academy Press.

NATIONAL RESEARCH COUNCIL. (1987). *Current issues in atmospheric change.* Washington, DC: National Academy Press.

NATIONAL RESEARCH COUNCIL. (1989a). *Improving risk communication.* Committee on Risk Perception and Communication. Washington, DC: National Academy Press.

NATIONAL RESEARCH COUNCIL. (1989b). *Alternative agriculture.* Washington, DC: National Academy Press.

NELSON, T. (1996). Closing the loop. *World Watch, 9* (6), 10–17.

NETTING, R.M. (1981). *Balancing on an Alp: Ecological change and continuity in a Swiss mountain community.* Cambridge, England: Cambridge University Press.

NETTING, R. M. (1986). *Cultural Ecology,* 2nd ed. Prospect Heights, IL: Waveland Press.

NEW YORK TIMES. (1994, October 8). Recycling business booming, driving up prices for waste. *Omaha World-Herald,* 1–2.

New York Times (1997, September 24). In U.S. car population has boomed since '69: United States Department of Transportation Person Transportation Survey. *Omaha World-Herald,* p. 4.

NEW YORK TIMES NEWS SERVICE (1997, March 3). No fix in sight for global warming.

NIXON, W. (1993). A breakfast among PEERS: Environmental whistleblowers have some stories to tell. *E-The Environmental Magazine, 4* (5), 14–19.

NORSE, D. (1992). A new strategy for feeding a crowded planet. *Environment, 43,* (5), 6–39.

OBERSCHALL, A. (1973). *Social conflict and social movements.* Englewood Cliffs, NJ: Prentice Hall.

ODUM, E. P. (1971). *Fundamentals of ecology,* 3rd ed. Philadelphia: W. B. Sauders.

ODUM, E. P. (1983). *Basic ecology.* Philadelphia: W. B. Sauders.

OELSCHLAEGER, M. (1994). *Caring for creation: An ecumenical approach to the environmental crisis.* New Haven, CT: Yale University Press.

OLSEN, M. E. (1968). *The process of social organization.* New York: Holt, Rinehart, & Winston.

OLSEN, M. E. (1981). Consumers' attitudes toward energy conservation. *Journal of Social Issues, 37,* 108–131.

OLSEN, M. E. (1991). *Societal dynamics: Exploring macrosociology.* Englewood Cliffs, NJ: Prentice Hall.

OLSEN, M. E., and CLUETT, C. (1979). *Evaluation of Seattle City light neighborhood conservation program.* Seattle, WA: Battelle Human Affairs Research Center.

OLSEN, M. E., LODEWICK, D. G., and DUNLAP, R. E. (1992). *Viewing the world ecologically.* Boulder, CO: Westview Press.

OLSHANSKY, S. J. CARNES, B., ROBERTS, R., and SMITH, L. (1997). Infectious diseases—New and ancient threats to world health. *Population Bulletin, 52,* 2.

O'MEARA, M. (1998). CFC production continues to plummet. In L. Brown, C. Flavin, and M. Renner (Eds.), *Vital signs, 1998: Environmental trends that are shaping our future,* (pp. 70–78). New York: W. W. Norton.

O'MEARA, M. (1999). Urban air taking lives. In L. Starke. (Ed.), *The state of the world 1999* (pp. 128–129). New York: W. W. Norton.

OPHULS, W., and BOYAN, A. S., Jr. (1992). *Ecology and the politics of scarcity revisited.* NY: W. H. Freeman.

ORLOV, B. S. (1980). Ecological anthropology. *Annual Review of Anthropology, 9*: 253–273.

OSTROM, E. (1990). *Governing the commons: Evolution of institutions for collective action.* Cambridge, England: Cambridge University Press.

OSTWALD, W. (1909). *Energetische Grundlagen der Kulturwissenshaften.* Leipzig, Germany: Vorvort.

PADDOCK, W., and PADDOCK, P. (1975). *Famine 1975!* Boston: Little Brown.

PAEHLKE, R. C. (1989). *Environmentalism and the future of progressive politics.* New Haven, CT: Yale University Press.

PALMER, G. (1989). Cited in *The Washington Post,* March 8.

PARK, R. E., BURGESS, E. W., and MCKENZIE, R. (1925). *The city.* Chicago: University of Chicago Press.

PARKIN, S. (1989). *Green parties: An international guide.* London, UK: Heretic Books.

PARRICK, D. W. (1969). An approach to the bioenergetics of rural West Bengal. In A. Vayda (Ed.), *Environment and cultural behavior* (pp. 29–46). Garden City, NJ: Natural History Press.

PARRY, M. L. (1988). *The impact of climatic variations on agriculture.* Reidel, Netherlands: Kluiver Academie.

PARSONS, T. (1951). *The social system.* Glencoe, IL: The Free Press.

PASSARINI, E. (1998). Sustainability and sociology. *The American Sociologist, 29* (3), 59-70.

PAYER, C. (1974). *The debt trap: The IMF and the Third World.* New York: Monthly Review Press.

PELLOW, D. N. (1994). *Environmental justice and popular epidemiology: Grassroots empowerment or symbolic politics.* Presented at the annual meeting of the Midwest Sociological Society, St. Louis, MO.

PELTO, P. J. (1973). *The snowmobile revolution: Technological and social change in the Arctic.* Menlo Park, CA: Cummings.

PELTO, P. J., and MULLER-WILLIE, L. (1972). Snowmobiles: Technological revolution in the Arctic. In H. R. Bernard and P. J. Pelto (Eds.), *Technology and cultural change.* New York: Macmillan.

PIMENTEL, D. (1999, October 16). Cited in J. Anderson, Budding invasions costly. *Omaha World-Herald,* 1–2.

PIMENTEL, D. (1992a, October). Rural populations and the global environment. *Rural Sociology,* 12–26.

PIMENTEL, D. (1992b). Land degradation and environmental resources. In T. Miller, *Living in the environment* (7th ed., pp. 330–331). Belmont, CA: Wadsworth.

PIMENTEL, D., and et al. (1992c). Genetic engineering in agriculture. *International Journal on the Unity of the Sciences, 5*: 77–96.

PIRAGES, D. (1977). *The sustainable society: Implications for limited growth.* New York: Praeger.

PODOBNIK, B., (1999, August) Towards a sustainable energy regime: Technological forecasting and social change. Presented at the 1998 meeting of the American Sociological Association, San Francisco, CA.

POINTING, C. (1991). *A green history of the world.* London: Sinclair Stevenson.

POMFRET, J. (1993). Exodus in Europe: Millions of immigrants put the burden of hope in the West. *Washington Post National Weekly Edition, 10* (40), 6–7.

POPULATION REFERENCE BUREAU (1998). *World population data sheet: Demographic data and estimates for the countries and regions of the world* [book edition]. Washington, DC: Population Reference Bureau.

PORTER, G., and BROWN, J. (1990). *Global environmental politics*. Boulder, CO: Westview Press.

POSTEL, S. (1989). *Water for agriculture: Facing the limits. Worldwatch Paper 93*. Washington, DC: Worldwatch Institute.

POSTEL, S. (1992a). Water scarcity. *Environmental Science and Technology, 26* (12), 2332–2333.

POSTEL, S. (1992b). *The last oasis: Facing water scarcity*. New York: W. W. Norton.

POSTEL, S. (1993). Water scarcity spreading. In L. R. Brown, H. Kane, and E. Ayers (Eds.), *Vital signs, 1993: The trends that are shaping our future* (pp. 106–107). New York: W. W. Norton.

POSTEL, S. (1994). Carrying capacity: Earth's bottom line. In L. Brown et al. *State of the World, 1994* (pp. 3–21). New York: W. W. Norton.

POSTEL, S. and CARPENTER, S. (1997). Freshwater ecosystem services. In G. Daily (Ed.), *Nature's services: Societal dependence on natural ecosystems* (pp. 195–214). Washington, DC: Island Press.

RADFORD E. and DRIZD, T. (1982). Blood carbon monoxide levels in persons 3–74 years of age. *Advance Data, 76*, 8. Hyattsville, MD: National Center for Health Statistics.

RAPPAPORT, R. A. (1968). *Pigs for the ancestors: Ritual in the ecology of a New Guinea people*. New Haven, CT: Yale University Press.

RAPPAPORT, R. A. (1971). The flow of energy in an agricultural society. *Scientific American, 224* (3), 121–32.

RATHJE, W. L. (1990). The history of garbage. *Garbage. 2* (5), 32–41.

RAUBER, P. (1998). Nations vs. corporations. *Sierra, 83* (3), 17.

RAVEN, P. H. (1990). Endangered realm. *The emerald realm: Earth's precious rain forests*. Washington, DC: National Geographic Society.

RAVENSTEIN, E. (1989). The laws of migration, I and II. *Journal of the Royal Statistical Society, 48*, 167–235; *52*, 241–305.

REDCLIFT, M. (1987). *Sustainable development: Exploring the contradictions*. London: Methuen.

REDDY, A. K. N., and GOLDEMBERG, J. (1990). Energy for the developing world. *Scientific American, 263* (3) 110–119.

REICH, R. (1991). *The wealth of nations: Preparing ourselves for 21st-century capitalism*. New York: Alfred A. Knopf.

RENNER, M. (1997, January/February). Chiapas: The fruits of despair, *World Watch*, 10 (1) 12–24.

RENNER, M., (1998), Pollution control markets expand, In L. Brown, M. Renner, and C. Flavin (Eds.), *Vital signs, 1998: The environmental trends that are shaping our future* (pp. 144–145). New York: W. W. Norton.

RENNER, M. (1999). Wars increase once again. In L. Brown, M. Renner, and B. Halweil, (Eds.), *Vital signs; 1999 The environmental trends that are shaping our future* (pp. 112–133). New York: W. W. Norton.

REPETTO, R. (1987). Population, resources, environment: An uncertain future. *Population Bulletin, 42*, 2.

RIDLEY, M., and LOW, R. S. (1993, September). Can selfishness save the environment? *Atlantic,*: 76–86.

RIJKSINSTITUUT VOOR VOLKSGESONDHEIT EN MILIEUHYGIENE. (1991). *National environmental outlook, 1990–2010*. Bilthoven, Netherlands: RIVM.

RISSLER, J. (1999, Fall). Killer corn, *Nucleus, 21* (3), 1–3.

ROGERS, R. A. (1994). *Nature and the crisis of modernity.* Montreal, Canada: Black Rose Books.

ROHR, D. (1992, June 1). Too much, too fast. *Newsweek,* 34–35.

ROODMAN, D., (1998). *The natural wealth of nations: Harnessing the market for the environment,* New York: W. W. Norton.

ROODMAN, M. (1999). Building a sustainable society. In L. Starke (Ed.). *State of the world, 1999* (pp. 169–188). New York: W. W. Norton.

ROSA, E. A. (1998, Winter). Risk and environmental sociology. *Environment, Technology, and Society, 88,* 8.

ROSA, E. A., KEATING, K. M., and STAPLES, C. (1981). Energy, economic growth and quality of life: A cross-national trend analysis. *Proceedings of the International Congress of Applied Systems Research and Cyberntetics* (pp. 258–264). New York: Pergamon.

ROSA, E. A., and KREBILL-PRATHER, R. (1993). *Mapping cross-national trends in carbon releases and societal well-being.* Discussion paper submitted to the committee on the Human Dimensions of Global Change, Commission on the Behavioral and Social Sciences, National Research Council. Washington, DC.

ROSA, E. A., MACHLIS, G. E. and. KEATING, K. M. (1988). Energy and society. *Annual Review of Sociology, 14,* 149–172.

ROSENBAUM, W. A. (1989). The bureaucracy and environmental policy. In J. P. Lester (Ed.), *Environmental politics and policy: Theories and evidence* (pp. 213–237). Durham, NC: Duke University Press.

ROSENZWEIG, C., and PARRY, M. (1993, September 21). Cited in Computer vision of global warming: Hardest on have-nots. *New York Times.* Reprinted in Union of Concerned Scientists, *Pledge Bulletin,* Cambridge, MA, February 1994.

ROSS, M. H., and WILLIAMS, R. H. (1981). *Our energy: Regaining control.* New York: McGraw-Hill.

ROSSET, P. (1997). Alternative agriculture and crisis in Cuba. *Technology and Society, 12* (2), 19–25.

ROSSIDES, D. W. (1993). *American society: An introduction to macrosociology.* Dix Hills, NY: General Hall.

ROTHMAN, B. K. (1991). Symbolic interactionism. In H. Etzkowitz and R. Glassman (Eds.), *The renascence of sociological theory: Classical and contemporary* (pp.151–176). Itasca, IL: F. E. Peacock.

ROYAL SOCIETY OF LONDON and THE U.S. NATIONAL ACADEMY OF SCIENCES (NAS). (1992). *Population growth, resource consumption, and a sustainable world.* Washington, DC: National Academy Press.

RUBENSTEIN, D., (1995, Winter/Spring). Environmental accounting for the sustainable corporation: Strategies and techniques, *Human Ecology Review,* 2 (1), 1–21.

RUCKELSHAUS, W. D. (1990). Toward a sustainable world. *Managing planet earth: Readings from Scientific American* (pp. 125–136). New York: W. H. Freeman.

RUNYAN, C. (1999). Action on the front lines. *World Watch, 12,* (6), 12–21.

RYAN, J. C. (1992). Conserving biological diversity. In L. Starke (Ed.), *State of the world, 1992* (pp. 9–26). New York: W. W. Norton.

RYAN, J. C. and DURNING, A. (1997). *Stuff: The secret lives of everyday things.* Seattle, WA: Northwest Environment Watch.

SABATIER, P. (1975). Social movement and regulatory agencies: Toward a more adequate—and less pessimistic—theory of "clientele capture." *Policy Sciences, 6,* 301–342.

SACHS, A. (1995). Population Growth Steady. In L. Brown et al. (Eds.), *Vital signs, 1995: The trends that are shaping our future.* New York: W. W. Norton.

SADIK, N. (1992, January). Poverty, population, and pollution. *The UNESCO Courier,* 18–21.

SAHLINS, M. (1972a). *Stone Age economics: Production, exchange, and politics in small tribal societies.* Chicago: Aldine.

SAHLINS, M. (1972b). The original affluent society. *Stone Age Economics* (pp.1–38). New York: Aldine.

SALE, K. (1980). *Human scale.* New York: Coward, McCann, and Geoghegan.

SALE, K. (1990). Planet water: The spirituality of water. *Annals of Earth 8* (2), 56–57. Reprinted in *Utne Reader,* 57, May/June, 1993.

SALE, K. (1993). *The green revolution: The American environmental movement, 1962–1992.* New York: Hill and Wang.

SAMDAHL, D. M., and ROBERTSON, R. (1989). Social determinants of environmental concern: Specification and test of the model. *Environment and Behavior, 21,* 57–81.

SAMUELSON, R. J. (1992, June 1). The end is not at hand. *Newsweek,* 43.

SANCTION, T. A. (1989, January 2). What on earth are we doing? *Time,* 28.

SANDERSON, S. (1995). *Macrosociology: An introduction to human societies* (3rd ed.). New York: HarperCollins.

SCHIPPER, L., and LICHTENBERG, A. J. (1976). Efficient energy use and well-being: The Swedish example. *Science, 194,* 1001–1013.

SCHNAIBERG, A. (1980). *The environment: From surplus to scarcity.* New York: Oxford University Press.

SCHNAIBERG, A. (1983, Spring). Redistributive goals versus distributive politics: Social equity limits in environmental and appropriate technology movements. *Sociological Inquiry, 53,* 200–219.

SCHNAIBERG, A, and GOULD, K. (1994). *Environment and society: The enduring conflict.* New York: St. Martin's Press.

SCHNEIDER, S. H. (1990a). The changing climate. In *Managing planet earth: Readings from Scientific American* (pp. 26–36). New York: W. H. Freeman.

SCHNEIDER, S. H. (1990b, July). Cooling it. *World Monitor,* 30–38.

SCHNEIDER, S. and LONDER, R. (1984). *The coevolution of climate and life.* San Francisco: Sierra Club Books.

SCHOR, J. (1992). *The overworked American: The unexpected decline of leisure.* New York: Basic Books.

SCHULZE, E., and MOONEY, H. (1993). *Biodiversity and ecosystem function.* Berlin: Springer-Verlag.

SCHUMACHER, E. F. (1973). *Small is beautiful: Economics as if people mattered.* New York: HarperCollins.

SCHUTZ, A. (1967). *The phenomenology of the social world.* Evanston, IL: Northwestern University Press. (Original work published 1932).

SELIGMAN C., BECKER, L. J., and DARLEY, J. M. (1981). Encouraging residential energy conservation through feedback. In A. Baum and J. Singer (Eds.), *Advances in environmental psychology, Vol. 3. Energy conservation: Psychological perspectives.* Hillsdale, NJ: Erlbaum.

SEN, A. (1981). *Poverty and famines.* New York: Oxford University Press.

SEN, A. (1993). The economics of life and death. *Scientific American, 208* (5), 40–47.

SHARER, R. (1983). *The ancient Maya.* Palo Alto, CA: Stanford University Press.

SHILS, E. (1991). The virtue of civil society. *Government and Opposition, 26* (1), 4–10.

SHIVA, V. (1988). *Staying alive*. London: Zed Books.

SIEGEL, S., and CASTELLAN, N. J. JR. (1988). *Nonparametric statistics for the behavioral sciences* (2nd ed.). New York: McGraw Hill.

SILVER, C. S. and DEFRIES, R. S. (1990). *One earth, one future: Our changing global environment*. Washington, DC: National Academy of Sciences, National Academy Press.

SIMKINS, R. (1994). *Creator and creation: Nature in the worldview of ancient Israel*. Peabody, MA: Hendrickson.

SIMON, J. L. (1990). Population growth is not bad for humanity. *National Forum: The Phi Kappa Phi Journal, 70*, 1.

SIMON, J. L. (1994, April). More people, greater wealth, more resources, healthier environment. *Economic Affairs*.

SIMON, J. L. (1996). *Ultimate Resources 2*. Princeton, NJ: Princeton University Press.

SIMON, J. L. (1998). There is no crisis of unsustainability. In T. Miller, *Living in the environment* (10th ed., pp. 26–27). Belmont, CA: Wadsworth.

SIMON, J. L. and KAHN, H. (1984). *The resourceful earth*. Oxford, England: Basil Blackwell.

SIMON, J. L., and WILDAVSKY, A. (1993, May 13). Facts, not species, are periled. *New York Times*.

SIMPSON, C., and RAPONE, A. (1996, Winter/ Spring). Rebellion in Chiapas: Ecological and cultural spaces in collision, *Human Ecology Review*, 2 (2), 157–169.

SINDING, S. W. (1992, October 23–25). *Getting to replacement: Bridging the gap between individual rights and demographic goals*. Presented at IPPF Family Planning Congress, Delhi, India.

SINGER, S. F. (1992). The benefits of global warming. *Society, 29* (3), 33–40.

SINGER, S. F. (1997). *Hot talk, cold science: Global warming's unfinished business*. Oakland, CA: Independent Institute.

SIVARD, R. L. (1993). *World military and social expenditures*. Washington, DC: World Priorities Publications.

SLOVIC, P. (1987). Perception of risk. *Science, 236*, 280–286.

SMITH, J. B., and TIRPAK, D. (Eds.) (1988). *The potential effect of a global climate change on the United States*. [Draft report to Congress.]

SMITH, T. W. (1985). The polls: Americans' most important problems: Part I. National and international. *Public Opinion Quarterly, 46*:38–61.

SNOW, D., and BENFORD, R. (1988). Ideology, frame resonance, and participant mobilization. In B. Klandfermans, H. Kriesi, and S. Tarrow (Eds.), *Structure to action: Comparing social movement research across cultures*. Greenwich, CT: JAI Press.

SNOW, D., and BENFORD, R. (1992). Master frames and cycles of protest. In A. D. Morris and C. M. Mueller (Eds.), *Frontiers in social movement theory*. New Haven, CT: Yale University Press.

SOCOLOW, R. H. (1978). *Saving energy in the home: Princeton's experiments at Twin Rivers*. Cambridge, MA: Ballinger.

SODDY, F. (1926). *Wealth, virtual wealth, and debt: The solution to the economic paradox*. London: Oxford University Press.

SOUTHWICK, C. H. (1996). *Global ecology in human perspective*. New York: Oxford University Press.

SPEARS, J., and AYENSU, E. S. (1985). Resources, development, and the new century: Forestry. In R. Repetto (Ed.), *The global possible: Resources, development, and the new century* (pp. 299–335). New Haven, CT: Yale University Press.

SPENCER, H. (1880). *First principles.* New York: A. L. Burt.

SPENCER, H. (1896). *The principles of sociology.* New York: Appleton.

STANISLAW, J., and YERGIN, D. (1993). Oil: Reopening the door. *Foreign Affairs, 72* (4), 81–93.

STARK, L. (1999). *State of the world, 1999.* New York: W. W. Norton.

STARK, R. (1994). *Sociology.* Belmont, CA: Wadsworth.

STAROBIN, P. (1993). Unequal shares. *National Journal, 25* (37), 2171–2226.

STARR, C. (1969). Social benefit versus technological risk. *Science, 165,* 1232–1238.

STEPHAN, E. G. (1970). The concept of concept of community in human ecology. *Pacific Sociological Review, 13,* 218–228.

STERN, P. C. (1992). Psychological dimensions of global environmental change. *Annual Review of Psychology, 43,* 269–302.

STERN, P. C. and ARONSON, E. (1984). *Energy use: The human dimension.* New York: W. H. Freeman.

STERN, P. C., DIETZ, T., and KALOF, L. (1993). Value orientations, gender, and environmental concern. *Environment and Behavior, 25* (3), 322–348.

STERN, P. C., and OSKAMP, S. (1987, March 28). Managing scarce environmental resources In D. Stokals and I. Altman (Eds.), *Handbook of environmental psychology* (vol. 2). New York: Wiley.

STERN, P. C., YOUNG, O. R., and DRUCKMAN, D. (Eds.). (1992). *Global environmental change: Understanding the human dimensions.* Washington, DC: National Academy of Sciences, National Academy Press.

STRANGE, M. (1994, May). Farmers for the next century. *Center for Rural Affairs Newsletter.*

STRAUSS, M. (1998a). Fish catch hits a new high. In L. Brown, M. Renner, and C. Flavin, (Eds.), *Vital signs 1998: Environmental trends that are shaping our future* (pp. 34–35). New York: W. W. Norton.

STRAUSS, M. (1998b). Trade remains strong. In L. Brown, M. Renner, and C. Flavin (Eds.), *Vital Signs 1998: Environmental trends that are shaping our future* (pp. 34–35). New York: W. W. Norton.

STROH, M., and RALOFF, J. (1992, April 4). New UN soil survey: The dirt on erosion. *Science News,* 215.

SULLIVAN, W. (1991, December 3). Antarctic ice buildup. *New York Times,* C5.

SWITZER, J. V. (1994). *Environmental politics: Domestic and global dimensions.* New York: St Martin's Press.

SZTOMPKA, P. (1993). *The sociology of social change.* Cambridge, MA: Blackwell Publishers.

TAINTER, J. A. (1988). *The collapse of complex societies.* Cambridge, England: Cambridge University Press.

TAKACS, D. (1996). *The idea of biodiversity: Philosophies of paradise.* Baltimore, MD: John Hopkins University Press.

TAYLOR, B. (1992). *Our limits transgressed: Environmental political thought in America.* Lawrence, KS: University of Kansas Press.

TAYLOR, D. (1993). Environmentalism and the politics of inclusion, In R. Bullard (Ed.), *Confronting environmental racism: Voices from the grassroots.* Boston: South End Press.

TETT, S., STOTT, P., ALLEN, M., INGRAM, W., and MITCHELL, J. (1999, June 10). Causes of twentieth-century temperature change near the earth's surface. *Nature, 399,* 569–572.

THOMAS, W. I. (1923). *The unadjusted girl.* Boston: Little Brown.

THOMAS, F. (1994, October 18). Report links water, risk of cancer. *Omaha World-Herald,* 1, p. 7.

THOMPSON, D. (1999). Capital hill meltdown. *Time*, Aug 9: 56–59.

TILMAN, D. (1997). Biodiversity and ecosystem functioning. In G. Daily (Ed.), *Nature's services: Societal dependence on natural ecosystems* (pp. 93–112). Washington, DC: Island Press.

TILLY, C. (1984). *Big structures, large processes, huge comparisons*. New York: Russell Sage Foundation.

TOFFLER, A. (1980). *The third wave*. New York: William Morrow.

TOUGH, A. (1992). *Crucial questions about the future*. Lanham, MD: University Press of America.

TOURAINE, A. (1978). *The voice and the eye: An analysis of social movements*. Cambridge, England: Cambridge University Press.

TSOUKALAS, T. (1994). Environmental whistleblowers in federal agencies: A brief organizational profile. *Environment, Technology, and Society, 74*, 1–5.

TUXILL, J. (1997). Death in the family tree. *World Watch, 10* (5), 13–21.

TUXILL, J. (1998). Vertebrates signal biodiversity losses. In L. Brown et al. (Eds.), *Vital signs, 1998: Environmental trends that are shaping our future* (pp. 128–129) New York: W. W. Norton.

TUXILL, J. (1999). Appreciating the benefits of plant biodiversity. In L. Brown et al. (Eds.), *State of the world, 1999* (pp. 96–114). New York: W. W. Norton.

UNGAR, S. (1992). The rise and (relative) decline of global warming as a social problem. *The Sociological Quarterly, 33* (4), 483–502.

UNGAR, S. (1998). Bringing the issue back in: Comparing the marketability of the ozone hole and global warming. *Social Problems, 48* (4), 510–527.

UNION OF CONCERNED SCIENTISTS. (1992). *Warning to humanity*. Washington, DC: Union of Concerned Scientists.

UNITED NATIONS. (1981). *World population prospects as assessed in 1980*. Population Studies Report no. 78. New York: United Nations.

UNITED NATIONS. (1982). *Estimates and projections of urban, rural, and city populations, 1950–2025*. United Nations Population Division New York: United Nations.

UNITED NATIONS. (1996). *Human development report 1996*. United Nations Human Development Programme. New York: Oxford University Press.

UNITED NATIONS. (1998a). *Human development report 1998*. United Nations Human Development Programme. New York: Oxford University Press.

UNITED NATIONS. (1998b). *World urbanization prospects: The 1996 revisions*. New York: United Nations.

UNITED NATIONS. (2000). *World population prospects*: The 1998 revisions. United Nations Population Division. New York: United Nations.

U. S. BUREAU OF THE CENSUS. (1995). *Statistical abstract of the United States*. Washington, DC: Government Printing Office.

U.S. BUREAU OF THE CENSUS (1979). Illustrative projections of world populations to the 21st century. *Current population reports*, Special Studies, Series P-23, no. 79. (Washington, DC: Government Printing Office).

U.S. BUREAU OF THE CENSUS (1991). *Statistical abstract of the United States: 1991*, 111th Ed. (Washington, DC: Government Printing Office).

U.S. DEPARTMENT OF AGRICULTURE (1993). *World grain situation and outlook*. February. (Washington, DC: Government Printing Office).

U.S. DEPARTMENT OF ENERGY (1989). *Energy conservation trends: Understanding the factors that affect conservation gains in the U.S. economy*. Washington, D.C : DOE/PE-0092.

U.S. ENVIRONMENTAL PROTECTION AGENCY (1988). *Environmental progress and challenges: EPA's update.* (Washington, DC: Government Printing Office).

U.S. ENVIRONMENTAL PROTECTION AGENCY. (1990). *Superfund: Environmental progress.* Washington, DC: EPA Office of Emergency and Remedial Response.

U.S. GEOLOGICAL SURVEY AND BUREAU OF MINES. (1980) Geological Survey Circular 831.

VAN DEN BERGHE, P. (1977–1978). Bridging the paradigms. *Society, 15*: 42–49.

VAN LIERE, K.D. and DUNLAP, R. E. (1980). The social bases of environmental concern: A review of hypotheses, explanations, and empirical evidence. *Public Opinion Quarterly, 44*: 43–59.

VITOUSEK, P. M., ET AL. (1986). Human appropriation of the products of photosynthesis. *BioScience*, 36, 368.

VOGLEY, W. A. (1985). Nonfuel minerals in the world economy. In R. Repetto (Ed.), *The global possible: Resources, development, and the new century* (pp. 457–473). New Haven, CT: Yale University Press.

WALLACE, R. A., and WOLF, A. (1991). Contemporary sociological theory: Continuing the classical tradition, 3rd ed. Englewood Cliffs, NJ: Prentice Hall.

WALLERSTEIN, I. (1980). *The modern world system, 2*. New York: Academic Press.

WALSH, E., WARLAND, R., and CLAYTON SMITH, D. (1993). Backyards, NIMBYs, and incinerator sitings: Implications for social movement theory. *Social Problems 40* (1), 25–38.

WALTHEN, T. (1993). A guide to trade and the environment. In D. Zalke, P. Orbuch, and R. F. Housman (Eds.), *Trade and environment: Law, economics, and policy* (pp. 3–21). Washington, DC: Island Press.

WALTON, J. (1993). *Sociology and critical inquiry: The work, tradition, and purpose* (3rd ed.). Belmont, CA: Wadsworth.

WASHINGTON POST, (1998, April 21). Big extinction under way, biologist says. *Omaha World-Herald*, p. 5.

WASHINGTON POST, (1999, December). EPA plans gas, emissions rules. Cited in *Omaha World-Herald*, p. 12.

WARREN, R., and PASSEL, J. (1987). A count of the uncountable: Estimates of undocumented aliens counted in the 1980 United States census. *Demography, 24* (3), 375.

WEBER, M. (1922/1958). *The Protestant ethic and the spirit of capitalism.* New York: Scribners.

WEEKS, J. R. (1994). *Population: An introduction to concepts and issues* (5th ed.). Belmont, CA: Wadsworth.

WEEKS, J. R. (1996). *Population: An introduction to concepts and issues* (6th ed.). Belmont, CA: Wadsworth.

WEINBERG, A., PELLOW, D., and SCHNIBERG, A. (1998). Ecological modernization in the internal periphery of the USA: Accounting for recycling's promises and performance. Presented to the American Sociological Association, August, San Francisco, CA.

Weinberg, C. J., and WILLIAMS, R. H. (1990). Energy from the sun. *Scientific American, 263* (3), 147–163.

WHITE, G. F. (1980). Environment. *Science, 209* (4), 183–189.

WHITE, L., (1949). Energy and the evolution of culture. In L. White, Jr., *The evolution of culture* (pp. 363–393). New York: Farrar, Straus, and Giroux.

WHITE, L., JR. (1967). The historical roots of our ecological crisis. *Science, 155*: 1203–1207.

WILLIAMS, J. A. and MOORE, H. A. (1994). The rural-urban continuum and environmental concerns. *Great Plains Research: A Journal of Natural and Social Sciences, 12*, 195–214.

WILSON, A. (1992). *The culture of nature: North American landscape from Disney to the Exxon Valdez*. Cambridge, MA: Blackwell.

WILSON, E. O. (1975). *Sociobiology: The new synthesis*. Cambridge, MA: Belknap Press of Harvard University.

WILSON, E. O. (1990). *Threats to Biodiveristy. Managing planet earth: Readings from Scientific American* (pp. 49–59.) New York: W. H. Freeman.

WINTERBOTTOM, R. (1990). *Taking stock: The tropical forestry action plan after five years.* Washington, DC: World Resources Institute.

WOLF, E. C. (1986). *Beyond the green revolution: New approaches for Third World agriculture.* Worldwatch Paper no. 73. Washington, DC: Worldwatch Institute.

WOLF, E. R. (1982). *Europe and the peoples without history*. Berkeley, CA: University of California Press.

WORLD BANK. (1991). *World development report*. New York: Oxford University Press.

WORLD COMMISSION ON ENVIRONMENT AND DEVELOPMENT. (1987). *Our common future.* Oxford: Oxford University Press.

WORLD ENERGY COUNCIL. (1993). *Energy for tomorrow's world: The realities, the real options and the agenda for achievement*. New York: St Martin's Press.

WORLD RESOURCES INSTITUTE. (1993). *Environmental almanac*. New York: Houghton Mifflin.

WORLDWATCH INSTITUTE. (1994). Matters of scale: The price of beef. *World Watch, 7* (4), 39.

WRIGHT, R. (2000, January 17). Continental drift, *The new republic*, 18–23.

WRIGLEY, T., SEMITH, R., and SANTER, D. (1998, November). Anthropogenic influence on the autocorrelation structure of hemispheric mean temperatures. *Science, 27.*

YANKELOVICH, D. (1981). *New rules*. New York: Random House.

YOUNG, G. L. (1994). Community with three faces: The paradox of community in postmodern life with illustrations from the United States and Japan. *Human Ecology Review, 1* (1), 137–146.

ZALKE, D., ORBUCH, P., and HOUSMAN, R. F. (Eds.). (1993). *Trade and environment: Law, economics, and policy*. Washington, DC: Island Press.

ZEY, M. G. (1994). *Seizing the future*. New York: Simon & Schuster.

ZWERDLING, D. (1993, October 20). Report on pesticide research by IRRI. *All Things Considered*, Washington, DC: National Public Radio.

Name Index

Subject Index